APPLIED COMPUTATIONAL MATERIALS MODELING

Theory, Simulation and Experiment

APPLIED COMPUTATIONAL MATERIALS MODELING

Theory, Simulation and Experiment

Edited by

Guillermo Bozzolo
Ohio Aerospace Institute
Cleveland, OH, USA

Ronald D. Noebe
NASA Glenn Research Center
Cleveland, OH, USA

Phillip B. Abel
NASA Glenn Research Center
Cleveland, OH, USA

Springer

Guillermo Bozzolo
Ohio Aerospace Institute
Cleveland, OH, USA

Ronald D. Noebe
NASA Glenn Research Center
Cleveland, OH, USA

Phillip B. Abel
NASA Glenn Research Center
Cleveland, OH, USA

Consulting Editor: D.R. Vij

Applied Computational Materials Modeling: Theory, Simulation and Experiment

ISBN 13: 978-1-4419-3575-5

ISBN 10: 0-387-34565-5 (e-book)
ISBN 13: 978-0-387-34565-9 (e-book)

Printed on acid-free paper.

Printed in the United States of America.

9 8 7 6 5 4 3 2 1

springer.com

CONTENTS

12. FRONTIERS IN SURFACE ANALYSIS: EXPERIMENTS AND MODELING

D. Farías, G. Bozzolo, J. Garcés, R. Miranda

13. THE EVOLUTION OF COMPOSITION AND STRUCTURE AT METAL-METAL INTERFACES: MEASUREMENTS AND SIMULATIONS

R.J. Smith

PREFACE

While it is tempting to label computational materials modeling as an emerging field of research, the truth is that both in nature and foundation, it is just as much an established field as the concepts and techniques that define it. It is the recent enormous growth in computing power and communications that has brought the activity to the forefront, turning it into a possible component of any modern materials research program. Together with its increased role and visibility, there is also a dynamic change in the way computational modeling is perceived in such a vast field as materials science with its wide range of length and time scales. As the pace of materials research accelerates and the need for often inaccessible information continues to grow, the demands and expectations on existing modeling techniques have progressed that much faster. Primarily because there is no one technique that can provide all the answers at every length and time scale in materials science, excessive expectations of computational materials modeling should be avoided if possible. While it is apparent that computational modeling is the most efficient method for dealing with complex systems, it should not be seen as an alternative to traditional experimentation.

Instead there is another option, which is perhaps the one that is most likely to become the defining characteristic of computational materials modeling. It is the role of modeling as a tool to connect the different and seemingly unrelated features across the various length scales and types of materials, in synergy with fundamental theory and applied experimentation. As in other areas of evolving modern technology, it is not immediately apparent when we can fully expect to achieve the necessary maturity, coherence, and completeness to fulfill this goal. But until that point is reached, it is necessary to document the progress, acknowledging both successes and limitations, not just with the objective of perfecting the underlying methods and techniques, but also with the goal of gaining a deeper understanding of the power and essence of the activity. In this regard, the current volume provides a necessarily limited yet broad view of computational materials modeling as applied to metallic materials, and lays the foundation for achieving greater understanding, acceptance, and the proper utilization of computational materials science.

Starting with a detailed description of the fundamental methods for the calculation of electronic structure and their role in the description of complex materials, the first few chapters concentrate on modeling of alloy phase equilibria, emphasizing the connection between first-principles methods and computational thermodynamics. Topics of substantial current interest are examined within this framework, including metallic glass formation and the transition between the nano- to the micro-scale regime by exploring the structure of clusters. The emphasis then shifts to the role of modeling methods in surface science, detailing currently accepted techniques including quantum approximate methods with proven effectiveness in bridging the gap between surface science and bulk solids. New approaches, as well as established techniques are discussed in subsequent chapters, including phase-field simulations, CALPHAD modeling, the quantum approximate BFS method for alloys, and the molecular orbital approach to alloy design. Having established a solid foundation for different modeling approaches, the focus is then on specific materials and properties with current examples and applications, including the intelligent design of age-hardenable aluminum alloys and intergranular modeling of deformation and fracture in metallic materials. The final chapters point to non-traditional areas of materials and surface modeling, with a detailed discussion of current topics in surface analysis, metal-metal interfaces, and nuclear materials.

Needless to say, progress in computational materials modeling is much broader than the contents of this book, but even if every current topic or development could be included, it would not change the fact that in many ways the discipline has come of age, becoming a vital and fundamental part in any modern materials research program. It is our hope that the detailed and comprehensive contributions of our guest authors will help reflect this promising conclusion and provide the proper perspective for the current status and future prospects of computational materials modeling.

Guillermo Bozzolo
Ohio Aerospace Institute

Ronald D. Noebe and Phillip B. Abel
NASA Glenn Research Center

Chapter 1

AB INITIO MODELING OF ALLOY PHASE EQUILIBRIA

Axel van de Walle, Gautam Ghosh and Mark Asta
Materials Science and Engineering Department, Northwestern University, Evanston IL, USA

Abstract: In this chapter we provide an overview of the methodologies underlying first-principles (or *ab initio*) calculations of alloy phase stability, and provide examples intended to illustrate their capabilities and accuracy. The calculations of the thermodynamic properties of elemental solids and ordered compounds, as well as alloy phases with dilute and concentrated compositional disorder are covered. The integration of first-principles methods with computational-thermodynamic methods is illustrated with an application to the development of multicomponent Nb-based alloy systems for potential high-temperature structural applications.

Keywords: First-principles (or ab initio) calculations, phase diagrams, thermodynamics, phonons, configurational disorder, point defects, computational-thermodynamic methods.

1. INTRODUCTION

In computational materials modeling, the terms *ab initio*, or "first principles" typically refer to methods involving applications of quantum-mechanical total-energy calculations performed within the framework of electronic density-functional theory (DFT).[1-4] DFT-based methods allow one to predict bulk and defect properties for elemental and multicomponent solids, starting from a knowledge of only the atomic numbers of the constituent atomic species. Over the past two decades DFT methods have demonstrated high accuracy in calculations of a wide variety of materials structural and energetic properties, including heats of formation, lattice parameters, elastic constants, and the formation energies of point and planar defects. By coupling DFT energy methods with statistical-mechanical models, first-principles methods for calculating finite-temperature alloy thermodynamic properties have also been extensively developed.[5-12] With continuing progress in method development, coupled with increases in computational power and the availability of highly developed software, first-principles methods are beginning to find growing applications to problems in materials design.

In modern approaches to alloy design,[13] computational methods for modeling the evolution of phase-transformation microstructure form a critical component. Computational-thermodynamic methods based on the CALPHAD framework[14] have become widely used as the basis for modeling phase stability and phase-transformation kinetics in complex multicomponent alloy systems.[15-17] The accuracy of the predictions derived from these methods depends critically upon the thermodynamic and kinetic models that form the basis for calculations of phase stability and phase-transformation driving forces. Thus, highly developed thermodynamic and kinetic databases are generally a prerequisite for successful applications of computational thermodynamic methods in materials design. For new, relatively unexplored systems, modeling efforts are often hindered by the need for extensive experimental measurements required in database development. Recent efforts have demonstrated the feasibility of augmenting experimental thermodynamic measurement efforts through the application of *ab initio* methods for the direct calculation of alloy thermodynamic properties.[18-23] Since *ab initio* methods yield calculated thermodynamic properties with very limited input required from experiment, they offer the potential for significantly limiting the extent of costly experimental measurements required in thermodynamic database development. First-principles methods are particularly advantageous as a framework for providing predictive estimates of thermodynamic properties in situations where direct experimental measurements are difficult due to constraints imposed by sluggish kinetics or metastability.

In this chapter we provide an overview of the methodologies underlying first-principles calculations of alloy phase stability, and provide examples intended to illustrate the capabilities and accuracy of these methods. The following section provides an overview of the theoretical formalism underlying such calculations. The methods are further described and applications discussed in the following three sections. These sections are devoted to thermodynamic calculations for elemental solids and ordered compounds, as well as alloy phases with dilute and concentrated compositional disorder. This chapter ends with a section describing the integration of first-principles and computational-thermodynamic methods in an application to multicomponent Nb-based alloy systems of interest for potential high-temperature structural materials applications.[24]

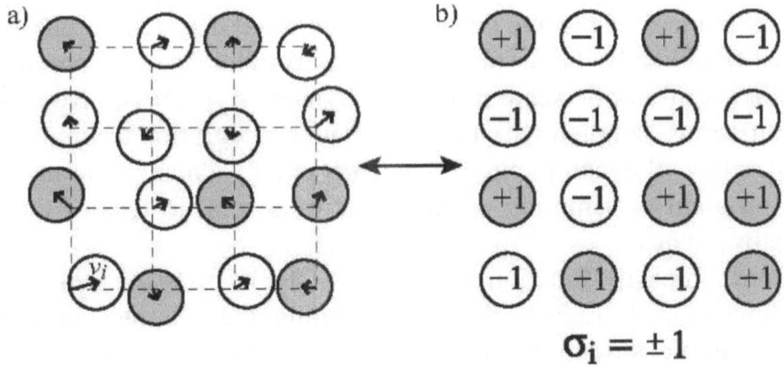

Figure 1. a) Disordered crystalline alloy. The state of the alloy is characterized both by the atomic displacements v_i and the occupation of each lattice site. b) Mapping of the real alloy onto a lattice model characterized by occupation variables σ_i describing the identity of atoms on each of the lattice sites.

2. FIRST-PRINCIPLES CALCULATIONS OF THERMODYNAMIC PROPERTIES: OVERVIEW

At finite temperature (T) and pressure (P) thermodynamic stability is governed by the magnitude of the Gibbs free energy (G):

$$G = E - TS + PV \tag{1}$$

where E, S and V denote energy, entropy and volume, respectively. In principle, the formal statistical-mechanical procedure for calculating G from first-principles is well defined. Quantum-mechanical calculations can be performed to compute the energy $E(s)$ of different microscopic states (s) of a system, which then must be summed up in the form of a partition function (Z):

$$Z = \sum_s \exp\left(-E(s)/k_B T\right) \tag{2}$$

from which the free energy is derived as $F = E - TS = -k_B T \ln Z$, where k_B is Boltzman's constant.

Figure 1.a illustrates, for the case of a disordered crystalline binary alloy, the nature of the disorder characterizing a representative finite-temperature atomic structure. This disorder can be characterized in terms of the configurational arrangement of the elemental species over the sites of the underlying parent lattice, coupled with the displacements characterizing positional disorder. In principle, the sum in Eq. 2 extends over all configurational and displacive states accessible to the system, a phase space that is astronomically large for a realistic system size. In practice, the methodologies of atomic-scale molecular dynamics (MD) and Monte-Carlo (MC) simulations, coupled with thermodynamic integration techniques (e.g., Ref. 25) reduce the complexity of a free energy calculation to a more tractable problem of sampling on the order of several to tens of thousands representative states.

Electronic density-functional theory (DFT) provides an accurate quantum-mechanical framework for calculating the relative energetics of competing atomic structures in solids, liquids and molecules for a wide range of materials classes. Due to the rapid increase in computational cost with system size, however, DFT calculations are typically limited to structures containing several-hundred atoms, while *ab initio* MD simulations are practically limited to time scales of less than a nanosecond. For liquids or compositionally-ordered solids, where the time scales for structural rearrangements (displacive in the latter case, configurational and displacive in the former) are sufficiently fast, and the size of periodic cells required to accurately model the atomic structure are relatively small, DFT-based MD methods have found direct applications in the calculation of finite-temperature thermodynamic properties.[12,26] For crystalline solids containing both positional and concentrated compositional disorder, however, direct applications of DFT to the calculation of free energies remains intractable; the time scales for configurational rearrangements are set by solid-state diffusion, ruling out direct application of MD, and the necessary system sizes required to accurately model configurational disorder are too large to permit direct application of DFT as the basis for MC simulations. Effective strategies have nonetheless been developed for bridging the size and time-scale limitations imposed by DFT in the first-principles computation of thermodynamic properties for disordered solids. The approach involves exploitation of DFT methods as a framework for parameterizing coarse-grained statistical models which serve as efficient "effective Hamiltonians" in direct simulation-based calculations of thermodynamic properties.

3. THERMODYNAMICS OF COMPOSITIONALLY ORDERED SOLIDS

In an ordered solid thermal fluctuations take the form of electronic excitations and lattice vibrations and, accordingly, the free energy can be written as $F = E_0 + F_{elec} + F_{vib}$, where E_0 is the absolute zero total energy while F_{elec} and F_{vib} denote electronic and vibrational free energy contributions, respectively. This section describes the calculation of the electronic and vibrational contributions most commonly considered in phase-diagram calculations under the assumption that electron-phonon interactions are negligible (i.e., F_{elec} and F_{vib} are simply additive).

To account for electronic excitations, electronic DFT can be extended to nonzero temperatures by allowing for partial occupations of the electronic states.[27] Within this framework, and assuming that both the electronic charge density and the electronic density of states can be considered temperature-independent, the electronic contribution to the free energy $F_{elec}(T)$ at temperature T can be decomposed as

$$F_{elec}(T) = E_{elec}(T) - E_{elec}(0) - TS_{elec}(T) \tag{3}$$

where the electronic band energy $E_{elec}(T)$ and the electronic entropy $S_{elec}(T)$ are respectively given by

$$E_{elec}(T) = \int f_{\mu,T}(\varepsilon)\varepsilon g(\varepsilon)d\varepsilon \tag{4}$$

$$S_{elec}(T) = -k_B \int (f_{\mu,T}(\varepsilon)\ln f_{\mu,T}(\varepsilon) + \\ +(1 - f_{\mu,T}(\varepsilon))\ln(1 - f_{\mu,T}(\varepsilon)))g(\varepsilon)d\varepsilon \tag{5}$$

where $g(\varepsilon)$ is the electronic density of states obtained from a density functional calculation, while $f_{\mu,T}(\varepsilon)$ is the Fermi distribution when the electronic chemical potential is equal to μ,

$$f_{\mu,T}(\varepsilon) = \left(1 + \exp\left(\frac{(\varepsilon - \mu)}{k_B T}\right)\right)^{-1} \tag{6}$$

The chemical potential μ is the solution to $\int f_{\mu,T}(\varepsilon)g(\varepsilon)d\varepsilon = n_e$, where n_e is the total number of electrons. Under the assumption that the electronic density of states near the Fermi level is slowly varying relative to $f_{\mu,T}(\varepsilon)$, the equations for the electronic free energy reduce to the well-known Sommerfeld model, an expansion in powers of T whose lowest order term is

$$F_{elec}(T) = -\frac{\pi^2}{6} k_B^2 T^2 g(\varepsilon_F) \tag{7}$$

where $g(\varepsilon_F)$ is the zero-temperature value of the electronic density of states at the Fermi level (ε_F).

The quantum treatment of lattice vibrations in the harmonic approximation provides a reliable description of thermal vibrations in many solids for low to moderately high temperatures.[28] To describe this theory, consider an infinite periodic system with n atoms per unit cell and let $u\begin{pmatrix} l \\ i \end{pmatrix}$ for $i = 1, \ldots, n$ denote the displacement of atom i in cell l away from its equilibrium position and let M_i be the mass of atom i. Within the harmonic approximation, the potential energy U of this system is entirely determined by (i) the potential energy (per unit cell) of the system at its equilibrium position E_0 and (ii) the *force constants tensors* $\Phi_{\alpha\beta}$ whose components are given, for $\alpha, \beta = 1,2,3$, by

$$\Phi_{\alpha\beta}\begin{pmatrix} l & l' \\ i & j \end{pmatrix} = \frac{\partial^2 U}{\partial u_\alpha \begin{pmatrix} l \\ i \end{pmatrix} \partial u_\beta \begin{pmatrix} l' \\ j \end{pmatrix}} \tag{8}$$

evaluated at $u\begin{pmatrix} l \\ i \end{pmatrix} = 0$ for all l, i. Such a harmonic approximation to the Hamiltonian of a solid is often referred to as a Born-von Kármán model.

The thermodynamic properties of a harmonic system are entirely determined by the frequencies of its normal modes of oscillations, which can be obtained by finding the eigenvalues of the so-called $3n \times 3n$ dynamical matrix of the system:

$$D(k) = \sum_l e^{i2\pi(k.l)} \begin{pmatrix} \frac{\Phi\begin{pmatrix} 0 & l \\ 1 & 1 \end{pmatrix}}{\sqrt{M_1 M_1}} & \cdots & \frac{\Phi\begin{pmatrix} 0 & l \\ 1 & n \end{pmatrix}}{\sqrt{M_1 M_n}} \\ \vdots & \ddots & \vdots \\ \frac{\Phi\begin{pmatrix} 0 & l \\ n & 1 \end{pmatrix}}{\sqrt{M_n M_1}} & \cdots & \frac{\Phi\begin{pmatrix} 0 & l \\ n & n \end{pmatrix}}{\sqrt{M_n M_n}} \end{pmatrix} \tag{9}$$

for all vectors k in the first Brillouin zone. The resulting eigenvalues $\lambda_b(k)$ for $b = 1 \ldots 3n$, provide the frequencies of the normal modes through $\nu_b(k) = \frac{1}{2\pi}\sqrt{\lambda_b(k)}$. This information for all k is conveniently summarized by $g(\nu)$, the phonon density of states (DOS), which specifies the number of modes of oscillation having a frequency lying in the infinitesimal interval $[\nu, \nu + d\nu]$. The vibrational free energy (per unit cell) F_{vib} is then given by

$$F_{vib} = k_B T \int_0^\infty \ln(2\sinh(\frac{h\nu}{2k_B T}))g(\nu)d\nu \tag{10}$$

where h is Planck's constant and k_B is Boltzman's constant. The associated vibrational entropy S_{vib} of the system can be obtained from the well-known thermodynamic relationship $S_{vib} = -\partial F_{vib}/\partial T$. The high temperature limit (which is also the classical limit) of Eq. 10 is often a good approximation over the range of temperature of interest in solid-state phase diagram calculations

$$F_{vib} = k_B T \int_0^\infty \ln\left(\frac{h\nu}{k_B T}\right)g(\nu)d\nu \tag{11}$$

The high temperature limit of the vibrational entropy difference between two phases is often used as convenient measure of the magnitude of the effect of lattice vibrations on phase stability. It has the advantage of being temperature-independent, thus allowing a unique number to be reported as a measure of vibrational effects. Fig. 2 (from Ref. 29) illustrates the use of the above formalism to assess the relative phase stability of the θ and θ' phases responsible for precipitation hardening in the Al-Cu system.[29] Interestingly, accounting for lattice vibrations is crucial in order for the calculations to agree with the experimentally observed fact that the θ phase is stable at typical processing temperatures ($T > 475K$).

A simple improvement over the harmonic approximation, called the quasiharmonic approximation, is obtained by using volume-dependent force constant tensors. This approach maintains all the computational advantages of the harmonic approximation while permitting the modeling of thermal expansion. The volume dependence of the phonon frequencies induced by the volume-dependence of the force constants is traditionally described by the Grüneisen parameter $\gamma_{kb} = -\partial \ln(\nu_b(k))/\partial \ln V$. However, for the purpose of modeling thermal expansion, it is more convenient to directly parameterize the volume dependence of the free energy itself. This dependence has two sources: the change in entropy due to the change in the phonon frequencies and the elastic energy change due to the expansion of the lattice:

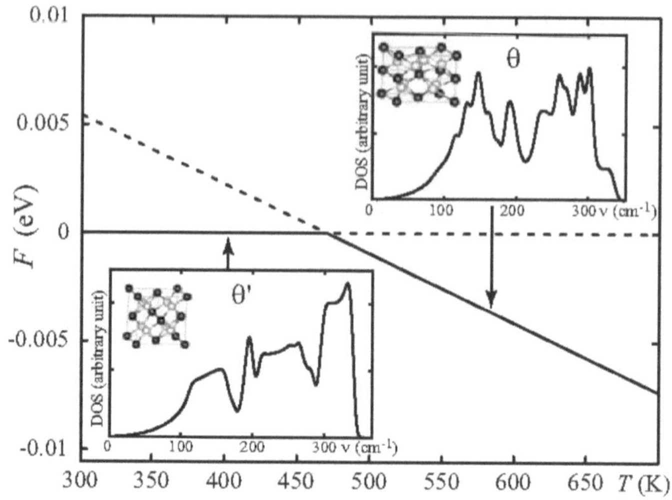

Figure 2. Temperature-dependence of the free energy of the θ and θ' phases of the Al_2Cu compound. Insets show the crystal structures of each phase and the corresponding phonon density of states. Dashed lines indicate region of metastability and the θ phase is seen to become stable above about 475 K. (Adapted from Ref. 29 with the permission of the authors).

$$F(T,V) = E_0(V) + F_{vib}(T,V) \tag{12}$$

where $E_0(V)$ is the energy of a motionless lattice whose unit cell is constrained to remain at volume V, while $F_{vib}(T,V)$ is the vibrational free energy of a harmonic system constrained to remain with a unit cell volume V at temperature T. The equilibrium volume $V^*(T)$ at temperature T is obtained by minimizing $F(T,V)$ with respect to V. The resulting free energy $F(T)$ at temperature T is then given by $F(T,V^*(T))$.The quasiharmonic approximation has been shown to provide a reliable description of thermal expansion of numerous elements up to their melting points, as shown in Fig. 3.

First-principles calculations can be used to provide the necessary input parameters for the above formalism. The so-called direct force method proceeds by calculating, from first principles, the forces experienced by the atoms in response to various imposed displacements and by determining the value of the force constant tensors that match these forces through a least-squares fit. Note that the simultaneous displacements of the periodic images of each displaced atom due to the periodic boundary conditions used in most *ab initio* methods typically requires the use of a supercell geometry, in order

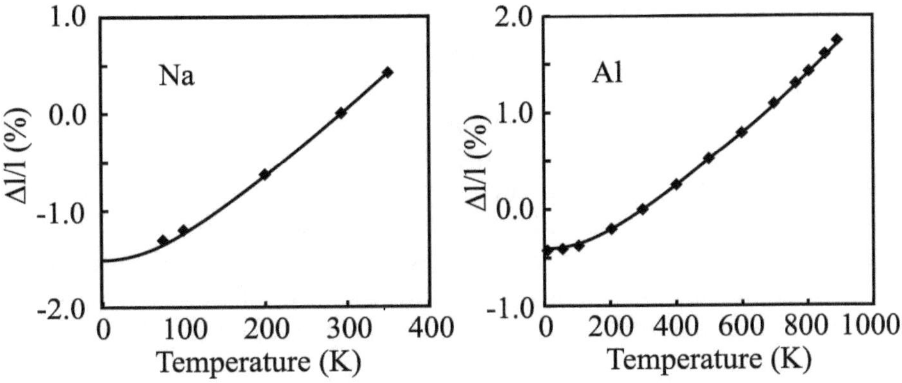

Figure 3. Thermal expansion of selected metals calculated within the quasiharmonic approximation (diamonds denote experimental measurements). (Reproduced from Ref. 30 with the permission of the authors).

to be able to sample all the displacements needed to determine the force constants. While the number of force constants to be determined is in principle infinite, in practice, it can be reduced to a manageable finite number by noting that the force constant tensor associated with two atoms that lie farther than a few nearest neighbor shells can be accurately neglected for many systems. Alternatively, linear response theory[31-39] can be used to calculate the dynamical matrix $D(k)$ directly using second-order perturbation theory, thus circumventing the need for supercell calculations. Linear response theory is also particularly useful when a system is characterized by nonnegligible long-range force-constants, as in the presence of Fermi-surface instabilities or long-ranged electrostatic contributions.

The above discussion has centered around the application of harmonic (or quasi-harmonic) approximations to the statistical modeling of vibrational contributions to free energies of solids. While harmonic theory is known to be highly accurate for a wide class of materials, important cases exist where this approximation breaks down due to large anharmonic effects. Examples include the modeling of ferroelectric and martensitic phase transformations where the high-temperature phases are often dynamically unstable at zero temperature, i.e., their phonon spectra are characterized by unstable modes. In such cases effective-Hamiltonian methods have been developed to model structural phase transitions from first principles.[40] Alternatively, direct application of *ab initio* molecular-dynamics offers a general framework for modeling thermodynamic properties of anharmonic solids.[12,26]

4. THERMODYNAMICS OF COMPOSITIONALLY DISORDERED SOLIDS

We now relax the main assumption made in the previous section, by allowing atoms to exit the neighborhood of their local equilibrium position. This is accomplished by considering every possible way to arrange the atoms on a given lattice. As illustrated in Fig. 1.b, the state of order of an alloy can be described by occupation variables σ_i specifying the chemical identity of the atom associated with lattice site i. In the case of a binary alloy, the occupations are traditionally chosen to take the values +1 or −1, depending on the chemical identity of the atom.

Returning to Eq. 2, all the thermodynamic information of a system is contained in its partition function Z and in the case of a crystalline alloy system, the sum over all possible states of the system can be conveniently factored as follows:

$$Z = \sum_{\sigma} \sum_{v \in \sigma} \sum_{e \in v} \exp[-\beta E(\sigma, v, e)] \tag{13}$$

where $\beta = (k_B T)^{-1}$ and where σ denotes a configuration (i.e. the vector of all occupation variables); v denotes the displacement of each atom away from its local equilibrium position; e is a particular electronic state when the nuclei are constrained to be in a state described by σ and v; and $E(\sigma,v,e)$ is the energy of the alloy in a state characterized by σ, v and e.

Each summation defines an increasingly coarser level of hierarchy in the set of microscopic states. For instance, the sum over v includes all displacements such that the atoms remain close to the undistorted configuration σ. Eq. 13 implies that the free energy of the system can be written as

$$F(T) = -k_B T \ln\left(\sum_{\sigma} \exp[-\beta F(\sigma, T)] \right) \tag{14}$$

where $F(\sigma,T)$ is nothing but the free energy of an alloy with a fixed atomic configuration, as obtained in the previous section

$$F(\sigma, T) = -k_B T \ln\left(\sum_{v \in \sigma} \sum_{e \in v} \exp[-\beta E(\sigma, v, e)] \right) \tag{15}$$

The so-called "coarse graining" of the partition function illustrated by Eq. 15 enables, in principle, an exact mapping of a real alloy onto a simple lattice model characterized by the occupation variables σ and a temperature-dependent Hamiltonian $F(\sigma,T)$.[6,41]

4.1 Cluster Expansion Formalism

Although we have reduced the problem of modeling the thermodynamic properties of configurationally disordered solids to a more tractable calculation for a lattice model, the above formalism would still require the calculation of the free energy for every possible configuration σ, which is computationally intractable. Fortunately, the configurational dependence of the free energy can often be parameterized using a convenient expansion known as a cluster expansion.[5,6,8,42] This expansion takes the form of a polynomial in the occupation variables

$$F(\sigma,T) = J_0 + \sum_i J_i \sigma_i + \sum_{i,j} J_{i,j} \sigma_i \sigma_j + \sum_{i,j,k} J_{i,j,k} \sigma_i \sigma_j \sigma_k + \ldots \qquad (16)$$

where the so-called effective cluster interactions (ECI) J_0, J_i, J_{ij}, ... need to be determined. The cluster expansion can be recast into a form which exploits the symmetry of the lattice by regrouping the terms as follows

$$F(\sigma,T) = \sum_\alpha m_\alpha J_\alpha \left\langle \prod_{i \in \alpha'} \sigma_i \right\rangle \qquad (17)$$

where α is a cluster (i.e. a set of lattice sites) and where the summation is taken over all clusters that are symmetrically distinct while the average $\langle \ldots \rangle$ is taken over all clusters α' that are symmetrically equivalent to α. The multiplicity m_α weight each term by the number of symmetrically equivalent clusters in a given reference volume (e.g. a unit cell). While the cluster expansion is presented here in the context of binary alloys, an extension to multicomponent alloys (where σ_i can take more than 2 different values) is straightforward.[42] Note that the presence of vacancies can also be accounted for within the cluster expansion formalism by merely treating "vacancies" as an additional atomic species.[43,44] The cluster expansion formalism can also be extended to allow for multiple coupled sublattices,[45] thus permitting the treatment of interstitials.

It can be shown that when all clusters α are considered in the sum, the cluster expansion is able to represent any function of configuration σ by an appropriate selection of the values of J_α. However, the real advantage of the cluster expansion is that, for many systems it is found to converge rapidly. An accuracy that is sufficient for phase diagram calculations can often be achieved by keeping only clusters α that are relatively compact (e.g. short-range pairs or small triplets), as illustrated in Fig. 5.b. The unknown parameters of the cluster expansion (the ECI J_α) can then be determined by fitting them to $F(\sigma,T)$ for a relatively small number of configurations σ obtained from first-principles computations (typically less than 50). Once the ECI have been determined, the free energy of the alloy for any given configuration can be quickly calculated, making it possible to explore a large number of configurations without recalculating the free energy of each of them from first principles.

Within the framework of the cluster expansion formalism, *ab initio* modeling of alloy free energies and phase diagrams proceeds through the following steps: (i) calculation of the free energies $F(\sigma,T)$ for a selected set of alloy configurations to fit the cluster-interaction coefficients in Eq. 15, (ii) determination of the optimal set of clusters to construct a predictive cluster expansion, (iii) determination of the "ground-state" structures, i.e., those minimum-energy structures representing the stable thermodynamic phases at low temperature, and (iv) Monte-Carlo simulations of configurational free energies for all phases of interest. The remainder of this section describes each of these steps in further detail.

While the methods for performing step (i) were described already in the previous section, construction of a cluster expansion typically requires calculation of $F(\sigma,T)$ for on the order of 50 structures, which represents a computationally demanding task and thus is commonly performed using more approximate yet computationally more efficient techniques. A popular way to handle this issue is to approximate $F(\sigma,T)$ by the corresponding energy at absolute zero $F(\sigma,0)$, which is readily provided by first-principles total energy methods. However, in recent years, the validity of this approximation has been scrutinized.[41,46-52] It is generally recognized that this approximation is one of the main causes of the widely reported systematic overestimation of transition temperatures in *ab initio* phase diagram calculations.[41]

In an effort to alleviate this problem, computationally efficient ways to account for lattice vibrations have been proposed. The simplest scheme is the Moruzzi-Janak-Schwarz (MJS) method,[53] in which the vibrational free energy is calculated within the Debye approximation and under reasonable assumptions regarding the elastic tensor in order to obtain an expression that solely depends on a compound's bulk modulus and average density. Agreement with the experimentally measured phase diagrams was substantially improved by including lattice vibrations in this way in a number of alloy systems.[54-56]

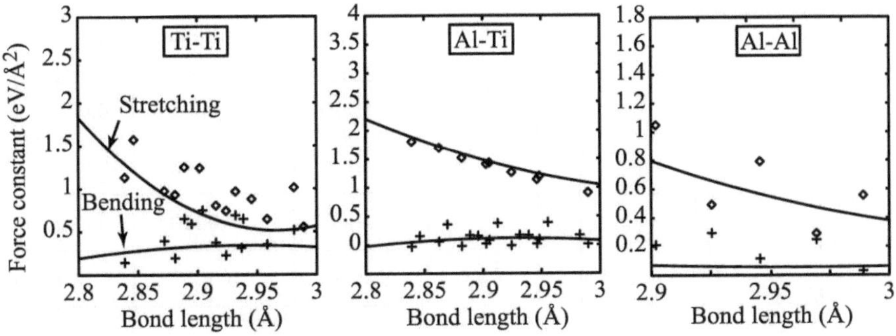

Figure 4. Determination of the bond stiffness vs. bond-length relationship in the hcp Ti-Al system the for construction of length-dependent transferable force constants (LDTFC).

Another popular scheme[57-59] is the use of a shorter-range cluster expansion to represent the configuration dependence of $F(\sigma,T) - F(\sigma, 0)$, thus necessitating fewer lattice dynamics calculations, while maintaining very accurate cluster expansion for the zero-temperature $F(\sigma,0)$ contribution.

The concept of length-dependent transferable force constants (LDTFC) has recently been suggested[41,60-62] to improve the computational efficiency of lattice dynamics calculations. The basic idea is to rely on the observation that bond length is a good predictor of bond stiffness (for a given lattice and a given type of chemical bond). The bond length-bond stiffness relationships can be determined by calculating, from first-principles, the stretching and bending force constants in a few ordered compounds as a function of volume (see Fig. 4). Once this relationship is known, the force constants needed for the calculation of the vibrational free energy of a given structure can be predicted solely from the knowledge of its relaxed geometry (which provides the bond lengths). Since it then becomes possible to quickly determine the lattice dynamics for a large number of configurations, the accuracy of the cluster expansion of the vibrational free energy is no longer limited by the number of terms that can be included in the cluster expansion, although it may be limited by the accuracy of the LDTFC approximation, which can fortunately be assessed during the determination of the bond length-bond stiffness relationships.

The second step in the construction of a cluster expansion is the optimal selection of the clusters to be included in the expansion. If too few terms are kept, the predicted energies may be imprecise because the truncated cluster expansion cannot account for all sources of energy fluctuations. If too many terms are kept, the least-squares fit becomes excessively noisy because the number of fitting parameters is too large relative to the amount of available

a)

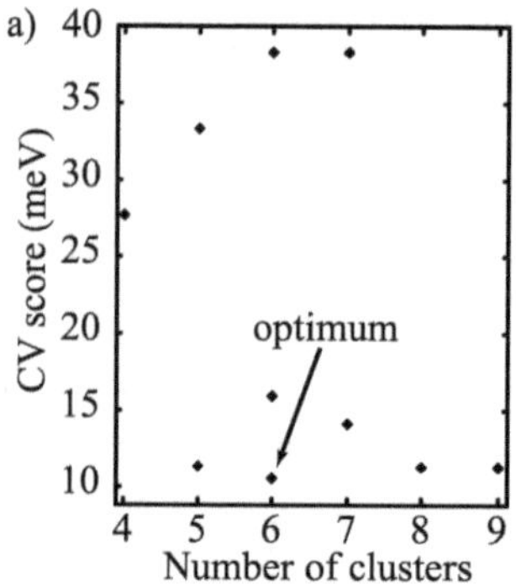

b)

	Energetic (eV)	Vibrational (eV)	Electronic (eV)
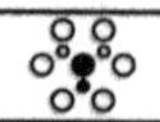	0.0479	-0.0042	-0.0004
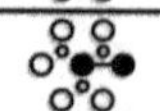	0.0417	-0.0048	-0.0008
	-0.0137	0.0020	0.0003
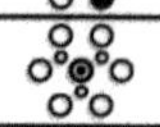	0.0089	0.0017	-0.0013

at T = 1000K (Vibrational and Electronic columns)

Figure 5. a) Optimal selection of clusters via crossvalidation (the optimal number of clusters includes the empty, the point cluster and 4 pairs). b) Values of the effective cluster interactions (ECI) obtained from a least-squares fit to a set of 14 Ti-rich structures. The energetic, vibrational and electronic components of the free energy at a representative temperature of 1000 K are separately given, to illustrate the typical relative magnitude of each contribution.

data. The best compromise between those two unwanted effects can be found by minimizing the cross-validation (CV) score,[63,64] defined as

$$(CV)^2 = n^{-1}\sum_{i=1}^{n}(E_i - E_{(i)}) \tag{18}$$

where E_i is the calculated energy of structure i, while $E_{(i)}$ is the predicted value of the energy of structure i obtained from a least-squares fit to the $(n - 1)$ other structural energies. Choosing the number of terms that minimizes the CV score has been shown to be an asymptotically optimal[65] selection rule. In contrast to the well-known mean squared error, the CV score is not monotonically decreasing in the number of fitting parameters. As the number of parameters to be fitted increases, the CV score first decreases because an increasing number of degrees of freedom are available to explain the variations in energy. The CV score then goes through a minimum before increasing, due to a decrease in predictive power caused by an increase in the noise in the fitted ECI. The best compromise between these two effects can then be found. Naturally, as more first-principles data is calculated and included in the fit, the minimum CV score will be achieved by including an increasing number of terms in the cluster expansion. The cluster selection procedure is illustrated in Fig. 5.a, while Fig. 5.b shows the ECI obtained through a least-squares fit to a database of *ab initio* 14 free energies determined using the LDTFC method for the vibrational contributions and Eq. 3 for the electronic contributions.

In some applications the development of a converged cluster expansion can be complicated by the presence of long-ranged interatomic interactions mediated by electronic-structure (Fermi-surface), electrostatic and/or elastic effects. Long-ranged interactions lead to an increase in the number of ECIs that must be computed, and a concomitant increase in the number of configurations that must be sampled to derive them. For metals it has been demonstrated how long-ranged electronic interactions can be derived from perturbation theory using coherent-potential approximations to the electronic structure of a configurationally-disordered solid as a reference state (see Ref. 7 for a review). Effective approaches to modeling long-ranged elastically-mediated interactions have also been formulated.[66,67] Such elastic effects are known to be particularly important in describing the thermodynamics of mixtures of species with very large differences in atomic "size".

In the limiting case when some of the atomic species in a multicomponent alloy are dilute, a so-called local cluster expansion can be used[44] and offers the advantage of reducing the number of ECI that need to be determined, since the interactions between dilute species do not need to be determined. The main difference with the conventional cluster expansion is that the symmetry group used to determine the equivalence between clusters is the local point group associated with a single dilute antisite defect (or vacancy) rather than the full space group of the parent lattice. Local cluster expansions have also found important application in the parameterization of the configurational-dependence of activation barriers for diffusion.[68,69]

4.2 Determining Ground-State Structures

The cluster expansion tremendously simplifies the third step of the analysis of phase stability: The search for the lowest energy configuration at each composition of the alloy system. Determining these so-called ground states is important because they determine the general topology of the alloy phase diagram: Each ground state is typically associated with one of the stable phases of the alloy system. There are four main approaches to identify the ground states of an alloy system.

With the enumeration method, all the configurations whose unit cell contains less than a given number of atoms are enumerated and their energy is quickly calculated using the value of $F(\sigma,0)$ predicted from the cluster expansion. The energy of each structure can then be plotted as a function of its composition (see Fig. 7) and the points touching the lower portion of the convex hull of all points indicate the ground states. While this method is approximate, as it ignores ground states with unit cell larger than the given threshold, it is simple to implement and has been found to be quite reliable, thanks to the fact that most ground states indeed have a small unit cell.

Simulated annealing offers another way to find the ground states. It proceeds by generating random configurations via Monte Carlo simulations using the Metropolis algorithm[70,71] that mimic the ensemble sampled in thermal equilibrium at a given temperature. As the temperature is lowered, the simulation should converge to the ground state. Thermal fluctuations are used as an effective means of preventing the system from getting trapped in local minima of energy. While the constraints on the unit cell size are considerably relaxed relative to the enumeration method, the main disadvantage of this method is that, whenever the simulation cell size is not an exact multiple of the ground state unit cell, artificial defects will be introduced in the simulation that need to be manually identified and removed. Also, the risk of obtaining local rather than global minima of energy is not negligible and must be controlled by adjusting the rate of decay of the simulation temperature.

There exists an exact, although computational demanding, algorithm to identify the ground states.[5] This approach relies on the fact that the cluster expansion is linear in the correlations $\sigma_\alpha \equiv \left\langle \prod_{i\in\alpha'} \sigma_i \right\rangle$. Moreover, it can be shown that the set of correlations σ_α that correspond to "real" structures can be defined by a set of linear inequalities. These inequalities are the result of lattice-specific geometric constraints and there exists systematic methods to generate them.[5] As an example of such constraints, consider the fact that it is impossible to construct a binary configuration on a triangular lattice where the nearest neighbor pair correlations take the value -1 (i.e., where all nearest-neighbors are between different atomic species). Since both the objective function and the constraints are linear in the correlations, linear programming techniques can be used to determine the ground states. The main difficulty associated with this method is the fact that the resulting linear programming problem involves a number of dimensions and a number of inequalities that both grow exponentially fast with the range of interactions included in the cluster expansion.

While the three above methods enable the determination of the ground states that are superstructures of a given parent lattice, a data mining technique[72,73] has been proposed to carry out ground state searches among crystal structures that do not share a common parent lattice. Instead of relying on a cluster expansion, this method relies on a large database of existing *ab initio* formation energy calculations in a variety of alloy systems to identify correlations between the formation energies of various crystal structures, through a so-called principal moment analysis. These correlations can then be used to guide the ground state search for a new alloy system not contained in the database as follows. First, the formation energies a few simple candidate compounds are calculated. Second, the correlations known from the database are used to predict the most likely ground states given the known formation energy of the few simple compounds considered. Next, the formation energies of the predicted ground states are calculated and used to further refine the ground state prediction. The process is iterated until no new ground states are predicted. While this method does not have the ability to discover ground states with a crystal structure that is not included in the database, its predictive capabilities steadily improve over time, as researchers contribute to enlarge the database of available formation energies.

4.3 Free Energy Calculations

Once the ground-state structures (i.e., the stable phases at low temperature) have been derived, construction of a solid-state phase diagram from first principles requires calculation of free energies for each of these ordered phases, as well as the different possible disordered solid solutions, and any other ordered phases that may appear at finite temperatures. Calculation of the required composition and temperature dependent free energies formally requires summing the configurational partition function given in Eq. 13. Historically, the infinite summation defining the alloy partition function has been approximated through various mean-field methods.[5,6] However, the difficulties associated with extending such methods to systems with medium to long-ranged interactions, and the increase in available computational power enabling Monte Carlo (MC) simulations to be directly applied, have led to reduced reliance upon these techniques more recently.

MC simulations readily provide thermodynamic quantities such as energy or composition by making use of the fact that averages over an infinite ensemble of microscopic states can be accurately approximated by averages over a finite number of states generated by "importance sampling". Moreover, quantities such as the free energy, which cannot be written as ensemble averages, can nevertheless be obtained via thermodynamic integration[25] using standard thermodynamic relationships to rewrite the free energy in terms of integrals of quantities that can be obtained via ensemble averages. For instance, since energy $E(T)$ and free energy $F(T)$ are related through $E(T)= \partial\,(F(T)/T)\,/\partial\,(1/T)$ we have

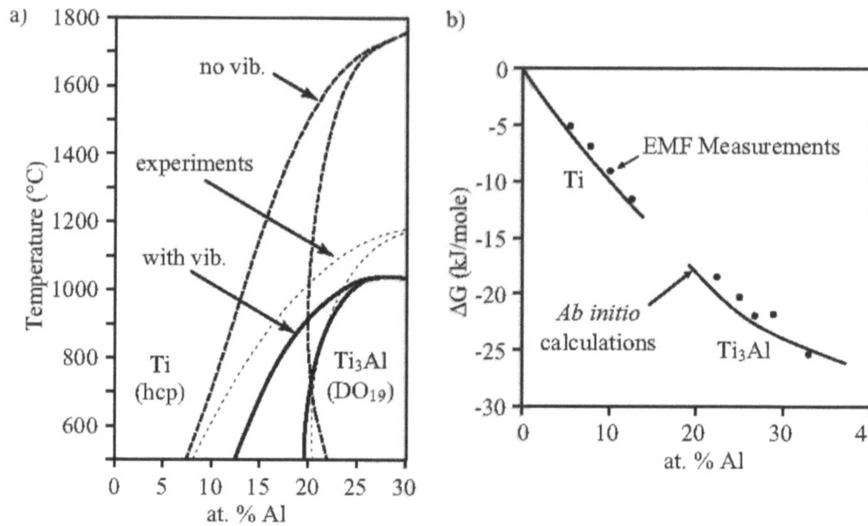

Figure 6. a) Comparison between an experimentally assessed phase diagram[74] and the *ab initio* phase diagram calculated including or excluding the effect of lattice vibrations. b) Comparison between calculated free energies with the ones obtain from electromotive force (EMF) measurements.[75]

$$\frac{F(T)}{T} - \frac{F(T_0)}{T_0} = \int_{T_0}^{T} \frac{E(T)}{T^2} dT \tag{19}$$

and free energy differences can therefore be obtained from MC simulations providing *E(T)*. Figures 6.a and 8 show two phase diagrams obtained by combining first principles calculations, the cluster expansion formalism and MC simulations, an approach which offers the advantage of handling, in a unified framework, both ordered phases (with potential thermal defects) and disordered phases (with potential short-range order).

Figure 6.a demonstrates how the inclusion of the effect of lattice vibrations in *ab initio* calculations dramatically improves the agreement between calculated and experimentally determined phase diagrams. Figure 6.b illustrates that *ab initio* predictions of free energies that include both configurational and vibrational effects can provide accuracy comparable with the scatter in experimental determinations of the same quantity via electromotive force measurements. A predicted phase diagram with a more complex topology is given in Fig. 8, along with the pattern of Li ion ordering corresponding to each of the ordered phases.

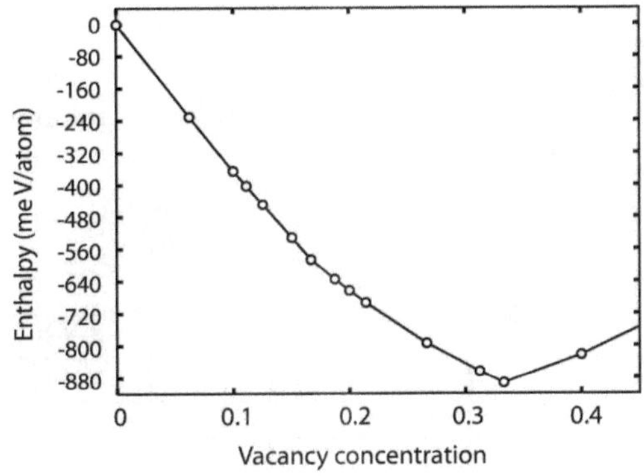

Figure 7. Ground state search using the enumeration method in the Sc_x $Vacancy_{1-x}S$ system. Diamonds represent the formation energies of about 3×10^6 structures, predicted from a cluster expansion fitted to LDA energies. The ground states, indicated by open circles, are the structures whose formation energy touches the convex hull (solid line) of all points. (Reproduced from Ref. 76, with the permission of the authors).

4.4 Free Energy of Phases with Dilute Disorder

To conclude this section we review a simplified framework for the calculation of configurational free energies in alloy phases with "dilute" compositions. For phases containing a sufficiently dilute concentration of defects (e.g., solute elements for a solid solution, or antisite defects for an intermetallic compound), such that defect interactions can be safely neglected, the free energy can be written in the following approximate form:

$$F = F_0 + \sum_i \sum_\alpha (x_i^\alpha (\Delta F_i^\alpha + k_B T \ln x_i^\alpha)) \tag{20}$$

where F_0 is the free energy of the non-defected phase (pure element for a dilute solid solution or stoichiometric compound for an intermetallic phase), the sums are over the sites i of the unit cell and the different defect types α, x_i^α is the site concentration of defect α, and ΔF_i^α is the defect free energy of formation, including vibrational and electronic entropy contributions. When alloy phases are sufficiently dilute, the calculation of configurational free energies and phase diagrams can thus be reduced to computation of the symmetry-distinct defect formation free energies ΔF_i^α entering Eq. 20. Such calculations can be accurately performed through the use of supercell

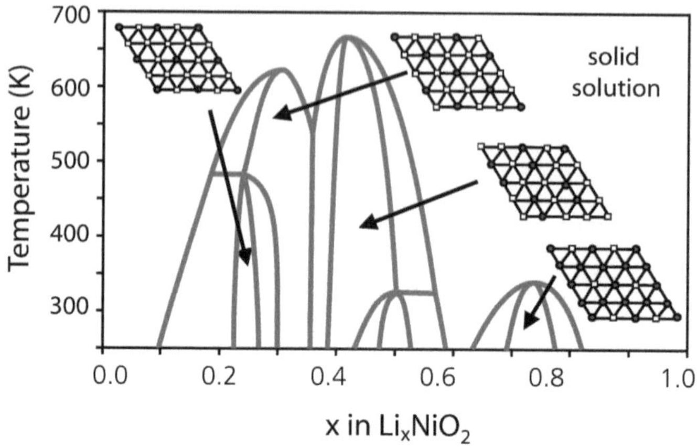

Figure 8. Phase diagram of the Li_xNiO_2 system calculated from first principles. The insets show the crystal structures of the predicted ordered phases. Reproduced from Ref. 77, with the permission of the authors.

geometries within standard first-principles codes. Recent examples illustrating the application of this formalism for the first-principles calculation of solvus boundaries in Al-Sc and Al-Si can be found in Refs. 78 and 79, respectively. For both systems, vibrational contributions to the calculated solute free energies were found to lead to sizeable corrections to calculated phase-boundary temperatures. The formalism has also been applied recently to investigations of solute partitioning in multicomponent two-phase alloy systems.[80,81]

The formalism outlined in the preceding paragraph has also found numerous applications in first-principles calculations of equilibrium vacancy and anti-site-defect concentrations, as well as solute site-selection preferences in multicomponent intermetallic compounds (see, e.g., Refs. 82-87, and references cited therein). In such applications, the free energy given in Eq. 20 is more commonly expressed as a grand-potential, written in terms of solute chemical potentials rather than the site fractions. In this formalism the equilibrium defect concentrations can be conveniently derived from the derivative of the grand potential with respect to the chemical potentials.

5. INTEGRATION OF *AB INITIO* AND CALPHAD METHODS FOR MULTICOMPONENT ALLOY DESIGN

In this section we describe recent applications of *ab initio* thermodynamics methods within an effort aimed at the accelerated design and development of Nb-based superalloys combining oxidation resistance, creep strength and ductility for aeroturbine applications at temperatures at or above 1573 K. The approach, detailed elsewhere,[24] involved the design of a Nb-based superalloy with coherent aluminide dispersion as a strengthening phase, analogous to classical γ-γ' microstructure in Ni-base superalloys. Aluminide-phase dispersions were chosen based on the desire to have the precipitates act not only as strengthening particles, but also as Al reservoirs to achieve Al_2O_3 passivation for oxidation resistance. In the initial stages of the project first-principles calculations were employed to calculate the total energies and cohesive properties of several aluminide compounds with ordered bcc-based Heusler-type crystal-structures,[24,88] in order to identify promising candidate coherent-precipitate phases in a bcc (Nb) matrix. From a consideration of the calculated formation energies and lattice parameters, it was found that Pd_2HfAl provided the optimum combination of phase stability and lattice misfit. This choice then defined the quaternary system Al-Hf-Nb-Pd as the basis for the design of creep-resistant two-phase coherent microstructures.[89,90] Below we focus on the modeling of phase stability in Al-Hf-Nb, which represents an important ternary sub-system. Since the experimental phase diagram and thermodynamic data for this system was very limited, *ab initio* methods were exploited to accelerate the development of the databases required to employ computational-thermodynamics methods, based on the CALPHAD approach, as a predictive framework for guiding the materials-design process. In this work, the application of the methods described in the previous sections was facilitated by employing the Alloy Theoretic Automated Toolkit (ATAT), which is a set of open-source programs for performing *ab initio* thermodynamics calculations.[63,91-93]

5.1 Overview of CALPHAD Approach

As background for describing the integration of *ab initio* and computational-thermodynamic methods, we provide here a brief overview of the free-energy models employed in the CALPHAD calculations for the Al-Hf-Nb system. The molar Gibbs energy of the solid-solution (fcc, bcc, and hcp) and liquid phases were expressed as:

$$G_m^{\phi} - H^{SER} = {}^{ref}G^{\phi} + {}^{id}G + {}^{xs}G^{\phi} \tag{21}$$

where φ denotes the phase, H^{SER} denotes a standard value of the enthalpy at room temperature, ${}^{ref}G^{\varphi}$ denotes the composition-weighted average of the free-energies of the pure-element reference states, and ${}^{id}G^{\varphi}$ is the ideal Gibbs energy of mixing. The quantity ${}^{xs}G^{\varphi}$ is the excess Gibbs energy of mixing, expressed by a Redlich-Kister-Muggianu polynomial:[94,95]

$$
{}^{xs}G^{\phi} = \sum_{i \neq j} \left\{ x_i x_j \left[L_{ij}^0 + (x_i - x_j) L_{ij}^1 + (x_i - x_j)^2 L_{ij}^2 + \ldots \right] \right\} +
$$

$$
+ \sum_{i \neq j \neq k} x_i x_j x_k \left[L_{ijk}^0 x_i + L_{ijk}^1 x_j + L_{ijk}^2 x_k \right] \tag{22}
$$

where i,j,k = Al, Hf, Nb, and both the binary and ternary interaction parameters, (L_{ij}) and (L_{ijk}), may be temperature dependent.

For the stoichiometric intermetallic phases (i.e., line compounds) the molar Gibbs energy is expressed as:

$$
G^{\psi} - H^{SER} = {}^{ref}G + \Delta^f G^{\psi} \tag{23}
$$

where $\Delta^f G^{\psi}$ is the Gibbs energy of formation. In the case of intermetallics with finite homogeneity range, a sub-lattice model[96] is used where the molar Gibbs energy is also expressed in the form given by Eq. 21. As a specific example, consider a binary intermetallic phase A_pB_q with a free energy represented by a two-sublattice model, denoted as $(\mathbf{A},B)_p(A,\mathbf{B})_q$ with p and q being the sublattice-site ratios. The reference free energy is written as:

$$
{}^{ref}G^{\psi} = Y_A^I Y_A^{II} \Delta G_{A:A}^{\psi} + Y_A^I Y_B^{II} \Delta G_{A:B}^{\psi} + Y_B^I Y_A^{II} \Delta G_{B:A}^{\psi} + Y_B^I Y_B^{II} \Delta G_{B:B}^{\psi} \tag{24}
$$

where Y_A^I and Y_A^{II} are the site fractions of A on the first and second sublattices, respectively, and similarly for Y_B^I and Y_B^{II}. The parameters $\Delta G_{A:A}$, $\Delta G_{A:B}$, $\Delta G_{B:A}$, and $\Delta G_{B:B}$ in Eq. 24 represent the Gibbs free energy of formation for the stable phase A_pB_q and the three virtual phases, A_pA_q (pure A), A_qB_p, and B_pB_q (pure B), all having the structure ψ. It is important to note that the construction of such a free-energy model requires values for the Gibbs energies of formation of all the virtual phases A_pA_q, A_qB_p and B_pB_q, even if A_pB_q is the only thermodynamically stable phase with structure ψ. The excess free energy in this two-sublattice model takes the form:

$$
{}^{xs}G^{\psi} = Y_A^I Y_B^I \left(Y_A^{II} L_{A,B:A}^{\psi} + Y_B^{II} L_{A,B:B}^{\psi} \right) + Y_A^{II} Y_B^{II} (Y_A^I L_{A,A:B}^{\psi} +
$$

$$
+ Y_B^I L_{B,A:B}^{\psi}) + Y_A^I Y_B^I Y_A^{II} Y_B^{II} L_{A,B:A,B}^{\psi} \tag{25}
$$

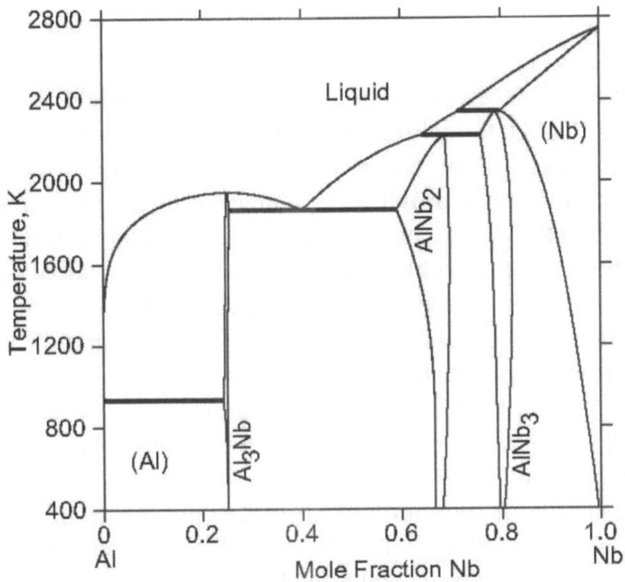

Figure 9. Calculated Al-Nb phase diagram obtained by integrating *ab initio* phase stability results and experimental data within CALPHAD formalism.

with $L^{\psi}_{A,B:A} = L^{0}_{A,B:A} + (Y^{I}_{A} - Y^{I}_{B})L^{1}_{A,B:A}$, and so forth, where again the polynomial coefficients (e.g., $L^{\psi}_{A:A,B}$) may be temperature dependent.

In practice, CALPHAD thermodynamic modeling requires values for the various polynomial coefficients and reference free energies entering Eqs. 21-25. These parameters are generally fit to available experimental data for measured phase boundaries and thermodynamic properties. In the absence of sufficient measured data to uniquely fit the model parameters, *ab initio* methods can be employed as a "virtual calorimeter" to augment thermodynamic databases with computed values of enthalpies and entropies of formation for solid solutions and intermetallic compounds.

5.2 Integrated *ab initio* and CALPHAD phase-stability modeling

5.2.1 The Al-Nb system

There are six phases in the equilibrium phase diagram: liquid, bcc, fcc, Al_3Nb (DO_{22}), $AlNb_2$ (σ) and $AlNb_3$ (A15). All of these intermetallic phases have measurable homogeneity ranges. Previous CALPHAD assessments,[97,98] and *ab initio* calculations of thermodynamic properties[99] for bcc and fcc phases have been reported in this system. Our own efforts were aimed at a somewhat more comprehensive study of phase stability of all solid phases in the system integrating *ab initio* calculations within CALPHAD as a basis for refining the assessed phase diagram and free-energy models.

In the CALPHAD modeling of Al-Nb intermetallics, a two-sublattice model for Al_3Nb and $AlNb_3$, and a three-sublattice model for the σ phase[100] were used. An advantage provided by integrating *ab initio* methods with CALPHAD is that the energy parameters of all the virtual phases, required in the construction of the sublattice models, can be calculated directly. For example, in a three-sublattice description of the σ phase, $(\mathbf{Al},Nb)_{10}(Nb)_4(Al,\mathbf{Nb})_{16}$, values are required for the formation free energies of the three virtual phases $Al_{26}Nb_4$, $Al_{16}Nb_{14}$ and Nb_{30} having the structure of the σ phase. Similarly, a two-sublattice description of the A15 phase, $(\mathbf{Al},Nb)(Al,\mathbf{Nb})_3$, gives rise to three virtual A15 phases Al_4, Al_3Nb and Nb_4. For the construction of the CALPHAD free-energy models we have calculated the energy of formation of all virtual phases in the Al-Nb system. Additionally, the formation entropies for all of these compounds were estimated based on a linear correlation between entropies and energies of formation which we have parameterized from our own and previously published calculations making use of the methods described in sections 2 and 3.[101] All *ab initio* and experimental data were then integrated within the CALPHAD formalism, and the resultant calculated Al-Nb phase diagram is shown in Fig. 9.

5.2.2 The Al-Hf system

There are seven intermetallic phases, three solid solutions (bcc, fcc and hcp) and the liquid phase in the equilibrium Al-Hf phase diagram.[102] In addition, three metastable phases have also been reported in this system. CALPHAD modeling of phase equilibria was reported twice, first by Kaufman[103] when there was no experimental thermodynamic data available, and later by Wang[104] using limited experimental data. As part of a recent study of the systematics of alloy energetics and to augment the calorimetric data of Al-TM (TM = Ti, Zr, Hf) intermetallics, we calculated the heats of formation of all the reported stable, metastable, and numerous virtual Al-Hf intermetallics.[105]

For CALPHAD modelling of phase diagrams, both enthalpies and entropies of formation of phases are required. While the enthalpy of formation of intermetallics is often measured by calorimetry, values for entropies of formation are reported only occasionally. As discussed in sections 2 and 3, both of these quantities can be computed by *ab initio* methods. However, for intermetallic compounds with complex crystal structures, the calculation of vibrational entropies of formation can be computationally demanding. The Al-Hf system falls in this category as the

Al_3Hf_2 and Al_2Hf_3 compounds have orthorhombic and tetragonal crystal structures, respectively, with 40 and 20 atoms per unit cell. To simplify the approach, entropy values were again estimated using the correlation between intermetallic enthalpies and entropies of formation (alluded to above) parameterized from extensive *ab initio* calculations.

Our CALPHAD modeling of the Al-Hf system combined *ab initio* calculations and calorimetric data for the formation enthalpies of Al-Hf intermetallics, with estimated values of the formation entropies, to derive the parameters in line-compound free-energy models. In addition, the modeling made use of *ab initio* thermodynamic data for solid solution phases calculated by the cluster expansion and Monte Carlo techniques.

5.2.3 The Hf-Nb system

There are only three phases in the equilibrium Hf-Nb phase diagram: bcc and hcp solid solutions and the liquid. The thermodynamic properties of the solid solution phases were calculated by the cluster-expansion method, and the results were then integrated within the CALPHAD formalism to model the phase diagram of this system, as has been described in detail by us elsewhere.[21] While the heat of mixing for the hcp solid-solution phase was found to be weakly positive, that of the bcc solid-solution was predicted to be weakly negative. In other words, the cluster expansion technique predicted a weak ordering tendency on the bcc lattice. It was not possible to anticipate this tendency in a previous CALPHAD assessment of the Hf-Nb system[106] due to the lack of measured enthalpy data, and the bcc phase was thus assigned a positive heat of mixing.

5.2.4 The ternary Al-Hf-Nb system

In modeling this ternary phase diagram, all Al-Hf intermetallics were treated as stoichiometric binary phases, i.e., without allowing any Nb solubility, with the single exception of Al_3Hf. Since both Al_3Nb and Al_3Hf have the DO_{22} structure at high temperature, they were treated as one phase, i.e., $Al_3(Hf,Nb)$, allowing random mixing of Nb and Hf on one sublattice. The sublattice models for $AlNb_3$ (A15) and $AlNb_2$ (σ) were extended to include Hf, $(\mathbf{Al},Hf,Nb)(Al,Hf,\mathbf{Nb})_3$ and $(Al,Hf,Nb)_{10}(Nb,Hf)_4(Al,Hf,Nb)_{16}$, respectively. This introduces additional virtual phases, the energy parameters of which were all calculated by *ab initio* methods. Modeling of ternary solid-solution phases was facilitated by cluster-expansion calculations of the thermodynamic properties of fcc and hcp solid solution in the Hf-Nb and Al-Nb systems, respectively. In a more traditional CALPHAD approach, relying on measured data entirely, such data would not be accessible due to the fact that fcc and hcp solid-solutions phases are absent in the respective

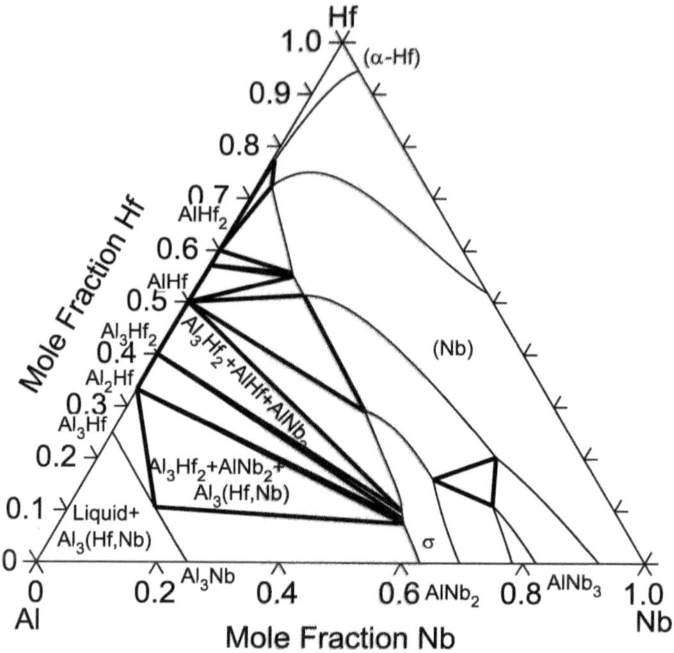

Figure 10. Calculated isothermal section of the Al-Hf-Nb system at 1573 K.

equilibrium phase diagrams. No ternary interaction was introduced for the solution phases (liquid, bcc, fcc and hcp) since, at the time of this study, there was no experimental or *ab initio* thermodynamic data available for ternary compositions.

Figure 10 shows the calculated isothermal section of Al-Hf-Nb system at 1573 K. In the context of the design of Nb-base superalloy, it turned out that a very important prediction is that the solubility of Al in (Nb) can be increased significantly by adding Hf. For example, at 1573 K the solubility of Al in (Nb) is about 8 atomic percent. However, our calculated results show that by adding 30 atomic percent Hf the solubility of Al can be increased by more than a factor of two, to 17 atomic percent. As will be shown below, based on the results of kinetics simulations, that were later verified experimentally,[107] this increase in Al solubility proves to be highly beneficial for significantly reducing oxygen transport kinetics in the (Nb) solid-solution phase.

5.3 Computational kinetics of the Al-Hf-Nb system: Oxygen in bcc solid solution

One of the most substantial challenges in the design of Nb-based superalloys for high-temperature applications is the need to very

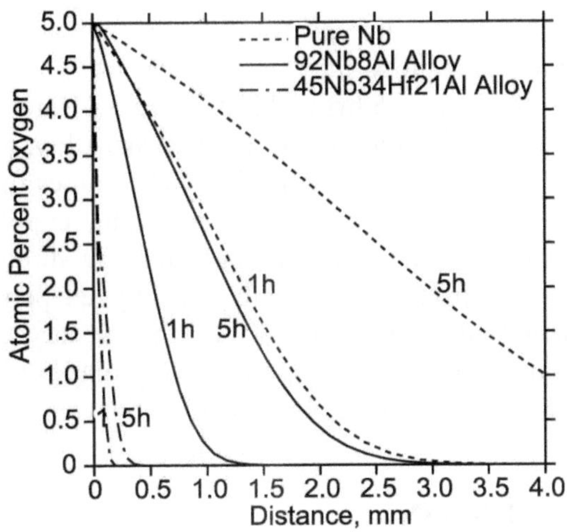

Figure 11. Intrinsic oxygen transport kinetics in pure Nb and Nb-based alloys at 1573 K, demonstrating the effect of alloy chemistry.

significantly improve the intrinsic oxidation resistance. As discussed in detail by Ref. 24, the formation of a self-protective external oxide scale requires a decrease in the oxygen solubility and a decrease in oxygen transport kinetics, both by several orders in magnitude compared to pure Nb. Within the framework of computational thermodynamics and kinetics, we demonstrated that Al-Hf-Nb solid solutions help in achieving these goals.

The modeling and simulation of oxygen transport was carried out using the DICTRA (DIffusion Controlled TRAnsformation)[108] package. This is a general software package to simulate diffusion controlled transformations in multicomponent systems involving multiple phases, in one dimension. DICTRA uses Thermo-Calc to calculate the thermodynamic factor of the phases to convert mobility into diffusivity, and also to compute the local equilibrium between the phases. In other words, to use DICTRA successfully a complete thermodynamic description of the participating phase(s) is needed first, which is then combined with a database comprising the diffusion parameters of the corresponding phase(s). In parallel with the thermodynamic-database development efforts described above, a multicomponent mobility database relevant to Nb-based alloys was also developed[107] within the formalism proposed by Andersson.[109]

Figure 11 compares the simulated oxygen profiles in pure Nb and two Nb-based alloys at 1573 K, clearly demonstrating the effect of alloy chemistry on the intrinsic transport kinetics of oxygen in the (Nb) solid-solution phase. In these simulations it was assumed that the oxygen concentration at the surface of the bulk alloy is 5 atomic percent, which is based on values derived from the equilibrium Nb-O phase diagram. This is equivalent to assuming that, irrespective of the type of oxide, the oxygen potential at the oxide/base metal interface is constant. The simulated oxygen penetration depths after 1 and 5 hrs at 1573 K are 3.21 and 6.01 mm in pure Nb, 1.26 and 2.51 mm in 92Nb8Al (atomic percent) alloy, and 0.14 and 0.35 mm in 45Nb34Hf21Al (atomic percent) alloy. The values for Nb-Al and Nb-Al-Hf alloys are in excellent agreement with the experimental data as verified by measured hardness profiles and also by direct observation of microstrucutres.[107] Compared to pure Nb, the oxygen penetration depth is reduced by a factor of more than 20 in the ternary Nb-Al-Hf alloy. Mechanistically, this is most likely due to strong binding energies of Al-O and Hf-O, and possibly Hf-O-Nb and Al-Hf-O in the ternary alloy.

6. CONCLUSION

Ab initio thermodynamic modeling of alloys typically proceed by employing a hierarchy of modeling techniques targeted at handling a specific type of thermal fluctuation. At the lowest level, electronic excitations can be modeled within standard finite temperature density functional theory while lattice vibrations can be treated in the quantum or classical regime within the quasiharmonic approximation. At the next level, configurational excitations are described via a lattice model in conjunction with a so-called temperature-dependent cluster expansion which conveniently summarizes the effective interactions between nearby atoms that implicitly account for electronic, vibrational as well as relaxation effects via a technique called the "coarse graining" of the "fast" degrees of freedom of the partition function.

The appeal of this approach is that the cluster expansion can be used to efficiently search for the ground states of the alloy system (which will typically determine the phases present in the phase diagram) and to carry out Monte Carlo simulations to determine the phase diagram and/or any thermodynamic property of interest. This formalism thus encompasses, at the atomic level, dilute or concentrated point defects in ordered phases as well as short-range order in solid solutions.

At the topmost level of the hierarchy, CALPHAD thermodynamic modeling relies on thermodynamic databases describing the relationships between various macroscopic quantities (e.g. free energies and compositions of each phase) in terms of simple functional forms in which the unknown parameters are traditionally fit to available experimental data for measured phase boundaries and thermodynamic properties. In the absence of sufficient measured data to uniquely fit the model parameters, *ab initio* methods can be employed as a "virtual probe" to augment thermodynamic databases with computed values of enthalpies and entropies of formation for solid solutions and intermetallic compounds.

ACKNOWLEDGEMENTS

Financial support was provided by the Air Force Office of Scientific Research, USAF, under grant number F49620-01-1-0529. The authors are grateful to Prof. G.B. Olson for his interest and support on the phase stability modeling of Al-Hf-Nb system. Supercomputing resources were provided by the National Partnership for Advanced Computational Infrastructure at the University of Michigan and also at the University of Illinois at Urbana-Champaign.

REFERENCES

1. P. Hohenberg and W. Kohn, Inhomogeneous electron gas, *Phys. Rev.* **136**, 864 (1964).
2. W. Kohn and L. J. Sham, Self-consistent equations including exchange and correlation effects. *Phys. Rev.* **140**, A1133 (1965).
3. R. M. Dreizler and E. K. U. Gross, Density Functional Theory: an Approach to the Quantum Many-Body Problem. Springer-Verlag, Berlin, 1990.
4. M. C. Payne, M. P. Teter, D. C. Allan, T. A. Arias, and J. D. Joannopoulos, Iterative minimization techniques for *ab initio* total-energy calculations: molecular dynamics and conjugate gradients. *Rev. Mod. Phys.* **64**, 1045 (1992).
5. F. Ducastelle, *Order and Phase Stability in Alloys. Elsevier Science*, New York, 1991.
6. D. de Fontaine, Cluster approach to order-disorder transformation in alloys, *Solid State Phys.* **47**, 33 (1994).
7. G. M. Stocks, D. M. C. Nicholson, W. A. Shelton, B. L. Gyorffy, F. J. Pinski, D. D. Johnson, J. B. Staunton, P. E. A. Turchi, and M. Sluiter, *First principles theory of disordered alloys and alloy phase stability*. In Turchi and Gonis, 1994, p. 305.
8. A. Zunger, *First principles statistical mechanics of semiconductor alloys and intermetallic compounds*. In Turchi and Gonis, 1994, p. 361.
9. P. E. A. Turchi, First principles theory of disordered alloys and alloy phase stability. Vol. 1, New York. John Wiley, 1995, p. 21.
10. G. Ceder, A. van der Ven, C. Marianetti, and D. Morgan, First-principles alloy theory in oxides, *Modelling Simul. Mater Sci Eng.* **8**, 311 (2000).
11. M. Asta, V. Ozoliņš, and C. Woodward, A first-principles approach to modeling alloy phase equilibria. *JOM - Journal of the Minerals Metals & Materials Society* **53**,16 (2001).
12. D. Alfé, G. A. de Wijs, G. Kresse, and M. J. Gillan, Recent developments in *ab initio* thermodynamics, *Int. J. Quantum Chem.* **77**, 871-879 (2000).
13. G. B. Olson, Computational design of hierarchically structured materials. *Science* **277**, 1237-1242 (1997).
14. L. Kaufman and H. Bernstein, *Computer Calculation of Phase Diagrams*, Academic Press, New York, 1970.
15. U. R. Kattner, G. Eriksson, I. Hahn, R. Schmid-Fetzer, S. Sundman, V. Swamy, A. Kussmaul, P. J. Spencer, T. J. Anderson, T. G. Chart, A. C. E. Silva, B. Jansson, B. J. Lee, and M. Schalin, Applications of computational thermodynamics: Groups 4 and 5: Use of thermodynamic software in process modelling and new applications of thermodynamic calculations. *CALPHAD* **24**, 55 (2000).

16. G. Ghosh and G. B. Olson, Computational thermodynamics and the kinetics of martensitic transformation. *J. of Phase Equilibria* **22**, 199 (2001).
17. J. Ågren, F. H. Hayes, L. Höglund, U. R. Kattner, B. Legendre, and R. Schmidt-Fetzer, Applications of computational thermodynamics. *Z. Metallkde.* **93**, 128-142 (2002).
18. B. P. Burton, N. Dupin, S. G. Fries, G. Grimvall, A. F. Guillermet, P. Miodownik, W. A. Oates, and V. Vinograd, Using *ab initio* calculations in the CALPHAD environment. *Z. Metallkde.* **92**, 514-525 (2001).
19. L. Kaufman, P. E. A. Turchi, W. M. Huang, and Z. K. Liu, Thermodynamics of the Cr-Ta-W system by combining the *ab initio* and CALPHAD methods, *CALPHAD* **25**, 419-433 (2001).
20. C. Wolverton, X.-Y. Yan, R. Vijayaraghavan, V. Ozoliņš, Incorporating first-principles energetics in computational thermodynamics approaches, *Acta Mater.* **50**, 2187-2197 (2002).
21. G. Ghosh, A. van de Walle, M. Asta, and G. B. Olson, Phase stability of the Hf-Nb system: From first-principles to CALPHAD. *CALPHAD* **26**, 491-511 (2002).
22. Y. Zhong, C. Wolverton, Y. A. Chang, Z. K. Liu, A combined CALPHAD//first-principles remodeling of the thermodynamics of Al-Sr: Unsuspected ground state energies by rounding up the (un)usual suspects. *Acta Mater.* **52**, 2739-2754 (2004).
23. K. Ozturk, Y. Zhong, L. Q. Chen, C. Wolverton, J. O. Sofo, and Z. K. Liu, Linking first-principles energetics to calphad: An application to thermodynamic modeling of the Al-Ca binary system, *Metall. Mater. Trans A* **36**, 5-13 (2005).
24. G. B. Olson, A. J. Freeman, P. W. Voorhees, G. Ghosh, J. Perepezko, M. Erberhart, and C. Woodward, Quest for noburnium: 1300°C cyberalloy, pp. 113-122, Warrendale, PA. TMS, 2004.
25. D. Frenkel and B. Smit, *Understanding Molecular Simulation: From Algorithms to Applications*, Academic Press, San Diego, second edition, 2002.
26. D. Alfé, M. J. Gillan, and G. D. Price, *Ab initio* chemiation potentials of solid and liquid solutions and the chemistry of the earth's core. *J. Chem. Phys.* **116**, 7127-7136 (2002).
27. N. D. Mermin, Thermal properties of the inhomogeneous electron gas. *Phys. Rev.* **137**, A1441 (1965).
28. A. A. Maradudin, E. W. Montroll, and G. H. Weiss, *Theory of Lattice Dynamics in the Harmonic Approximation*, Second Edition. Academic Press, New York, 1971.
29. C. Wolverton, V. Ozoliņš, Entropically favored ordering: The metallurgy of Al_2Cu revisited, *Phys. Rev. Lett.* **86**, 5518 (2001).
30. A. A. Quong and A. Y. Lui, First-principles calculations of the thermal expansion of metals. *Phys. Rev. B* **56**, 7767 (1997).
31. S. Baroni, P. Giannozzi, and A. Testa, Green's-function approach to linear response in solids. *Phys. Rev. Lett.* **58**, 1861 (1987).
32. P. Giannozzi, S. de Gironcoli, P. Pavone, and S. Baroni, *Ab initio* calculation of phonon dispersions in semiconductors. *Phys. Rev. B* 43, 7231 (1991).
33. X. Gonze, D. C. Allan, and M. P. Teter, Dielectric tensor, effective charges, and phonons in α-quartz by variational density-functional perturbation theory. *Phys. Rev. Lett.* **68**, 3603 (1992).
34. X. Gonze, First-principles responses of solids to atomic displacements and homogeneous electric fields: Implementation of a conjugate-gradient algorithm, *Phys. Rev. B* 55, 10337 (1997).
35. X. Gonze, and C. Lee, Dynamical matrices, Born effective charges, dielectric permittivity tensors, and interatomic force constants from density-functional perturbation theory, *Phys. Rev. B* **55**, 10355 (1997).

36. J. T. Devreese, and P. Camp, eds., *Electronic Structure, Dynamics and Quantum Structural Properties of Condensed Matter*. Plenum, New York, 1985.
37. S. de Gironcoli, Lattice dynamics of metals from density-functional perturbation theory, *Phys. Rev. B* **51**, 6773 (1995).
38. U. V. Waghmare, *Ab Initio Statistical Mechanics of Structural Phase Transitions (Lattice Dynamics, Piezoelectric, Ferroelectric)*. PhD thesis, Yale University, 1996.
39. V. Ozoliņš, *Structural and Vibrational Properties of Transition Metal Systems from ab initio Electronic-Structure Calculations*. PhD thesis, Royal Institute of Technology, Stockholm, Sweden, 1996.
40. W. Zhong, D. Vanderbilt, and K. M. Rabe, Phase transition in $BaTiO_3$ from first principles, *Phys. Rev. Lett.* **73**, 1861 (1994).
41. A. van de Walle, and G. Ceder, The effect of lattice vibrations on substitutional alloy thermodynamics, *Rev. Mod. Phys*. **74**, 11 (2002).
42. J. M. Sanchez, F. Ducastelle, and D. Gratias, Generalized cluster description of multicomponent systems. Physica A, **128**, 334 (1984).
43. A. Van der Ven, M. K. Aydinol, G. Ceder, G. Kresse, and J. Hafner, First-principles investigation of phase stability in Li_xCoO_2, *Phys. Rev. B* **58**, 2975 (1998).
44. A. Van der Ven, and G. Ceder, Vacancies in ordered and disordered binary alloys treated with the cluster expansion. *Phys. Rev. B* **71**, 054102 (2005).
45. P. D. Tepesch, G. D. Garbulsky, and G. Ceder, Model for configurational thermodynamics in ionic systems. *Phys. Rev. Lett.* **74**, 2272 (1995).
46. G. D. Garbulsky, and G. Ceder, Effect of lattice vibrations on the ordering tendencies in substitutional binary alloys, *Phys. Rev. B* **49**, 6327 (1994).
47. G. D. Garbulsky, Ground-state structures and vibrational free energy in first-principles models of substitutional-alloy thermodynamics. PhD thesis, Massachusetts Institute of Technology, 1996.
48. G. J. Ackland, Vibrational entropy of ordered and disordered alloys. In G. Stocks and P. Turchi, editors, Alloy Modelling and Design, p. 149, Pittsburgh, PA. The Minerals, Metals and Materials Society, 1994.
49. J. D. Althoff, D. Morgan, D. de Fontaine, M. Asta, S. M. Foiles, and D. D. Johnson, Vibrational spectra in ordered and disordered Ni_3Al, *Phys. Rev. B* **56**, R5705 (1997).
50. L. Anthony, J. K. Okamoto, and B. Fultz, Vibrational entropy of ordered and disordered Ni_3Al, *Phys. Rev. Lett.* **70**, 1128 (1993).
51. L. Anthony, L. J. Nagel, J. K. Okamoto, and B. Fultz, Magnitude and origin of the difference in vibrational entropy between ordered and disordered Fe_3Al, *Phys. Rev. Lett.* **73**, 3034 (1994).
52. L. J. Nagel, *Vibrational entropy differences in Materials*, PhD thesis, California Institute of Technology, 1996.
53. V. L. Moruzzi, J. F. Janak, and K. Schwarz, Calculated thermal properties of metals, *Phys. Rev. B* **37**, 790 (1998).
54. M. Asta, R. McCormack, and D. de Fontaine, Theoretical study of alloy stability in the Cd-Mg system, *Phys. Rev. B* **48**, 748 (1993).
55. J. M. Sanchez, J. P. Stark, and V. L. Moruzzi, First-principles calculation of the Ag-Cu phase diagram. *Phys. Rev. B* **44**, 5411, 1991.
56. C. Colinet, J. Eymery, A. Pasturel, A.T. Paxton, et al., A first-principles phase stability study on the Au-Ni system, *J. Phys.: Condens. Matter* **6**, L47 (1994).
57. G. D. Garbulsky and G. D. Ceder, Contribution of the vibrational free energy to phases stability in substitutional alloys: methods and trends, *Phys. Rev. B* **53**, 8993 (1996).
58. P. D. Tepesch, A. F. Kohan, G. D. Garbulsky, G. Ceder, C. Coley, H. T. Stokes, L. L. Boyer, M. J. Mehl, B. P. Burton, K. J. Cho and J. Joannopoulos, A model to compute

phase diagrams in oxides with empirical or first-principles energy methods and application to the solubility limits in the CaO-MgO system, *J. Am. Ceram.* **49**, 2033 (1996).
59. V. Ozoliņš, C. Wolverton, C., and A. Zunger, First-principles theory of vibrational effects on the phase stability of Cu-Au compounds and alloys, *Phys. Rev. B* **58**, R5897 (1998).
60. A. van de Walle, and G. Ceder, First-principles computation of the vibrational entropy of ordered and disordered Pd_3V, *Phys. Rev. B* **61**, 5972 (2000).
61. E. Wu, G. Ceder, and A. van de Walle, Using bond-lengthdependent transferable force constants to predict vibrational entropies in Au-Cu, Au-Pd, and Cu-Pd alloys. *Phys. Rev. B* **67**, 134103 (2003).
62. A. van de Walle, Z. Moser, and W. Gasior, First-principles calculation of the Cu-Li phase diagram, *Archives of Metallurgy and Materials* **49**, 535 (2004).
63. A. van de Walle and G. Ceder, Automating first-principles phase diagram calculations, *Journal of Phase Equilibria* **23**, 348 (2002).
64. M. Stone, Cross-validatory choice and assessment of statistical predictions. *J. Roy. Stat. Soc. B Met.* **36**, 111 (1974).
65. K.–C. Li, Asymptotics optimality for C_p, C_l, cross-validation and generalized cross-validation: discrete index set, *Ann. Stat.* **15**, 956 (1987).
66. D. B. Laks, L. G. Ferreira, S. Froyen, and A. Zunger, Efficient cluster expansion for substitutional systems, *Phys. Rev. B* **46**, 12587 (1992).
67. C. Wolverton and A. Zunger, An Ising-like description of structurally-relaxed ordered and disordered alloys, *Phys. Rev. Lett.* 75, 3162 (1995).
68. A. van der Ven and G. Ceder, First principles calculation of the interdiffusion coefficient in binary alloys, *Phys. Rev. Lett.* **94**, 045901 (2005).
69. A. van der Ven, G. Ceder, M. Asta, and P. D. Tepesch, First principles theory of ionic diffusion with non-dilute carriers, *Phys. Rev. B* **64**, 184307 (2001).
70. K. Binder and D. W. Heermann, *Monte Carlo Simulation in Statistical Physics*, Springer-Verlag, New York, 1988.
71. M. E. J. Newman and G. T. Barkema (1999). Monte Carlo Methods in Statistical Physics. Clarendon Press, Oxford, 1999.
72. S. Curtarolo, D. Morgan, K. Persson, J. Rodgers, and G. Ceder, Predicting crystal structures with data mining of quantum calculations, *Phys. Rev. Lett.* **91**, 135503 (2003).
73. D. Morgan, J. Rodgers, and G. Ceder, Automatic construction, implementation and assessment of Pettifor maps, *J. Phys.: Condens. Matter* 15, 4361 (2003).
74. I. Ohnuma, Y. Fujita, H. Mitsui, K. Ishikawa, R. Kainuma, and K. Ishida, Phase equilibria in the Ti-Al binary system, *Acta Mater.* **48**, 3113 (2000).
75. V. V. Samohval, A. A. Vecher, and P. A. Poleshchuk, Thermodynamic properties of aluminum-titanium and aluminum-vanadium alloys, *Russ. J. Phys. Chem.* **45**, 1174 (1971).
76. Gus L. W. Hart, and Alex Zunger, Origins of nonstoichiometry and vacancy ordering in $Sc_{1-x}\square_xS$, *Phys. Rev. Lett.* **87**, 275508 (2001).
77. M. E. Arroyo y de Dompablo, A. Van der Ven, and G. Ceder, First-principles calculations of lithium ordering and phase stability on the Li_xNiO_2, *Phys. Rev. B* **66**, 064112 (2002).
78. V. Ozoliņš, and M. Asta, Large vibrational effects upon calculated phase boundaries in Al-Sc. *Phys. Rev. Lett.* **86**, 448 (2001).
79. V. Ozoliņš, B. Sadigh, and M. Asta, Effects of vibrational entropy on the Al-Si phase diagram. *J. Phys. Cond. Matter* **17**, 1-14 (2005).

80. E. A. Marquis, D. N. Seidman, M. Asta, C. Woodward, and V. Ozoliņš, Equilibrium mg segregation at Al/Al_3Sc heterophase interfaces on an atomic scale: Experiments and computations, *Phys. Rev. Lett.* **91**, 036101 (2003).
81. R. Benedek, A. van de Walle, S. S. A. Gerstl, M. Asta, D. N. Seidman, and C. Woodward, Partitioning of impurities in multi-phase TiAl alloys, *Phys. Rev. B* **71**, 094201 (2005).
82. C. L. Fu, Y. Y. Ye, M. H. Yoo, K. M., and Ho, Equilibrium point defects in intermetallics with the B2 structure -NiAl and FeAl, *Phys. Rev. B* **48**, 6712-6715 (1993).
83. C. L. Fu, and G. S. Painter, Point defects and the binding energies of boron near defect sites in Ni_3Al: A first-principles study, *Acta Mater.* **45**, 481-488 (1997).
84. C. Woodward, and S. Kajihara, Density of thermal vacancies in γ-TiAl-M, M=Si, Cr, Nb, Mo, Ta or W, *Acta Mater.* 47, 3793-3798 (1999).
85. C. Woodward, M. Asta, G. Kresse, J. and Hafner, Density of constitutional and thermal defects in $L1_2$ Al_3Sc, *Phys. Rev. B* **63**, 094103 (2001).
86. M. Fahnle, Atomic defects and diffusion in intermetallic compounds: The impact of the *ab initio* electron theory, *Defects and Diffusion Forum* **203**, 37-46 (2002).
87. R. Drautz, I. Schultz, F. Lechermann, and M. Fahnle, Ab-initio statistical mechanics for ordered compounds: Single-defect theory vs. cluster-expansion techniques, *Phys. Stat. Solid. B* **240**, 37-44 (2003).
88. W. Lin, and A. J. Freeman, Cohesive properties and electronic structure of Heusler $L2_1$-phase compounds Ni_2XAl (X=Ti, V, Zr, Nb, Hf, and Ta), *Phys. Rev. B* **45**, 61-68 (1992).
89. A. Misra, R. Bishop, G. Ghosh, and G. B. Olson, Phase equilibria in prototype Nb-Pd-Hf-Al alloys, *Metall. Mater. Trans. A* **34**, 1771-1782 (2003).
90. A. Misra, G. Ghosh, and G. B. Olson, Phase relations in the Nb-Pd-Hf-Al system. *J. Phase Equilibria and Diffusion* **25**, 507-514 (2004).
91. A. van de Walle, The alloy theoretic automated toolkit. Software package. http://cms.northwestern.edu/atat/, 2001.
92. A. van de Walle, M. Asta, and G. Ceder, The alloy theoretic automated toolkit: A user guide, *CALPHAD* **26**, 539 (2002).
93. A. van de Walle, M. and Asta, Self-driven lattice-model monte carlo simulations of alloy thermodynamic properties and phase diagrams, *Modelling Simul. Mater. Sci. Eng.* **10**, 521 (2002).
94. O. Redlich and A. Kister, Algebraic representation of thermodynamic properties and classification of solutions, *Ind. Eng. Chem.* **40**, 345-348 (1948).
95. Y.–M. Muggianu, M. Gambino, and J. P. Bros, Enthalpies of formation of liquid alloys bismuth-gallium-tin at 723 K - Choice of an analytical representation of integral and partial thermodynamic functions of mixing for this ternary system, *J. Chem. Phys.* **72**, 83-88 (1975).
96. B. Sundman and J. Ågren, Regular solution model for phases with several components and sublattices, suitable for computer-applications, *J. Phys. Chem. Solids* **2**, 297-301 (1981).
97. U. R. Kattner and W. J. Boettinger, Thermodynamic calculation of the ternary Ti-Al-Nb system, *Mater. Sci. Eng. A* **152**, 9-17 (1992).
98. C. Servant and I. Ansara, Thermodynamic assessment of the Al-Nb system. *J. Chem. Phys.* **94**, 869-888 (1997).
99. C. Colinet, A. Pasturel, D. Manh, Nguyen, D. Pettifor, and P. Miodownik, Phase stability of the Al-Nb system, *Phys. Rev. B* **56**, 552-565 (1997).

100. I. Ansara, Chart, T. G., Guillermet, A. F., Hayes, F. H., Kattner, U. R., Pettifor, D. G., Saunders, N., and Zeng, K. Thermodynamic modeling of selected topologically close-packed intermetallic compounds. *CALPHAD* **21**, 171–218 (1997).
101. Ghosh, G., Liu, J. Z., van de Walle, A., and Asta, M. (2005). Unpublished research. Northwestern University.
102. J. Murray, A. J. McAlister, D. J. and Kahan, The Al-Hf system, *J. Phase Equilibria* **19**, 376-379 (1998).
103. L. Kaufman, and H. Nesor, Calculation of the Ni-Al-W, Ni-Al-Hf and Ni-Cr-Hf systems, *Canadian Metall. Quarterly* **14**, 221-232 (1975).
104. T. Wang, Z. Jin, and J. C. Zhao, Thermodynamic assessment of the Al-Hf binary system, *J. Phase Equilibria* **23**, 416-423 (2002).
105. G. Ghosh, and M. Asta, First-principles calculation of structural energetics of Al-TM (TM = Ti, Zr, Hf) intermetallics. *Acta Mater.* **53**, 3225–3252 (2005).
106. Guillermet, A. F. Gibbs energy coupling of phase stability and thermochemistry in the Hf-Nb system. *J. Alloys and Compounds* **234**, 111–118 (1996).
107. Misra, A. (2005). Noburnium: Systems Design of Niobium Superalloys. PhD thesis, Northwestern University, Evanston, Illinois.
108. Borgenstam, A., Engström, A., Höglund, L., and Ågren, J. Dictra, a tool for simulation of diffusional transformations in alloys. *J. Phase Equilibria* **21**, 269–280 (2000).
109. Andersson, J.-O. and Ågren, J. Models for numerical treatment of multicomponent diffusion in simple phases. *J. Appl. Phys.* **72**, 1350–1355 (1992).

Chapter 2

USE OF COMPUTATIONAL THERMODYNAMICS TO IDENTIFY POTENTIAL ALLOY COMPOSITIONS FOR METALLIC GLASS FORMATION

Y. Austin Chang
Dept. of Mat. Sci. and Eng., University of Wisconsin, 1509 University Ave, Madison, WI 53706, USA, chang@engr.wisc.edu

Abstract: A simple thermodynamic analysis illustrates that the formation of a eutectic, e.g. in a binary, is due to the greater thermodynamic stability or the more negative excess Gibbs energy of the liquid versus those of the solid phases. Further decreasing the excess Gibbs energy of the liquid extends its stability to even lower temperatures. Decreases in a liquid's viscosity with concomitant increases in its diffusivity when the temperature is lower, creates a favorable kinetic condition for the formation of materials in the glassy or amorphous state. A calculated isopleth of $Zr_{56.28-c}Ti_cCu_{31.3}Ni_{8.7}Al_{8.5}$, from 0.0 to 15 mol% Ti shows that the liquidus temperature of the quaternary alloy decreases rapidly with the addition of Ti, reaching a minimum at 4.9 mol% Ti, and then increases again. These results indicate that the glass forming ability of the alloys would increase with the addition of Ti, reaching a maximum at or near 4.9 mol% Ti, and then decrease again. This was indeed confirmed experimentally. In a second example, we used a similar thermodynamic analysis to predict alloy compositions of sputter-deposited Al-Zr thin-films in the amorphous state. The predicted alloys compositions were also in accord with those obtained experimentally. In this analysis, we assumed intermetallic compounds do not form due to kinetic constraint since sputter-deposition is an extremely rapid process. In addition we assumed the Gibbs energies of the amorphous phase to be the same as those of the undercooled liquid.

Keywords: phase diagrams, thermodynamics, metallic glasses

1. INTRODUCTION

Bulk metallic glasses (BMGs) exhibit unique properties such as high strength (≈ 300 ksi or 2 GPa), excellent wear and corrosion resistance, high fracture toughness (50 MPa-m$^{1/2}$), outstanding castability, and low cost for alloy synthesis and fabrication.[1-4] All of these properties make them extremely attractive for practical applications as structural and functional materials. The success in making metallic glasses stimulated from the earlier work of Clement et al. in synthesizing Au-Si glass foils by rapid quenching, 10^5-10^6 K/s, from the melts.[5] Subsequent advances were made with decreasing cooling rates first to 10^3 K/s and eventually to 1 to 10 K/s in the nineteen eighties and nineties.[1,2,6-8] The ability to make these novel metallic materials at cooling rates approaching that of conventional casting conditions affords opportunities not only for investigating the fundamental behavior of deeply undercooled metallic melts but also for manufacturing potential near net-shaped BMG components for practical applications. However, as pointed out by Gottschall,[10] there remains an urgent need to formulate a method for predicting families of alloy compositions with a greater tendency for glass formation. In a recent review, Löffler made the following statement, "-----*the search for new bulk metallic glass compositions is somewhat a 'trial-and-error' method, involving in many cases the production of hundreds to thousands of different alloy compositions*".[11] However, with advancement made in computational thermodynamics, we can indeed calculate multicomponent phase diagrams, from which potential alloy compositions can be identified with tendencies to form BMGs.[12,13] In this chapter, we will first show that the topological features of a simple binary phase diagram are governed by the relative thermodynamic stabilities of the liquid phase versus the competing solid phases. When the liquid phase is much more stable than the solid phases, a eutectic liquid forms. This liquid favors glass formation. We will then next give two examples to demonstrate the success in the use of computational thermodynamics to identify potential alloy compositions for glass formation. While the first example focuses on bulk multicomponent glasses for primarily structural applications, the second one focuses on amorphous binary thin-films as sensors for disk drive applications.

2. PHASE DIAGRAM FEATURES FAVORING GLASS FORMATION

It has been known that the existence of a eutectic in a pseudo-binary oxide system favors glass formation. One example is a soda silicate glass that occurs at the eutectic composition of approximately 76 mol% SiO_2 in Na_2O-SiO_2.[14] The formation of such a eutectic in Na_2O-SiO_2 has its thermodynamic origin in that the excess Gibbs energy of the liquid phase is more negative than that of the competing solid phases. In the following, we will use a simple regular solution model to represent the excess Gibbs energy of the liquid and solid phases of a simple binary *A-B*. Both *A* and *B* exhibit the same crystal structure such as bcc and their entropies of fusion differ slightly. By varying the regular solution parameters of the solid and liquid phases, different types of phase diagrams can be obtained. The regular solution model is represented by the following equation,

$$G^E = x_A x_B L_0 \tag{1}$$

where G^E is the excess Gibbs energy, x_A and x_B the mole fractions of the component elements and L_0 the regular solution parameter. The Gibbs energy of a phase is,

$$\Delta G = x_A x_B L_0 + RT[x_A \ln x_A + x_B \ln x_B] \tag{2}$$

where ΔG is the Gibbs energy of a solution relative to those of pure *A* and *B* respectively, *R* the gas constant, and *T* the absolute temperature. The Gibbs energies of fusion of *A* and *B* are,

$$\Delta_{fus} G(A) = 32500 - 11.3T \text{ J mol}^{-1} \tag{3.a}$$

$$\Delta_{fus} G(A) = 38429 - 10.4T \text{ J mol}^{-1} \tag{3.b}$$

In the following, we present three figures with the following sets of solution parameters: Fig. 1.a $L_0(L) = 0$, $L_0(S) = -20$ kJ mol^{-1}, Fig. 1.b $L_0(L) = 0$, $L_0(S) = 20$ kJ mol^{-1}, Fig. 1.c $L_0(L) = 0$ and $L_0(S) = 50$ kJ mol^{-1}; $L_0(L) = -50$ kJ and $L_0(S) = 50$ kJ mol^{-1}. The term one mol means one mole of atoms, i.e. $A_{1-x}B_x$. In all three figures, the solid/liquid phase boundaries (or the solidus/liquidus curves) are also calculated using $L_0(L) = L_0(S) = 0$, i.e. both liquid and solid behave like an ideal solution. The temperatures of the calculated liquidus/solidus curves under this condition, shown as dashed lines, vary monotonically from pure A to pure B. The temperatures of the calculated liquidus/solidus curves using $L_0(L) = 0$ and $L_0(S) = -20$ kJ mol^{-1} shown in Fig. 1.a increase with composition, reaching a maximum and

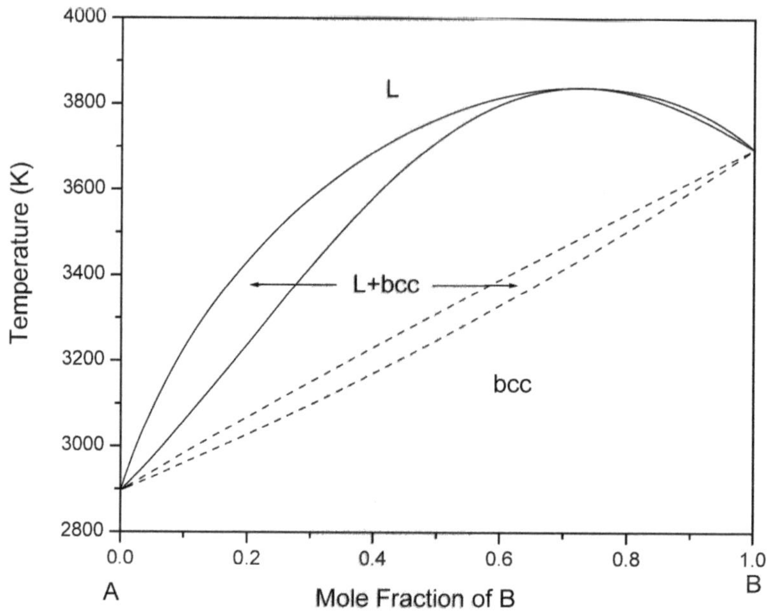

Figure 1.a. Solid lines calculated using $L_0(L) = 0$, $L_0(S) = -20$ kJ mol^{-1}. Dashed lines calculated using $L_0(L) = L_0(S) = 0$.

decrease again. Moreover, the maximum melting temperature occurs at a single composition, i.e. the compositions of the liquid and the solid at the melting point are the same. This kind of melting is referred to as congruent melting. It is similar to the melting of a pure component *A* or *B*. On the other hand, other two-phase alloys melt over a range of temperature with corresponding composition changes. The resulting higher melting temperatures are reasonable since the solid phase becomes thermodynamically more stable with respect to the liquid phase. When both the liquid and solid behave ideally, the liquidus/solidus curves change smoothly with composition. The results shown in Fig. 1.b are the reverse since the regular solution parameter of the solid phase is less exothermic than that of the liquid phase. A minimum congruent melting occurs in this case. In addition to exhibiting minimum congruent melting, something else also happens. The solid phase undergoes phase separation or the formation of a miscibility gap at lower temperatures. According to the regular solution model, the critical point for phase separation is $T_c = L_0(S)/2R = 1203$ K with T_c being the critical point.

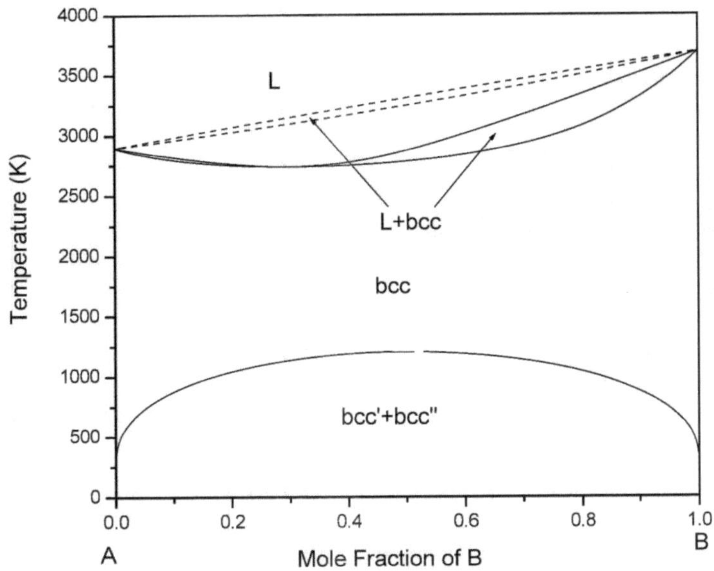

Figure 1.b. Solid lines calculated using $L_0(L) = 0$, $L_0(S) = 20$ kJ mol^{-1}. Dashed lines calculated using $L_0(L) = L_0(S) = 0$.

Figure 1.c shows two calculated phase diagrams. The solid lines are calculated using $L_0(L) = 0$ and $L_0(S) = 50$ kJ mol^{-1}, yielding a eutectic phase diagram. Let us compare the values of $L_0(S)$ used here with that used in calculating the phase diagram in Fig. 1.b since the liquid is ideal in both cases. The thermodynamic parameters indicate that the liquid in this case is much more stable than the solid phase. This condition favors the liquid extending its stability to a lower temperature. When we next increase the value of $L_0(L)$ from 0 to –50 kJ mol^{-1} and keep that of $L_0(S) = 50$ kJ mol^{-1}, the eutectic point is lower by 815 K, frequently referred to as a deep eutectic. This is due to the fact that the liquid becomes even more stable when compared to the solid phase. Binary alloys if indeed exhibiting such a feature would have great tendency to form glass when solidified from the melt at or near the eutectic composition. Since the viscosity of a liquid increases (with a corresponding decrease in diffusivity) when the temperature is lowered, it becomes kinetically favored for this liquid to form glass upon cooling. Moreover, it is known that glasses normally form over a range of composition in the compositional vicinity of the eutectic but requiring larger undercooling at compositions away from the eutectic composition. In the following we will present two examples to show that thermodynamically calculated phase diagrams can indeed predict alloy compositions with great tendencies to form metallic glasses.

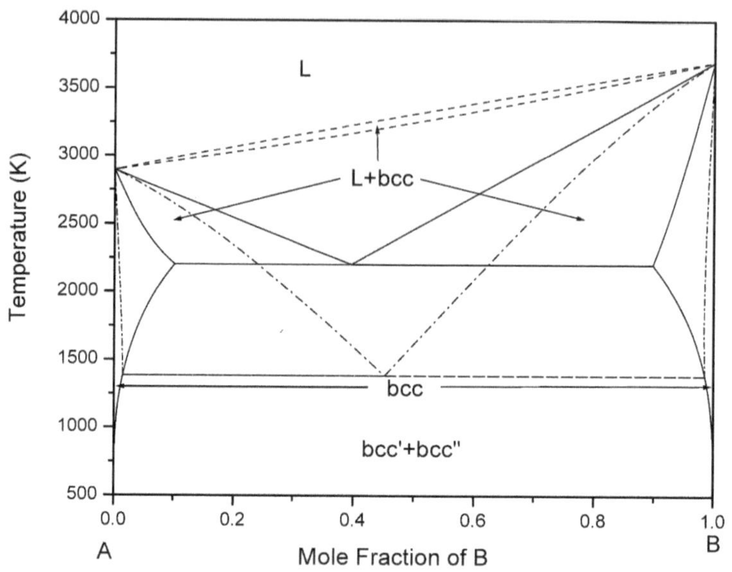

Figure 1.c. Solid lines calculated with $L_0(L) = 0$, $L_0(S) = 50$ kJ mol^{-1} and dashed-dotted line with $L_0(L) = -50$ kJ mol^{-1}, $L_0(S) = 50$ kJ mol^{-1}. Dashed lines calculated using $L_0(L) = L_0(S) = 0$.

However in order to be able to calculate multicomponent phase diagrams, it is necessary to have first a robust phase diagram calculation software, which does not require the supply of initial values by the user. Otherwise the calculation of multicomponent phase diagrams becomes extremely challenging. Often the calculated phase diagrams are not the stable one due to the supply of inappropriate initial values by the users. In addition to the phase diagram calculation software, it is necessary to have secondly thermodynamic descriptions of the multicomponent systems in question.

3. EXAMPLES USING COMPUTATIONAL THERMODYNAMICS TO IDENTIFY ALLOY COMPOSITIONS FOR GLASS FORMATION

3.1 Addition of Ti to improve the glass forming ability (GFA) of a known glass-forming Zr-Cu-Ni-Al alloy

In an attempt to improve the glass forming ability (GFA) of a known quaternary glass-forming alloy, $Zr_{56.28}Cu_{31.3}Ni_{4.0}Al_{8.5}$, Ma et al.[15] recently calculated an isopleth in terms of temperature as a function of the composition of Ti keeping the mol% of Cu, Ni and Al constant at 31.3, 4.0

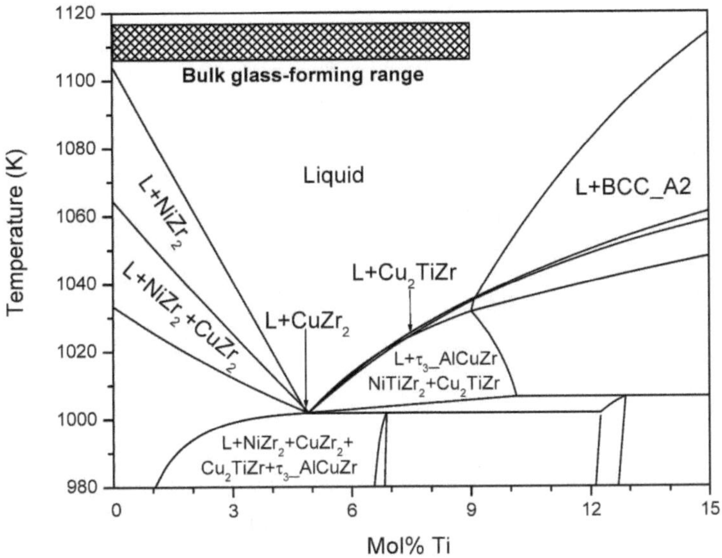

Figure 2. A calculated isopleth, T vs mol% Ti, keeping the mol% of Cu, Ni and Al at 31.3, 4, and 8.5. The alloy composition at the origin is $Zr_{56.2}Cu_{31.3}Ni_{4.0}Al_{8.5}$.

and 8.5 mol%, respectively. Figure 2 shows such a calculated isopleth using the software Pandat.[16-18] It is evident from this figure that the liquidus temperatures from the quaternary alloy $Zr_{56.28}Cu_{31.3}Ni_{4.0}Al_{8.5}$, i.e. without Ti, decrease rapidly reaching a minimum at 4.9 mol% Ti and then increases again. The experimental results obtained by Ma et al. as given below indeed showed that this is the case.[15] A series of $Zr_{56.28-c}Ti_cCu_{31.3}Ni_{8.7}Al_{8.5}$ alloys with values of c varying from 0 to 10 mol% Ti, were prepared with the expectation that the alloy with 4.9 mol% Ti would exhibit the highest GFA. The quaternary $Zr_{56.28}Cu_{31.3}Ni_{4.0}Al_{8.5}$ alloy was found to be a bulk glass-forming alloy based on the calculated low-lying liquidus surface of the quaternary Zr-Cu-Ni-Al system.[19] Alloy ingots with the nominal compositions $Zr_{56.28-c}Ti_cCu_{31.3}Ni_{8.7}Al_{8.5}$ (c = 0 ~ 10.0 mol%) were prepared by arc melting pieces of high purity metals, with Zr being 99.95 wt% and the rest Ti, Cu, Ni and Al being 99.99 wt%, in a Ti-gettered argon atmosphere. Each of the ingot samples was remelted several times to assure good mixing and then suction-cast (or drop-cast), under a purified Ar (or He) atmosphere, into a copper mold with an internal cylindrical cavity with diameters ranging from 1 to 5 mm (or 6 to 14 mm). The amorphous nature of the as-cast rods was examined by analyzing the central part of their cross-sections using X-ray diffraction (XRD) with a Cu-Kα source, and Scanning Electron Microscopy (SEM) in the Backscattered Electron Imaging (BEI) mode. The glass transition and crystallization behaviors of these alloys upon reheating

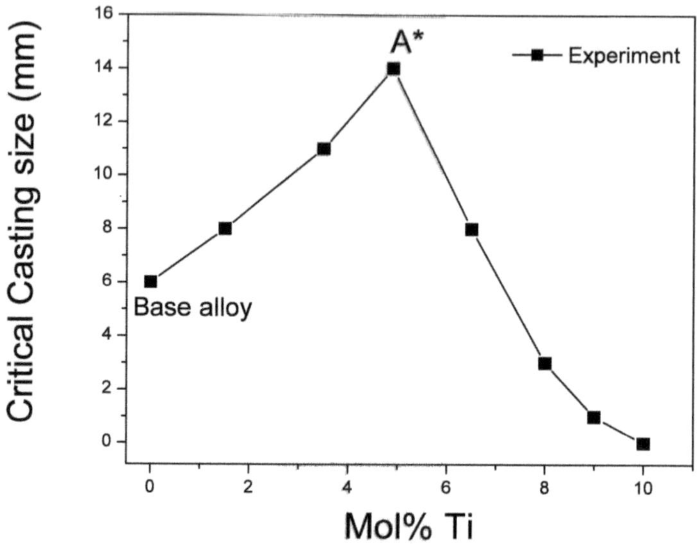

Figure 3.a. Critical diameters of the cast glassy rods vs mol% of Ti. The 14 mm diameter rod at 4.9% is marked A*.

were characterized using a Perkin-Elmer DSC7 (Differential Scanning Calorimeter) at a heating rate of 20 K/min.

As shown in Fig. 3.a the GFA of the quaternary base alloy increases with the addition of Ti in terms of the critical diameters of amorphous rods formed, reaching a maximum at 4.9 mol% Ti and then decrease again. At 10 mol% Ti, it was no longer possible to achieve bulk glass formation. Also shown in Fig. 3.b are the 6 mm-diameter glass rod formed by casting the base alloy and the > 14 mm diameter glass rod formed with the alloy containing 4.9 mol% Ti. Since the technique used by Ma et al.[15] is not capable of casting a rod larger than 14 mm diameter, it was concluded that larger diameters than 14 mm could be obtained. The inset in Fig. 3.b shows an arc-melted 20 gm-button used for casting the amorphous rod of the alloy A*. These results are consistent with the liquidus temperatures shown in Fig. 4. This is anticipated since a minimum amount of undercooling is required at the lowest temperature.

Figure 4 shows the XRD patterns obtained from the as-cast rods of four representative alloys, i.e. $Zr_{56.2-c}Ti_cCu_{31.3}Ni_{4.0}Al_{8.5}$ with c = 0, 1.5, 4.9, 6.5 mol% respectively. They are denoted as the base alloy, A1, A* and A3, respectively. The base alloy exhibits two typical amorphous halos in its 6 mm-diameter sample. On the other hand, the 7 mm-diameter rod shows two crystalline peaks due to the presence of $CuZr_2$ and $NiZr_2$ respectively, indicating that the critical casting diameter for this alloy is ~6 mm.

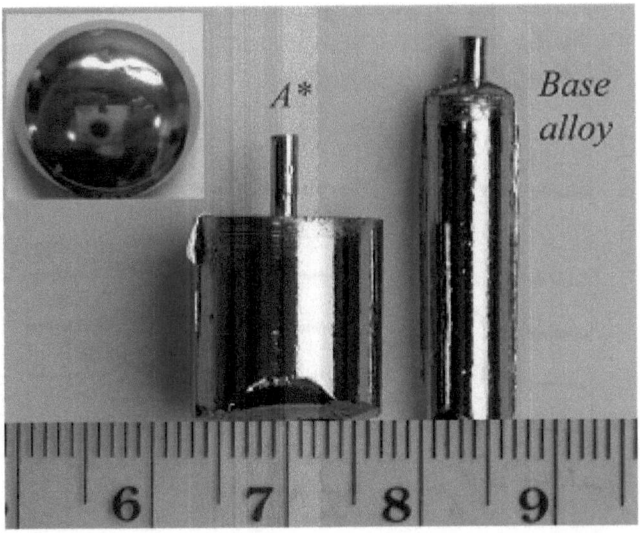

Figure 3.b. Pictures of the 6 and 14 mm diameter glass rods cast from the base alloy and the 4.9 mol% Ti alloy, marked A*. The inset shows an arc-melted 20 gm-button used for casting A*.

The diffraction patterns of the cast 10 mm-diameter rod, from the alloy containing 1.5 mol% Ti and denoted as A1, also exhibit two similar crystalline peaks superimposed on the main halo. These peaks show that this rod is only partially glass. However, it is abundantly clear that there is no crystalline peaks discernible in the XRD patterns of alloy A* obtained from its 5 mm diameter sample. This means that the rod is a monolithic glass. For alloys containing more than 6.5% Ti such as A3, their XRD patterns reveal even more and sharper peaks, indicating the presence of a considerable amount of crystalline phases in their 10 mm-samples. However, with increasing Ti contents beyond this critical composition of 4.9 mol% Ti, the critical casting diameter diminishes rapidly reducing close to nothing when the Ti content reaches 10 mol%.

The DSC curves of the cast amorphous rods presented in Fig. 5 exhibit endothermic inflection characteristic of a glass transition at a temperature, T_g, ranging from 656 to 675 K, followed by one or two pronounced exothermic peaks corresponding to crystallization events. Values of T_x and T_g for each amorphous alloy obtained from the DSC traces with T_x being the onset crystallization temperature are summarized in Table 1. The thermodynamically calculated liquidus temperatures T_l are also given in this table as well as the frequently used GFA criteria, $\Delta T_x = (T_x - T_g)$,[20] T_{rg} [21] and γ.[22] It is found that the value of T_{rg} peaks at 4.9 mol% Ti, which corresponds exactly to the best glass-forming alloy, i.e. A*.

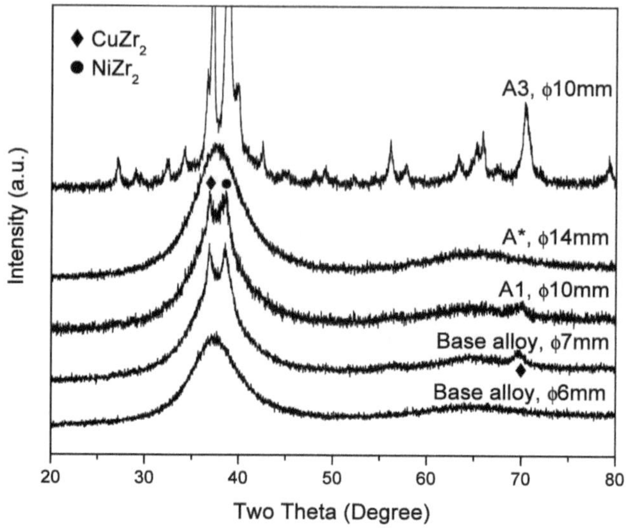

Figure 4. XRD patterns of the 6 and 7 mm dia. rods cast from the base alloy, 10 mm rod from A1 (1.5 mol % Ti), 14 mm rod from A*, and 10 mm from A3 (6.5 mol% Ti).

This is not surprising. Since values of Tg are insensitive to alloy composition, the shape of the compositional dependence of the reduced-glass temperatures is governed by the sharp decreases in the liquidus temperatures. On the other hand, the experimental measured results appear to be somewhat inconsistent with the other two criteria, i.e. ΔT_x and γ.

Except for the above mentioned alloy series (A), Ma et al.[15] also calculated the isopleths in terms of T versus the compositions of Cu, Ni, Al, and ($Zr_{0.5628}Cu_{0.313}Ni_{0.040}Al_{0.085}$) respectively. In other words, each of the elements of Cu, Ni, Al or ($Zr_{0.5628}Cu_{0.313}Ni_{0.040}Al_{0.085}$) was replaced with Ti. First, all the experimentally determined values of the GFA are consistent with the calculated liquidus temperatures. Moreover, the minimum liquidus temperature calculated at 4.9 mol% Ti when replacing Zr is by far the lowest. One can thus conclude that the strategy using the thermodynamically calculated liquidus temperatures has been proven to be robust in locating the bulkiest BMG-former with optimum minor-alloying additions.

3.2 Synthesis of Precursor Amorphous Alloy Thin-films of Oxide Tunnel Barriers Used in Magnetic Tunnel Junctions

We will present in this section how computational thermodynamics can also be used to facility processing innovation for synthesizing precursor

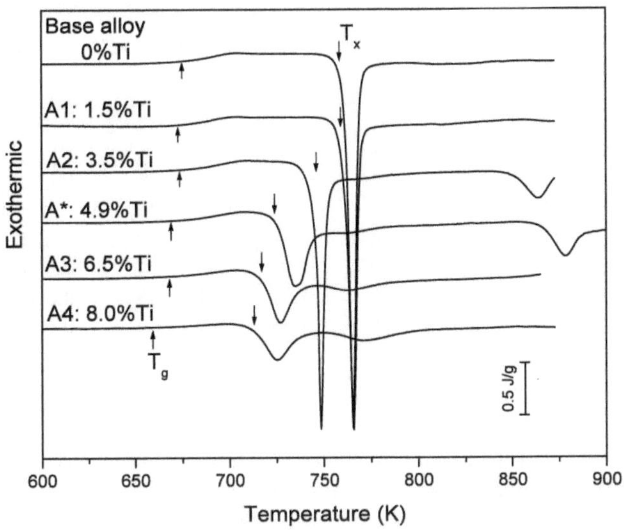

Figure 5. DSC traces of $Zr_{56.2-c}Ti_cCu_{31.3}Ni_{4.0}Al_{8.5}$ alloys (c= 0, 1.5, 3.5, 4.9, 6.5, 8.0) obtained from 2-mm cast rods. The upward arrows indicate T_g and the downward ones T_x.

amorphous alloy thin-films of oxide tunnel barriers used in magnetic tunnel junctions (MTJs). These junctions are being considered as sensitive magnetic sensors and nonvolatile storage cells in magnetic access memories.[23-26] A MTJ consists of two ferromagnetic metal electrodes (e.g. Co) separated by a thin tunnel barrier such as aluminum oxide with a thickness around 1 nm. One of the major challenges is to fabricate MTJs with both high tunneling magnetoresistance (TMR) and low product of junction resistance and area (RA) for practical applications.[27] The quality of a tunnel barrier such as aluminum oxide plays a critical role in the performance of such a device.[28,29] The current practice is to obtain an aluminum oxide barrier by oxidizing a thin crystalline Al layer.[30-34] However, since polycrystalline aluminum films have grain boundaries, the thin oxide barrier formed tends to exhibit non-uniform surfaces as well as other types of defects. An alternative approach is to oxidize an amorphous thin-film of an alloy such as (Al,Zr). The oxide films thus formed tend to exhibit smooth interfaces with fewer defects, thus leading to higher performance of the MTJs, i.e. with greater TMR.[28]

Since an amorphous phase was found in the Al-Zr system with different alloy preparation methods,[35-40] Yang et al.[33] adopted this binary as a model system for their thermodynamic and experimental study.

Table 1. Glass-forming ability and thermal properties of a series of (Zr,Ti,Cu,Ni,Al) alloys (denoted as series A) whose compositions are obtained by replacing Zr with Ti in a base alloy $Zr_{56.2}Cu_{31.3}Ni_{4.0}Al_{8.5}$.

Alloys	Ti Replacement (mol% Ti)	d_{max} (mm)	T_l (K)	T_g (K)	T_x (K)	T_x-T_g (K)	T_g/T_l	$T_x/(T_g+T_l)$
Base alloy	0.0	6	1104	675	761	86	0.611	0.428
A1	1.5	8	1073	673	762	89	0.627	0.436
A2	3.5	11	1030	674	746	72	0.654	0.438
A*	4.9	>14	1002	669	724	55	0.668	0.433
A3	6.5	8	1018	668	717	49	0.656	0.425
A4	8.0	3	1029	659	713	54	0.640	0.422
A5	9.0	1	1035	656	711	55	0.634	0.420
A6	10.0	0	1052	–	–	–	–	–

d_{max}: experimentally attained maximum diameter of glassy rods using copper mould casting.
T_l: the liquidus temperature calculated thermodynamically.
T_g: the glass transition temperature measured using DSC.
T_x: the onset temperature of crystallization measured using DSC.

They presented a thermodynamic formulation to predict alloy compositions, which show tendencies to form amorphous thin-films when fabricated by a rapid quenching process such as sputter-deposition. TEM and XRD were used to confirm the formation of amorphous alloy films. Based on the methodology proposed by Yang et al.,[33] other alloys with higher amorphous-forming ability could be obtained as additional candidates for precursor metals of oxide tunnel barriers.

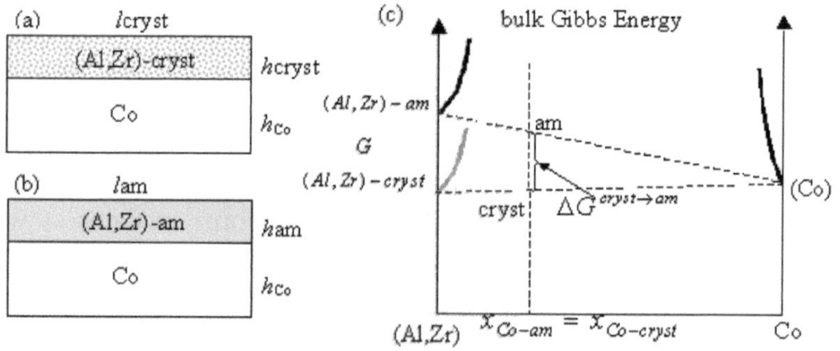

Figure 6. Schematic diagrams of the bilayer structures with the (Al, Zr) layer in (a) crystalline state, (b) amorphous state; (c) the bulk Gibbs energy vs the composition of Co of the bilayer in (a) and (b) for the case that (Al, Zr)-cryst is more stable than (Al, Zr)-am. The symbols l and h refer to the length and height, respectively. The Co layer is in the crystalline state.

Figures 6.a and 6.b show two bi-layered structures, one with an alloy of (Al,Zr) in the crystalline state on top of a thin layer of Co and the other with the same alloy in the amorphous state. The dimensions of the bi-layered structure are given in these figures, i.e. l_{cryst}, l_{am}, h_{am}, and h_{Co}, with the subscripts denoting crystalline and amorphous respectively. The total Gibbs energy of such a bi-layer structure is taken to include three parts: bulk Gibbs energy, interfacial energy and surface energy. The bulk Gibbs energies of these two (Co)/(Al,Zr) bi-layer structures are shown schematically in Fig. 6.c as a function of the Co composition for the structure given in Figs. 6.a and 6.b, respectively. The assumption made is that all the intermetallic compounds normally stable in Al-Zr do not form due to kinetic constraints when using sputter-deposition to fabricate these thin metallic films. In other words, the sputter-deposition process is so rapid that nucleation of these intermetallics becomes unfavorable. In deriving their thermodynamic model, Yang et al.[33] considered only the existence of amorphous and crystalline phases of (Al,Zr). Moreover, the Gibbs energy curves of (Co), (Al, Zr)-am and (Al, Zr)-cryst are represented by sharp curves at the two ends of the diagram because the mutual solubilities between Co and (Al,Zr)-am or (Al,Zr)-cryst are negligible at low temperatures (<500°C). The tangent line between Co and (Al, Zr)-cryst represents the metastable phase equilibrium between Co and (Al,Zr)-cryst. Similarly, the tangent line between Co and (Al,Zr)-am represents the metastable phase equilibrium between Co and (Al,Zr)-am. The Gibbs energy of the amorphous phase is approximated to be that of the undercooled liquid and the volume Gibbs energies of (Al, Zr)-cryst can be either lower or higher than those of the amorphous phase depending on which state is more stable. Figure 6.c shows the Gibbs energies of a mixture of (Al,Zr)-am and (Co) as well as that of (Al,Zr)-cryst and (Co) as a function of composition of Co at a constant T. For the case shown in this figure, the Gibbs energy of the two-phase mixture consisting of (Al,Zr)-am and (Co) is higher than that of (Al,Zr)-cryst and (Co). At a specific composition of x_{Co}, the Gibbs energy difference between the two states is shown in Fig. 6.c as $\Delta G^{cryst \rightarrow am}$. The symbol $\Delta G^{cryst \rightarrow am}$ denotes the Gibbs energy of transformation from a crystalline state for an (Aly_{Al},Zry_{Zr}) alloy to an amorphous state of an (Aly_{Al},Zry_{Zr}) alloy since (Co) in the structures shown in Fig. 6.a and 6.b remain in the crystalline state. The symbols y_{Al} and y_{Zr} denote the mole fractions of Al and Zr in the binary (Al;Zr) alloys. This transformation energy is the barrier to be overcome for the formation of the amorphous state. The analytical equation of $\Delta G^{cryst \rightarrow am}$ at the as-deposited temperature is described as,[33]

$$\frac{\Delta G^{cryst\to am}_{(Al,Zr)}}{(1-x_{Co})} = y^{am}_{Al}{}^{0}\Delta H^{fus}_{Al}\left(1-\frac{T_{as}}{T^{cryst}_{m,Al}}\right) + y^{am}_{Zr}{}^{0}\Delta H^{fus}_{Zr}\left(1-\frac{T_{as}}{T^{cryst}_{m,Zr}}\right)$$
$$+\left({}^{ex}G^{am}_{(Al,Zr)} - G^{cryst}_{(Al,Zr)}\right) \quad (4)$$

In Eq. 4, x_{Co} is the overall composition of Co in the bi-layer structure; T_{as} the film deposition temperature; y^{am}_{Al} and y^{am}_{Zr} have been defined above. Since the (1-x_{Co}) term is always positive and does not affect the sign of $\Delta G^{cryst\to am}$, the Co layer in the model can be replaced by other materials without changing the validity of the conclusions. $T^{cryst}_{m,Al}$ and $T^{cryst}_{m,Zr}$ refer to the transition temperatures from the pure crystalline Al and Zr to their pure liquid, respectively. Similarly, $\Delta H^{cryst\to am}_{Al}$ and $\Delta H^{cryst\to am}_{Zr}$ represent the enthalpies of fusion of Al and Zr at their respective melting temperatures. ${}^{ex}G^{am}_{(Al,Zr)}$ and ${}^{ex}G^{cryst}_{(Al,Zr)}$ denote the excess Gibbs energies of the (Al,Zr) alloys exhibiting the amorphous and crystalline state, respectively. In order to evaluate the energy barrier from the (Al,Zr) crystalline to (Al,Zr) amorphous structure, a prerequisite is to know which crystalline structure is the most stable structure. Based on the Gibbs energy of solution phases calculated from the thermodynamic description of the Al-Zr system developed by Wang et al.,[41] the Gibbs energy of the (Al, Zr) solution with the fcc structure was found to be the most stable among the common crystal structures, consistent with the experimental data to be presented later. Using the SGTE lattice stabilities of Al and Zr,[42] and the excess Gibbs energies of the undercooled liquid (Al, Zr)-am and the fcc (Al, Zr)-cryst phases,[41] the values of $\Delta G^{cryst\to am}$ versus the composition of Zr in the top layer are shown in Figure 7. It is evident from the values of $\Delta G^{cryst\to am}$ shown in this figure that in the mid-part of the diagram, amorphous alloys are likely to form during sputter-deposition.

It is indeed somewhat surprising that in view of the simplicity of the thermodynamic formulation, the calculated compositions of the (Al,Zr) alloys for amorphous phase formation are in reasonable agreement with the experimental data presented in Figure 8. These data were obtained from TEM micrographs and SAD patterns of co-sputtered deposited alloy films.[33] As shown in this figure, alloys with compositions denoted as A and B exhibit crystal grains with dotted SAD patterns. However, as the composition approaches point C, most of the grains disappear in the micrograph and multiple diffraction rings fade with a halo ring becoming clear in the SAD patterns. This suggests a transitional region from a

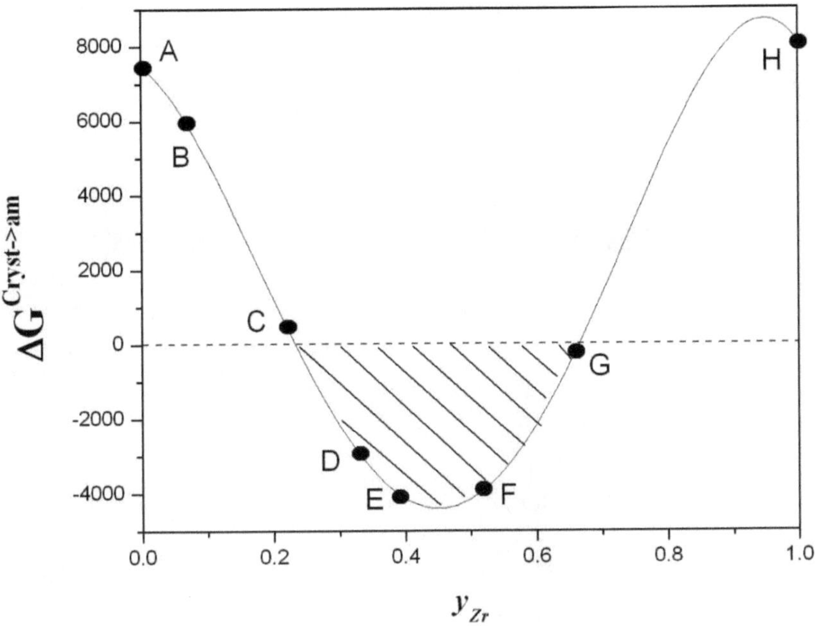

Figure 7. The Gibbs energy difference between the fcc and the amorphous phase of (Al,Zr) vs mol fraction of Zr. Co-sputter deposited films were made at the compositions: A, B, etc. to H.

polycrystalline structure to an amorphous state, in accord with the calculation. With increases in the Zr concentrations, the Al-Zr alloy films appear to be amorphous, which can be seen from the single diffuse ring in SAD patterns and the typical amorphous micrographs[43] (defocused to enhance the contrast) at composition points D, E and F. At point G both the micrograph and the SAD pattern experienced an appreciable change from those of point F, suggesting the film transforms from an amorphous state to a crystalline structure again. At point H, i.e. pure Zr, a polycrystalline fcc structure can be observed from both the micrograph and SAD pattern, indicating the fcc Zr exist. Thicker films with typical compositions were deposited on glass for the XRD structure characterization, as shown in Figure 9. In the three XRD diffraction patterns, the big humps at about 24° result from the glass substrate, which was adopted to exclude any possible peaks from the substrate. This exercise indicates that the thermodynamic approach presented here can be used to make similar predictions for many other alloys and can identify alloy compositions for forming amorphous phases via sputter-deposition, provided thermodynamic descriptions of the alloys in question are available.

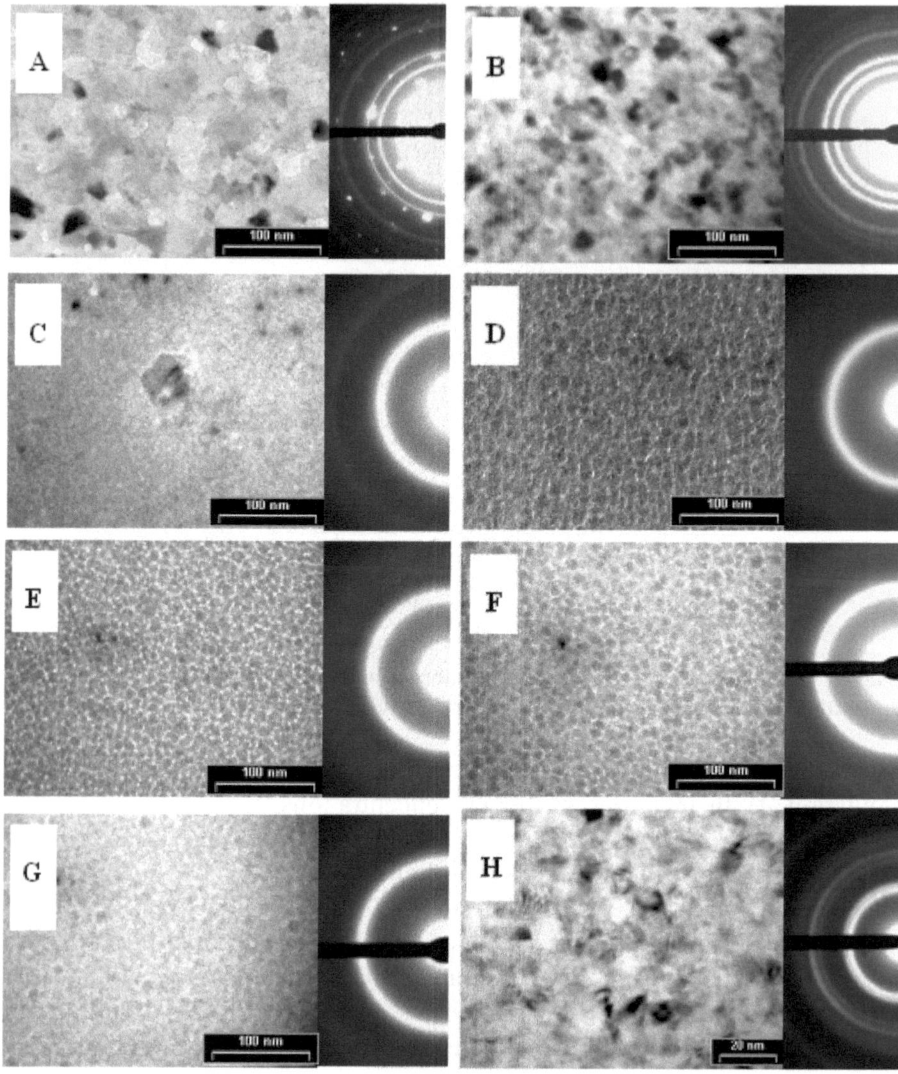

Figure 8. TEM micrographs and SAD patterns of co-sputter deposited (Al, Zr) films at the compositions: A 0, B 7, C 22, D 33, E 39, F 52, G 66, and H 100, all in mol% of Zr.

4. CONCLUSIONS

We have shown that it is possible to use computational thermodynamics to predict alloy compositions with tendencies to form materials in the glassy or amorphous state. Two examples were given. The first one is to add Ti to a known glass-forming quaternary $Zr_{56.28}Cu_{31.3}Ni_{8.7}Al_{8.5}$ alloy in order to

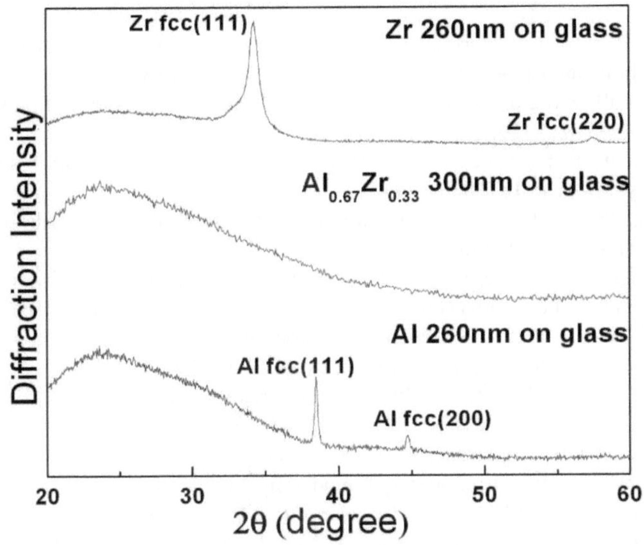

Figure 9. XRD patterns obtained from three typical films on glass: Zr 260 nm, $Al_{0.67}Zr_{0.33}$ 300 nm and Al 260 nm. Note the Zr film exhibits the fcc structure.

improve its glass forming ability. In the second one, we calculated the Gibbs energy difference between the liquid and the fcc phase to predict alloy compositions with tendencies to form amorphous thin-films prepared via sputter-deposition. In both cases, the calculations were confirmedly experimentally.

ACKNOWLEDGEMENTS

I wish to thank my current graduate students, J. Zhu, J. Yang, H. Cao, and C.-H. Ji and my post doc D. Ma, my recent former graduate students Drs. Y. Yang of CompuTherm, LLC, Madison, WI and P. Ladwig of Hutchinson Technology Inc., Hutchinson, MN, for all of their help. The financial support from the DOE-BES (DE-FG-02-99ER45777), the DARPA (ARO Contract No. DAAD 19-01-1-525), and the Wisconsin Distinguished Professorship is gratefully acknowledged.

REFERENCES

1. A. Inoue., High strength bulk amorphous alloys with low critical cooling rates. *Mater. Trans. JIM*, **36**, 866 (1995).
2. W.L. Johnson, Bulk Glass-Forming Metallic Alloys: Science and Technology. MRS Bulletin, **24**, 42 (1999).

3. C.T. Liu, L. Heatherly, D.S. Easton, C.A. Carmichael, J.H. Schneibel, C.H. Chen, J.L. Wright, M.H. Yoo, J.A. Horton, A. Inoue, Test environments and mechanical properties of Zr-base bulk amorphous alloys. *Metall. Mater. Trans.*, **29A**, 1811 (1998).
4. R.J. Gottschall, Structure and properties of bulk amorphous alloys: Foreword address, *Metall. Mater. Trans.*, **29A**, 1777 (1998).
5. W. Clement Jr., R. H. Willens, and P. Duwez, Noncrystalline structure in solidified gold-silicon alloys, *Nature*, **187**, 869 (1960).
6. H.S. Chen, Thermodynamic considerations on the formation and stability of metallic glasses. *Acta Metall.*, **22**, 1505 (1974).
7. A.J. Drehman, A. L. Greer, and D. Turnbull, Bulk formation of a metallic glass: palladium-nickel-phosphorus (Pd40Ni40P20), *Appl. Phys. Lett.* **41**(8), 716 (1982).
8. A. Inoue, T. Zhang, and N. Nishiyama, Preparation of 16 mm diameter rod of amorphous $Zr_{65}Al_{7.5}Ni_{10}Cu_{17.5}$ alloy, *Mater. Trans. JIM.*, **34**, 1234 (1993).
9. A. Peker and W. L. Johnson, A highly processable metallic glass: $Zr_{41.2}Ti_{13.8}Cu_{12.5}Ni_{10.0}Be_{22.5}$. *Appl. Phys. Lett.*, **63**, 2342 (1993).
10. R.J. Gottschall, First Conference on Bulk Metallic Glasses, Singapore, 25 October 2000. Sponsored by the United Engineering Foundation, New York, N. Y., USA, 2000.
11. J. F. Löffler, Bulk metallic glasses. *Intermetallic*, **11**, 529 (2003).
12. Y. A. Chang, S.–L. Chen, F. Zhang, X.–Y. Yan, F.–Y. Xie, R. Schmid-Fetzer, and W. A. Oates, Phase Diagram Calculation: Past, Present and Future. *Prog. Mater. Science*, **49**, 313 (2004).
13. Y. A. Chang, Phase Diagram Calculation in Teaching, Research and Industry. *Metall. Mater. Trans.*, to be published.
14. F. C. Kracek, The system sodium oxide-silica, *J. Phys. Chem.*, **34**, 1583 (1930).
15. D. Ma, H. Cao, K.–C. Hsieh, L. Ding, Y. Pan, and Y.A. Chang, Bulkier Glass Formation Enhanced by Minor Alloying Additions, Appl. Phys. Lett, under review, 2005.
16. S.–L. Chen, S. Daniel, F. Zhang, Y.A. Chang, W.A. Oates, and R. Schmid-Fetzer, On the Calculation of Multicomponent Stable Phase Diagrams. *J. of Phase Equilb.*, **22**, 373 (2001).
17. S.–L. Chen, S. Daniel, F. Zhang, Y.A. Chang, X.–Y. Yan, F.–Y. Xie, R. Schmid- Fetzer, and W.A. Oates, The PANDAT Software Package and its Applications. *CALPHAD*, **26**, 175 (2002).
18. S.–L. Chen, F. Zhang, S. Daniel, F.–X. Xie, X.–Y. Yan, Y.A. Chang, R. Schmid-Fetzer, and W. A. Oates, Calculating Phase diagrams Using PANDAT and PanEngine, *J. of Metal*, **55**, 48 (2003).
19. X.Y. Yan, Y.A. Chang, Y. Yang, F.–Y. Xie, S.–L. Chen, F. Zhang, S. Daniel, and M. H. He, Thermodynamic Approach for Predicting the Tendency of Multicomponent Metallic Alloys for Glass Formation" *Intermetallics*, **9**, 535 (2001).
20. A. Inoue, Stabilization of metallic supercooled liquid and bulk amorphous alloys, *Acta Mater.* **48**, 279 (2000).
21. D. Turnbull, Under what conditions can a glass be formed. *Contemp. Phys.*, **10**, 473 (1969).
22. Z. P. Lu and C.T. Liu, Glass formation criterion for various glass-forming systems. *Phys. Rev. Lett.*, **91**, 115505/1 (2003).
23. S.A. Wolf, D.D. Awschalom, R. A. Buhrman, J.M. Daughton, S. von Molnár, M.L. Roukes, A.Y. Chtchelkanova, and D.M. Treger, Spintronics: A spin-based electronics vision for the future. *Science*, **294**, 1488 (2001).
24. J.S. Moodera, L.R. Kinder, T.M. Wong, and R. Meservey, Large magnetoresistance at room temperature in ferromagnetic thin film tunnel junctions. *Phys. Rev. Lett.*, **74**, 3273 (1995).

25. I. Zutic, J. Fabian, and S. D. Sarma, Spintronics: Fundamentals and applications. *Rev. Mod. Phys.*, **76**, 323 (2004).
26. G. A. Prinz, Magnetoelectronics. *Science*, **282**, 1660 (1998).
27. H. Shim, B.K. Cho, J. Kim, T.W. Kim, and W J. Park, Effect of nitrogen plasma treatment at the Al_2O_3/Fe interface in magnetic tunnel junction. *J. Appl. Phys.*, **93**, 7026 (2003).
28. J.S. Moodera, J. Nassar, and G. Mathon, Spin-tunneling in ferromagnetic junctions. *Annu. Rev. Mater. Sci.*, **29**, 381 (1999).
29. E.Y. Tsymbal, O.N. Mryasov, and P.R. LeClair, Spin-dependent tunnelling in magnetic tunnel junctions. *J. Phys.: Condens. Matter*, **15**(4), R109 (2003).
30. K. Ohashi, K. Hayashi, K. Nagahara, K. Ishihara, K. Fukami, J. Fujikata, S. Mori, M. Nakada, T. Mitsuzuka, K. Matsuda, H. Mori, A. Kamijo, and H. I. Tsuge, Low-resistance tunnel magnetoresistive head, *IEEE Trans. Magn.*, **36**(5), 2549 (2000).
31. Z.G. Zhang, P.P. Freitas, A.R. Ramos, N.P. Barradas, and J.C. Soares, Resistance decrease in spin tunnel junctions by control of natural oxidation conditions. *Appl Phys Lett.*, **79**, 2219 (2001).
32. J.R. Childress, M.M. Schwickert, R.E. Fontana, M.K. Ho, P.M. Rice, and B.A. Gurney, Low-resistance IrMn and PtMn tunnel valves for recording head applications., *J. Appl. Phys.*, **89**, 7353 (2001).
33. J.J. Yang, P.F. Ladwig, Y. Yang, C.–X. Ji, F.X. Liu, B.B. Pant, A.E. Schultz, and Y.A. Chang, Oxidation of tunnel barrier metals in magnetic tunnel junctions. *J. Appl. Phys.*, in press, 2005; J.J. Yang and Y.A. Chang, Amorphous alloy thin films as precursor metals of oxide tunnel barrier used in magnetic tunnel junctions. *J. Appl Phys.*, in press, 2005.
34. P.F. Ladwig, J.J. Yang, Y. Yang, F. Liu, B.B. Pant, A.E. Schultz, and Y.A. Chang, Selective Oxidation of Tunnel Barrier Metals in Magnetic Tunnel Junctions. *Appl. Phys. Lett.*, in press, 2005.
35. S. Lee, C. Choi, and Y. Kim, Effect of Zr concentration on the microstructure of Al and the magnetoresistance properties of the magnetic tunnel junction with a Zr-alloyed Al–oxide barrier. *Appl. Phys. Lett.*, **83**, 317 (2003).
36. E. Ma, and M. Atzmon, Calorimetric evidence for polymorphous constraints on metastable zirconium-aluminum phase formation by mechanical alloying. *Phys. Rev. Lett*, **67**(9), 1126 (1991).
37. E. Ma, F. Brunner, and M. Atzmon, Stability and thermodynamic properties of supersaturated solid solution and amorphous phase formed by ball milling in the zirconium-aluminum system. *J. Phase Equil.*, **14**(2), 137 (1993).
38. H.J. Fecht, G. Han, Z. Fu, and W.L. Johnson, Metastable phase formation in the zirconium-aluminum binary system induced by mechanical alloying, *J. Appl. Phys.*, **67**, 1744 (1990).
39. H. Yoshioka, H. Habazaki, A. Kawashima, K. Asami, and K. Hashimoto, Anodic polarization behavior of sputter-deposited aluminum-zirconium alloys in a neutral chloride-containing buffer solution. *Electrochimica Acta,* **36**, 1227 (1991).
40. J. Ho and K. Lin, The metastable Al-Zr alloy thin films prepared by alternate sputtering deposition. *J. Appl. Phys.*, **75**, 2434 (1994).
41. T. Wang, Z. Jin, and J C. Zhao, Thermodynamic assessment of the Al-Zr binary system. *J Phase Equil.*, **22**, 544 (2001).
42. A.T. Dinsdale, SGTE data for pure elements, *CALPHAD*, **15**, 317 (1991).
43. D. Williams, and C.B. Carter, *Transmission Electron Microscopy: A Textbook for Materials Science*, Chapt. 28, Plenum Press, New York, 1996.

Chapter 3

HOW DOES A CRYSTAL GROW? EXPERIMENTS, MODELS AND SIMULATIONS FROM THE NANO- TO THE MICRO-SCALE REGIME

J.L. Rodríguez-López[1], J.M. Montejano-Carrizales[2], M. José-Yacamán[3]
[1] Advanced Materials Department, IPICYT, Camino Presa San José 2055, 78216 San Luis Potosí, S.L.P., Mexico; [2] Instituto de Física, Universidad Autónoma de San Luis Potosí, 78000 San Luis Potosí, S.L.P., Mexico; [3] Department of Chemical Engineering and Texas Materials Institute, The University of Texas, 78712-1063 Austin Texas, USA

Abstract: Modern research in the field of small metallic systems has confirmed that many nanoparticles take Platonic and Archimedean solids related shapes. A Platonic solid looks the same from any vertex, and intuitively they appear as good candidates for atomic equilibrium shapes. A good example is the icosahedral (I_h) particle that only shows {111} faces that produce a more rounded structure. Indeed, many studies report the I_h as the most stable particle at the size range $r \leq 20$ Å for noble gases and for some metals. In this chapter, we discuss the structure and shape of mono- and bimetallic nanoparticles in the size range from 1–300 nm. First, AuPd nanoparticles (1–2 nm) that show dodecahedral atomic growth packing. Next, in the range of 2–5 nm, we discuss a surface reconstruction phenomenon observed also on AuPd and AuCu nanoparticles. These binary alloy nanoparticles show the fivefold edges truncated, resulting in {100} faces on decahedral structures, an effect largely envisioned and reported theoretically, with no experimental evidence in the literature before. Next, we review a monometallic system (≈ 5 nm) that we termed the *decmon*. Finally, we present icosahedrally derived star gold nanocrystals (100–300 nm) which resemble the great stellated dodechaedron, a Kepler-Poisont solid. We conclude that the shape or morphology of some mono- and bimetallic particles evolves with size following the sequence from atoms to the Platonic solids. As the size increases, they tend to adopt Archimedean related shapes and then beyond the Archimedean (Kepler-Poisont) solids, up to the bulk structure of solids.

Keywords: Nanoscale materials, Synthesis and characterization, Spectroscopy and geometrical structure of clusters, Molecular dynamics and other simulation methods

1. INTRODUCTION

Philosophers, mathematicians, and artists have long been fascinated by regular polyhedra, and their aesthetically pleasing symmetry appeals to the academician and layman alike. Plato in his landmark work *Timaeus* deduced the nowadays known *Platonic solids*[1] which are polyhedra formed by the faces made of same regular polygon. In total there exist five Platonic solids, *i.e.*, the tetrahedron, the cube, the octahedron, the dodecahedron, and the icosahedron. Composed of regular triangles there exist the tetrahedron, the octahedron, and the icosahedron, with 4, 8, and 20 faces, respectively. The cube, with 6 faces, is the only Platonic solid composed of squares, and the dodecahedron, with 12 faces, is the only one composed of pentagons. No other solid can be constructed from regular polygons. This is the reason that the classical Greeks attributed five shapes of special significance to the five Greek elements like fire, earth, water, air, and cosmos, believing they must constitute the basic building blocks of the universe. Also, the discovery of regular concave polyhedra, commonly known as the Kepler-Poisont solids, dates from the Renaissance.[2,3] Interestingly, regular star polyhedra were first envisioned by artists who embellished their renderings of the Platonic solids by replacing their surfaces with pyramids. For example, a marble inlay of the small stellated dodecahedron is found in the floor of St. Mark's Basilica in Venice, Italy, which is attributed to Paolo Uccello, a fifteenth-century florentine artisan. The German artist Wentzel Jamnitzer produced a conceptual drawing of the great stellated dodecahedron in his book *Perspectiva Corporum Regularium*, published in 1568.

It was more than fifty years later when the regular concave polyhedra were explained from a mathematical standpoint, when in 1619 Johannes Kepler described the small and great stellated dodecahedra in the second volume of his landmark work *Harmonices Mundi*. The dual structures of the small and great stellated dodecahedra (the great dodecahedron and great icosahedron, respectively) were subsequently described by Louis Poinsot in 1809.[8]

Interestingly, modern technology has confirmed that many nanoparticles take on some Platonic and Archimedean solids related shapes. Platonic solids look the same from any vertex, and intuitively they appear as good candidates for atomic equilibrium shapes. A very clear example is the icosahedral (I_h) particle that only shows {111} faces that contribute to produce a more rounded structure. Indeed, many studies report the I_h as the most stable particle at the size range $r \leq 20$ Å for noble gases and for some metals.[9-11] However, in the icosahedron Platonic solid cluster, an internal strain builds up with size; and at some point the surface tension is suppressed

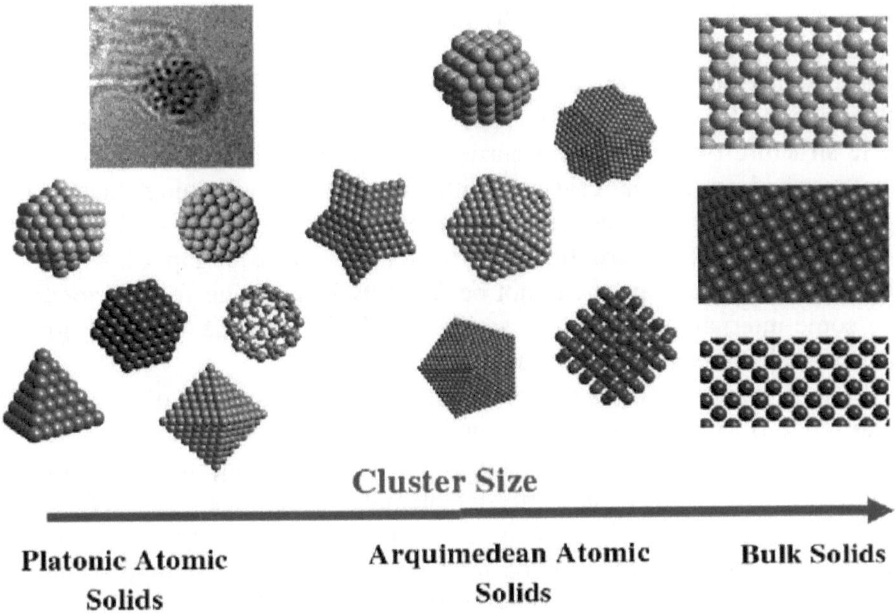

Figure 1. Proposed atomic growth sequence that would follow the metallic and bimetallic nanoparticles. In the left region, we illustrate the five Platonic solids, with an experimental image that illustrates the dodecahedral Platonic atomic solid.[4] The other platonic solid structures such as icosahedra, tetrahedra, cubes, octahedral[5-7] have widely been reported. For the icosahedron, for example, as the cluster size increases, internal strain builds up with size; and at some point the surface tension is suppressed in expense of strain energy and a new shape is favored. In that limit two different faces appear and an Archimedean solid is generated (central region). This transition is size dependent and for noble metals $r \approx 20$ Å will correspond to the optimum value for the transition. Then, the shape of mono- and bimetallic particles evolves with size following the sequence from atoms to the Platonic solids to Archimedean related shapes; then to the Kepler-Poisont solids (beyond the Archimedean solids), reaching at the very end the bulk structure of solids.

in expense of strain energy and a new shape is favored. In that limit two different faces appear and an Archimedean solid is generated. A very elegant analytical treatment of the transition between an icosahedron (Platonic) to a Wulff (Archimedean) polyhedron has been presented by Nagaev,[12] and we refer the reader to that article for more details. This transition is size dependent and for noble metals $r \approx 20$ Å will correspond to the optimum value for the transition. The other platonic solid structures such as tetrahedra, cubes, octahedral[5-7] and the dodecahedral[4] have also been reported.

Pentagonal arrangement in multi-twinned particles (MTP) is quite known. MTP nanoparticles of transition metals with fcc lattice[13,14] and some related materials such as carbon[15] have been reported. Nanoparticles with fivefold symmetry have produced a long attraction during the last 50 years. The pioneering work of Ino and Ogawa[16] and Allpress and Sanders[17] described gold nanoparticles with icosahedral and decahedral structures in terms of a multi-twinned model. Later on, Marks described icosahedral nanoparticles in terms of a multi-twinned structure composed of 20 tetrahedrons joined on their {111} faces.[18] Based on those studies, the basic structure of a decahedral particle can be described as the junction of five tetrahedron single crystals with twin-related adjoining faces. The theoretical angle between two (111) planes is ≈ 70.5°, so by joining 5 tetrahedrons, which are bounded by {111} faces, a gap of ≈ 7.5° is generated. Thus, the space can not be filled by just joining five tetrahedrons and some internal strain is necessarily introduced into the structure, giving place to dislocations and other structural defects.[13,14]

Metallic nanostructures exhibit unique properties and are of interest in a wide range of applications including catalysis, electronics, and sensors.[19,20] Also, the morphology of a nanomaterial has a strong influence on its electromagnetic, optical, and physicochemical properties.[21] Therefore, the synthesis of polyhedral crystals is a prized pursuit in the field of nanoscience, both because of the intellectual curiosity that they stimulate, and because of the desirable properties that they possess.

In this work we review some of our recent results in mono- and bimetallic nanoparticle research in the size range from 1–300 nm. The next section contains some essential concepts about the tetrahedral packing. Then, we review AuPd nanoparticles in the 1–2 nm size range that show dodecahedral atomic growth packing, one of the Platonic solid shapes that have not been identified before in this small size range for metallic particles.[4] Next, with particles in the size range of 2–5 nm, we present an energetic surface reconstruction phenomenon observed also on bimetallic nanoparticle systems of AuPd and AuCu, similar to a re-solidification effect observed during cooling process in lead clusters.[22] These binary alloy nanoparticles show the fivefold edges truncated, resulting in {100} faces on decahedral structures, an effect largely envisioned and reported theoretically, with no experimental evidence in the literature. Next nanostructure we review is a monometallic system in the size range of ≈ 5 nm, termed the *decmon*.[23] We present some detailed geometrical analysis and experimental evidence that supports our models. Finally, in the size range of 100–300 nm, we present icosahedrally derived star gold nanocrystals which resemble the great stellated dodechaedron, a Kepler-Poisont solid.[24] Then, we conclude that the shape or morphology of some mono- and bimetallic particles evolves with size following the sequence from atoms to the Platonic solids, and with slightly greater particle size, they tend to adopt Archimedean related shapes. If the particle size is still greater, they tend to adopt shapes beyond the Archimedean (Kepler-Poisont) solids, reaching at the very end the bulk structure of solids. We demonstrate both experimentally and by means of computational simulations for each case that this structural atomic growth sequence is followed in such mono- and bimetallic nanoparticles.

2. THEORY OF ATOMIC PACKING

It has been known for a long time that the addition of a second metal significantly changes some properties of the nanoparticles. For example, the addition of a second metal to some catalysts increases its chemical activity and selectivity. However, very little is known about the crystal structure and shape of bimetallic or multimetallic nanoparticles. Another issue very poorly studied related to bimetallic systems is what difference exists between the alloy states of the bimetallic or multimetallic system in the bulk form and in the nanoparticle form. Some systems that show miscibility gaps at certain compositions in the bulk have been observed to form alloys when they are in the form of nanoparticles.[25] Also, when two or more different elements compose one particle the differences in atomic radii are going to influence the way the atoms accommodate. And because there exist evidence[26] that shows that nanoparticles have a shell periodicity and that they grow by the successive accommodation of layers, this difference in the atomic radii of the species in the alloy is going to have a great impact on the final structure and shape observed in the bimetallic nanoparticles. Thus, their growth is related to the mathematical problem of packing spheres in the most dense arrangement.

Monometallic nanoparticles are assumed to have three basic shapes, *i.e.* icosahedral, cubo-octahedral and decahedral, plus truncations of these basic shapes. The icosahedron, presenting (111) faces, has the lowest surface energy, but it implies a large internal strain. The cubo-octahedron has a large surface energy since its (111) and (100) faces are exposed, but it does not present internal strain. On the other hand, the decahedron has moderate internal strain and smaller (111) and (100) faces. In this sense it is not surprising that in some pure gold and palladium nanoparticles the truncated decahedron is often observed for sizes < 3 nm, while for the case of Pt the fcc structure seems to prevail. For the case of bimetallic nanoparticles, composed by a noble metal and a transition metal, recent experimental results indicate that the icosahedral phase structure is preferred over the rest of the different structures. For example, when alloying iron with gold nanoparticles, the formation rate of the icosahedral phase is increased to around 80% of the cluster population, just by adding iron to approximately 25% of the total concentration.[27]

When packing four contacting spheres in a tetrahedral arrangement, the regular tetrahedron formed has 0.7796 of its volume occupied by portion of

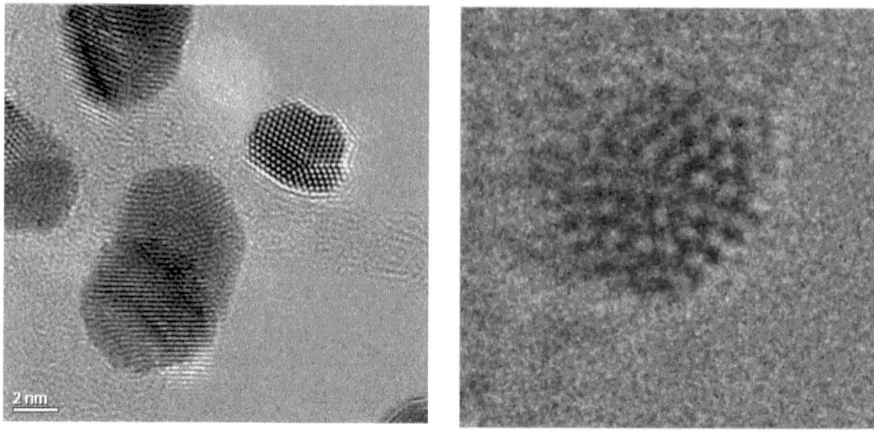

(a) (b)

Figure 2. (a) Example of asymmetric, multiple-twinned decahedral binary-alloy nanoparticle, where structural defects caused the evident assymmetry. (b) binary alloy nanoparticle with polytetrahedral atomic packing.

the spheres, which is the upper bound of the density of any sphere packing. Despite of the fact that arrays of tetrahedrons lead to a densely packed structure, they will not produce closed structures, unless some defects are introduced; for instance, in a decahedral structure, a gap of 7.5° is produced which can be filled by the introduction of a star disclination. The tetrahedral packing will thus result in gaps that can be filled by a network of disclinations. When the network is regular, the structure corresponds to the Frank-Kasper phases.[28,29] However, it is expected that the presence of a second metal will induce the formation of polytetrahedral structures. This tells us about the enhancement on the control of the structure in the bimetallic nanoparticles relative to the single metal particles.

Therefore, the multiple-twinned structure of these nanocrystals is a direct result of their fivefold symmetry. It is not possible to fill space by joining five perfect tetrahedra to form a center with fivefold rotational symmetry, because a gap of 7.5° remains between the final two tetrahedral units.[30] When crystals possessing fivefold symmetry are formed, such as the decahedron or the icosahedron, defects must be introduced at the tetrahedral boundaries to accomodate the structure.[13]

3. DISCUSSION OF EXPERIMENTAL RESULTS, SIMULATIONS, AND ATOMIC MODELS

3.1 The Dodecahedral Particle

In a very small size range of a AuPd system, we have studied colloidal (PVP) AuPd nanoparticles with Au to Pd atomic concentration ratio of 1:5, 1:1, and 5:1. Previously, we have seen that the addition of a second metal introduces a new degree of freedom and helps the tetrahedral packing to be more efficient as was predicted by Doye and Wales[29] and Leary and Doye.[31] Also, it has been reported that poly-tetrahedral packing can generate exciting novel structures.[22,32,33]

We located these small (1-2 nm) nanoparticles at the edges of the carbon films which allowed their complete characterization, as in those positions, the particles are stuck to some particular structure, avoiding structural variations in a shallow energy landscape. In the HREM images, we could observe nanoparticles of 1-2 nm size range with about 85% in rounded shape. HREM images of such rounded shape particles are obtained at different orientations by tilting the samples inside the microscope and recording the images at optimum focus condition. The shape of the particles did not match well with the shape of theoretically calculated images using an icosahedral particle model.

To identify the profiles revealed in experimental images (see Figs. 3 and 4 below), we searched for some small and almost spherical structures theoretically. We considered five geometries with a central site (Fig. 1 in Ref. 4): round capped dodecahedron (RCD); round TIC-based polyhedron (RTB); modified round capped dodecahedron (MRCD); the triacontrahedron (T_h); and the icosahedron (I_h) for comparison.

The study of the geometrical characteristics of these structures is based on the concepts of *equivalent sites* and on the *distance between sites in the cluster*. It is said that equivalent sites are those sites which keep the same distance with the central site and have similar environments. Equivalent sites are joined together in a shell, and various shells can form a layer that is considered to form onion-like clusters. On the other hand d_{NN} is the nearest neighbor distance, that is, the distance between nearest neighbor atoms in a cluster.

An onion-like regular structure is formed by covering a minimum cluster with a given geometry, with a layer formed by some shells preserving the original geometry. After that, the resulting cluster is covered by another layer, preserving the previous geometry, and so on. In contrast, another onion-like growth can be considered, where the cluster that is covered with a layer does not preserve the underlying geometry. The icosahedron (I_h)

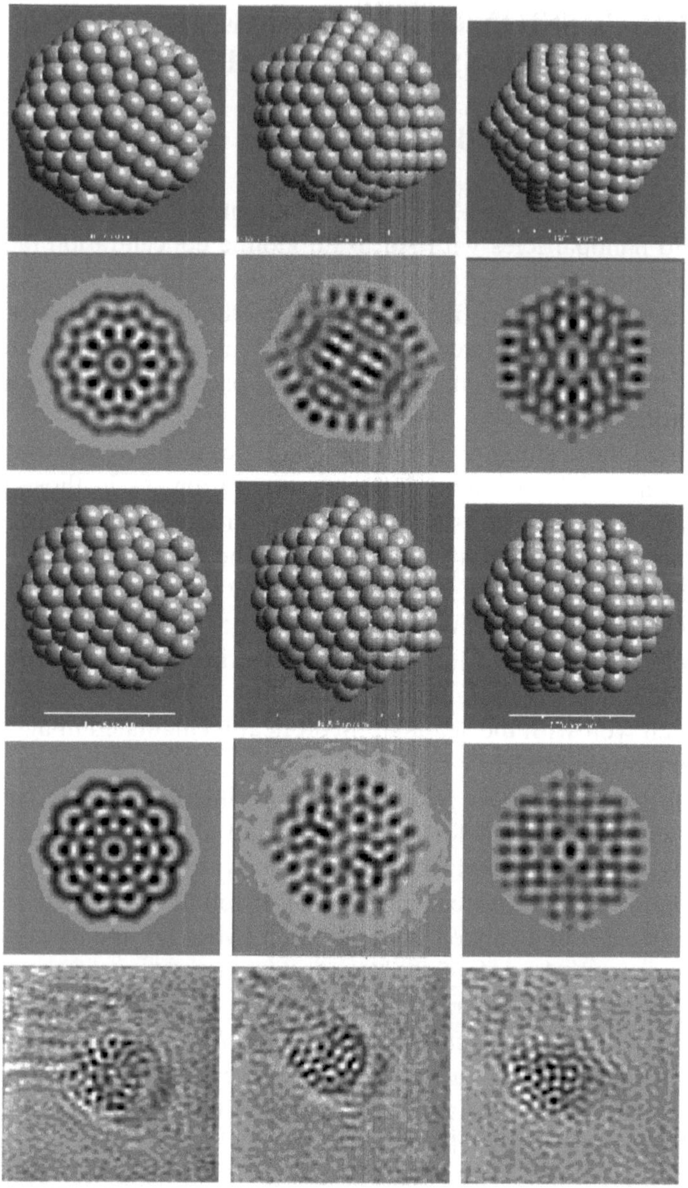

Figure 3. Different orientations [(001)/left, (010)/center, and (100)/right] for the model with 287 atoms of the round capped dodecahedron (RCD) in hard ball (first row), and the calculated HRTEM (second row). Third and fourth rows contains the same but for the modified round capped dodecahedron (MRCD) model with following TEM conditions: $E = 200$ kV and $C_s = 0.5$ mm at the Scherzer defocus. Last row shows experimental HRTEM images. Reprinted figure with permission from Ref. 4. Copyright 2006 by Wiley-VCH.

and the triacontrahedron (T_h) are onion-like regular structures, whereas RCD, RTB, and MRCD are onion-like structures that do not preserve the initial geometry of the clusters.

Of the five structures considered, the RTB and the I_h are polyhedra based on a 13-atom icosahedron, whereas RCD, MRCD, and T_h are polyhedra based on a 20-atom dodecahedron at the core (12 pentagonal faces formed with 20 atoms without a central atom). Adding atoms at each one of the central sites of the pentagonal faces of the 20-atoms dodecahedron we obtain the 32-atom triacontrahedron (30 rhombohedral faces formed with 32 atoms).

For a systematic comparison of the calculated images of the models with the experiment, we obtain several images at different tilting angles of the particles in the microscope and compared them with the images of the model using multi-slice dynamical diffraction theory. From the five models tested here, it is clear that the icosahedral structures do not explain the observed profile of the particles of sizes < 2 nm. Only for larger particles ≈ 3-5 nm the HREM were consistent with the well know contrast from I_h particles. A much better agreement was found with both, the round capped dodecahedron (RCD) and the modified round capped dodecahedron (MRCD) as shown in Fig. 3. Both structures have five fold axis with a *round* profile in the (001) direction and also appear with spherical profile in the (010) direction and somewhat faceted in the (100) direction, in agreement with the experimental data.

Moreover, two critical experiments revealed that the round shape particles are only consistent with the dodecahedral structures initially seen with elongated (ellipsoidal) shape (Fig. 4.a). The other shapes observed in that figure could be obtained by tilting away 15 degrees from the five fold axis in the *x* and the *y* directions. Finally, in Fig. 4.b, starting with the five fold axis and tilting away up to 60 degrees in the *x* direction, we recover the initial fivefold axis. This result is totally inconsistent with the icosahedron structure and consistent with the proposed dodecahedron structures.

Since we selected the model structures guided by geometrical considerations to fit the observed structures of the images, we calculated the stability of these models by first-principles density functional (DFT) calculations.[34] We used SIESTA[35] for gold monometallic clusters with the generalized gradient approximation (GGA).[36] Core electrons were replaced by scalar-relativistic norm conserving pseudopotentials[37] and valence electrons were described with a basis set of double-ξ, *s*, *p*, and *d* numerical pseudo-atomic orbitals. In each case, the geometries were relaxed until the maximum forces were smaller than 0.01 eV/Å. The calculated ICO is the most stable structure ($E_{ico}^{147} = -2.3451$ eV/atom), followed by the CO ($E_{co}^{147} = -2.3218$ eV/atom), the RTB ($E_{rtb}^{137} = -2.3162$ eV/atom), the RCD

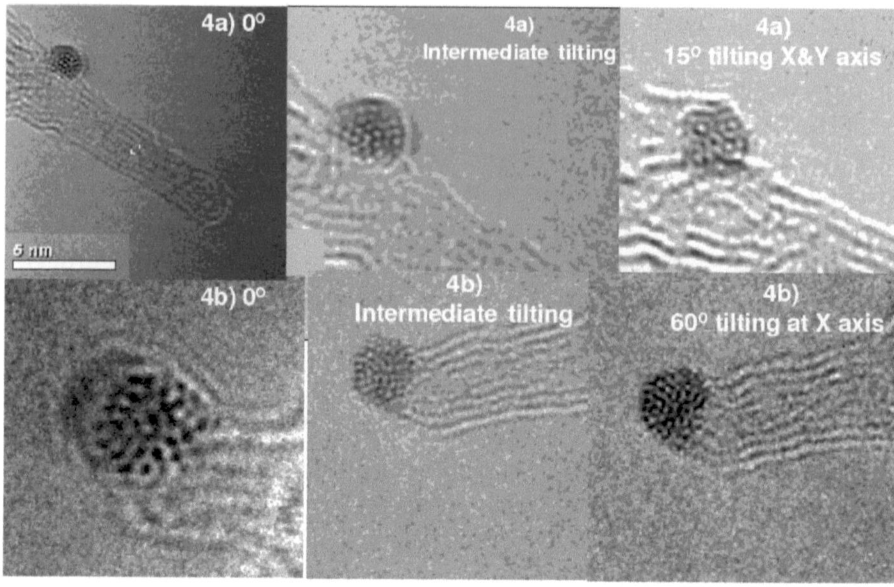

Figure 4. (a) HREM images of a 1.7 nm AuPd nanoparticle (supported on an extremely thin carbon film) showing a nearly spherical shape but with a pointed end (midle figure) which is consistent with the dodecahedral structure. The second particle contrast was obtained on tilting by 15 degrees in *x* and *y* respectively; (b) HREM images of a 2 nm AuPd particle showing the typical contrast in five fold symmetry. After tilting of 60 degrees along *x* axis, the five fold symmetry is recovered. The intermediate contrast is shown. The latter behavior of the particle is not consistent with the I_h and can only be explained by dodecahedral particle. Reprinted figure with permission from Ref. 4. Copyright by 2006 Wiley-VCH.

($E_{rcd}^{124} = -2.3036$ eV/atom), and the MRCD ($E_{mrcd}^{104} = -2.2184$ eV/atom).[38] Additionally, this calculation was repeated using the EAM[39] fitted for monometallic Pd and the main trends above described for DFT calculations still remain. The results are presented in Fig. 5. We compare the cohesive energies but for families of icosahedra (ICO) and cubo-octahedra (CO) for reference, and families of the models proposed in this study. It can be seen that ICO is the most stable structure in the range up to 310 atoms; T_h is the less stable structure. In the range between 100 and 150 atoms, the RTB competes with the stability of ICO. As a general trend, the RTB and RCD are the most stable among MRCD, and other structures. In the range between 250 and 310 atoms RCD, RTB, and MRCD are energetically favorable structures.

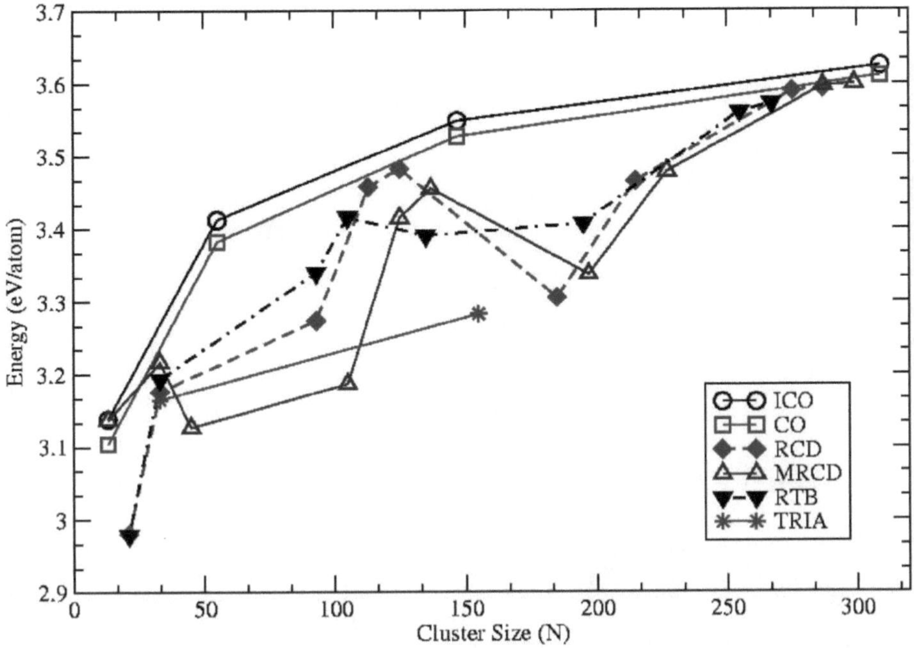

Figure 5. Energy stability EAM calculations for the five different family models in the small size range (1–2 nm) we have studied: the round capped dodecahedron (RCD), the modified round capped dodecahedron (MRCD), the round TIC-based polyhedron (RTB), the triacontrahedron (T_h), and the icosahedron (I_h). For comparison, we have also calculated the curve for the cubo-octahedral family. Reprinted figure with permission from Ref. 4. Copyright 2006 by Wiley-VCH.

We conclude for this section that we have explored for the first time the structure of bimetallic AuPd nanoparticles using HREM at a series of tilting angles with respect to the electron beam for the size range between 1–2 nm. From the experiment, we observe that addition of a second metal allow the nanoparticles to reach a quasi-spherical shape. That shape can be explained in terms of a dodecahedron shape, which is very close to the spherical shape in the fivefold axis. The majority of the particles in the experiment correspond to dodecahedron shape. Also, mono-metallic DFT (Au) and EAM (Pd) calculations demonstrate that the structure is stable.

3.2 Surface Reconstructed Decahedron

In this section we present experimental evidence that two of the well known defects in solid state materials that influence surface stability *i.e.*,

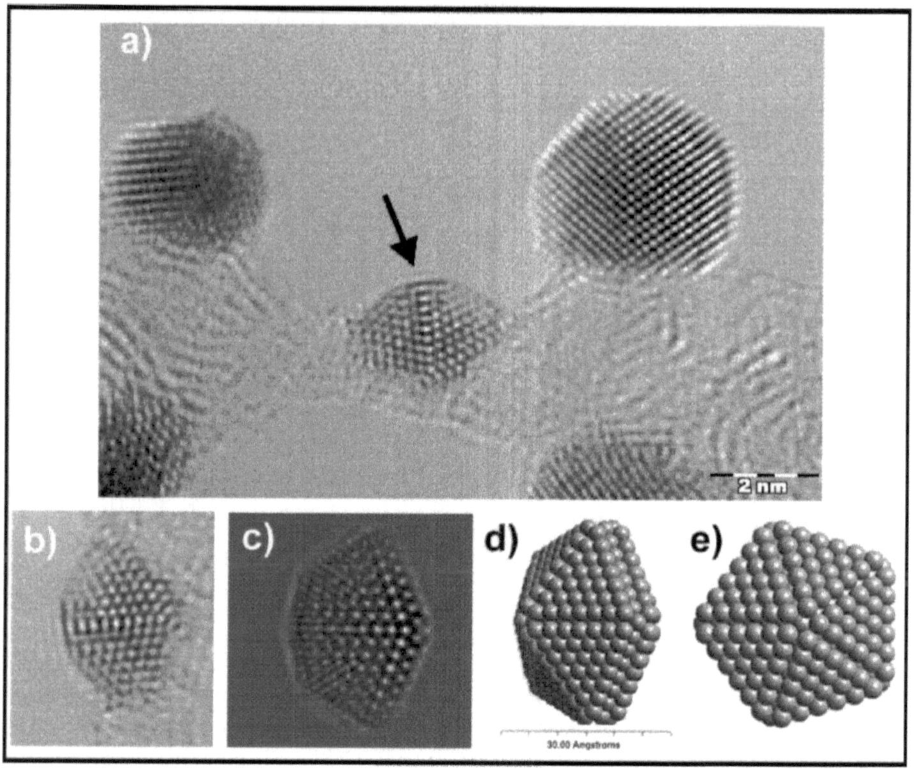

Figure 6. (a) pointed by the arrow and (b) shows high resolution images for a decahedral AuPd particle presented in our experiments. This kind of particle is compared with the simulated HRTEM image at the proper orientation in (c) using the model in (d). A model for the particle where surface reconstruction occurs is presented in (e). The model is a regular pentadecahedra where the last layer of atoms does not follow the fcc (...*ABC*) sequence. The faulted packing (HCP) in the last layer is reconstructed due to energy considerations of stability. Reprinted figure with permission from Ref. 22. Copyright 2004 by the American Physical Society.

vacancies and surface reconstruction that tend to reduce the free energy of the systems, are present in our particles and thus these defects stabilize the structures of our bimetallic nano-systems. The formation of vacancies in the interior and on the surface of the cluster stabilize icosahedral structure, which has been demonstrated theoretically earlier for pure noble metal clusters.[40] These vacancies imply important modifications on the diffusion profiles in metallic clusters (due to the migration of vacancies) and the consequence is noticeable in the chemical ordering in bimetallic clusters.[41] Surface reconstruction is induced by the binary nature of the system, mainly due to this chemical ordering. Our extensive experiments on AuPd and AuCu

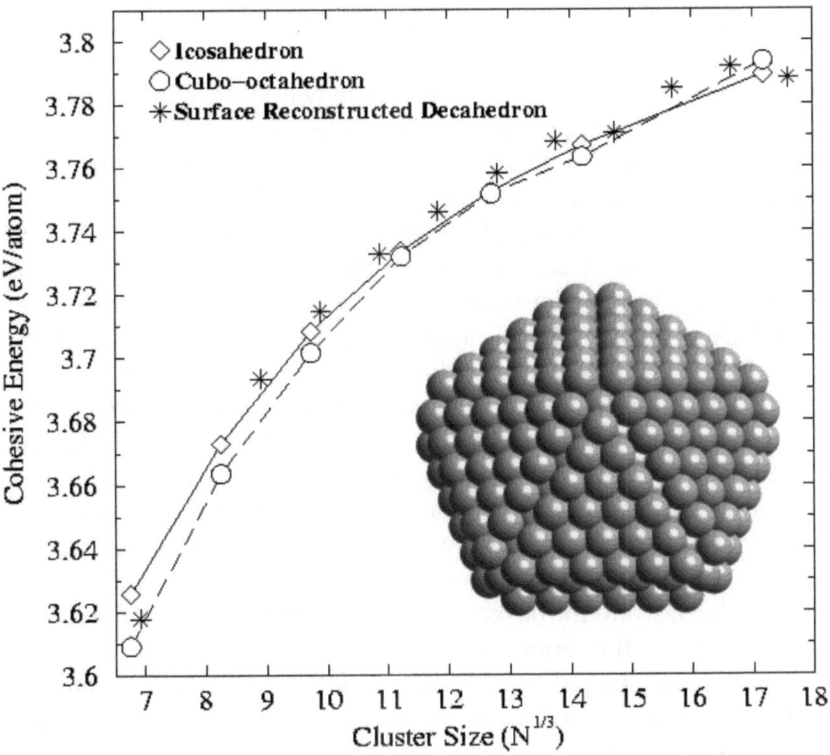

Figure 7. Cohesion energy *versus* particle size for the structure with surface reconstruction shown in Fig. 6 and (inset) calculated by means of the EAM using Pd parameters. For reference, we have plotted the results for the most known stable structure families at this size range *i.e.,* the icosahedral and cubo-octahedral structures. Reprinted figure with permission from Ref. 22. Copyright 2004 by the American Physical Society.

nanoclusters reveal that at a given composition (Au50/Pd50 in case of AuPd clusters), we obtain a high rate of formation of these unusual structures with fivefold although non-perfect symmetry and report on surface reconstruction phenomenon in these systems. We performed molecular dynamics simulations (MD) on bimetallic systems and proposed a new structure with fivefold symmetry and reconstructed surface layer which corresponds to the experimentally observed non-perfect pentagonal structures.[22]

For experimental and simulation analysis we considered clusters of mono-metallic Au, Pd, and bimetallic nanoparticles of Au_5Pd, AuPd, $AuPd_5$ nominal compositions. It was found that the decahedral structure was the most common type of particle observed, although with significant deviations and defects with respect to the perfect decahedral structure. First of all, we remark on the missing atom in the center of these bimetallic particles, being this a characteristic fact in our experiments. For larger particles, several surface vacancy sites are observed, (see Ref. 22 for details) and the TEM images evidence the bimetallic contrast and the decahedral structure.

One of the main results in this work is shown in the experimental image for a AuPd particle supported on amorphous carbon shown in Figs. 6.a and 6.b. The simulated HRTEM image (Fig. 6.c) corresponds with the model for a particle with surface reconstruction, Fig. 6.d. This type of structural arrangement has been obtained by means of MD by Hendy *et al.*[42] on lead clusters. A model for a decahedral particle with surface reconstruction is presented in Fig. 6.e. The structure results with the outer layer of atoms in hexagonal close packed (HCP) surface sites, a surface reconstruction that improves the energetic stability of these binary particles. In these particles, the normal FCC atomic growth sequence ...*ABC* becomes ...*ABCB*, which is a surface reconstruction pointed out by Hendy *et al.*[42] on lead clusters after a re-solidification process. The first point to be remarked is that the above mentioned fact results in a missing atom in the central region where all the tetrahedrons meet (Fig. 6.e), mainly due to the loosely bonds this atom has because of the lower coordination at this site. And secondly, this hcp packing in the last atomic layer results on {100} faces distributed in the surface. Vacancies formation favoring icosahedral structures on metallic clusters had been studied by Mottet *et al.*[40] and surface reconstruction (truncated edges on the icosahedron) that improves the energy stability of multi-twinned particles on a wide range of sizes by Hendy *et al.*,[42] both studies using MD simulations.

The cohesion energy for that structure using the embedded atom method (EAM) in the Foiles' version[39] was calculated for palladium clusters. The results are presented in Fig. 7. As it can be seen, the relative stability of this structural growth pattern is as high as the cubo-octahedral and the icosahedral growth patterns. This high stability is explained by the large surface vacancy sites and the truncated five fold edges, as has been reported theoretically before by Marks[18] and Hendy *et al.*,[42] and justify the large number of particles with these features in our bimetallic systems.

We have found that binary nanoparticles can present a number of non-perfect decahedral structures and the most novel is the one reported in Fig. 6 that corresponds to a decahedral structure with an atom (or a number of atoms[43]) missing on the center, and truncated fivefold edges. The structure results from the formation of a fault on the particle packing, where in the last layer a stacking fault is produced. The hcp surface atomic arrangement (...*ABCB*) of the tetrahedrons forming the decahedral particle (which regularly follows an fcc tetrahedral packing in mono-atomic clusters) evidenced some {100} faces along the fivefold edges. This packing minimizes the surface free energy contribution in the small size regime ($\leq$ 3800 atoms) according to our EAM calculations.

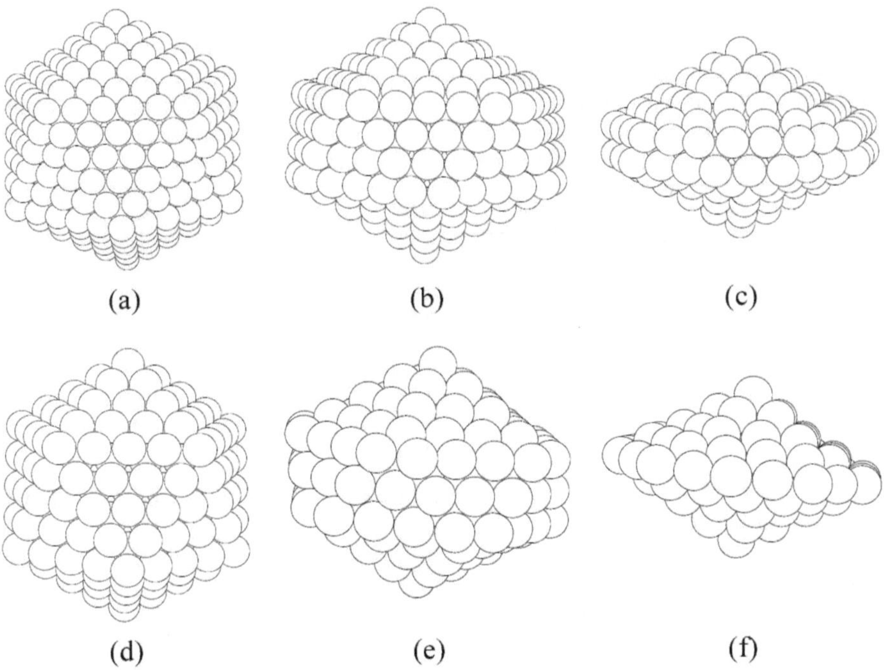

Figure 8. a) Icosahedron, I_h of 561 atoms; b) truncated icosahedron, TI_h, of 409 atoms; c) TI_h of 257 atoms; d) TI_h of 309 atoms; e) TI_h of 207 atoms; f) TI_h of 105 atoms.

3.3 The Montejano's Decahedron

In a previous paper it was presented a study of the structural stability, within the EAM scheme, between icosahedral and cubo-octahedral clusters.[44] In that study, it was found that the icosahedra family is more stable than the cubo-octahedra family for the small cluster sizes, and the opposite behavior is found as the cluster size increases. These results were confirmed by the experimental evidence.[45-47] For small sizes of transition metals clusters, they show the icosahedral and truncated decahedral (Mark's decahedrals) shape.[10,18,48]

Recently, using HRTEM characterization in samples of Au-nanoparticles, Ascencio *et al.*[49] observed images with constrast similar to the icosahedral or the truncated decahedra in gold nanoparticles samples.

Table 1. Geometrical characteristics for the truncated icosahedra TI_h family. The parameter ν (1st column) denotes the *order* of the icosahedron, N_ν is the number of atoms in the icosahedron of order ν, n_t is the number of truncations made. The fourth column lists the (*mnr*) characteristic numbers for each one of the TI_h. Next column shows the number of atoms eliminated (n_-) from the icosahedron of order ν, and the number of atoms on the surface (N_σ) and the total number of atoms of the TI_h are listed in columns six and seven.

ν	N_n	n_t	*mnr*	n_-	N_S	N
1	13	0	212	0	12	13
2	55	0	313	0	42	55
		1	221	32	22	23
3	147	0	414	0	92	147
		1	322	62	62	85
4	309	0	515	0	162	309
		1	423	102	122	207
		2	331	204	82	105
5	561	0	616	0	252	561
		1	524	152	202	409
		2	432	304	152	257
6	923	0	717	0	362	923
		1	625	212	302	711
		2	533	424	242	499
		3	441	636	182	287

However, they also observed squares faces of type (110) together to the triangular faces of type (111); then they proposed a new structure, they named it the *truncated icosahedron* (TI_h). These type of structures also were observed by Yacamán *et al.*[50] in palladium clusters.

In this section we review the geometrical characteristics of another structure very similar to the truncated icosahedron proposed by Ascencio *et al.*,[49] and we termed it the *decmon* (Montejano's decahedron). Also, we carried out a study of the cohesive energy as a function of the cluster size, within the Foiles' version[39] of the EAM, using the parameters of cooper as a first step to characterized the stability of these structures respect to the icosahedral and cubo-octahedral clusters.

3.3.1 The Symmetric Truncated Icosahedron

The truncated icosahedron (TI_h) is derived from the regular icosahedron, is a polyhedron that results from an adequated symmetric truncation of the icosahedron (this truncation can also be made in a non-symmetric way, as was presented originally by Ascencio et al.[49]

Two icosahedra with their respective truncations are presented in Fig. 8. The 561 (309) atoms icosahedron, shown initially at Fig. 8.a (Fig. 8.d) is truncated once (by elimination of a complete pentagonal cap) resulting in the TI_h of 409 (207) atoms, Fig. 8.b (Fig. 8.e). Truncating once again this last structure we obtain the TI_h of 257 (105) atoms, Fig. 8.c (Fig. 8.f). Notice from these figures that the truncated icosahedron can adopt two different shapes: a) the TI_{h30} and b) the TI_{h40}. The TI_{h30} is formed by 22 vertices joined by 50 edges to form 30 faces of three types: one type is a rectangular face (R) that shows (100) crystal planes, and two triangular faces showing (111) crystal planes, one of these faces is an equilateral triangle (ET), and the other one is an isosceles triangle (IT) (see Fig. 8.c), they count 10 faces of each type and note that the ET faces coincide in the pole vertices. Each side for the ET-face has a length of ($1.05ad_{NN}$), and two sides of the IT-face have a length of (bd_{NN}) and the other ($1.05ad_{NN}$), with a and b integers and d_{NN} the distance between first neighbors; consequently each R-face has two sides of length ($1.05ad_{NN}$), and two of (bd_{NN}). The TI_{h40} is formed by 32 vertices joined by 80 edges to form 40 faces of four types: the same three than for the TI_{h30} and one additional trapezoidal face (Tr) (see Figs. 8.b, 8.c, and 8.e. The aditional 10 trapezoidal faces and 20 edges form the waist of the TI_{h40}, *i.e.*, they are 10 lateral trapezoidal faces and 20 lateral edges. Notice that for the equator or waist of the TI_{h30}, an R face coincides with an IT face, while in the TI_{h40} the R and IT faces coincide with Tr faces.

Three numbers (*mnr*) are going to be used to geometrically characterize the TI_h; index m means the number of atoms, including the vertices, along the edge between two ET faces, index n means the same but between R and IT faces, and index r is the number of atoms along the Tr faces. For example, the (524) TI_{h40} structure is presented in Fig. 8.b, the (432) TI_{h40} in Fig. 8.c, the (423) TI_{h40} in Fig. 8.e, and the (331) TI_{h30} in Fig. 8.f.

It must be noted several things: First, for the TI_{h30} $m=n$ and $r=1$; second, the icosahedron corresponding with this notation has $m=r$ and $n=1$; third, the length of the edges of the ET-faces cannot be smaller than the half part of the length of the edge of the icosahedron, that is $m \geq (\nu+1)/2$; and fourth, $m \geq n$, r, that is the length of the edges of the IT- and R-faces can not be bigger than the length of the edges of the ET-faces. Some of the geometrical properties for the truncated icosahedra are presented in Table 1. The parameter ν (listed in the first column of the table) denotes the *order* of the icosahedron that eventually will be truncated, N_n is the number of atoms in the icosahedron of order ν, n_t is the number of truncations made to the icosahedron for to get the (*mnr*) truncated icosahedron. The fourth column lists the three characteristic numbers for each one of the TI_h, the number of atoms eliminated (n_-) from the icosahedron of order ν is presented in column

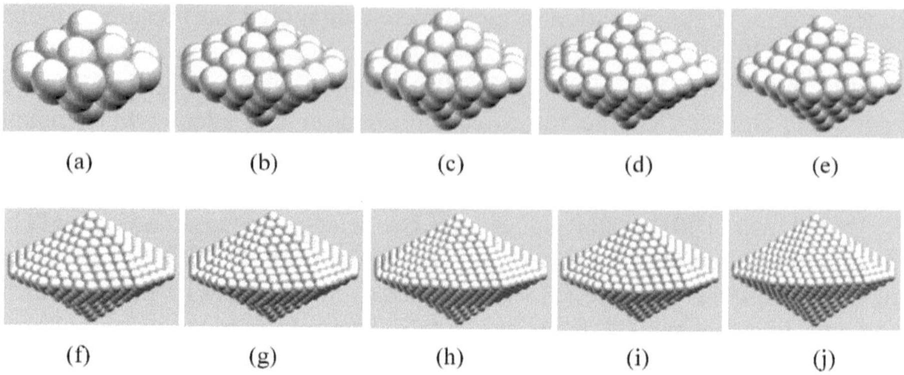

Figure 9. Decmon-like polyhedra of order *mn* (a) decm22 of 23 sites; (b) decm23 and (c) decm32 of 54 sites; (d) decm24 and (e) decm42 of 105 sites; (f) decm55 of 609 sites; (g) decm56 and (h) decm65 of 835 sites; (i) decm57 and (j) decm75 of 1111 sites.

five, the number of atoms on the surface (N_s) and the total number of atoms N of the TI_h are listed in columns six and seven, respectively. We mention as an example the TI_h (432), Fig. 8.c, of type TI_{h40}, which is generated from the 561-atoms icosahedron of order 5, 304 atoms were eliminated in 2 truncations resulting a 257-atoms truncated icosahedron with 152 surface atoms.

3.3.2 The Decmon-like Polyhedron

The decahedron looks like two inverted pentagonal pyramids joined by the bases. An Ino decahedron is built in the same way but the pyramids are separated by some layers of atoms forming five lateral rectangular faces. An icosahedron is built with the pyramids separated and rotated 36° between them in such away that causes the formation of ten equilateral triangular faces. From Fig. 8, it can be seen that the truncated icosahedra TI_{h30} and TI_{h40}, look like to be constructed in the same way that the polyhedra above described but the pyramids are different. It is possible to consider a decmon type pyramid with an irregular decagonal base. The truncated icosahedra has not specular symmetry with respect to the equator.

The *decmon*-like polyhedron, shown in Fig. 9, is a structure that results by reflection of the decmon pyramid with respect to the base, it looks like the truncated icosahedron TI_{h30} in Fig. 8.c, but on the equator or waist of the polyhedron, an R-face coincides with an R-face, and an IT-face coincides with an IT-face, that is, it is a symmetric polyhedron respecto to the equator. Also, notice that in the decmon polyhedra the length of the edges is not limited.

Table 2. Number of sites aggregated for each value of m (surface atoms N_σ), and the total number of sites for the *decmon* families $[m,m]$ with a central site, $[m,(m+1)]$ with a central decahedron of 7 sites, and $[m,(m+2)]$, with a central decahedron with a central site of 23 sites.

$m = n$			$n = m+1$				$n = m+2$			
m	N_σ	N	m	n	N_σ	N	m	n	N_σ	N
2	22	23	2	3	7	7	2	4	82	105
3	82	105	3	4	47	54	3	5	182	287
4	182	287	4	5	127	181	4	6	322	609
5	322	609	5	6	247	428	5	7	502	1111
6	502	1111	6	7	407	835	6	8	722	1833

In the decmon polyhedron there are only two types of edges and their lengths are characteristic for each particular decmon. The letters *mn* can be considered as the order of the decmon (decm*mn*), and the values taken by m and n are not limited. The indexes m and n are considered as above in the TI_h. Indeed, every decmon has inside it another decmon with lower order, *i.e.*, it growths up in an onion-like structure; the decm22 is covered by the decm33 one, which is covered by the decm44, and so on. Moreover, the decm23 (decm32) is covered by the decm34 (decm43), which is covered by the decm45 (decm54) and so on.

In this work we present the geometrical characteristics of three families of the decmon polyhedra: a) family with $m = n$, which have a central site, b) family $n = m+1$, with a central decahedron, and c) the family $n = m+2$, with a central decahedron plus a central site. It is worth to notice that interchanging the values of m and n we also obtain decmon polyedra with the same characteristics that the first but with different edge lengths, thus it takes place a large and rich variety of clusters.

Some geometrical characteristics of the three decmon families are listed in Table 2. $m=n$ family is described by three columns: m values in the first column, and the number of sites aggregated to the last cluster (or the number of surface sites, N_σ, for this value of m), and the total number of atoms, N, for the clusters in the second and third columns, respectively. The other families are described by four columns: m and n values in the first and second columns, and the number of surface sites, N_σ, and the total number of atoms, N, for the clusters in the third and fourth columns, respectively.

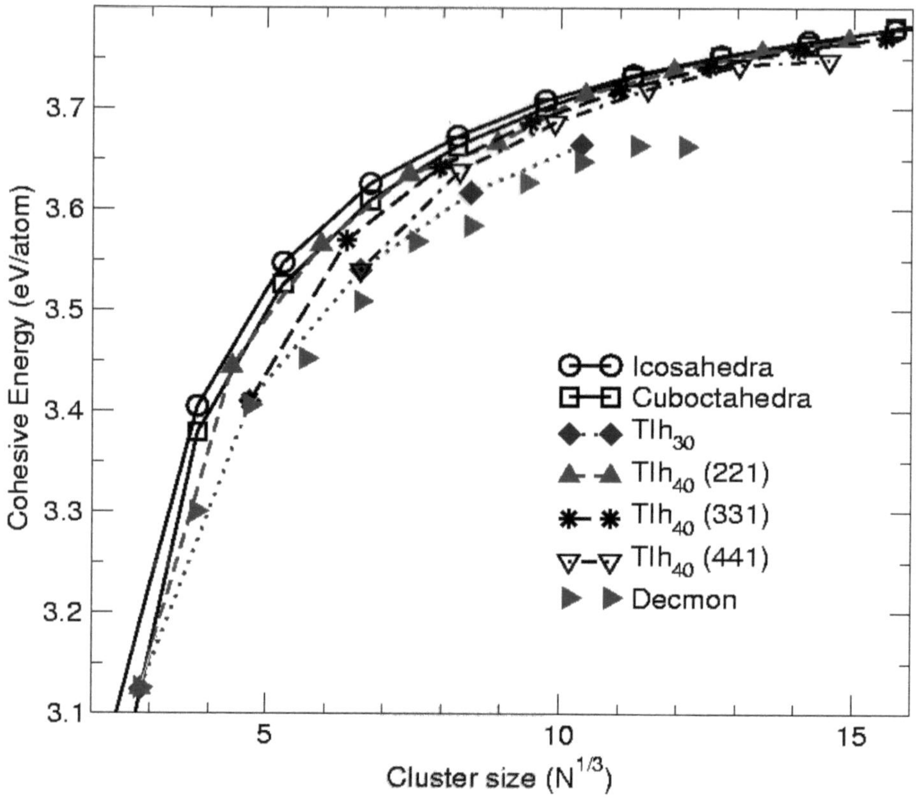

Figure 10. Cohesive energy per atom as function of the cluster size $N^{1/3}$, for the families of the cubo-octahedra (O_h), regular icosahedra (I_h), the truncated icosahedra TI_{h30}, the TI_{h40} (221), (331) and (441), and the decmon. Parameters used were for palladium. See the text below for details.

The decmon polyhedra for the three families are shown in Fig. 9. The decm22, Fig. 9.a, is a cluster with a central atom and 22 surface atoms; the decm33 is the decm22 covered by 82 atoms for a 105-atom decmon cluster (see Table 2); the decm44 is the decm33 covered by 182 atoms for a 287-atom decmon cluster; the decm55, shown in Fig. 9.f, is a cluster formed by the decm11 plus three layers to get a 609-atoms decmon cluster, 322 of these atoms form the cluster's surface. The decm23 (Fig. 9.b), is the smallest cluster of the $n=m+1$ family, is a 7-atoms decahedron covered by 47 atoms for to give a 54-atom decmon cluster; the decm32 (Fig. 9.c), has the same number of atoms but interchanging the number of atoms in the edges; the decm45 (decm54) is the decm34 (decm43) plus 127 atoms to form the 181-atoms decmon cluster (see Table 2); so the decm56 (Fig. 9.g) (decm65, Fig. 9.h), is the decm23 (decm32) plus three layers to get a 835-atoms

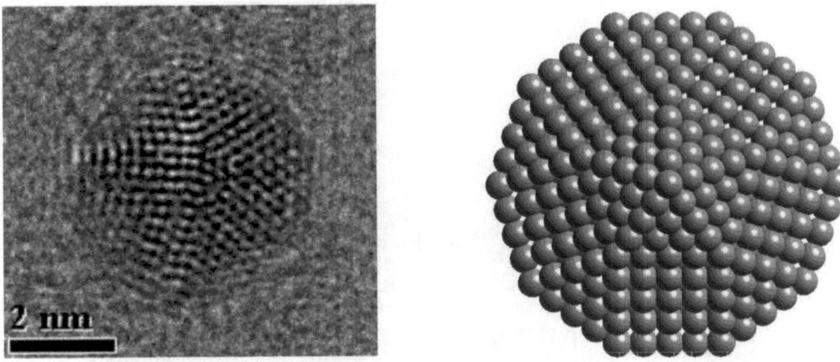

Figure 11. (left) Experimental HRTEM image for palladium nanoparticles where the decmon structure is appreciated. (right) shows the model for a decmon55.

decmon cluster. In the same way the decm24 and decm42, shown in Figs. 9.d and 9.e respectively, are the smallest of the $n=m+2$ or $n=m-2$ families and they are formed by a decahedron with a central site of 23 atoms covered with 82 atoms for a total number of 105 atoms; so the decm57 (Fig. 9.i), is the decm24 plus three layers to get a 1111-atoms decmon cluster, as well as for the dec75, Fig. 9.j.

The EAM calculations for the different sizes of Pd-decmon clusters are shown in Fig. 10, where the cohesive energy is plotted as a function of the cubic root of the number of sites (cluster size). We have selected palladium only as an example. Results for icosahedral and cubo-octahedral clusters are plotted for comparison. In Ref. 44 it was found that icosahedral cluster are more stable than cubo-octahedral ones for small sizes, and cubo-octahedral are the more stable for large sizes. Results for TI_{h30}, three TI_{h40} families and decmon clusters for different sizes are shown in Fig. 10. The $TI_{h40}(221)$ family corresponds to the family that begins with the (221) cluster which is covered by the (322) cluster, and this one, is covered by the (423) cluster and so on. The $TI_{h40}(331)$ family corresponds to the family that begins with the (331) cluster which is covered by the (432) cluster, followed by the (533) cluster and so on. Finally, the $TI_{h40}(441)$ family corresponds to the family that begins with the (441) cluster which is covered by the (542) cluster, this followed by the (643) cluster and so on. From the figure, we observe that all of them are energetically below the icosahedral and cubo-octahedral clusters curves, thus they are certainly less stable than these structures, but due to the energy difference shown in the figure, we deduce that they could compete in stability with icosahedral and cubo-octahedral clusters and thus decmon and TI_h are good candidates for stable structures for any other transition metal

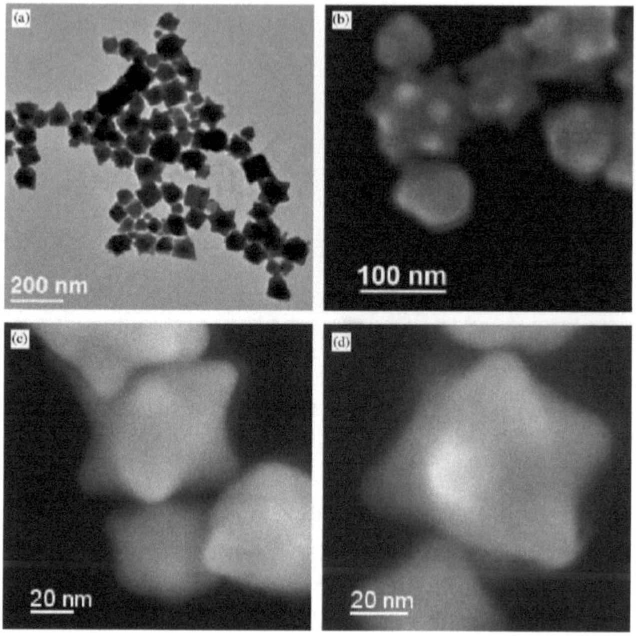

Figure 12. Star polyhedral gold nanocrystals. (a) Low-magnification TEM image of the as-synthesized product. (b) SEM image of a multiple-twinned star nanocrystal, revealing pyramids emerging around the perimeter of the crystal and from its upper surface. (c) and (d) SEM images of monocrystalline stars demostrating the presence of eight pyramids each oriented 90° from its nearest neighbors. Reprinted figure with permission from Ref. 24. Copyright 2005 by Elsevier B.V.

cluster. Comparison between TI_h, and decmon clusters shows that truncated icosahedral clusters are the more stable for small sizes and the decmon clusters are the more stables for larger sizes. Indeed, the stability transition between them occurs approximately at the same size transition that happens for icosahedral and cubo-octahedral clusters.

A HRTEM image of a palladium particle is shown in Fig. 11, where it can be noticed the good agreement with decmon clusters. A top view of the model for the decmon55 cluster with 609 atoms is shown at the right. We can conclude that the decmon structure is a good candidate for structural stable shapes to be adopted by transition metal clusters.

3.4 Star Polyhedral Gold Nanocrystals

Now we will review the structural features of star polyhedra gold nanocrystals, sinthesized by colloidal reduction by ascorbic acid. We have identified two distinct types of star polyhedra nanocrystals, corresponding to

Figure 13. The two classes of star polyhedral gold nanocrystals. (a) TEM image of a multiple-twinned star nanocrystal, and (b) proposed model, an icosahedron with each of its 20 {111} surfaces replaced by a tetrahedral pyamid. (c) TEM image of a monocrystalline star, and (d) propsed model, a cuboctahedron with each of its eight {111} truncations replaced by a tetrahedral pyramid. Reprinted figure with permission from Ref. 24. Copyright 2005 by Elsevier B.V.

icosahedra and cubo-octahedra with preferential growth of their exposed {111} surfaces. Specifically, the {111} faces of the original Arquimedean solids grow into tetrahedral pyramids, being the base of each pyramid the original polyhedral face. Figure 12 shows synthesized nanocrystals with enlarged pyramids on their surface. Approximately 60% of the as-syntesized product consist of star polyhedral gold nanocrystals, with 20% of them being multitwinned crystals with fivefold symmetry, and 40% being monocrystalline star polyhedra. The size of this crystals is on the order of 100–200 nm, but desspite of this broad variation, the geometric proportions of a given tipe of star nanocrystal are remarkably consistent from one crystal to another, regardless of size.

In the case of the icosahedrally derived star nanocrystals, the crystal shape closely resembles a great stellated dodecahedron, which is known as one of the Kepler-Poisont solids.

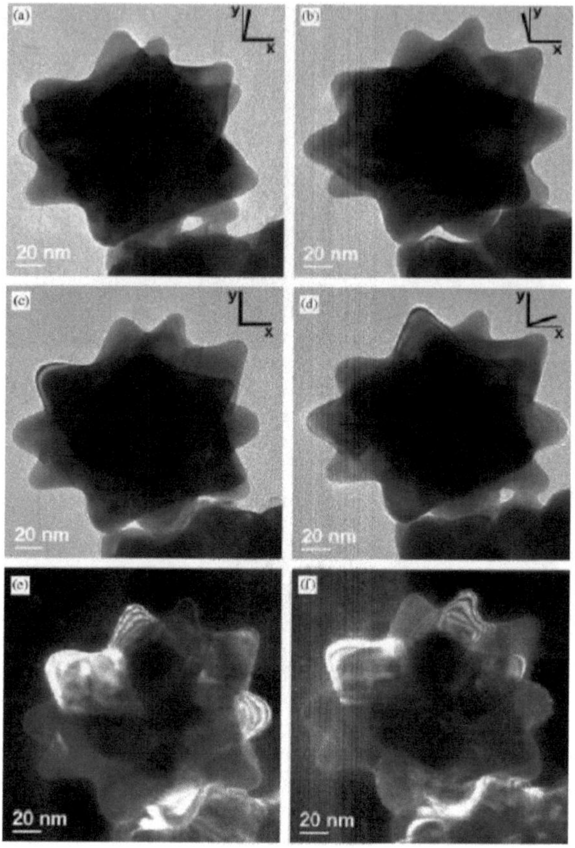

Figure 14. TEM tilt sequence for a multiple-twinned star nanocrystal, with [x°,y°] tilting angles (a) [0°,-15°]; (b) [0°,15°]; (c) [0°,0°]; and (d) [15°,0°]. (e,f) Weak beam dark field (WBDF) images illustrating some of the individual monocrystalline domains present within the crystal. Reprinted figure with permission from Ref. 24. Copyright 2005 by Elsevier B.V.

Figure 13.a is a representative TEM image of the first specific type of star polyhedral gold nanocrystal we have characterized. Tilting sequences of TEM images, as in Fig. 14, were performed, and by tracking the emergence and disappearance at various tilting angles, three dimensional reconstructions were obtained. Twenty distinct pyramids were observed, and their spatial arrangement suggested that they emanate from an icosahedral core. Based on the number of peaks and the overall symmetry of the structure, we propose a structural model for this nanocrystal, illustrated in Fig. 13.b, corresponding to an icosahedron with tetrahedral pyramids extending from each of its 20 {111} faces. The structure of these

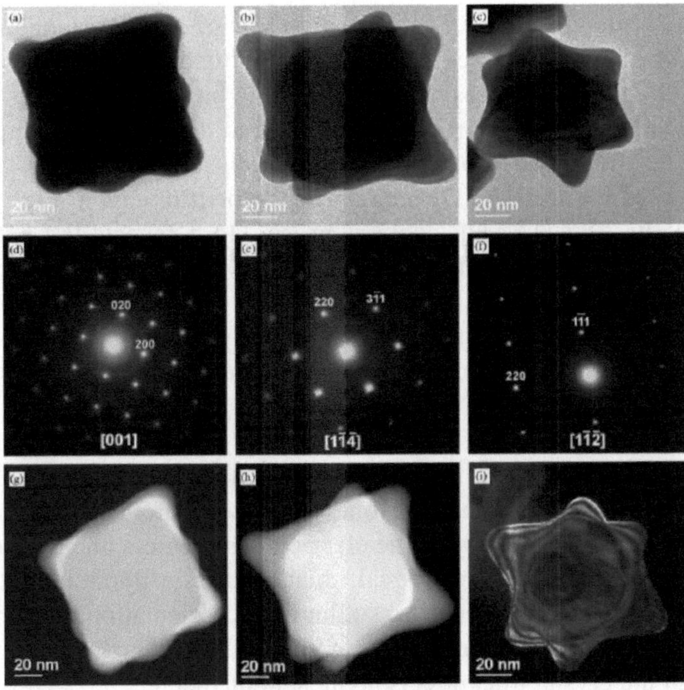

Figure 15. Monocrystalline star polyhedral gold nanocrystals. (a) TEM image, (d) SAED, and (g) HAADF image for the [001] zone axis.(b) TEM image, (e) SAED, and (h) HAADF iamge for the [114] zone axis. (c) TEM image and (f) SAED for the [112] zone axis. (i) Representative WBDF image. Reprinted figure with permission from Ref. 24 Copyright 2005 by Elsevier B.V.

icosahedrally derived star nanocrystals closely resembles the great stellated dodecahedron, a Kepler-Poisont solid.

However, a great stellated dodecahedron posseses elongated pyramids, the edge length of each pyramid being related to the edge length of the icosahedral base by the golden ratio.[51] In contrast, our proposed model was constructed by growing tetrahedral pyramids on the icosahedral faces, both to satisfy fcc stacking and to match the experimentally observed shape of the pyramids. Properly speaking, the proposed model is a 60-faced deltahedron.[52] It should be noted that the pyramids of the gold nanocrystals are not perfect tetrahedra, as the peaks of the pyramids appear to be rounded. Since the corners of faceted nanocrystals generally terminate in a series of terraced steps,[6,53] we expect that surface steps give a rounded appearance to the peak of a tetrahedral pyramid.

Figure 13.c is representative of the second type of star polyhedral gold nanocrystal we have characterized. Based on 3D reconstruction from TEM tilting sequences, and combined with SEM analysis discussed in Ref. 24

propose a structural model, seen in Fig. 13.d corresponding to a cuboctahedron with tetrahedral pyramids extending from each of its eight {111} faces. Figures 15.a-c present three orientations for the cuboctahedrally derived star nanocrystals, and the corresponding SAED patterns (Figs. 15.d-f) demonstrate the monocrystallinity of the structures. Figures 15.g and h are HAADF images of the orientations in Figs. 15.a and b, respectively. The HAADF contrast in Fig. 15.h reveals the presence of two pyramids in the central region of the crystal, in addition to the six pyramids visible at the perimeter of the crystal.

The orientation of the star nanocrystal in Fig. 15.a is such that the four pyramids at the upper surface of the crystal are aligned directly above the four pyramids at the lower crystal surface. In the corresponding HAADF image (Fig. 15.g), the contrast is brighter at the corners of the crystals than in the center. This contrast inversion occurs because in the center of the crystal, the electron beam travels through one continuous monocrystalline region. On the other hand, at the corners the electron beam enters one pyramid at the upper crystal surface, exit the crystal, then reenters a pyramid at the lower crystal surface with a different crystallographic orientation. This increases the local inelastic dispersion, resulting in brighter contrast at the corners even though the crystal thickness is greatest at the center.

Figure 15.i is a WBDF image of a cuboctahedrally derived star nanocrystal. The continuity of thickness fringes over the entire crystal structure demonstrates its monocrystallinity. In addition to illuminating the six pyramids around the perimeter of the crystal, WBDF contrast also reveals the two pyramids emerging in projection from the upper and lower surfaces of the crystal. These two pyramids are not visible in conventional TEM images at this orientation, due to the thickness of the central region of the crystal. The thickness fringe contours also illustrates rounding of the pyramid peaks, as noted and discussed previously. Thus, in both monocrystalline and multiple-twinned star polyhedra, local minimization of energy dictates rounded peaks for the tetrahedral pyramids.

Gold nanocrystals increase the proportion of their exposed {111} surfaces by assuming star polyhedral morphologies, since each exposed {111} surface of the original Archimedean solid is replaced by a tetrahedral pyramid presenting three exposed {111} surfaces. For fcc crystals the relative surface energies of the low-index crystallographic planes are $\gamma_{111} < \gamma_{100} < \gamma_{110}$. Thus, assuming a star morphology lowers the surface energy of a growing crystal, and this could be the driving force for star nanocrystal formation in our system.

The sizes of our icosahedrally derived star nanocrystals are on the order of 100–200 nm. However, icosahedral gold nanocrystals of this size are typically not observed. Recently, Yang and coworkers[5] demonstrated that by careful regulation of the growth rate for each crystallographic direction, icosahedral gold nanocrystals on the order of 100–300 nm can be synthesized. For our reaction conditions, in the presence of ascorbic acid and its oxidazed derivatives, icosahedral nanocrystals are stabilized by preferential growth of their exposed {111} surfaces, allowing for growth of icosahedrally derived star polyhedral nanocrystals.

Although the vast majority of monocrystalline stars are free from structural defects, a sizable number of multiple-twinned stars contain defects and have an imperfect shape. This is to be expected, since a multiple-twinned structure inherently posseses more lattice strain, and thus a higher potential for defects. However, characteristic defects were observed for both class of star polyhedral gold nanocrystals.

4. CONCLUSIONS

In conclusion, we have explored for the first time the structure of bimetallic AuPd nanoparticles using HREM at a series of tilting angles with respect to the electron beam for the size range between 1–2 nm. From the experiment, we observe that addition of a second metal allow the nanoparticles to reach a quasi-spherical shape. That shape can be explained in terms of a dodecahedron shape, which is very close to the spherical shape in the five fold axis. The majority of the particles in the experiment correspond to dodecahedron shape. Also, mono-metallic DFT (Au) and EAM (Pd) calculations demonstrate that the structure is stable.

Also, non-perfect decahedral shape and structure stabilization by surface reconstruction are reported in bimetallic AuPd and AuCu nanoparticles (1.5 nm-3 nm). A number of new structures are found when two different metallic atoms are present. These novel structures present surface vacancy sites and the formation of a fault on the particle packing. The hcp surface atomic arrangement (...*ABCB*) evidenced some {100} faces along the fivefold edges. This packing minimizes the surface free energy contribution and favors the formation of these structures, according to our energy calculations and some others results reported previously. The formation of faces {100} additional to the faces {111} present in decahedral particles would have important implications in heterogeneous catalysis, because these new faces are chemically reactive and bonding sites for molecules, being specially important on catalytic systems such as AuPd, AuCu, and PdPt.

We conclude that the shape for very small particles corresponds to a Platonic solid and larger sizes correspond to Archimedean solids. Thus, going from the very small size range of both mono- and binary metallic nanoparticles, the complete sequence of structures which is formed corresponds to Platonic-solids → Archimedean-solids, Kepler-Poisont solids, defected structures (twins, stacking faults, etc.) and finally bulk-like structures.

ACKNOWLEDGEMENT

This chapter was originally published as a brief review in the journal Modern Physics Letters B, Vol. 20, No. 13 (2006) 725-751, Low dimensional non-crystallographic metallic nanostructures: HRTEM simulation, models, and experimental results. Copyright World Scientific Publishing Company. The authors acknowledge financial support from CONACYT-Mexico through grant J42645-F.

REFERENCES

1. H. S. M. Coxeter, *Regular Polytopes*; Dover Publications Inc., New York, 1973.
2. M. J. Wenninger, *Polyhedron Models*, University Press, Cambridge, 1971.
3. A. Pugh, *Polyhedra: A Visual Approach*, University of California Press, Berkeley, 1976.
4. J. M. Montejano-Carrizales, J. L. Rodrguez-López, Umapada Pal, Mario Miki-Yoshida, and Miguel José-Yacamán, The Completion of the Platonic Atomic Polyhedra: The Dodecahedron, Small **2**, 351 (2006).
5. F. Kim, S. Connor, H. Song, T. Kuykendall, and P. Yang, Angew. *Chem. Int. Ed.* **43**, 3673 (2004).
6. Z. L. Wang, T. S. Ahmad, and M. A. El-Sayed, *Surf. Sci.* **380**, 302 (1997).
7. C. R. Henry, *Prog. Surf. Sci.* **80**, 92 (2005).
8. P. R. Cromwell, *Polyhedra*, Cambridge University Press, New York, 1997.
9. O. Echt, K. Sattler, E. Recknagel, *Phys. Rev. Lett.* **47**, 1121 (1981).
10. T. P. Martin, *Phys. Rep.* **273**, 199 (1996).
11. T. P. Martin, U. Näher, H. Schaber, U. Zimmermann, *Phys. Rev. Lett.* **70**, 3079 (1993).
12. E. L. Nagaev, *Phys. Rep.* **222**, 199 (1992).
13. K. Heinemann, M. J. Yacamán, C. Y. Yang and H. Poppa, *J. Crystal Growth* **47**, 177 (1979).
14. A. Howie and L. D. Marks. *Phil. Mag. A* **49**, 95 (1984).
15. B. Pauwels, D. Bernaerts, S. Amelinckx, G. Van Tendeloo, J. Joutsensaari, and E. I. Kauppinen, *J. Crystal Growth* **200**, 126 (1999).
16. S. Ino and S. Ogawa, *J. Phys. Soc. Japan* **22**, 1365 (1967).
17. J. G. Allpress and J. V. Sanders, *Philos. Mag.* **13**, 609 (1966).
18. L. D. Marks and D. J. Smith, *J. Crystal Growth* **54**, 425 (1981); L. D. Marks, Philos. Mag. A **49**, 81 (1984); L. D. Marks, *Rep. Prog. Phys.* **57**, 603 (1994).
19. M. José-Yacamán, J.A. Ascencio, H. B. Liu, J. Gardea-Torresdey, *J. Vac. Sci. Tehcnol. B* **19**, 1091 (2001).
20. H. Bönnemann and R. M. Richards, *Eur. J. Inorg. Chem.* **2001-10**, 2455 (2001).
21. L. M. Liz-Marzán, *Mater. Today* **7**, 26 (2004).

22. J. L. Rodrguez-López, J. M. Montejeno-Carrizales, U. Pal, J. F. Sánchez-Ramrez, H. Troiani, D. Garca, M. Miki-Yoshida, M. José-Yacamán, *Phys. Rev. Lett.* **92**, 196102 (2004).
23. J. M. Montejano-Carrizales, J. L. Rodriguez-López, and M. José-Yacamán, *Geometrical properties of the decmon structure*, submitted to *J. Crystal Growth*, (2006).
24. J. L. Burt, J. L. Elechiguerra, J. Reyes-Gasga, J. M. Montejano-Carrizales, and M. José-Yacamán, *J. Crystal Growth* **285**, 681 (2005).
25. T. Yonezawa and N. Toshima, *J. Mol. Catalysis* **83**, 167 (1993); J. Luo *et al.*, *Chem. Mater.* **17**, 3086 (2005).
26. J. M. Montejano-Carrizales, J. L. Rodrguez-López, C. Gutiérrez-Wing, M. Miki-Yoshida, and M. José-Yacamán; *Crystallography and shape of nanoparticles and clusters: geometrical analysis, image, diffraction simulation and high resolution images*, Chapter in: The Encyclopedia of Nanoscience and Nanotechnology. 2, 237-282 (2004); edited by H. S. Nalwa, American Scientific Publishers, Los Angeles, USA.
27. D. K. Saha, K. Koga, and H. Takeo, *Eur. Phys. J. D* **9**, 539 (1999).
28. F. C. Frank amd J. S. Kasper, *Acta Crystall.* **11**, 184 (1958); idem. **12**, 483 (1959).
29. J. P. K. Doye, D. J. Wales, *Phys. Rev. Lett.* **86**, 5719 (2001).
30. Y. Gao,L. Song, P. Jiang, L. F. Liu, X. Q. Yan, Z. P. Zhou, D. F. Liu, J. X. Wang, H. J. Yuan, Z. X. Zhang, X. W. Zhao, X. Y. Dou, W. Y. Zhou, G. Wang, S. S Xie, H. Y. Chen, and J. Q. Li, *J. Crystal Growth* **276**, 606 (2005).
31. R. H. Leary, J. P. K. Doye, Phys. Rev. E **60**, 6320 (1999).
32. D. R. Nelson and Frans Spaepen, in Solid State Physics: Advances in Research and Applications vol. 42, pp. 1, eds. H. Ehrenreich and D. Turnbull, Academic Press, 1989.
33. F. Dassenoy, M. -J. Casanove, P. Lecante, M. Verelst, E. Snoeck, A. Mosset, T. Ould Ely, C. Amiens, B. Chaudret, *J. Chem. Phys.* **112**, 8137 (2000).
34. W. Khon, L. J. Sham, *Phys. Rev.* **140**, 1133 (1965).
35. a) P. Ordejón, E. Artacho, M. J. Soler, *Phys. Rev. B* **1996**, 53 (R10441); b) D. Sánchez-Portal, P. Ordejón, E. Artacho, M. J. Soler, *Int. J. Quantum Chem.* **65**, 453 (1997); c) ibid, *Phys. Status Solidi* (b) **215**, 809 (1999).
36. J. P. Perdew, K. Burke, M. Ernzerhof, *Phys. Rev. Lett.* **77**, 3865 (1996).
37. N. Troullier, J. L. Martins, *Phys. Rev. B* **43**, 1993 (1991).
38. J. L. Rodrguez-López, (to be published 2006).
39. S. M. Foiles, M. I. Baskes, and M. S. Daw, *Phys. Rev. B* **33**, 7893 (1986).
40. C. Mottet, G. Tréglia, and B. Legrand, *Surf. Sci.* **383**, L797 (1997).
41. J. L. Rodrguez-López, J. M. Montejano-Carrizales, and M. José-Yacamán, *Appl. Surf. Sci.* **219**, 56 (2003).
42. S. C. Hendy and B. D. Hall, *Phys. Rev. B* **64**, 085425 (2001); S. C. Hendy and J. P. K. Doye, *ibid* **66**, 235402 (2002).
43. K. Koga and K. Sugawara, *Surf. Sci.* **529**, 23 (2003).
44. J. M. Montejano-Carrizales, M. P. Iñiguez, and J. A. Alonso, *J. Cluster Sci.* **5**, 287 (1994).
45. M. José-Yacamán, in *Catalytic Materials, Relationship between structure and reativity*, p. 341, edited by T. E. Whyte, R. A. Betta, E. G. Deroduane, and R. T. K. Baker, American Chemical Society, New York, 1983.
46. A. Renou and M. Gillet, *Surf. Sci.* **106**, 27 (1987).
47. J. M. Dominguez and M. José-Yacamán, in *Growth and properties of metal clusters*, p. 493, edited by J. Bourdon, Elsevier, The Netherlands, 1980.
48. C. L. Cleveland, U. Landman, T. G. Schaaff, M. N. Shafigullin, P. W. Stephan, and R. L. Whetten, *Phys. Rev. Lett.* **79**, 1873 (1997).
49. J. A. Ascencio, M. Pérez, and M. José-Yacamán, *Surf. Sci.* **447**, 73 (2000).

50. M. José-Yacamán, (*private communication, 2005*).
51. E. M. Weisstein, *Great Stellated Dodecahedron* from MathWorld - A Wolfram Web Resource. http://mathworld.wolfram.com/GreatStellatedDodecahedron.html.
52. E. M. Weisstein, *Spiky* from MathWorld—A Wolfram Web Resource. http://mathworld.wolfram.com/Spiky.html.
53. Z. L. Wang, R. P. Gao, B. Nikoobakht, and M. A. El-Sayed, *J. Phys. Chem.* B **104**, 5417 (2000).
54. Z. L. Wang, *J. Phys. Chem. B* **104**, 1153 (2000).

Chapter 4

STRUCTURAL AND ELECTRONIC PROPERTIES FROM FIRST-PRINCIPLES

X. Q. Wang
Department of Physics and Center for Theoretical Studies of Physical Systems, Clark Atlanta University, Atlanta, Georgia 30314

Abstract: First-principles calculation has emerged as one of the popular approaches in material science and engineering. In this article we provide a comprehensive, though not exhaustive, overview of the current status of first-principles calculations of materials. We categorize the different approaches using the local and orthogonal bases, and we discuss in some detail representative methods in each category. Finally, we offer some views on the strength and weakness of the approaches that are discussed, and touch upon some of the challenging problems that need to be addressed in the future.

Keywords: *ab initio* calculations, shape memory alloys, nanowires, clusters

1. INTRODUCTION

Recent advances in emerging technologies have created tremendous opportunities for improving the performance of existing materials and for designing new materials. Exploiting these opportunities, however, requires knowledge of structural and electronic properties on scales ranging from atomic to macroscopic. The development of this knowledge, especially at the atomic scale, is proving to be both a challenging and rewarding field of study. In this regard, high performance computer simulations are playing an increasing role as these simulations provide a basis for understanding material properties in detail. In many cases, these simulations can reliably provide information on material properties too difficult or expensive to undertake in the laboratory.

Several of theoretical and computational approaches have been devised to study the properties of advanced materials. First-principles methods involve the solution of quantum-mechanical equations (within certain well-established approximations) to determine the electronic structure, and thereby extract the forces on the atoms. Semi-empirical methods involve the use of simpler models for interatomic bonding such as tight-binding molecular-dynamics for semiconductors[1] and "pair-functional" many-atom potentials for metals.[2]

Although first-principles methods are generally more reliable than the empirical methods, they are currently limited to small systems (a few hundred atoms). Recently, several groups[3–12] have developed methods for performing first-principles electronic structure calculations that scale linearly with system size (the O(N) methods). These methods are now applicable to systems that could only be studied by means of empirical and semi-empirical methods a decade ago. Nowadays, the relative reduction in computational cost enables the molecular-dynamics simulations and therefore the investigation of complicated physical and chemical systems.

This chapter is organized as follows: In Section 2 we briefly discuss the density functional theory and various approximations involved in first-principles calculations, along with the basis sets used for solving the self-consistent equations. In Section 3 we present examples of several first-principles calculations, including the fullerenes, metallic clusters, metallic surfaces, semiconductor nanowires, and shape memory alloys. Section 4 contains conclusions on the application of the various first-principles approaches. The examples presented in this overview to some extent reflect our own research interests and they are by no means exhaustive. Nevertheless, we hope that they give a satisfactory cross-section of the current state of the first-principles calculation methods.

2. FIRST-PRINCIPLES METHODS

Recent advances in *ab initio* methods have experienced a considerable amount of success in predicting ground-state structural and cohesive properties of condensed-matter systems.[4–12] The pioneering work of Car and Parrinello[15] based on dynamical simulated annealing promoted a new type of approach applicable to density-functional theory within the local-density approximation (LDA).[15,18] Density-functional molecular dynamics (Car-Parrinello)[15] and other iterative methods[18] based on plane-wave basis have made such calculations possible for systems consisting of several hundred atoms.[5, 6, 8, 9, 11–14]

Most of contemporary LDA calculations are based on the Kohn-Sham formulation.[13,14,17] LDA provides structural and elastic data in good agreement with experiment; lattice constants, bulk moduli, elastic constants and phonon frequencies are usually predicted within 5% of experimental values.[14] For binding energies, LDA consistently overestimates experimental values by approximately 10-20%. This error is attributed to the incomplete cancellation of errors within the LDA method. While these methods have been very successful, several difficulties arise when they are extended to systems with large length scales or those containing transition metal atoms.

The first-principles methods based on plane-wave basis sets require many components in the expansion to keep track of the locality of the electronic wave functions. In the plane-wave basis, the kinetic energy operator is diagonal while the potential energy matrix is not sparse. In contrast, both the kinetic energy and potential energy matrices are approximately band-diagonal in the wavelet basis. Moreover, the wavelet transform, along with the associated multiresolution analysis, does not involve long-range operations and is thus particularly suitable for parallelization and wave-function-based O(N) algorithms, since every operation can be partitioned into hierarchical real-space domains. In the following, we briefly discuss density functional theory, first-principles molecular dynamics, code developments, and the wavelet bases for electronic structure calculations.

2.1 Density Functional Theory

To properly handle a many-electron system so that one can derive its various properties from fundamental quantum mechanics is a constant challenge in theoretical physics and chemistry. Although the interaction between electrons is well known, the facts that electrons, with a spin quantum number of 1/2, have to obey specific statistical rules and that one normally has to deal with quite a few of them at the same time make this problem immensely formidable. One approach that has become the standard one for large-scale electronic simulations is the density functional theory[17] in the so-called Kohn-Sham framework.[13,14] It is based on a theorem stating that the ground-state energy of a many-electron system can be represented as a functional of the electron density only. Thus one can hope to obtain the electronic energy without dealing with the many-body wave function which is highly multidimensional with the notorious property of being antisymmetric with respect to particle exchange. Being a scalar in the real space, the electron density is a much simpler quantity to manage, making it possible to investigate more complex systems. By minimizing the energy functional with respect to possible density distributions one can then determine the ground-state electronic energy for a given atomic arrangement.

The energy minimization procedure is most conveniently carried out by a mapping of the truly interacting system to an auxiliary system of noninteracting particles with the same density distribution.[13] The resulting total-energy functional

$$E[n] = T_0[n] + \int d^3r \; v_{ext}(r) n(r) + E_h[n] + E_{xc}[n] \qquad (1)$$

includes the kinetic energy functional of the noninteracting system $T_0[n]$, external potential energy, Hartree energy $E_h[n]$, and the so-called exchange-correlation energy functional $E_{xc}[n]$. $E_{xc}[n]$ includes all the many-body effects as well as the difference in the kinetic energies of the interacting and noninteracting systems.

The direct variation of energy with respect to the density is replaced by finding the noninteracting orbitals ψ self-consistently in the local Kohn-Sham equations

$$\left(-\frac{\hbar^2}{2m} \nabla^2 + V_{ext} + V_{eff} \right) \psi_i = \varepsilon_i \psi_i \qquad (2)$$

where the effective one-particle potential V_{eff} includes the Hartree potential and the exchange-correlation potential derived from a functional derivative $V_{xc} = \delta E_{xc}/\delta n$. The density is calculated from all occupied one-particle orbitals. The fact that the effective potential is a simple local function makes a tremendous difference in practical calculations. Other quantum-chemistry schemes such as the Hartree-Fock method commonly involves nonlocal operators which require much more computational resources. It is not surprising that the density-functional theory has become the prevailing approach in modern electronic-structure calculations with wide applications in quantum chemistry and materials physics.

Inarguably one could not have solved the exact many-body problem by regrouping energy terms. As a matter of fact, although the existence of the exchange-correlation energy functional E_{xc} is fully established, its exact form remains unknown and contains integrals of nonlocal quantities. It is therefore a challenging many-body problem to investigate this important quantity in real materials.[14] In practical calculations, approximations to the energy functional E_{xc} are required. Commonly used ones include the local-density approximation (LDA),[13] in which the density is assumed to be locally uniform and the result for a homogeneous electron gas is used point by point based on the local density, and the generalized gradient approximation (GGA), in which the gradient correction to the LDA is added.

In order to study systems of hundreds of atoms, one focuses on the properties of the valence electrons and employ norm-conserving pseudopotentials to model the effects of core electrons. The one-particle orbitals will be expanded in terms of plane waves to eliminate any bias in the basis functions. The self-consistent solution of the corresponding Kohn-

Sham orbitals will be carried out by evaluating relevant quantities in either the real space or momentum space.

2.2 Molecular Dynamics with *Ab Initio* Forces

The density-functional molecular-dynamics (Car-Parrinello) method[15] involves a fictitious molecular dynamics solution of electronic wave functions on the Bohn-Oppenheimer surface,

$$\mu \frac{\partial^2}{\partial t^2} \psi_i(rt) = -H\psi_i(rt) + \sum_j \Lambda_{ij}\psi_j(rt) \tag{3}$$

where μ is a fictitious mass and Λ_{ij} are the Lagrange constraints. The Gram-Schmidt orthogonalization scheme is employed to keep the wave functions of the occupied states orthogonal.

2.3 Algorithm development and coding improvement

The fundamental idea behind the O(N) techniques is the utilization of the spatial locality or near-sightedness[3] of the system, a consequence of the sparse density matrix. For instance, the density-matrix for insulators can be constructed in terms of Wannier functions[8,11] that are characterized by a spatial decay length and a dominant Fourier component. The density-matrix method[4,7,10] uses explicitly the sparsity of the density matrix to achieve a linear size scaling. Also notable are the divide-and-conquer approach and the fast multipole method in quantum chemistry calculations. Recently, several groups[11,12] proposed the approach using wavelets as basis functions. The theory of wavelets allows one to apply a multiscale (multiresolution) analysis to problems that exhibit widely varying length scales. Furthermore, the dual localization property of the wavelet bases will be useful for improving the existing O(N) methods that are yet based solely on the real-space locality.

Wavelet analysis is a relatively new mathematical discipline that has generated much interest in mathematics, physics, computer science, and engineering. Crucial to wavelets is their ability to analyze different parts of a function at different scales and to represent a function minimally at each scale.[12] Wavelets are basis functions that are particularly well suited for representing piecewise smooth functions. The reason is that wavelets are well localized in space as well as in scale and that they can represent polynomials up to a certain order exactly. These properties are used in signal analysis for compression of digital images, reduction of noise, and edge detection. For practical implementations, the compactly supported orthogonal wavelets are particularly useful for employing the pyramid algorithm in the discrete wavelet transform. An orthonormal wavelet is generated by an analyzing wavelet and an auxiliary function called the scaling function. The basis is constructed from the analyzing wavelet and scaling function by the operations of translations and dyadic dilations. Unlike the sines and cosines that comprise the bases in Fourier analysis, which are localized in Fourier space but delocalized in real space, both the mother wavelet and the scaling function are dual localized in real and reciprocal spaces. We are motivated to use these desirable properties of wavelets for electronic structure calculations.

In comparison with the widely used plane-wave basis, an orthonormal wavelet basis is advantageous for electronic structure calculations in that it yields a naturally optimized non-uniform grid with a drastic decrease in the work needed in the calculation. We have been developing and implementing an efficient compression scheme[11] in connection with the compactly supported wavelets for the application in self-consistent electronic structure calculations. This work is significant as it can lead to an effective O(N) algorithm for large-scale electronic structure calculations of novel materials.

2.4 Wavelet Bases

Recently, it has been recognized[11,12] that the wavelet bases with dual localization characteristics in real and reciprocal spaces are desirable for electronic structure calculations. By employing non-uniform grids, it is possible to add resolution locally. For instance, for a cluster or surface system one can employ a basis that uses a high density of grid points corresponding to the regions where the atoms are located and a low density of points for the vacuum regions. This leads to drastic savings in the basis size and thereby total computational workload. Orthonormal wavelet bases provide such a desirable representation for electronic structure calculations. One of the important features of the orthonormal wavelet bases is its ability to reduce the number of components needed for solving the Kohn-Sham equation,[17,13] reminiscent of compression in applying wavelets to image processing. Furthermore, multiresolution analysis provides automatic preconditioning[18] on all length scales, which increases the rate of convergence of the electronic wave functions.

2.4.1 Orthonormal wavelet bases for electronic structure calculations

We have devised an approach utilizing compactly supported, orthonormal wavelet bases for quantum molecular dynamics (Car-Parrinello) algorithm.[15] A wavelet selection scheme was developed and tested for various atomic and molecular systems. The method showed systematic convergence with the increase of wavelet-selected grid size, along with improvement on compression rates. This method yields an optimal grid for self-consistent electronic structure calculations, and offers a realistic approach for the study of transition metal clusters.

A few years ago it was optimistic that the nice properties of wavelets would automatically lead to efficient solution methods for electronic structure calculations. However, this early optimism remains to be honored. Wavelets have not had the expected impact on electronic structure calculations. While the solution is sparsely represented in a wavelet basis, applying wavelet compression alone does not necessarily reduce the computational work. One needs to reduce the overhead to make the algorithm efficient. In the following we discuss some of the most promising approaches and shed light on the obstacles that need to be overcome in order to obtain successful wavelet-based algorithms for electronic structure calculations. Schematically, wavelet-based methods for electronic structure calculations can be classified into the three categories described in the following subsections.

2.4.2 Methods based on scaling function expansions

The unknown solution is expanded in scaling functions at some chosen level J and is solved using a Galerkin approach. Because of their compact support, the scaling functions can be regarded as alternatives to splines or the piecewise polynomials used in finite element scheme. It is worthwhile to point out that continuous wavelet transforms can be used for quantum chemistry calculations based on LCAO (linear combination of atomic orbitals) in that one can use Gaussian or Slater functions for scaling functions. While this approach is important in its own right, it cannot exploit wavelet compression. Hence methods in this category are not adaptive.

2.4.3 Wavelets and finite difference

In this approach wavelets are used to derive adaptive finite difference methods. Instead of expanding the solution in terms of wavelets, the wavelet transform is used to determine where the finite difference grid must be refined or coarsened to optimally represent the solution. The computational overhead can be low because one works with point values in real space representation. One such approach for electronic structure calculations is developed using Wavelet Optimized Finite Difference Method. It works by using wavelets to generate an irregular grid that is then exploited for the finite difference method. This approach is analogous to various real-space algorithms.

2.4.4 Methods based on wavelet expansions

The unknown solution is expressed in terms of a wavelet instead of scaling functions so the wavelet compression can be applied, either to the solution, the differential operator, or both. Several different approaches have been considered for exploiting the sparsity of the wavelet representation. One is to perform all operations in the wavelet domain. The operators are sparse and the number of significant coefficients scale as log(N).

Wei and Chou[12] have devised a method for using compactly supported orthogonal wavelets, developed by Daubechies, for self-consistent electronic structure calculations. This approach takes advantage of the existence of a fast, discrete wavelet transform (FWT) associated with the Daubechies wavelets. They have demonstrated that the resulting Hamiltonian matrix in wavelet space can be reduced by a factor of hundreds, making it feasible for obtaining the eigenvalues and eigenfunctions through standard diagonalization. While the preliminary application of this method in atomic and molecular systems is promising, there remain several important issues relevant to practical applications. These include the selection of the wavelet components, the effective construction of the matrix elements of the local pseudopotential, and the convergence of the electronic wave functions. The selection of the wavelet components is a key issue in applying wavelet bases to electronic structure calculations.

In principle, the selection should be based on the eigenfunctions of the Hamiltonian matrix, i.e., the outcome of the diagonalization process. However, the eigenfunctions are not known *a priori*. Approximate methods, e.g. using the wavelet transform of the local potential as a guide,[11] may not guarantee the selected wavelet components being optimal. This may further lead to problems in convergence of the electronic wave functions. It is also worth pointing out that the representation of the local pseudopotential in the wavelet space is of a complicated form and amounts to an O(N) process for a system of N grid points. This severely hampers the practical applicability of the method.

In an effort to improve this situation, we devised the application of the Daubechies wavelet bases in the Car-Parrinello (CP) algorithm.[11] Our approach preserves the advanced features of the CP method, and the transform between the real and wavelet spaces can be efficiently carried out using the FWT. The FWT associated with the wavelets of compact support has many similarities to FFT. Both are unitary and scale as NlogN. The primary advantage of FWT over FFT is the capability of representing the wave functions with a minimal number of components that scales as $N_w \log_N$ with N_w wavelet components, a desirable feature for inhomogeneous systems. Using the CP method, the calculations associated with the local pseudopotential–a predominant part in the self-consistent calculations–scale linearly with system size for each occupied state. In addition, the construction of the separable non-local pseudo-potential matrix elements of the Kleinman-Bylander form scales as $O(N_W^2)$, while the matrix elements of the kinetic energy can be obtained through transformation of a band-diagonal matrix into wavelet space.

The use of orthonormal wavelet bases is capable of reducing the cost for the bottlenecks in plane-wave based calculations. In fact, for calculations of both local and non-local pseudopotential contributions, the compression associated with the wavelet bases is most beneficial in that an $O(N^2)$ process can be reduced to $O(N^2)$. The key point for combining speed and precision of the calculation is to select adequately the most significant wavelet coefficients to be kept. In order to do so, one may exploit the self-similar behavior of wavelet coefficients: from each scale to the next finer one, all coefficients are multiplied by a common small factor (which gets smaller for an increasing number of vanishing moments of the wavelet). Hence those coefficients that are negligible at a given scale lead to negligible ones at finer scales; no significant coefficients reemerge. Then combining the selection of significant wavelet coefficients and the CP algorithm at each successive scale, from coarse to fine, one obtains fast convergence to the eigenvalues and eigenfunctions. Both the convergence and the compression rate (percentage of coefficients being kept) increase with the grid size, and so does the advantage of the method over the conventional plane wave approach.

3. APPLICATIONS

3.1 Structure and dynamics of carbon fullerenes

The discovery of the icosahedral C_{60} molecule[21] and the subsequent synthesis of C_{60} solid[22] have stimulated a veritable flurry in the study of fullerenes, fullerites, and carbon nanotubes. The observation of superconductivity in alkali-metal-doped C_{60} crystals[23] produces further excitement in the interesting properties of this new form of all-carbon material. Phonon-pairing theories based on both low-frequency and high-frequency intramolecular vibrations have been proposed[24,25] to explain the mechanism responsible for superconductivity in such a system. Recent investigations tend to agree that electron-phonon interactions play an important role. In fact, experimental measurements by neutron inelastic scattering[26] and Raman scattering[27,28] have revealed that the positions and intensities of the peaks corresponding to intramolecular bond-stretching modes of alkali-metal-doped C_{60} change considerably with regard to those of pristine C_{60}, which suggests that these modes are strongly coupled to the electrons. A detailed information about the vibrational properties would be desirable for an accurate evaluation of deformation potentials and the electron-phonon coupling strength important for understanding the mechanism of superconductivity in these compounds. Because of the molecular nature of the bonding in fullerene solids, a study of the vibrational modes of the isolated molecule offers a good guide to the phonon spectrum in the solid. Furthermore, the characterization of the vibrational modes for this unique, highly-symmetrical structure is an interesting problem in its own right.

A wealth of experimental data has been accumulated for the vibrational frequencies of the fullerene molecules. The vibrational spectra were measured and identified over a broad range of frequencies by the methods of Raman and infrared spectroscopies,[28] as well as neutron inelastic scattering measurements. In order to examine the efficiency and accuracy of first-principles methods for the vibrational frequencies of C_{60}, a systematic study of the effect of basis sets on the vibrational spectrum was carried out. The results using large basis sets are in excellent accord with the experimental data available from Raman, infrared, and neutron inelastic scattering measurements,[28] which provides a benchmark to identify conclusively the modes observed in various experiments. A detailed analysis of effects of basis set reveals some of the deficiencies of previous theoretical studies. Moreover, the constructed force constant matrix yields complete information on the vibrational properties at T=0 which, together with the associated eigenfrequencies and eigenvectors, is useful for testing the reliability of empirical methods and for calculating electron-phonon-coupling matrix elements for various phonon modes.

The calculation was carried out using a first-principles density-functional approach for molecules with analytical energy gradients.[15] The Hedin-Lundqvist form was used for the exchange correlation energy of the electron within the local-density-approximation. The calculation was performed with three different LCAO basis sets: minimal (MIN), double-numerical (DN), and DN with polarization functions (DNP), i.e., functions with angular momentum one higher than that of the highest occupied orbital in the free atom. It amounts to five, nine, and fourteen atomic orbitals for each carbon atom, for the calculation using MIN, DN, and DNP basis sets, respectively.

The geometry of the icosahedral C_{60} molecule is characterized by two distinct bond lengths: the bond of the pentagon edge d_1 (single bond) and the hexagon-hexagon bond d_2 (double bond). For C_{60}, the highest occupied molecular orbital (HOMO) is five-fold degenerate with H_u symmetry, while the lowest unoccupied molecular orbital (LUMO) is three-fold degenerate with T_{1u} symmetry. Both HOMO and LUMO are predominantly π molecular orbitals. The geometry optimizations using conjugate-gradient procedures give distinct bond lengths of 1.391 and 1.444 Å for double and single bonds, respectively, in good agreement with other first-principles results and the experimentally measured values.

The calculated vibrational frequencies are shown in Table 1 and compared with the experimental frequencies of the optically active infrared (T_{1u}) and Raman (A_g+H_g) modes. The overall agreement with the measured spectra is very good, with a maximum deviation of ~5% and a mean square deviation (MSD) of 2.5% with respect to the whole set of 14 modes.

As seen in Table 1, the calculation by use of DN and DNP basis sets yields very close vibrational frequencies. The computed vibrational frequencies agree in detail with experimental data available from Raman, infrared, and neutron inelastic scattering measurements. Further improvement of the first-principles results can be achieved by using time-dependent density-functional calculations.

3.2 Shell structures of metal clusters

The atomic and electronic structures of metal clusters have been a subject of intensive theoretical and experimental studies. For metallic clusters with predominant *sp* valence electrons, the shell structure of the jellium model provides a useful guideline for describing the electronic structures.[29–32,44,37] The electronic shell structure has been verified for medium sized Al clusters both experimentally and theoretically.[38] Theoretical calculation on the Al_{13} cluster revealed that the stability can be substantially enhanced by closing the electronic shell through doping or charging.[33] As a result, the impurity-doped Al_{13} cluster has been viewed as a building block for the cluster assembled solid.[33] However, it remains unclear if the electronic shell structure is still valid for large clusters. It is expected that triple charged ionic cores of Al clusters lead to strong perturbations to the shell structures. In this regard, recent experiments[29] on the photoelectron spectroscopy of Al clusters have shown that the electronic shell structure diminishes for $n \approx 75$. Therefore, accurate first-principles study for large metallic clusters is clearly desirable.

Table 1. Comparison between the calculated 46 distinct vibrational frequencies, $\omega_{LDA}^{(0)}$ and ω_{LDA} (in cm^{-1}), and the experimental Raman and infrared data, $\omega_{expt.}$ (in cm^{-1}) along with the corresponding irreducible representations R of the icosahedral group. $\omega_{LDA}^{(0)}$ and ω_{LDA} are results using DN and DNP basis sets, respectively.

	Even-parity				Odd-parity		
	$\omega_{LDA}^{(0)}$	ω_{LDA}	$\omega_{expt.}$		$\omega_{LDA}^{(0)}$	ω_{LDA}	$\omega_{expt.}$
A_g	484	483	496	A_u	980	947	
	1547	1529	1470				
				T_{1u}	539	533	528
T_{1g}	566	566			566	548	577
	848	825			1216	1214	1183
	1299	1292			1501	1485	1429
T_{2g}	571	550		T_{2u}	353	344	
	772	771			724	717	
	820	795			997	987	
	1371	1360			1234	1227	
					1566	1558	
G_g	492	484					
	579	564		G_u	353	356	
	765	763			765	752	
	1127	1117			782	784	
	1335	1326			984	977	
	1535	1528			1346	1339	
					1472	1467	
H_g	272	263	273				
	439	432	437	H_u	406	396	
	720	713	710		544	534	
	782	778	774		680	663	
	1121	1111	1099		750	742	
	1285	1282	1250		1234	1230	
	1478	1469	1428		1368	1360	
	1606	1598	1575		1596	1588	

Ecker and coworkers[35] successfully synthesized a compound containing a large metallic Al_{77} cluster unit. The Al_{77} cluster is presumably the largest metallic cluster synthesized hitherto, whose structure has been measured by X-rays. The Al_{77} cluster constitutes a fascinating prototype for studying the crossover between the metallic cluster and the bulk metal of main-group elements. The experimentally characterized Al_{77} cluster is onion-like, having three layers of atoms covering the central atom. The three layers consist of 12, 44 and 20 atoms, respectively. The central atom is coordinated by

12 neighbors similar to the bulk arranged in a distorted icosahedral (I_h) environment. The coordination number for the atoms in outer shells decreases from the center outward, and undergoes a transition from the metallic center to a molecular surface. The atoms in the second shell have a coordination number of 10, while the atoms in the outmost layer arrange in a distorted I_h symmetry, with only four neighbors from the inner shell. The nearest-neighbor distance is about 2.7 Å for the central atom and 2.6 Å for the outer atoms. The latter indicates a small compression for the outer atoms, similar to the Al surface. However, little is known about the electronic structure of such a large metallic cluster. It is important to understand the role played by the ligands in the synthesized Al_{77}-containing compound. An outstanding question concerns the stability of Al_{77} as to whether the Al_{77} cluster itself is the stable core unit, or the chemical interaction between Al_{77} and its ligands is crucial.

Inspired by the experimental work,[35] we carried out a successful density-functional study on the electronic and structural properties of Al_{77}. Our attention was directed to its atomic and electronic shell structures. Through a detailed analysis on the electronic structures, we find that there exist strong interatomic shell interactions and significant charge transfer between Al_{77} and its ligands. Moreover, the electronic shell structure for Al_{77} can no longer be identified beyond the 1*h* shell, in agreement with recent experimental observations.

The calculations are based on the *ab initio* molecular dynamics method.[15] The electronic density is expressed in Kohn-Sham orbitals that are expanded in plane waves. The local-density approximation to the density-functional theory is used for the exchange correlation potential.[18]

3.2.1 Atomic shells

The electronic structure calculations show that Al_{77} and its inner core Al_{13} are stable. We find that Al_{13} has all the features of electronic structure of I_h-Al_{13}. Since the highest occupied orbital of Al_{13} is not completely occupied, the structure of I_h-Al_{13} is unstable against Jahn-Teller distortion. The structural distortions are reflected in the electronic structure in which the peaks in the density of states are split. The existence of a gap just above the Fermi level indicates that the structure of the inner core is very stable with an electron donation. For Al_{57}, we find that more peaks appear below the Fermi level. The Fermi level sitting on a high peak implies that this cluster is unstable against structural relaxation. However, with another layer of 20 atoms added, Al_{77} becomes stable as the Fermi level sitting at a dip indicates a higher stability of the structure. It is instructive to compare the electronic structure of Al_{57} and Al_{77}. Both are quite similar at the low energy part, the only difference appears around the Fermi energy.

The onion-like Al_{77} is stable against the geometry optimizations. The relaxation only slightly changes the structure of Al_{13} with an energy gain of 0.06 eV/atom. While the structure of Al_{77} is quite stable, in the sense that the relaxation slightly changes the structure of the outmost shell with a small energy gain of 0.04 eV/atom, the structure of the inner shell (Al_{57}) undergoes little distortion. This is consistent with the results of low density states at the Fermi level. The electronic structure of Al_{77} cluster remains almost the same after the relaxation. On the contrary, Al_{57} is quite unstable against the relaxation. Not only the structure of the outmost shell changes, but also the structure of the inner shell does, with a large energy gain of ~0.18 eV/atom due to the geometry optimization. Based on the above analysis, Al_{77} can be considered as a two-layer atomic shell covering a very stable inner core of Al_{13}.

3.2.2 Charge transfer

The fundamental features of the electronic structure of Al_{77} remain intact with ligands and charging. The experimentally synthesized compound consists of the Al_{77} cluster unit and ligands. The ligand is composed of 20 nitrogen atoms, 30 silicon atoms, and 120 methyl groups. To study the effect of ligands on the electronic structure of Al_{77}, we include a layer of 20 N atoms at the experimentally determined positions. The calculated electronic structure of $Al_{77}N_{20}$ is shown in the top panel of Fig. 1. It is readily observable that the predominant contribution from 20 N atoms to DOS is featured around -2 eV. Apart from this, the main feature in DOS of Al_{77} remains. As a matter of fact, our calculation on the charged cluster confirmed the rigid band behavior of Al_{77} electronic structure.

3.2.3 Electronic shells

The electronic structure of Al clusters can be well described by the spherical jellium model up to a certain cluster size where the separate *s* and *p* levels overlap. The most recent experiment of photoelectron spectroscopy on size-selected Al clusters showed[34] that the critical size is around N=9, above which the jellium model starts to be valid. The first-principles results confirmed that the jellium model holds for Al_{13}.[32,33] However, the experiment also found that the shell effect diminishes above Al_{75}.

The Al_{77} cluster characterized by the experiment provides a desired system to examine the shell structure of the spherical jellium model for large Al clusters. It is worth noting that the density of states of Al_{77} is much different from that of the free electron-like. Certain electronic shell structures from jellium model can be clearly identified. For instance, 1*s*, 1*p*, 1*d*(2*s*) states that are present in small Al_{13} clusters remain visible. The large gaps just above 1*g*, 1*h* and 1*i* shell states are evident. This is in remarkable agreement with the photoelectron spectroscopy observation on a close shell at 1*g* and 1*i* for the observed Al clusters. Because of strong interaction between +3 ionic core and valence, beyond the 1*i* shell state, all the shell states overlap together and the gap between the states disappears as labeled in Fig. 1. These results show that the shell structure in Al cluster as large as Al_{77} disappears, which is in accordance with the photoelectron spectroscopy experimental observations.[34]

The local electronic density of states from the central atom to outer shell, as shown in Fig. 1, can help to understand if the electronic shell and atomic shell structure are correlated. We get those local electronic density of states for each atom by projecting the wavefunction in the Wigner-Seitz cell into atomic-like orbitals. Although the space division is somewhat arbitrary, the obtained results can qualitatively locate the states for different atomic shells. We find that the local electronic density of states for all the atomic shells are quite similar to the total density of states. Although the electronic shells with lower kinetic energy are mainly distributed in the inner atomic shells and the electronic shells with higher kinetic energy have a trend to distribute in the outer atomic shells, basically all the electronic shells are delocalized on all atomic shells, there are no localized electronic shells which validate the jellium model. Thus each atomic shell has a significant contribution to the density of states at the Fermi energy.

The first-principles results show that the onion-like Al_{77} cluster is quite stable, while the Al_{57} cluster with one less layer is not. From a structural stability point of view, Al_{77} can be considered as a two atomic layer shell of 64 atoms on a stable inner core of Al_{13}. There exists strong interaction between atomic layers in the cluster. The interaction between Al_{77} and its ligands is ionic-like. Our results on the electronic structure of Al_{77} show that the shell model for the electronic structure of the Al cluster is valid around the *1i* shell state for large clusters. Beyond the 1*i* shell state the gap becomes very small for a number of atoms larger than 70, in close agreement with experimental results.

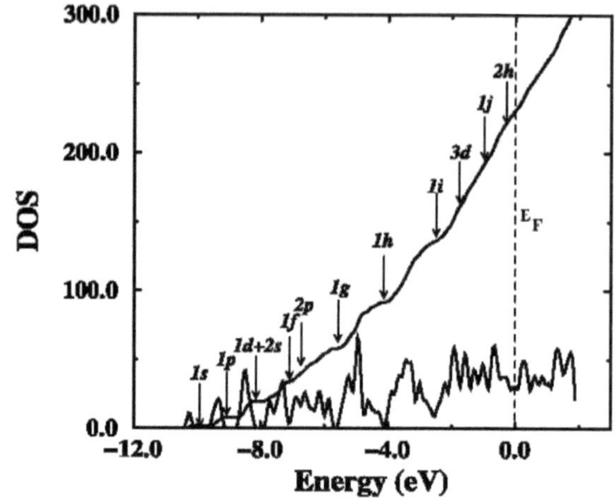

Figure 1. The electronic density of states (DOS) and integrated DOS of the Al_{77} cluster. The shell structures of the spherical jellium model are labeled. The shell structures up to 1*i* are clearly observed.

3.3 Microfacets of metal surfaces

The "missing-row" reconstruction of noble metal (110) surfaces has been the subject of much theoretical and experimental attention.[39–43,45] The (1×2) missing-row reconstruction is found on clean Au, Pt, and Ir, while Cu and Ag are dormant cases in which the reconstruction can be provoked only by a modest alkali coverage.[47] Through the removal of every other row, the flat (110) surface is converted into a sequence of tiny (111) facets, which are favored by the lower (111) surface energy relative to that of (110). Experimental studies of the (110) configurations have revealed that the reconstruction is accompanied by heavy distortions (multilayer relaxations).[44,37] Another characteristic feature of the missing-row reconstructed (110) surfaces is the discovery of roughening and deconstruction transitions.[41,42,45] For Au(110), the deconstruction of the (1×2) missing-row occurs at about 650 K, followed by a roughening transition at about 700 K. The critical exponents at the deconstruction transition temperature T_C are in accordance with the two-dimensional (2D) Ising universality class;[43] while the roughening transition is of the Kosterlitz-Thouless (KT) type. For Pt(110), the two processes proceed at the same temperature.[48] The intriguing interplay between the roughening and destruction led to the proposal of several phenomenological models.[40,49–51]

Many implications of the complicated relaxation patterns of the missing-row surfaces on the roughening and deconstruction transitions are as yet unexplored. Recent experimental studies[41,42] have evidently demonstrated that the (1×1) configuration plays little role in the deconstruction and roughening transitions, contrary to the early models based on order-disorder transitions via lattice gas formation.[45] Instead, antiphase domains consisting of (1×3) steps are identified, of crucial importance for Au(110). While the stability of the (1×2) missing-row configuration against the (1×1) unreconstructed structure has been understood theoretically,[39] the relative stability between the (1×2) and (1×3) configurations has remained a paucity of theoretical study. Existing semi-empirical models are not yet capable of determining conclusively the energy orders and structural features.[2, 52]

Based on a systematic first-principles analysis, the relative stability of the (1×2) and (1×3) missing-row reconstructed configurations has been clarified. The first-principles calculation results show that the delocalization of *d*-electrons and the accumulation of *sp*-electrons near the surface associated with the multilayer relaxations, are important factors in stablizing the experimentally observed (1×2) missing-row configurations for Au and Pt(110). For Ir(110), the missing-row configurations show weaker top layer distortion. As a result, a faceted (331) structure is shown to be energetically favorable. The study reveals important correlation between the top-layer distortion and the relative stability of the missing-row configurations. Moreover, the analysis provides useful guidelines for the construction of the empirical models for the noble metals.

With use of LDA and GGA, both the (1×2) and (1×3) missing-row reconstructed structures of Au(110) undergo a large topmost interlayer contraction, a slight row pairing in the second layer, and notable buckling in the third layer. The topmost-layer rows move straight downward into the bulk region while the second layer atoms move slightly upward. The calculation of the surface energies shows that the larger the value of n, the lower the surface energy of the unrelaxed (1×n) configuration. Our calculations for the unrelaxed missing-row configurations of other 5*d* noble metals Pt and Ir have also confirmed this trend, which is in contrast to the 4*d* metals. This generic feature is attributed to the relativistic effect.[47] In contrast to Au and Pt, Ir(110) favors a faceted (331) configuration. This is confirmed by our calculation to be the structure of the lowest surface energy.

The relative surface-energy order for (1×2) and (1×3) is reversed after taking into account the multilayer relaxations. Our calculations have confirmed that the experimentally observed (1×2) missing-row structure has the lowest surface energy for Au and Pt. In this respect, Ir stands for the only case among the noble metals that failed to achieve this reverse.

The study of the lowest energy configuration for Ir deserves its own right. The faceted (331) configuration is slightly lower in energy as compared with

($1\times\infty$). The different trend of Ir as compared with Au and Pt is attributed to the much weaker *d*–electron localization effects.[46]

The first-principles calculation results for the relative stability between the (1×2) and (1×3) missing-row configurations show that the *d*-electron delocalization and *sp*-electron spilling-out associated with the multilayer relaxation account for the stability of the experimentally observed (1×2) missing-row configurations. The calculated surface energies for Au, Pt, and Ir provide useful information on the electronic structure effects on the relative stability of the missing-row configurations. These results are useful for the construction of empirical models capable of large-scale molecular dynamics simulations.

3.4 Nanotechnology: nanowires

Nanometer-sized, quasi one-dimensional materials such as semiconductor nanowires and carbon nanotubes represent ideal building blocks for the bottom-up assembly of complex nanoscale architectures for integrated electronic and photonic devices. One of the novel properties of such nanostructures stands for the simultaneous functionality both as device elements and as the interconnecting wires for the devices. In the past few years much effort has focused on the synthesis, fabrication and characterization of core-shell structured semiconductor heterostructures. Those composite nanocrystals constitute a new class of semiconductor materials with unique chemical, electronic, and optical properties not present in single-component compounds. Substantial tunability of those properties make this kind of semiconductor systems very attractive for the real-world applications such as optoelectronic and telecommunication devices, high-performance field-effect transistors, and lasers.

Recently, Lauhon, Gudiksen, Wang, and Lieber[56,58] reported the heteroepitaxial growth of crystalline germanium-silicon and silicon-germanium core-shell structures with band-offsets injecting holes onto either germanium core or shell regions. The experimental synthesis of core-shell and core-multishell structures provides intriguing opportunities for the development of nanowire-based devices. For Si-core and Ge-shell nanowires, the experimental study demonstrated that the Ge shell is fully crystallized for low-temperature growth. A single diffraction peak is observed along the axial direction, indicating compressively strained Ge and tensile-strained Si. On the other hand, for Ge-core nanowires, an amorphous Si-shell is formed initially and then becomes crystallized after thermal annealing. Associated with 4% mismatch between the Ge and Si lattice parameters, the epitaxial SiGe core-shell nanowire heterostructures are prototype systems for the study of strained silicon effect.

Motivated by the experimental accomplishments, we have carried out a detailed first-principles density functional theory study of the composition dependence of the structural and electronic properties of the core-shell type SiGe nanowire heterostructures. Our attention is directed towards an analysis of the relationship between structural characteristics and chemical composition in these core-shell structured systems, quantitative estimation of the degree of deviations from the linearity assumed by the Vegard's law, as well as to shed light on the peculiarities of electronic band structures in these systems.

During the past decade there has been considerable discussion in the literature about changes in the local (bond lengths and bond angles) and global (lattice parameters, cell volume) structures with the variation of the chemical composition in various composite semiconductor materials. For alloys, a relationship between structural parameters and chemical composition is generally assumed to follow the Vegard's law, which states that the relaxed lattice parameter of a two-component system is a linear function of the composition. Examples in which the linear behavior is obeyed are very rare in semiconductor alloys, as illustrated by the numerous experimental and theoretical studies. Nevertheless, it is generally recognized that the analysis of the deviations from linearity for structural parameters can provide important information on physical and chemical characteristics of the systems.

Experimental studies of nanowire heterostructures have shown a strong orientation dependence of the optical behavior in these systems. Direct and indirect gaps have been observed experimentally for Si and Ge nanowires along the [111] direction, respectively. The direct and indirect energy gaps for Si and Ge nanowires can be readily understood from the point of view of projecting the bulk conduction bands onto [111]-oriented nanowire conduction bands. The Γ and X points of the bulk are projected onto the conduction-band center, while the L point is projected onto the conduction-band edge of the nanowires. Since the bulk Ge has the conduction-band minimum at the L point, the thin Ge nanowires oriented along [111] direction have indirect band gaps. For the bulk Si the conduction-band minimum has predominantly the X-character, thus the corresponding [111]-oriented thin Si nanowires have direct band gaps. It is therefore of particular interest to understand the composition dependence of the band gap and the implied direct-to-indirect and indirect-to-direct transition in the SiGe core-shell structured nanowires. It is well known that the LDA-type calculations underestimate systematically the band gaps for Si and Ge systems (the LDA approach predicts that bulk Ge is a semimetal). However, the calculations using the GW approximation indicate that the qualitative behavior of the band gaps remains intact.

The analysis of the changes in the local and global geometry and electronic structure of the epitaxial SiGe hydrogen-passivated core-shell structured nanowires reveal the strong core-type specific features for both structural and electronic properties. There exist negative and positive deviations from the linear behavior for the calculated structural parameters of the nanowires with the tensile strained Si-core and compressive strained Ge-core, respectively. Direct-to-indirect transition for the fundamental band gap is found in Ge-core/Si-shell nanowires, reminiscent of the L $\rightarrow$ X crossing in Ge quantum dots. Although the synthesized SiGe nanowires are about 40 nm in diameter, we believe that the results of theoretical study for the thin nanowires would shed light on the peculiarities of structural and electronic properties of the large core-shell structured nanowires. For instance, the obtained weak size dependence of the critical composition for direct-to-indirect gap transition suggests that the synthesized[59] Ge-core/Si-shell nanowires with different shell sizes (5 and 15 nm) have direct and indirect fundamental gaps, respectively. One can expect a smooth synthesis process by adding Ge-shell onto Si-core, as the calculation results indicate a gradual increase of the tensile strain, whereas the strain relaxation for compressively strained Ge-cores is much more involved. In fact, the experimental observation on initial formation of amorphous Si-shells around Ge-core indicates the importance of annealing conditions for the crystallized core-shell nanowires. The preservation of the direct band gap practically over the whole compositional range predicted for Si-core/Ge-shell nanowire heterostructures along with the possibility of the band gap energy variation with change of composition, diameter of nanowire, and/or shell thickness suggests a promising perspective for band gap engineering and fabrication a novel nanoarchitectures for optoelectronic applications.

3.5 Shape memory alloys

Shape memory alloys (SMAs) have attracted much attention in recent years as smart and functional materials due to the unique properties of shape memory smart functions and super-elasticity.[53,54] Unlike the current genre of strained silicon micromechanical devices, NiTi-based SMAs do not suffer from extreme fragility and are ideal candidates for extreme critical applications in the automotive, aerospace, electronics industries, and mechanical engineering, such as couplings, fasteners, connectors, and actuators.

From a metallurgical perspective, the principal properties of SMA are explained by the phase transition called martensitic transformation (from the austenitic phase to the martensitic phase and vice versa). In spite of the

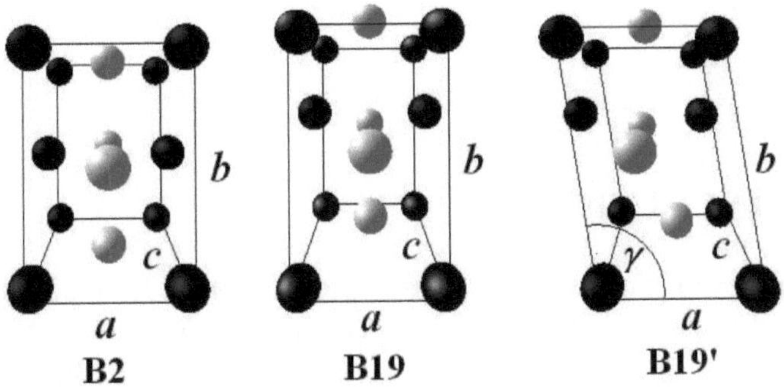

Figure 2. Unit cells for the cubic B2, the orthorhombic B19, and the monoclinic B19' crystal structures. Ni and Ti atoms are represented by black and grey balls, respectively.

technological importance of this type of materials, there is a striking lack of understanding of thermoelastic shape memory phenomena at the atomistic level. Several first-principles studies[53,54] have been performed for NiTi, TiPd, and TiPt. Despite the progress made, some questions remain regarding the relative stability of the martensite phases, as well as identifying the mechanisms that promote the shape memory behavior. Consequently, it is desirable to build on the current body of work based on first-principles calculations focusing on the structures and pathways associated with the martensite transitions for several binary systems.

We have carried out calculations of various relevant SMA structures for NiTi, TiPd, and TiPt. In all cases, the high-temperature austenite phase has a simple cubic B2 structure (space group *Pm*3*m*), while the room temperature martensite phase has either an orthorhombic B19 structure (space group *Pmma*), or a monoclinic B19' structure (space group *P*21/*m*).

Theoretical predictions of shape memory behavior, which rely on properly describing the stable martensite phase, deserve further examination. To this end, we have determined the crystalline structure of these alloys and optimized the ground state energy by means of constant pressure molecular dynamics. Defining γ as a rotation on the ab plane, we have systematically calculated the unit-cell parameters, atomic positions, and differences of the B19 and B19' structures with respect to the B2 phase for NiTi, TiPd, and TiPt. The calculated results for the relative stability of the three phases are in agreement with experimental observations for all three binary alloys. In addition, our first-principles calculation revealed the existence of a twinned martensite structure for NiTi. The existence and understanding of the twinned martensite structure is crucial for the shape memory effect, in that this is the structure that will be "memorized" after the loading and heating/cooling cycles. Based on this, we are further investigating the effect of ternary substitution effects in the high temperature shape memory alloys.[60]

4. CONCLUSIONS

In this review article we have attempted to provide a comprehensive overview of the current status of first-principles simulations methods and their applications in materials science. Because of the tremendous and continuing progress in first-principles calculation strategies, this review is by no means exhaustive. We hope that we have conveyed the message that first-principles calculations are a truly vibrant enterprise of multi-disciplinary nature. They combine the skills of physicists, materials scientists, chemists, mechanical and chemical engineers, applied mathematicians and computer scientists. The marriage of disciplines and the concomitant dissolution of traditional barriers between them, represent the true power and embody the great promise of first-principles approaches for enhancing our understanding of, and our ability to control complex physical phenomena.

ACKNOWLEDGMENTS

This work was supported in part by NASA under grant No. NAG3-2833, by National Science Foundation by Grant No. DMR-0205328 and EEC-9731680. The author is grateful to M. Y. Chou, C. J. Tymczak, X. G. Gong, C. Z. Wang, K. M. Ho, R. Musin for contributions during the various collaborative projects involved in this review. The author thanks G. Bozzolo and A. Msezane for useful comments and a critical reading of the manuscript.

REFERENCES

1. W. A. Harrison, Electronic Structures and the Properties of Solids (Dover, New York, 1989); C. Z. Wang, B. L. Zhang, K. M. Ho, and X. Q. Wang, *Int. J. Mod. Phys. B* **7**, 4305 (1993).
2. M. S. Daw, S. M. Foiles, and M. I. Baskes, *Materials Science Reports* **9**, 251 (1993); A. E. Carlsson, Solid State Physics **43**, 1 (1990); A. M. Raphuthi, X. Q. Wang, F. Ercolessi, and J. B. Adams, *Phys. Rev. B* **52**, R5554 (1995); F. Ercolessi and J. B. Adams, *Europhys. Lett.* **26**, 583 (1994).
3. W. Kohn, *Phys. Rev. Lett.* **76**, 3168 (1996).
4. X. P. Li, R. W. Nunes and D. Vanderbilt, *Phys. Rev. B* **47**, 10891 (1993).
5. E. L. Briggs, D. J. Sullivan, and J. Bernholc, *Phys. Rev. B* **54**, 14362 (1996).
6. F. Gygi, *Europhys. Lett.* 19, 617 (1993), *Phys. Rev. B* **51**, 11190 (1995).

7. S. Goedecker and M. Teter, *Phys. Rev. B* **51**, 9455 (1995).
8. J. N. Kim, F. Mauri, and G. Galli, *Phys. Rev. B* **52**, 1640 (1995).
9. E. Hernandez and M. J. Gillian, *Phys. Rev. B* **51**, 0157 (1995).
10. A. F. Voter, J. D. Kress, and R. N. Silver, *Phys. Rev.* ***B*** **53**, 12733 (1996).
11. C. J. Tymczak and Xiao-Qian Wang, *Phys. Rev. Lett.* **78**, 3654 (1997).
12. S. Wei and M. Y. Chou, *Phys. Rev. Lett.* **76**, 2650 (1996).
13. W. Kohn and L. J. Sham, *Phys. Rev. A* **140**, 1133 (1965).
14. See, for instance, R. G. Parr and W. Yang, Density Functional Theory of Atoms and Molecules (Oxford Univ. Press, Oxford, 1989).
15. R. Car and M. Parrinello, *Phys. Rev. Lett.* **55**, 2471 (1985).
16. C. J. Tymczak, G. S. Japaridze, C. R. Handy, and Xiao-Qian Wang, *Phys. Rev. Lett.* **80**, 3673 (1998).
17. P. Hohenberg and W. Kohn, *Phys. Rev.* **136**, B864 (1964).
18. M. C. Payne, M. P. Teter, D. C. Allan, T. A. Arias, and J. D. Joannopoulos, *Rev. Mod. Phys.* **64**, 1045 (1992).
19. S. Goedecker, *Rev. Mod. Phys.* **71**, 1085 (1999).
20. X. Q. Wang, C. Z. Wang, and K. M. Ho, *Phys. Rev. B* **51**, R8656 (1995).
21. H. W. Kroto, J. R. Heath, S. C. O'Brien, R. F. Curl, and R. E. Smalley, *Nature* (London) **318**, 162 (1985).
22. W. Krätschmer, L. D. Lamb, K. Fostiropoulos, and D. R. Huffman, *Nature* (London) **347**, 354 (1990).
23. A. F. Hebard *et al.*, *Nature* (London) **350**, 660 (1991).
24. C. M. Varma *et al.*, *Science* **254**, 989 (1991).
25. M. A. Schlüter, M. Lannoo, M. Needels, G. A. Baraff, and D. A. Tománek, *Phys. Rev. Lett.* **68**, 526 (1992).
26. K. Prassides, C. Christides, M. J. Rosseinsky, J. Tomkinson, D. W. Murphy, and R. C. Haddon, *Europhys. Lett.* **19**, 629 (1992).
27. M. G. Mitch, S. J. Chase, and J. S. Lannin, *Phys. Rev. Lett.* **68**, 883 (1992).
28. P. Zhou, K. Wang, P. C. Eclund, M. S. Dresselhaus, G. Dresselhaus, and R. A. Jishi, *Phys. Rev. B* **46**, 2595 (1992).
29. W. D. Knight *et al.*, *Phys. Rev. Lett.* **52**, 2141 (1984).
30. W. Ekardt, Phys. Rev. B **29**, 1558 (1984); M. Y. Chou et al., *Phys. Rev. Lett.* **52**, 2141 (1984).
31. W. A. de Heer, *Rev. Mod. Phys.* **65**, 611 (1993); M. Brack, *ibid*, **65**, 677 (1993).
32. M. Y. Chou and M. L. Cohen, *Phys. Lett. A* **113**, 420 (1987).
33. X. G. Gong and V. Kumar, *Phys. Rev. Lett.* **70** 2074 (1993); S. N. Khanna and P. Jena, *Phys. Rev. Lett.* **69**, 1664 (1992).
34. Xi Li, H. Wu, X. B. Wang and L. S. Wang, *Phys. Rev. Lett.* **81**, 1909 (1998).
35. A. Ecker et al., Nature, **387**, 379 (1997).
36. W. A. de Heer, P. Milani, and A. Chatelain, *Phys. Rev. Lett.* **63**, 2834 (1989).
37. M. F. Jarrold, J. E. Bower, and J. S. Kraus, *J. Chem. Phys.* **86**, 3876 (1987); D. M. Cox *et al., ibid*, **84**, 4651 (1986).
38. A. Nakajima, T. Kishi, T. Sugioka and K. Kaya, *Chem. Phys. Lett.* **187**, 239 (1991).
39. K. M. Ho and K. P. Bohnen, *Phys. Rev. Lett.* **59**, 1833 (1987); *Europhys Lett.* **4**, 345 (1987).
40. Marcel den Nijs, *Phys. Rev. Lett.* **66**, 907 (1991).
41. M. Sturmat, R. Koch, and K. H. Rieder, *Phys. Rev. Lett.* **77**, 5071 (1996).
42. C. Höfner and J. W. Rabalais, *Phys. Rev. B* **58**, 9990 (1998).
43. J. C. Campuzano, M. S. Foster, G. Jennings, R. F. Willis, and W. N. Unertl, *Phys. Rev. Lett.* **54**, 2684 (1985).

44. W. Moritz and D. Wolf, *Surf. Sci.* **163**, L655 (1985).
45. J. C. Campuzano, A. M. Lahee, and G. Jennings, *Surf. Sci.* **152/153**, 265 (1978).
46. A. Nduwimana, X. G. Gong, and X. Q. Wang, *Appl. Surf. Sci.* **219**, 129 (2003).
47. C. L. Fu, K. M. Ho, *Phys. Rev. Lett.* **63**, 1617 (1989).
48. I. K. Robinson, E. Vlieg, and K. Kern, *Phys. Rev. Lett.* **63**, 2758 (1989).
49. I. Vilfan and J. Villain, *Phys. Rev. Lett.* **65**, 1830 (1990); J. Villain and I. Vilfan, *Surf. Sci.* **199**, 165 (1988).
50. G. Santoro, M. Vendruscolo, S. Prestipino, and E. Tosatti, *Phys. Rev. B* **53**, 13169 (1996).
51. L. Balents and M. Kardar, *Phys. Rev. B* **46**, 16031 (1992).
52. M. Garofalo, E. Tosatti, and F. Ercolessi, *Surf. Sci.* **188**, 321 (1987).
53. Y. Y. Ye, C. T. Chan and K. M. Ho, *Phys. Rev. B* **56**, 3678 (1997).
54. X. Huang, G. J. Ackland, and K. M. Rabe, *Nature Materials* **2**, 307 (2003).
55. Y. Kudoh, M. Tokonami, S. Miyazaki and K. Otsuka, *Acta Metall.* **33**, 2049 (1985).
56. C. M. Lieber, Solid State Commun. **107**, 607 (1998); M. Nirmal, L. Brus,. *Acc. Chem. Res.* **32**, 407 (1999); D. H. Cobden, *Nature* **409**, 32 (2001); Y. Cui, C. M. Lieber, *Science* **291**, 851 (2001).
57. X. Zhao, C. M. Wei, L. Yang, M. Y. Chou, *Phys. Rev. Lett.* **92**, 236805 (2004).
58. S. J. Tans, R. M. Verschueren, C. Dekker, *Nature* **393**, 49 (1998); Y. Cui, Q. Q. Wei, H. K. Park, C. M. Lieber, *Science* **293**, 1289 (2001); X. Duan, Y. Huang, R. Agarwal, M.C. Lieber, *Nature* **421**, 241 (2003).
59. L. J. Lauhon, M. K. Gudiksen, D. Wang, C. M. Lieber, *Nature* **420**, 57 (2002).
60. D. Cupid, G. Bozzolo, and X. Q. Wang, unpublished.

Chapter 5

SYNERGY BETWEEN MATERIAL, SURFACE SCIENCE EXPERIMENTS AND SIMULATIONS

C. Creemers, S. Helfensteyn, J. Luyten and M. Schurmans
K.U.Leuven, Chemical Engineering Department, W. de Croylaan 46, B-3001 Leuven, Belgium

Abstract: This chapter examines the link between bulk alloys and surfaces and the necessary conditions that effective simulation techniques must meet in order to properly describe, in a consistent manner, the varied phenomena that characterize each field of research. After a brief description of the thermodynamics of alloy formation and surface segregation in ordered and disordered alloys, model calculations combining the Monte Carlo method and suitable energy models are presented as fully complementary to experiments in order to gain insight in segregation and other surface phenomena, deployed by materials in order to minimize their surface free energy and hence their total Gibbs free energy. In the confrontation between experiments and modeling, it is assumed that the surface, as observed experimentally, corresponds to the equilibrium situation of minimal Gibbs free energy. The entropy part is modeled by the stochastic nature of Monte Carlo simulations, while the energy part is taken into account by appropriate energy models. In this chapter, we describe the (modified) embedded atom method, the derivation of its parameters, and its applications to several cases of interest.

Keywords: Surface segregation; Surface thermodynamics; Computer modeling and simulation; Single crystal surfaces

1. INTRODUCTION

This chapter means to highlight the synergism that exists between surface science experiments, simulations or model calculations and finally, the intrinsic material properties. The origin of this synergism, that is illustrated in Figure 1, lies of course in the fact that all three 'partners' are governed by the same 'inter-particle-interactions', i.e. interatomic interactions between

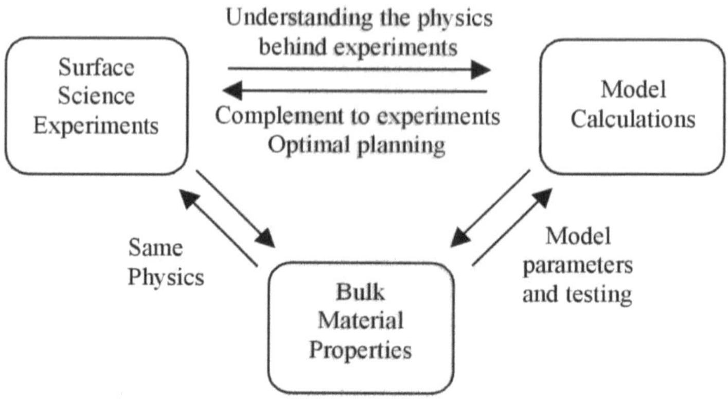

Figure 1. Synergism between (surface) experiments, simulations and material properties.

the atoms of a pure metal or an alloy. Both experimentalists and modelers should exploit this synergism to their maximum advantage. From simulations, people engaged in experimental (surface) research can gain an improved understanding of the physics behind their observations and, at the same time, simulations can help to fill in the blanks in their experiments and to plan and steer experimental work more economically. Furthermore, their experimental findings should always be in agreement with established material properties, as they emerge from e.g. the phase diagram. On the other hand, modelers need reliable values of material properties for the material under study in order to determine the model parameters and to check the reliability of their calculations. The agreement with experimental observations (on the surface behavior) is the ultimate touchstone for simulations and the primary condition for reliable predictions and extrapolations of trends.

As mentioned, this synergy stems from the fact that both the formation of alloys and the existence of distinct bulk phases on one hand, and the different surface phenomena on the other hand are based on the same interatomic interactions, that also determine the basic material properties. Therefore in the next sections a short survey is presented on the elementary thermodynamics of alloy formation and on the behavior of surfaces. These principles remain the basis for the sound interpretation of both experimental results and model calculations. For the sake of simplicity, only binary alloys are treated in the simple framework of the constant bond energy model (CBE). This approach is by far insufficient to give a correct description of phenomena in alloys, but still valuable trends and conclusions emerge.

As the deviations from ideal solutions become more important, the simplifying assumptions of CBE do no longer hold. In the first place, substantial deviations from the ideal mixing entropy can be expected.

Varying degrees of short and long range order will occur and make it impossible to treat bulk effects and surface segregation by simple analytical equations. Therefore in the next section, the Monte Carlo (MC) method is presented as an improved way to account for configurational entropy in alloy systems.

More accurate energy models are needed in order to correctly describe the effects of the detailed atomic surrounding and of the interatomic distances. Several energy models with varying degree of sophistication are able to describe the bonding in alloys much more accurately and are presented briefly. In conjunction with MC, these energy models allow for a far more accurate and realistic description of materials, both with respect to their bulk properties as to the surface behavior.

This approach is finally applied to a few case studies that exemplify the merits of the method and the added value brought about by reflecting on the synergy.

2. THERMODYNAMICAL BASIS

In our universe, nothing happens spontaneously unless it is accompanied by an increase of the total entropy. Viewed from a particular system, spontaneous processes try to disperse matter and to level energy between system and environment, or to realize an optimal compromise between these two entropic aspects. Any system therefore strives to attain the equilibrium situation of *minimum* Gibbs free energy and unless kinetic limitations prevent it, as is often the case in for example ceramics, this situation will be the endpoint of the spontaneous evolution. In a piece of material, this is accomplished by the subtle interplay of inter-particle (for an alloy: interatomic) interactions. The very same effects are thus governing alloy formation, bulk phase behavior of alloys, surface segregation, and other effects at alloy surfaces. The thermodynamic basis of these processes is therefore briefly outlined.

For the sake of simplicity, only binary alloys in the CBE approximation are considered. This simple pair-potential model assumes that the configurational energy of an alloy is determined solely by the pair-wise interactions between its constituent atoms. The constant bond energy between two atoms is taken as independent of the distance between them. This model further assumes that the atoms reside on fixed lattice sites and interact only with their nearest neighbors. The bond strength between two identical atoms is directly derived from the heat of sublimation for the pure metal

$$\varepsilon_{ii} = \frac{2\Delta h_i^{sub}}{Z} \tag{1}$$

with Z the number of nearest neighbors of an atom in the pure i lattice.

The energetic interactions between unlike atoms are crudely characterized by one single parameter, the regular solution parameter α. This constant is defined as

$$\alpha = \varepsilon_{AB} - \frac{\varepsilon_{AA} + \varepsilon_{BB}}{2} \tag{2}$$

and can be calculated from the molar heat of mixing of an alloy A_xB_y as

$$\alpha = \frac{\Delta h_{mix}}{N_{Av} \cdot Z \cdot x \cdot (1-x)} \tag{3}$$

The CBE model only differs from regular solution theory in that the latter also assumes ideal entropy of mixing

$$\Delta s_{mix}^{id} = -R\sum_i x_i \ln x_i \tag{4}$$

while CBE combined with Monte Carlo simulations can handle various degrees of short and long range order as deviations from perfectly random solutions.

2.1 Thermodynamics of alloy formation

The cohesion of a solid is of course the result of the overall attractive interactions between the constituting particles, be it atoms in a metal, alloy or covalent network crystal, molecules in a molecular solid or chains in a polymer. These interactions determine the properties of a pure solid: the stronger these interactions, the higher the melting and boiling points. The particular crystal lattice is thereby determined by the chemical identity of the atoms and by the balance in bond strength between first and second nearest neighbors.

When different metals are alloyed, new interatomic interactions between unlike particles arise. In ideal solutions and alloys the energy of these new AB interactions is simply the average of AA- and BB-bonds, causing a zero enthalpy of formation ($\Delta h_{mix} = 0$, $\alpha = 0$) when preparing the alloy from the juxtaposition of the pure elements. As a consequence, the atomic arrangement

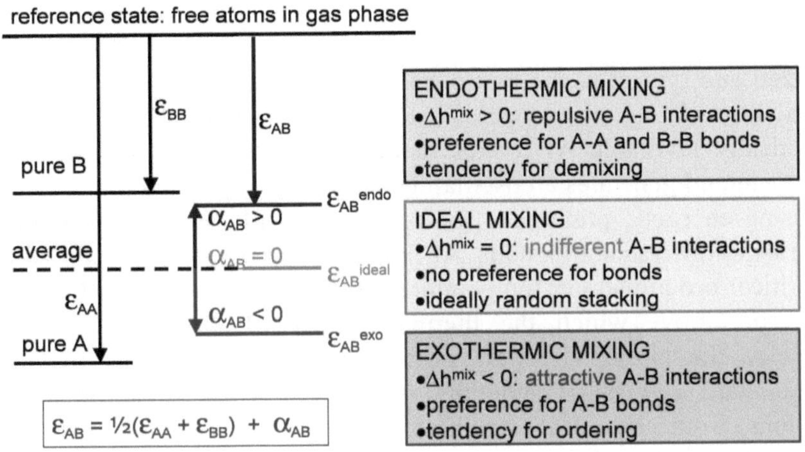

Figure 2. Energy scheme illustrating the three possibilities upon alloying: indifferent AB interactions lead to ideal solutions with zero heat of mixing and ideal entropy of mixing; repulsive A-B interactions cause endothermic mixing and a tendency towards demixing; preferential attractions between A and B lead to exothermic mixing and a tendency to form ordered compounds.

will be ideally random and the molar entropy of mixing will be given by the well known expression for the ideal entropy of mixing (Eq. 4).

In non-ideal alloys preferential interactions exist, either between unlike neighbors AB or between like atom pairs AA and BB. Inevitably, the atomic stacking will then no longer be random and the assumption of ideal mixing entropy will fail. A preference for AB bonds ($\alpha < 0$) leads to varying degrees of short range order and eventually long range order while repulsive AB interactions ($\alpha > 0$) will favor A-A and B-B bonds and will eventually induce phase separation. Figure 2 presents a scheme in which the energy effects upon alloying are indicated together with the main consequences for the phase diagram. As responsible for these preferential interactions, firstly *chemical* attractions or repulsions must be recognized, caused by the degree of 'chemical likeness' of the constituting elements. These effects are caused by the behavior of the valence (shell) electrons that to a large extent also determine the crystallography and the lattice of the pure elements. In addition to this chemical resemblance, the *size of the atoms* is a very important factor. Unlike in liquid mixtures, where size differences between the molecules are easily accommodated by a change in the local composition or number of neighbors, in a solid with a more or less fixed lattice, size differences create strain fields in the lattice and only limited size differences can be accommodated before phase separation occurs.

If the overall interactions are more attractive than in the juxtaposition of the pure elements, alloying will be exothermic ($\Delta h_{mix} < 0$, $\alpha < 0$). A limited degree of exothermicity will cause some degree of short range order σ ($0 \leq \sigma \leq 1$). By this favorable energy effect, the solid phase will be stabilized and typically, the two-phase melting region has a tendency to be curved upward (e.g. the systems Au-Pd and Cu-Pt). The preference for energetically favorable AB bonds reduces the segregation of the minority component and generates an oscillating subsurface segregation depth profile. Increasing energetic preference for AB bonds will cause the formation of (long range) ordered compounds. At first this long range order only holds up to a critical order/disorder temperature that is much lower than the melting point, and above which the thermal energy of the atoms will again randomize the alloy. The new ordered phase α' will also be able to accommodate significant amounts of excess A or B, allowing large deviations from the ideal stoichiometry. The domain for this ordered compound is separated from the domain of the random solid solution α by rather narrow two-phase domains in which the ordered and the disordered phases coexist. As this intermetallic compound formation becomes more and more exothermic, the order-disorder critical temperature will increase, making the ordered compound stable up to the point of (distectic or peritectic) melting. The phase boundaries for the ordered phase will merge with the melting region. At the same time, the domain of existence of the ordered compound in the phase diagram will become (much) narrower and the correct stoichiometry will be better respected. Also, the two-phase regions (ordered compound β coexisting with the random solid solution) become broader. Eventually more and more ordered compounds with different well-defined stoichiometries will emerge from the phase diagram, separated by two-phase regions in which two immiscible compounds coexist (e.g. the Pt-Sn system). This evolution with increasing magnitude of α is shown in Figure 3.

If, on the other hand, the overall interactions in the alloy are less attractive than in the juxtaposition of the pure elements, AB interactions will be more repulsive (due to chemical repulsions or to strongly different atomic sizes) than the average of AA and BB bonds and alloying will be endothermic ($\Delta h_{mix} > 0$, $\alpha > 0$, e.g. the systems Ni-Pd and Au-Ni). There will be a preference for like neighbors causing another form of short range order or deviation from ideal randomness. These repulsions will somewhat 'destabilize' the solid phase, causing a tendency to lower the melting point and a downward curvature of the melting region. Sufficiently strong repulsions will cause a miscibility gap and a two-phase region in the existence domain of the otherwise single phase solid solution. At temperatures higher than the critical temperature for demixing, the repulsions are overcome and the alloy becomes a single solid solution again, in which the preference for like neighbors still subsists. As the repulsions

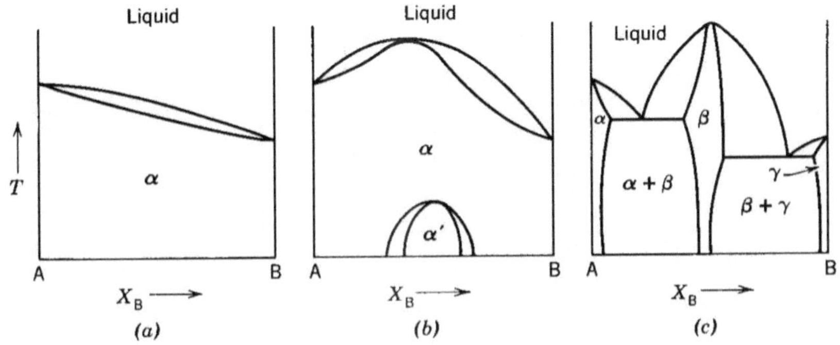

Figure 3. Evolution of the phase diagram of an exothermic alloy with increasing ordering tendency due to preferential attractions between A and B.[1]

become stronger, the melting region becomes 'azeotropic' with a common minimum for the solidus and the liquidus, as is shown in Figure 4. At the same time the two-phase region of immiscibility becomes broader and higher and the critical temperature for overruling the repulsions by thermal motions becomes higher. In the case of even stronger repulsions, the region of phase separation and the melting region coalesce into an eutectic phase diagram. Extreme repulsions will ultimately result in complete immiscibility and the disappearance of the 'side'-phases with α' and α'' solid solutions.

With respect to strain energy induced by large atomic size differences, it is perhaps useful to recall the principal trends in the systematic variation of the atomic radii throughout the periodic system, especially for the transition elements where the majority of the engineering materials and the catalytically active elements are found. For the main group elements, due to the increasing nuclear charge at constant principal quantum number, there is a pronounced decrease in atomic radius from left to right in a period. For the transition elements, the *d*-subshell is filled and the strong shielding of these deeper lying energy levels counters the increase in nuclear charge, causing a broad and shallow minimum in atomic radius at the triads, the Fe-Co-Ni groups (groups 8-9-10 or VIIIB). Moving downward in a group, the atomic radius increases due to the appearance of a new electronic shell, with increasing principal quantum number in each new period. However, in the second and third transition series, vertical neighbors have almost identical atomic radii as a consequence of the so-called lanthanide-contraction: the expected increase in atomic radius between 4*d* and 5*d* elements is perfectly balanced by the slow but altogether clear decrease over the 14 lanthanides, the 1st group of the inner transition elements, that precede the 5*d*-series. As a consequence, the elements Zr to Cd have almost identical sizes as their lower

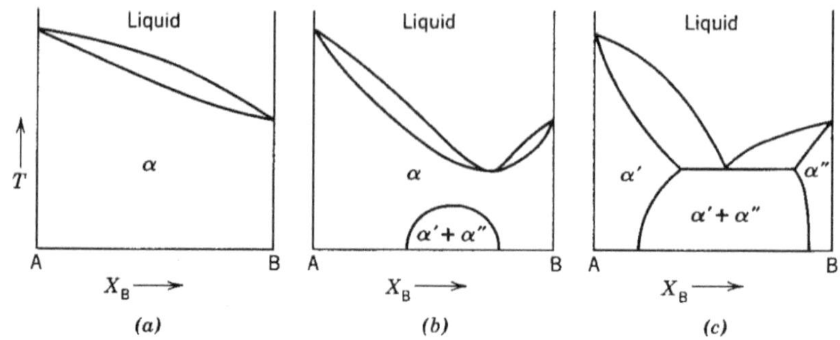

Figure 4. Evolution of the phase diagram of an endothermic alloy for increasing repulsive interactions between A and B.[1]

neighbors Hf to Hg, causing for example Zr and Hf and Nb and Ta always to appear together in their ores. These simple rules allow for a first quick estimation of the size differences and the associated repulsions between alloying partners and therefore of the tendency towards immiscibility and eventually phase separation.

With respect to the curvature of the two-phase melting region, one last effect can be pointed out: also in the molten alloy, similar chemical interactions exist. But unlike the rigid lattice, the liquid phase is able to accommodate the size difference strain by a local reorganization of the atoms. Only the chemical interactions subsist. But these non-idealities in the liquid can also influence the curvature of the two-phase melting region: a significant downward (upward) curvature of the solidus and liquidus in the melting region, if chemical attractions (repulsions) stabilize (destabilize) the liquid phase at the expense of the solid.

In summary, Table 1 shows the four different possible combinations, depending on whether alloy formation is endothermic or exothermic and depending on the temperature being higher or lower than the critical temperature.

2.2 Thermodynamics of surface segregation

In addition to the bulk behavior, also surface processes can contribute significantly in achieving the minimal Gibbs free energy for a piece of material by attenuating its surface energy. Surface segregation, surface relaxations in bond energy and in interlayer spacing and possibly surface reconstructions play a role in this respect. In a first instance, it is the altered surface composition after segregation that is explored in view of improved catalytic properties of an alloy. Surface segregation can be defined as the

Table 1. Possible combinations for non-ideal alloys.

	Exothermic mixing ($\alpha < 0$)	Endothermic mixing ($\alpha > 0$)
$T > T_c$	Disordered solid solution *e.g. Ni-Pd, Au-Pd, ...*	
$T < T_c$	Stoichiometric ordered intermetallic compounds *e.g. Pt-Sn, Co-Mo, Mo-Ni,...*	'Cherry model' Demixing into two phases *e.g. Au-Pt, Pt-Re, Pt-Rh...*

spontaneous enrichment of the surface, at equilibrium, in one of the components, caused by the mere presence of the surface. Because of the altogether different nature in origin and extent, a clear distinction will be made between segregation in disordered alloys on the one hand and in ordered intermetallic compounds on the other hand.

2.2.1 Surface segregation in disordered alloys

In randomly disordered alloys, surface segregation is essentially a partial demixing of the surface layers. This demixing causes a lowering of the configurational entropy. For segregation to be spontaneous and assuming reasonable values for the excess entropy, it must necessarily be exothermic. As a consequence, segregation in disordered alloys is always less pronounced at higher temperatures.

In the concept of the quasi-chemical approximation[2-4], atomic exchanges in segregation can be approximated as a 'quasi-chemical' reaction: surface segregation is considered as a process where atoms B from the bulk of an AB alloy swap with atoms A from the surface. Formally written as a chemical reaction this becomes

$$A_{surface} + B_{bulk} \Leftrightarrow A_{bulk} + B_{surface} \quad \text{(segregation of B)} \tag{5}$$

The equilibrium constant for this 'reaction' can be written in the usual way

$$K_{segr} = \frac{[A_{bulk}][B_{surface}]}{[A_{surface}][B_{bulk}]} = \frac{y(1-x)}{(1-y)x} \tag{6}$$

where y and x are the surface and bulk concentrations respectively in at.% B. This traditional way of writing the equilibrium constant is only valid for minor deviations from the assumed ideal entropy of mixing in disordered alloys. The classical thermodynamics of chemical equilibrium leads to the corresponding segregation equation

$$RT\ln\frac{y(1-x)}{(1-y)x}+\Delta G^0_{segr}=0 \tag{7}$$

The Gibbs free energy for segregation can be written as

$$\Delta G^0_{segr}=\Delta H^0_{segr}-T\Delta S^{ex}_{segr} \tag{8}$$

with ΔH^0_{segr} the enthalpy and ΔS^{ex}_{segr} the excess entropy. The condition for disordered alloys is clearly incorporated in the formalism since $RT\ln K_{segr}$ is the pure translation of the ideal mixing *entropy* in a stochastic disordered system (Eq. 4).

Equation 7 allows the calculation of the surface concentration y at temperature T for an alloy with bulk composition x, when the change in Gibbs free energy upon the exchange of atoms between bulk and surface can be evaluated. For a regular solution, only negligible entropy effects are assumed, $\Delta S^{ex}_{segr}\approx 0$, and thus

$$\Delta G^0_{segr}=\Delta H^0_{segr}-T\Delta S^{ex}_{segr}\approx\Delta H^0_{segr} \tag{9}$$

and

$$RT\ln\frac{y(1-x)}{(1-y)x}+\Delta H^0_{segr}=0 \tag{10}$$

If the driving force for segregation can be assumed to be constant, independent of the bulk composition, which is only strictly true for ideal alloys, Figure 5 gives the solutions to this equation for different values of the (dimensionless) segregation enthalpy. Three different situations can be observed. First, when the heat of segregation equals zero (K_{segr} = 1), no segregation occurs and the surface composition remains identical to the bulk composition. Next, when the segregation enthalpy is negative, B atoms segregate spontaneously by an exothermic process. At higher temperatures, segregation becomes less pronounced. Finally, when the enthalpy of segregation is positive, segregation of B atoms is endothermic and not spontaneous. The reaction will proceed in the opposite direction and atoms A will segregate instead.

Looking at the nature of the segregation enthalpy in non-ideal alloys, up to four different contributions can be distinguished, each representing a possible driving force for segregation

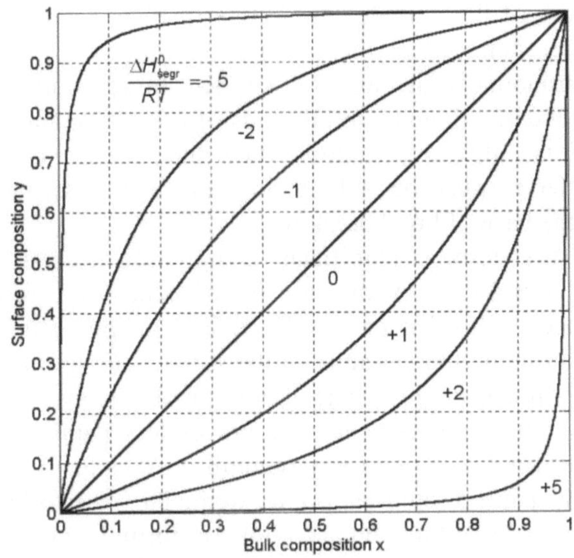

Figure 5. Graphical illustration of the segregation in a binary disordered alloy for constant values of the reduced, dimensionless segregation energy (Eq. 7).

$$\Delta H^0_{segr} = \Delta H^0_1 + \Delta H^0_2 + \Delta H^0_3 + \Delta H^0_4 \tag{11}$$

where ΔH^0_1 is the lowering of the surface energy, ΔH^0_2 the lowering of the total bond energy, ΔH^0_3 comes from the relaxation of elastic strain energy around the solute atoms, and ΔH^0_4 is the energy change due to preferential chemisorption on the surface.

The first two energy contributions can be straightforwardly evaluated in the constant bond energy model.

For the calculation of the *first* energy contribution, the assumption that the surface energy equals half the energy of the missing bonds at the surface leads to

$$\Delta H^0_1 = a_{\text{atom}}(\gamma_B - \gamma_A) = -\frac{m}{4}(\varepsilon_{BB} - \varepsilon_{AA}) \quad \text{with } \varepsilon_{ii} < 0 \tag{12}$$

with a_{atom} the surface area covered by one atom at the face concerned, γ_i the surface energy of a pure metal *i*, *m* the number of out-of-plane bonds and ε_{ii} the binding energy of an *i-i* bond. The element with the lower surface energy or heat of sublimation will normally segregate to the surface. The segregation is face-dependent because the surface energy γ_i and the area of an atom a_{atom} or alternatively the number of out-of-plane neighbors *m* depend on the orientation of the surface. As a consequence, segregation in an fcc lattice is most pronounced at the (110) face, followed by the (100) and finally is less pronounced at the (111) surface. In addition to the previous effects, it has to be mentioned that all kinds of relaxation and reconstruction processes diminish the segregation since the energy release in these spontaneous effects must be subtracted from the first estimate of the segregation on the base of bond breaking.

If only this first term is significant and if only nearest neighbors are accounted for, the influence of the segregation behavior is restricted to one layer. With respect to this first driving force for segregation it is perhaps also useful to recall a rule on the variation of the surface energy of pure transition elements. From first principles calculations, it is shown[5] that the surface energy of transition elements behaves more or less parabolically with the number of *d*-electrons (N_d), with a maximum at $N_d = 5$. In agreement with this observation, using a rigid-band-like approach,[6] the following rule for surface segregation in (dilute) binary alloys is formulated: (i) when the *d*-band filling of the solvent is greater than 5, the alloy minority component will segregate to the surface if it has more *d*-electrons than the solvent, (ii) likewise, when the *d*-band filling of the solvent is less than 5, the minority component will have a lower surface energy and tend to segregate if it has fewer *d*-electrons than the solvent.

The *second* energy term becomes appreciable in regular solutions and the segregation enthalpy can be expressed in terms of the regular solution parameter α

$$\Delta H_2^0 = \alpha[2l(x-y) + m(x-0.5)] \qquad (13)$$

with l the number of in-plane neighbors.

Figure 6 illustrates the effect of the second energy term for different values of α/RT. It is easily seen from Eq. 13 that for $x = 0.5$ the solution is $y = x$, independent of the value of the regular solution parameter. Positive values of the regular solution parameter and the corresponding A-B repulsions enhance the segregation of the minority component and give rise to a monotonous depth profile. In case of a negative value for the regular solution parameter, segregation of the minority component is reduced and an oscillating depth profile appears. The effect of this energy term must be evaluated together with the other driving forces, especially the first one, and can reach up to four layers in depth.

A *third* energy contribution pertains to the relaxation of elastic strain energy due to the difference in atomic size of A and B. The solution of

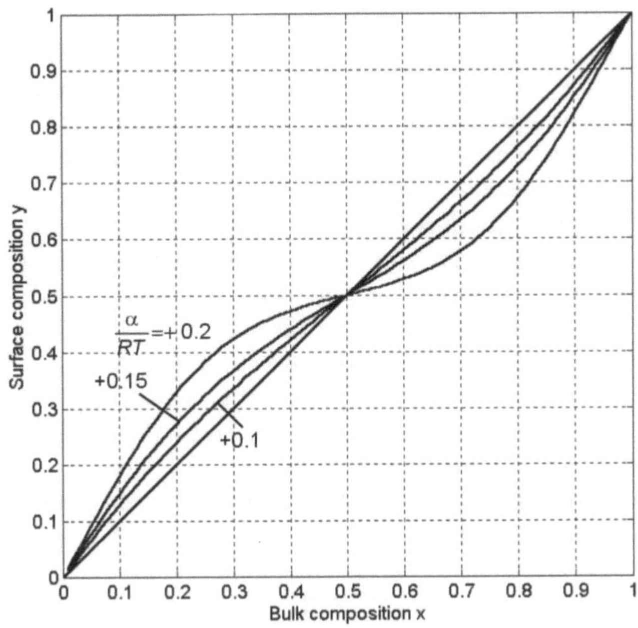

Figure 6. Graphical illustration of the effect of the second driving force for segregation to the (111) surface (m = l = 6) of a binary disordered alloy (Eqs. 10 and 13).

atoms with different sizes in an alloy creates local elastic deformations in the crystal lattice around these atoms. Upon segregation, the accompanying elastic strain energy can be (partially) released and this contributes a third driving force for segregation. This effect favors the segregation of the minority component when the difference in size exceeds 10% and is commonly believed to be of importance only for larger solute atoms. Moreover, this is always an exothermic effect. A coarse approximation of this elastic strain energy is obtained by extrapolating macroscopic elasticity theory to the atomic scale.[7] The elastic strain energy associated with a sphere of radius r_A of solute A incorporated into solvent B with a spherical cut-away of radius r_B is classically calculated as

$$E_{el} = \frac{24\pi K_A r_A (r_A - r_B)^2 G_B r_B}{3K_A r_A + 4G_B r_B} = -\Delta H_3^0 \tag{14}$$

with K_A the bulk modulus of solute A and G_B the shear modulus of solvent B. The extrapolation of elasticity theory for macroscopic continua down to the discrete atomic level is very questionable. Furthermore, this estimation is only valid for very dilute solutions, since it is assumed that the regions of elastic deformation do not influence each other. Next, it is not entirely correct to treat this effect separately from the second driving force, ΔH_2^0, since the elastic strain is generated during the alloying process. The associated energy is an essential part of the mixing energy for the solid alloy and is already taken into account by the regular solution parameter. Finally, it is possible that the elastic strain around larger solute atoms is not entirely relaxed upon segregation.

Up to now, segregation has been considered for a surface in vacuum. In reality, surfaces (e.g. of a catalyst) are often in contact with a (reacting) gas phase. Chemisorption of gaseous components can take place to a different extent on different atomic species at the surface. When a gaseous component X adsorbs preferably on one of the constituent atoms A or B (due to e.g. a different heat of adsorption), this preferential chemisorption of X may enhance or reduce the surface enrichment caused by the other driving forces. In some cases the segregation can even be reversed. Continuing the parallel with classical chemical equilibrium, this effect can formally be accounted for by a set of consecutive reactions

$$\text{Segregation } A_{surface} + B_{bulk} \Leftrightarrow A_{bulk} + B_{surface} \qquad \Delta H_1^0 + \Delta H_2^0 + \Delta H_3^0$$

$$\text{Chemisorption } B_{surface} + X_{gas} \Leftrightarrow (B-X)_{surface} \qquad \Delta H_4^0 \tag{15}$$

The enthalpy difference due to the preferential chemisorption is given by

$$\Delta H_4^0 = z_\mathrm{B}\varepsilon_\mathrm{BX}\theta_\mathrm{BX} - z_\mathrm{A}\varepsilon_\mathrm{AX}\theta_\mathrm{AX} \tag{16}$$

with z_i the surface coordination of i with X, $\varepsilon_{i\mathrm{X}}$ the bond energy between i and X, and θ_i the fractional coverage of i by X. An exothermic *fourth* driving force for segregation causes a preference for B-X bonds and a further surface enrichment in B atoms. In the opposite case, segregation of B is diminished and eventually reversed to segregation of A. In this case, the segregation of A that is not spontaneous ($\Delta G > 0$), but is entrained by another process that causes an even larger decrease of the Gibbs energy. The species X adsorbed on the surface acts as a 'chemical pump' to make the more reactive component A or B migrate to the surface where it can react (more) exothermically. An example of this behavior is observed for the system Pt-Rh. In Ultra High Vacuum (UHV) conditions Pt is the segregating component. Traces of carbon at the surface or of oxygen in the surrounding gas phase make Rh segregate instead.[8]

2.2.2 Surface segregation in ordered alloys

2.2.2.1 Stoichiometric ordered compounds

In ordered alloys, the situation is altogether different and somewhat more complex. Ordered alloys are formed in an exothermic process and exist because of the energetic preference for bonds between unlike atoms. In an ideally ordered intermetallic compound the maximum number of such A-B bonds is realised, therefore yielding a situation of minimum energy. Both the long and short range order are maximum and equal to one, corresponding to zero configurational entropy. Higher temperatures tend to first disturb and later destroy this order. Also segregation disturbs this ideal situation. First the influence of a non-zero temperature will be examined and it will be shown that, as long as $T<T_c$, the order is largely preserved. Next we will evaluate which element is likely to segregate to the surface and to what extent.

2.2.2.1.1 Effect of temperature on the order in stoichiometric ordered compounds

The configurational entropy can be evaluated using Boltzmann's well-known expression $S = k\ln W$, with k Boltzmann's constant and W the number of possible microscopic arrangements yielding the same macroscopic situation, combined with Stirling's approximation to the factorial function $\ln N! \approx N\ln N - N$, valid for large N values. When applied to a randomly disordered mixture, Boltzmann's formula directly yields the expression for the ideal mixing entropy $\Delta s_{mix}^{id} = -R\sum x_i \ln x_i$ (Eq. 4). When calculating the change in mixing entropy generated by a chemical reaction in this way, the classical equilibrium constants directly emanate from the minimization of the Gibbs energy.

For ordered alloys, the entropy can be elegantly calculated making use of the concept of sublattices. This will result in a slightly different expression for the equilibrium constants. The changes in configurational entropy can be evaluated by artificially splitting up the entire lattice of the ordered alloy in a number of sublattices.[9-11] Each of these sublattices is mainly filled with one species, with the minority species randomly distributed over the sites of the sublattice, hence again forming an ideal mixture. The total configurational entropy is then calculated as the sum of the entropy contributions in each sublattice. This approach will be demonstrated next on the basis of a rather simple example.

An ideally ordered alloy $A_{75}B_{25}$ is assumed with a face centered cubic (fcc) lattice ($L1_2$ structure) in which two sublattices can be discerned.

Sublattice α comprises the midplane sites, i.e. 75% of the total number of sites and is exclusively occupied by A atoms. Sublattice β groups the corner sites, represents 25% of the sites, and is exclusively occupied by B atoms. This ideal situation of zero configurational entropy can now be disturbed by either a deviation from the exact stoichiometry (see next section), by the thermal randomizing tendency at higher temperatures, or by segregation.

Next, assume that, as a consequence of thermal randomizing at higher temperatures, a fraction Δ of A atoms from the α sublattice has exchanged places with an equal number of B atoms from the β sublattice

$$A_\alpha + B_\beta \xleftrightarrow{\Delta} A_\beta + A_\alpha \tag{17}$$

The entropy for both sublattices can then be written as

$$\begin{aligned} S_\alpha(\Delta) &= -R\left(0.75\left(\frac{0.75-\Delta}{0.75}\ln\frac{0.75-\Delta}{0.75}+\frac{\Delta}{0.75}\ln\frac{\Delta}{0.75}\right)\right) \\ S_\beta(\Delta) &= -R\left(0.25\left(\frac{0.25-\Delta}{0.25}\ln\frac{0.25-\Delta}{0.25}+\frac{\Delta}{0.25}\ln\frac{\Delta}{0.25}\right)\right) \end{aligned} \tag{18}$$

The total configurational entropy is the sum of the entropy contributions in each sublattice

$$S_{total}^{config}(\Delta) = S_\alpha(\Delta) + S_\beta(\Delta) \tag{19}$$

Rearrangement gives

$$S_{total}^{config}(\Delta) = -R\begin{bmatrix} (0.75-\Delta)\ln(0.75-\Delta)+\Delta\ln(\Delta)+ \\ (0.25-\Delta)\ln(0.25-\Delta)+\Delta\ln(\Delta)+ \\ -0.75\ln(0.75)-0.25\ln(0.25) \end{bmatrix} \tag{20}$$

In this expression, the first two lines represent the usual expressions for the ideal molar entropy for each of the two sublattices forming ideal mixtures. The two terms on the third line account for the entropy decrease due to the order in the alloy: one mole of the highly ordered alloy consists of 0.75 mole of a sublattice α and 0.25 mole of sublattice β. The change in configurational entropy upon the atomic rearrangement at higher temperature is next given by

$$\Delta S_{thermal}^{0}(\Delta) = S_{total}^{config}(\Delta) - S_{total}^{config}(\Delta = 0) \tag{21}$$

The second term equals zero as it corresponds to the perfectly ordered lattice.

In the procedure of minimizing G by putting its derivative equal to zero, the two last constant terms of Eq. 20 will vanish. These considerations justify the form of the adapted equilibrium constants for the quasi-chemical processes in ordered alloys that will be used below.

In order to estimate the equilibrium situation corresponding to minimum Gibbs energy, this entropy balance must be combined with an energy balance. Straightforwardly summing the changes in bond energy leads to an antisite defect energy

$$\Delta E^0_{thermal}(\Delta) = -8\alpha\left(1 - \frac{16}{3}\Delta\right) \tag{22}$$

Continuing the quasi-chemical treatment, the equilibrium of the thermal rearrangement is described by

$$RT \ln K_{thermal} + \Delta E^0_{thermal}(\Delta) = 0 \tag{23}$$

with

$$K_{thermal} = \frac{[A_\beta]\cdot[B_\alpha]}{[A_\alpha]\cdot[B_\beta]} = \frac{\Delta^2}{(0.75-\Delta)\cdot(0.25-\Delta)} \tag{24}$$

It appears that the order in the bulk lattice is not appreciably degraded as long as the temperature does not approach too closely the critical order-disorder temperature. As an example, $Pt_{80}Fe_{20}$ can be considered (although this composition is off-stoichiometric). CBE predicts 1075 K as the order-disorder temperature for this alloy. Even at 900 K, Δ amounts to only 2%. In the simplified quasi-chemical treatment, one can then safely assume perfect bulk order as the starting situation for segregation.

2.2.2.1.2 Segregation in stoichiometric ordered compounds

As already mentioned, a stoichiometric ordered compound is a system at minimum energy and with perfect order. Segregation disturbs this ideal situation, hence is normally endothermic and causes partial atomic mixing at the surface with an increase in the configurational entropy. Whereas segregation in disordered alloys is *always* energy-driven, segregation in stoichiometric ordered alloys is most often (only) entropy-driven. It takes an unusually large difference in surface energy between the components A and B to turn the segregation into an exothermic process.

For the estimation of the segregation energy, in the QCA/CBE approach, four possibilities have to be considered. Again considering an fcc compound $A_{75}B_{25}$, both atomic species (A and B) can segregate to the surface. Moreover, this segregation can occur by a nearest neighbor exchange between the outermost layer and the second atomic layer or an exchange of atoms between the surface layer and a deeper bulk layer:

$$A_{\alpha,bulk} + B_{\beta,surface} \xleftrightarrow{\Delta} A_{\beta,surface} + B_{\alpha,bulk} \qquad \Delta E_1^0(\Delta) = -\frac{3}{2}(\varepsilon_{AA} - \varepsilon_{BB}) - 5\alpha$$

$$A_{\alpha,l_2} + B_{\beta,surface} \xleftrightarrow{\Delta} A_{\beta,surface} + B_{\alpha,l_2} \qquad \Delta E_2^0(\Delta) = -\frac{3}{2}(\varepsilon_{AA} - \varepsilon_{BB}) - 3\alpha$$

$$B_{\beta,bulk} + A_{\alpha,surface} \xleftrightarrow{\Delta} B_{\alpha,surface} + A_{\beta,bulk} \qquad \Delta E_3^0(\Delta) = +\frac{3}{2}(\varepsilon_{AA} - \varepsilon_{BB}) - 9\alpha$$

$$B_{\beta,l_2} + A_{\alpha,surface} \xleftrightarrow{\Delta} A_{\beta,surface} + B_{\alpha,l_2} \qquad \Delta E_4^0(\Delta) = +\frac{3}{2}(\varepsilon_{AA} - \varepsilon_{BB}) - 7\alpha$$

(25.1-4)

where l_2 denotes the second layer.

The first term in the four energy balances corresponds to the change in surface energy. This change in surface energy is realized at the 'cost' of respectively 3, 5, 7 or even 9 times ($-\alpha$). In ordering alloys $\alpha<0$, and, unless the difference in surface energy is unusually large, all four energy balances represent endothermic processes. These processes can then only occur because of the disorder (entropy) they entail.

Van Santen and Sachtler argued that at low to moderate temperatures, the least endothermic of these four possible processes will most likely occur and dominate the segregation.[3,4] This would result in segregation of the component with the lower surface energy by nearest neighbor exchange from the second to the first atomic layer. In this simple approach, the two outermost atomic layers form a closed system. The second atomic layer is depleted in the segregating element to the same extent as the enrichment in the first monolayer, leading to an oscillating concentration profile with the bulk concentration present from the third layer on. This is well illustrated by the $Pt_{75}Sn_{25}$ (111) alloy, where the minority component Sn segregates.[3,4] This system is also discussed in more detail in section 5.5. The oscillatory composition variation over the first few atomic layers is often observed for exothermic alloys in quantitative low energy electron diffraction (LEED) experiments.

However, this reasoning is primarily based on energy considerations and overlooks one particular entropy aspect: the number of possibilities for swapping atoms between the surface layer and the bulk of the alloy is so very much larger than the number of possible nearest neighbor exchanges between the first and the second layer. As a consequence, reaction (1) features a significantly more favorable reaction entropy. While it remains valid that the component with the lower surface energy will segregate, the segregation will be dominated by process (1) at all realistic temperatures. This is confirmed by more elaborate multilayer segregation models.

Continuing the above example and supposing that A features the lower surface energy (or heat of sublimation), reaction (1) will determine the surface composition, which can be calculated from the equilibrium equation

$$RT \ln K_{segr} + \Delta H_1^o = 0 \tag{26}$$

with

$$K_{segr} = \frac{[A_{\beta,\text{surface}}][B_{\alpha,\text{bulk}}]}{[A_{\alpha,\text{bulk}}][B_{\beta,\text{surface}}]} = \frac{\Delta^2}{(0.75-\Delta)(0.25-\Delta)} \tag{27}$$

The release of elastic strain energy upon segregation is not accounted for by CBE. In ordered compounds, however, elastic strain energy is normally not a major driving force, as the lattice parameter automatically adjusts to the ordered alloy stacking. It then follows that the mostly endothermic segregation in stoichiometric ordered alloys is in general less pronounced than in disordered alloys: if the phase diagram of a particular alloy system shows one or more true intermetallic compounds, which are proof of strong A-B interactions, the surface situation often corresponds to a truncated bulk without appreciable segregation enrichment Δ.

2.2.2.2 Segregation in off-stoichiometric ordered compounds

In stoichiometric compounds, segregation in the perfectly ordered lattice is necessarily accompanied by the creation of anti-site defects. This is not the case for off-stoichiometric alloys: new 'channels' for segregation become possible in which the excess amount of one component, accommodated in the 'wrong' sublattice, segregates to the surface from either the second layer or from the bulk. These processes are exothermic, since they somewhat restore the ideal situation of a maximum number of A-B bonds in the bulk where the atoms are fully coordinated. In this case, segregation of the excess component from the bulk is even more exothermic than segregation from the second layer. At first sight, it would seem that these processes would then occur more likely and more abundantly than the (almost always) endothermic segregation in the stoichiometric compound. However, by segregation according to this mechanism, the disorder caused by the excess amount on the 'wrong' sublattice is partially eliminated and the associated entropy decreases. This would again lead to typical equilibrium processes. Nevertheless, more abundant segregation is commonly observed in LEED or LEIS analyses of these alloys, often up to 100% A for e.g. $A_{80}B_{20}$ alloys.[12]

Summarizing the analysis of all the possible segregation paths in an $L1_2$ ordered fcc alloy $A_{75\pm\delta}B_{25\mp\delta}$ leads to the following conclusion: purely on the basis of the alloying parameter α and thus making abstraction of the difference in surface energy, the majority component A will always segregate. However, this surface enrichment in A occurs to a different extent for *under*- or *over*stoichiometric compositions. In $A_{75-\delta}B_{25+\delta}$, A segregates to the surface up to the stoichiometric composition of 75%; in $A_{75+\delta}B_{25-\delta}$, the same component A tends to enrich the surface up to 100%, irrespective of the magnitude of δ. At 0 K, this leads to a discontinuity: the surface concentration changes from 75% to 100% for an infinitesimal change in the bulk composition. This is confirmed by calculations by Ruban.[13] At higher temperatures the transition is still discontinuous, but becomes less abrupt as becomes evident from our simulations.

In addition to these effects induced by the alloying parameter α, the change in surface energy, that amounts to $-3/2(\varepsilon_{AA}-\varepsilon_{BB})$ must also be taken into account.

Finally, it should be noted that in off-stoichiometric alloys, the relaxation of strain energy can again become an appreciable issue and this effect also is always exothermic. Suppose that in $A_{75+\delta}B_{25-\delta}$ the A atoms are the larger ones. The excess A atoms are accommodated on the 'wrong' β lattice and indeed experience strain that can be released upon segregation to the surface. These A atoms replace the smaller B ones on β lattice sites where B atoms normally fit without appreciable strain. In the $L1_2$ structure these β lattice sites are completely surrounded by the larger A atoms on α sites. The approximation of the strain energy based on continuum-mechanical theory (Eq. 14) can then again cautiously be applied, but both elastic constants now pertain to material A

$$E_{el} = \frac{24\pi K_A r_A (r_A - r_B)^2 G_A r_B}{3K_A r_A + 4G_A r_B} = -\Delta H_3^0 \tag{28}$$

More sophisticated energy models will however automatically and much more correctly account for this effect.

3. MONTE CARLO SIMULATIONS

3.1 Introduction

In the previous part, surface segregation was treated by an analytic segregation equation in the Quasi-Chemical approximation. Although this way of working facilitates generalization, a macroscopic picture partly obscures processes at the atomic level. There are at least four reasons for a more detailed atom-by-atom approach. First, it is difficult to treat other than ideal entropies in a macroscopic approach. In an atomistic simulation a more accurate description of configurational entropy based on statistical mechanics is easier to embody. Secondly, the generalization to ternary or higher alloys, or the inclusion of lattice vacancies is more straightforward. It is also easier to explore the influence on segregation of other surface morphologies than atomically flat surfaces (e.g. step edges, small particles…). Lastly, complex energy models do not allow analytical treatment of e.g. segregation or ordering because of mathematical restrictions.

The detailed atomic configuration is a natural result of a MC simulation[14] and the actual state of order can easily be extracted from it. Two distinct quantities characterize the order, the long range order parameter (S) and the short range order parameter (σ).[15] The long range order quantifies the departure of the lattice as a whole from the perfectly ordered state. It is a measure for the number of atoms that reside on the correct sublattice

$$S = \max\left(\frac{\frac{N_{A,\alpha}}{N_\alpha} - x_A}{1 - x_A}, \frac{\frac{N_{B,\beta}}{N_\beta} - x_B}{1 - x_B} \right) \tag{29}$$

with N_j the number sites on sublattice j, $N_{i,j}$ the number of atoms i on sublattice j and x_i the mole fraction of species i. In order to evaluate S for

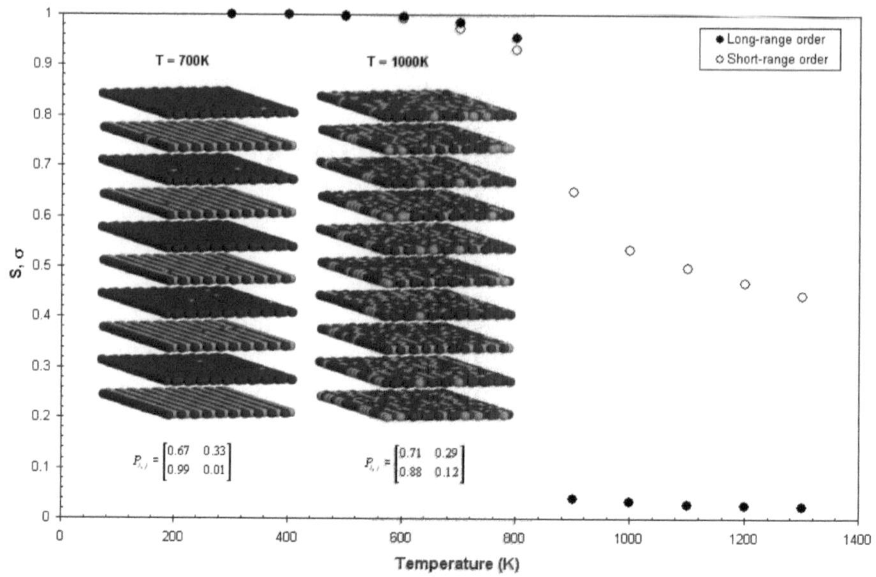

Figure 7. MC/CBE calculation of the order-disorder transition temperature for an $A_{75}B_{25}$ alloy ($L1_2$-type) and the associated variation of σ (○) and S (●). The insets show the actual state of order of (a part of) the simulation slab at the end of the simulation run. The matrices correspond to the nearest neighbor correlation matrix defined in the text.

a sampled equilibrium state, a template for perfect order is placed in all non-with N_j the number sites on sublattice j, $N_{i,j}$ the number of atoms i on sublattice j and x_i the mole fraction of species i. In order to evaluate S for a sampled equilibrium state, a template for perfect order is placed in all non-equivalent lattice points. Counting the number of atoms that reside on the correct sublattice, allows to calculate the long range order parameter from Eq. 29. The short-range order quantifies the departure from the perfectly ordered state in terms of the number of nearest neighbors only (an alternative, frequently used SRO parameter is the one by Warren-Cowley[16])

$$\sigma = \frac{N_{AB} - N_{AB}^{random}}{N_{AB}^{max} - N_{AB}^{random}} \tag{30}$$

with N_{AB} the actual number of AB bonds, N_{AB}^{random} the number of AB bonds in the perfectly random state and N_{AB}^{max} the maximum number of AB bonds in the ordered state. When the atoms are on fixed lattice points, the probability to find an atom j as a neighbor of atom i can be represented in a matrix notation[17] $P_{i,j}$. σ is directly related to the off-diagonal elements $P_{A,B}$ and $P_{B,A}$ in this matrix since

$$N_{AB} = \frac{x_A \cdot P_{A,B} + x_B \cdot P_{B,A}}{2} \tag{31}$$

Both the long-range order parameter S and the short-range order parameter σ vary between 1 (perfect order) and 0 (perfect disorder). Figure 7 shows the calculation of the order-disorder transition temperature and the associated variation of σ and S with MC/CBE. While at this transition temperature the long range order S falls abruptly from 1 to 0, a substantial degree of short range order subsists even at higher temperatures.

3.2 Statistical mechanics

The large number of particles ($\sim 10^{14} \ldots 10^{20}$) that are typically involved in realistic metallic alloys means that one cannot straightforwardly apply the laws of motion on the individual atoms to derive macroscopic properties of real alloy systems.

For most macroscopic properties, the detailed evolution of microscopic states over time is unimportant. The average of the macroscopic properties of these states remains constant although the microscopic states change continuously. The collection of different microscopic states with constant macroscopic properties is called an ensemble. Statistical mechanics provide ways to calculate the expectation values of the characteristics for these ensembles. An important theorem[17] in statistical mechanics is that the time average of a physical property (or measuring result) Q is equal to the ensemble average

$$\langle Q \rangle = \int P_j \cdot Q_j \tag{32}$$

P_j is the probability of finding a system in state j (with energy E_j). The determination of P_j depends on the type of ensemble that is used. In the investigation of surface segregation, two ensembles are of special importance: the canonical ensemble and the grand-canonical ensemble.

In the canonical ensemble, systems at constant volume and temperature have a specified number of particles N_j of each species j. The equilibrium occupation probability P_j follows the Boltzmann probability distribution

$$P_j = \frac{e^{\frac{-E_j}{kT}}}{\int e^{\frac{-E_j}{kT}} dE_j} = \frac{e^{-\beta E_j}}{\int e^{-\beta E_j} dE_j} \tag{33}$$

The grand canonical ensemble considers systems at constant volume, temperature, total amount of particles and chemical potential μ for every particle species. In analogy to the canonical ensemble the probability P_j can be calculated for a total number of particles N:

$$P_j = \frac{e^{\frac{-E_j-\mu N}{kT}}}{\int e^{\frac{-E_j-\mu N}{kT}} dE_j} = \frac{e^{-\beta(E_j-\mu N)}}{\int e^{-\beta(E_j-\mu N)} dE_j} \tag{34}$$

Only in systems with limited complexity and number of particles, can an analytical solution of the previous equations be obtained. Rather than concentrating on analytical solutions, we will adopt Monte Carlo Simulations. With Monte Carlo simulations, it is possible to find solutions to the integral in Eq. 32 without restriction of complexity or number of particles, even when it is impossible to find solutions with analytical methods.

3.3 Monte Carlo Simulations: the basics

Monte Carlo simulations sample the state space to find an estimate of Eq. 32. Just randomly picking out states with equal probability and estimating the properties in Eq. 32 would not work well: the exponential factor in the occupation probabilities implies that, especially at low temperatures, only a limited number of states contribute significantly to the average of macroscopic properties. This is also the case in actual experiments, because there too, according to the Boltzmann distribution, the ratio between the energy range of the sampled states during the experiment and the total energy of the system is very small. If the different states are selected with a probability p_j according to the probability of occurrence, then Eq. 32 modifies to

$$\langle Q \rangle = \int P_j \cdot p_j^{-1} \cdot Q_j \tag{35}$$

The procedure by which more probable states are selected with a higher frequency than others is called 'importance sampling'. If p_j is exactly equal to P_j and Eq. 35 is calculated over M discrete simulation steps than the equation further simplifies to

$$\langle Q \rangle = \frac{1}{M} \sum_{i=1}^{M} Q_i \tag{36}$$

The most essential part of correctly performing Monte Carlo simulations is to generate a random set of consecutive states according to the Boltzmann distribution. Markov processes are most frequently used to obtain such random set starting from an original state i of a system. A Markov process generates a new state j of the same system in a random fashion. The new random states are chosen according to a transition probability $P(i \rightarrow j)$. A succession of Markov processes is called a Markov chain. To guarantee that transition probabilities of Markov processes generate a Markov chain according to their Boltzmann probability, the transition probabilities $P(i \rightarrow j)$ have to fulfill a number of conditions:[14]

- they have to be constant in time.
- they should depend on properties of the initial state i and final state j only and not on any other state.
- the equation: $\sum P(i \rightarrow j) = 1$ has to be valid: every initial state should result in some final state. Note that $P(i \rightarrow i) > 0$ is possible.
- ergodicity: it should be possible to reach any final state in a finite number of steps.
- $p_i \cdot P(i \rightarrow j) = p_j \cdot P(j \rightarrow i)$ is sufficient to guarantee that state i is in equilibrium with state j. This is called the condition of detailed balance.

For the canonical ensemble, the condition of detailed balance yields an expression to calculate the transition probabilities from Eq. 33:

$$\frac{P(j \rightarrow i)}{P(i \rightarrow j)} = \frac{p_i}{p_j} = \frac{P_i}{P_j} = e^{-\beta \cdot (E_i - E_j)} \tag{37}$$

In a simulation step, the transition probability is equal to the selection probability times the acceptance probability. $g(i \rightarrow j)$ is the selection probability and $A(i \rightarrow j)$ is the probability of acceptance of state j when coming from state i. If all transitions are selected with the same probability g, than Eq. 37 states:

$$\frac{g(j \rightarrow i) \cdot A(j \rightarrow i)}{g(i \rightarrow j) \cdot A(i \rightarrow j)} = \frac{A(j \rightarrow i)}{A(i \rightarrow j)} = \frac{p_i}{p_j} = e^{-\beta (E_i - E_j)} \tag{38}$$

It is important to notice that this equation determines the ratios of the acceptance probabilities only. An efficient algorithm should maximize the acceptance probabilities to reach equilibrium as fast as possible. An efficient algorithm was published by Metropolis et al.[18] The acceptance probabilities are maximized by adopting Eq. 38 as follows:

$$A(i \to j) = 1 \qquad if\ E_j - E_i \le 0$$
$$= e^{-\beta(E_i - E_j)} \qquad if\ E_j - E_i > 0 \tag{39}$$

For the grand-canonical ensemble, analogous reasoning leads to:

$$A(i \to j) = 1 \qquad if\ E_j - E_i - \Delta\mu \le 0$$
$$= e^{-\beta(E_i - E_j - \Delta\mu)} \qquad if\ E_j - E_i - \Delta\mu > 0 \tag{40}$$

with $\Delta\mu = \mu_j - \mu_i$.

This enables one to generate a series of states according to their Boltzmann probability. The states generated in this way can be sampled in the course of a simulation to calculate the average $\langle Q \rangle$ according to in Eq. 36. This average is most accurate if the number of samples is high enough and if the sampled states are not correlated.

3.4 Monte Carlo simulations: practical issues

Next, a number of practical issues are discussed. A general algorithm[18,19] would be:

1. Choose an initial configuration X_i.
2. Generate a new configuration X_j.
3. Calculate $A(i \to j)$
   a) If $A(i \to j) = 1$ accept new configuration
   b) If $A(i \to j) < 1$ generate a random number (RN) between 0 and 1
      - If $RN \ge A(i \to j)$ keep old configuration
      - If $RN < A(i \to j)$ accept new configuration
4. Set $X_j = X_i$ and return to step 2

A simulation starts by designing and filling a simulation slab. To eliminate size effects, a simulation slab of 1000-10000 lattice sites is used. However, as this slab still does not have realistic dimensions, a quasi infinite lattice is approximated by applying periodic boundary conditions. In this respect, it is important to choose the dimensions in such a way that the periodicity is respected when order possibly develops. Apart from that, the choice of the the initial configuration is free. Three choices are commonly adopted: the slab at $T^{ini} = \infty$ (complete disorder), at $T^{ini} = 0$ (perfect order) or at $T^{ini} \approx T^{simulation}$ (e.g. in the course of an annealing). The choice is free because the actual sampling can only start after full equilibrium has first been reached, i.e. when $P(i,t)$ has become equal to $p(i)$. One can check if full equilibrium has indeed been reached by plotting a quantity (e.g. heat of formation) for successive simulation steps and by checking its constancy, or by starting at different initial compositions and checking after how many steps the results converge to the same value.

Once an initial configuration is chosen, new configurations are generated in the canonical ensemble by exchanging the positions of two different atoms that are randomly selected. As long as $T>0$, this procedure will fulfill the condition of ergodicity. However, especially at low temperatures, the simulation slab can possibly get trapped in some metastable state. An indication for a metastable state (= a local optimum) is that macroscopic properties remain dependent on the initial configuration, even after long equilibration. To eliminate metastable states, the exchange rate should be sufficiently high. Typically, equilibrium is reached after changing atoms ~1000 times the number of lattice sites in a simulation slab (1000 sweeps). However, this number is merely indicative and there are in fact cases where equilibrium is reached much faster or, on the contrary, much slower (e.g. in the occurrence of a metastable state, or when approaching critical transition temperatures). Only after equilibrium has been attained can the sampling of the simulation slab start. In order to get reliable statistics, the sampled microstates should be uncorrelated. Therefore at least 2-3 sweeps between two samples are necessary. But even this does not guarantee that the correlation will be zero, especially at low temperatures. It is possible to calculate the autocorrelation function and to adjust the sampling frequency of the simulation, however at an extra calculation cost. If there is no correlation between successive values of the quantity Q_i in the sampling process, the standard deviation σ of $<Q>$ can be calculated as

$$\sigma = \frac{1}{n-1}\sqrt{\sum (Q_i - <Q>)^2} \tag{41}$$

So far, a generally applicable simulation procedure in the canonical ensemble was outlined. But in many cases the efficiency can be enhanced by tayloring the Monte Carlo procedure to the specific problem at hand. This is rewarding since typically between 10^6 to 10^7 Monte Carlo steps are usually needed even in moderate simulations. One such technique is of special importance here. For the study of the segregation processes in binary alloys, either the canonical ensemble or the grand canonical ensemble can be chosen. In the *canonical ensemble* ($V, T, n_i = constant$ for all species i), simulations are quite fast. However, since the number of atoms of each element remains unchanged and since the simulation slab is of very limited dimensions, the surface to bulk ratio is unrealistically large and compositional rearrangements at the surface may have a large and unrealistic influence on the bulk composition. This can be remedied by defining a larger slab, or by iteratively adjusting the initial bulk composition, both at the cost of longer simulation times. Alternatively, the *grand canonical ensemble* (V, T, n_{tot} = *constant*) can be used. The constant bulk composition is then maintained by imposing the correct difference in chemical potential $\Delta\mu(T)$ between the components. In general, and all the more so when segregation is abundant, the results are then more accurate than with the canonical ensemble.

In the grand canonical ensemble $\Delta\mu(T)$, which depends on the temperature, is an alternative way to fix the bulk composition, but its value is not a priori known. In principle, iterative Monte Carlo simulations with a well-chosen value of $\Delta\mu(T)$ must be performed until the desired bulk composition is reached. In order to limit the number of these initial simulations, it is of great importance to have a good starting guess and an efficient updating algorithm.

For a disordered phase, Figure 8 presents an iterative scheme, proposed by Creemers et al.[20] The chemical potential of each species i in the alloy can be written as

$$\mu_i = \mu_i^* + RT\ln x_i + RT\ln\gamma_i \tag{42}$$

As long as the the nth iteration of the composition does not satisfy the desired composition x_i , a correction is made according to

$$\begin{aligned} \gamma_i^{(n)} &= \gamma_i^{(n\text{-}1)}\frac{x_i}{x_i^{(n)}} \\ \mu_i^{(n+1)} &= \mu_i^* + RT\ln\gamma_i^{(n)}x_i \end{aligned} \tag{43}$$

The correction is based on the assumption that, for a limited composition range, γ_i varies gradually and in a continuous way. This algorithm also meets the condition that for a thermodynamically stable phase, μ_i increases monotonically with the composition x_i.

Reasonable initial guesses $\mu_i^{(0)}$ are readily established whether T is higher or lower than the critical temperature T_c for order-disorder transition. When $T > T_c$, the alloy is disordered (no long range order) and the value for an ideal solution can be used: $\mu_i^{(0)} = \mu_i^* + RT\ln x_i$. When $T < T_c$, a high degree of long-range order exists in the alloy. A good value for $\mu_i^{(0)}$ can now be found, based on the theory of sublattices, both for stoichiometric and for slightly off-stoichiometric compounds. The details of this procedure are discussed in Ref. 20.

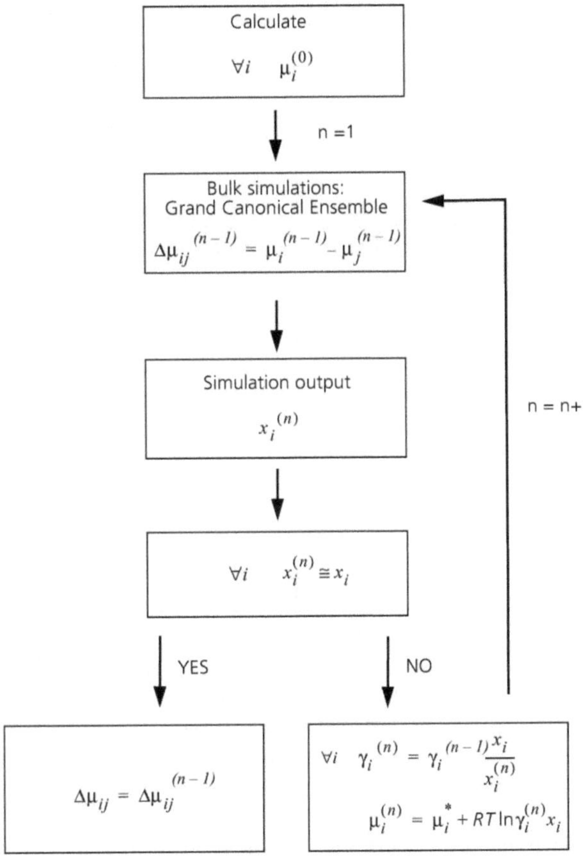

Figure 8. Algorithm for the evaluation of $\Delta\mu$.

A generally valid procedure for the determination of $\Delta\mu(T)$ is outlined in Figures 8 and 9. First, the order-disorder transition temperature is determined by Monte Carlo simulations in the canonical ensemble. Subsequently, knowing the critical order-disorder temperature, $\Delta\mu(T)$ can be determined, starting from the appropriate initial guess and following the scheme of Figures 8 and 9. As the chemical potential is a state property, the value of $\Delta\mu$ uniquely determines the bulk composition. This same value of $\Delta\mu$ can then be used in Monte Carlo simulations in the grand canonical ensemble to simulate several bulk and surface effects as segregation, relaxation, reconstruction and (dis)ordering. The great advantage of MC simulations is that no a priori assumption concerning the configurational aspects of the system needs to be made: bulk and/or surface order, (bulk) demixing, rearrangements upon segregation etc., all these phenomena follow naturally from the microscopic atomic swaps.

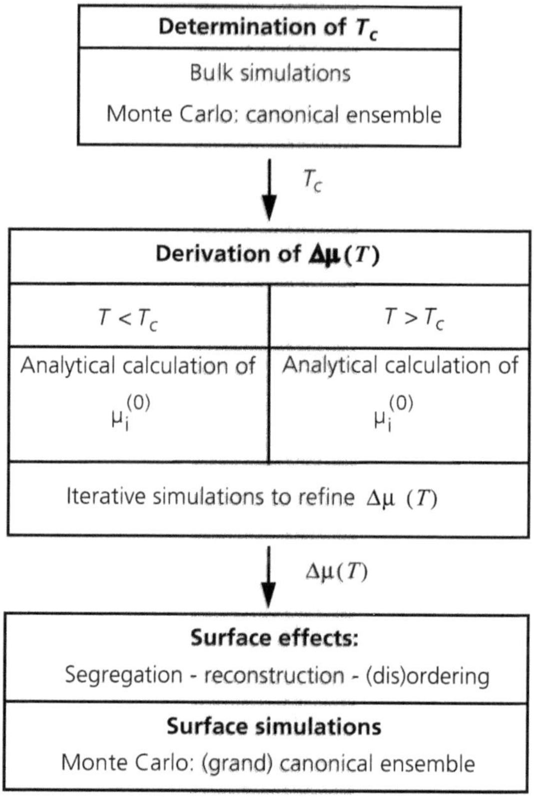

Figure 9. Setup for Monte Carlo simulations in the grand canonical ensemble.

4. BEYOND PAIR POTENTIALS

Although a lot of phenomenological aspects can be described by simple pair potentials (order-disorder transition, demixing, surface segregation, surface phase transitions, etc.), certain assumptions and premises make them improper for an accurate description of the behavior of real materials, such as transition metals and their (binary) alloys. A lot of the physics of the metallic bond is ignored, when considering (constant) pairwise interactions only between the atoms in the solid. Four major problems of pair-potentials for metallic systems are depicted by Ercolessi et al.[21] The first deals with the assumption that all bonds are independent of each other. This gives rise to a cohesive energy that scales with the number of neighbors Z. Based on experimental (or theoretical) values for structures with varying coordination number, it appears that the cohesive energy scales like $\sqrt{Z}$, rather than like Z. As an example, Table 2 shows the ratio of vacancy formation energy to

Table 2. Some physical properties of metals, revealing the weaknesses of simple pair-potentials (i.e. Lennard-Jones) methods.[24]

	E^{vac}/E^{coh}	C_{12}/C_{44}	E^{coh}/k_BT_m
PP (LJ)	1	1	13
Ni	0.31	1.2	30
Cu	0.37	1.6	30
Pd	0.36	2.5	25
Ag	0.39	2.0	27
Pt	0.26	3.3	33
Au	0.23	3.7	34

the cohesive energy, which would be equal to 1 if the metal obeyed a pair potential behavior. Another manifestation of the many-body character in metals becomes apparent from the mechanical properties. Two distinct elastic constants, C_{12} and C_{44}, are equal in the case of pure pairwise interactions,[22] resulting in a Cauchy ratio of 1. However, fcc metals have a Cauchy relation that is closer to 2 or even higher. Furthermore, in a two-body system, the melting temperature T_m is usually near $0.1E_{coh}/k_B$, but in reality, it is significantly lower. Finally, surface properties of a pair-potential model system are usually very different from those of a real metal. Often, first layer relaxations are erroneously predicted to be outward by pair-potential methods and reconstructions on noble metal surfaces can not be explained. Table 2 gives an overview of several material properties for the Ni- and Cu-group metals, compared to predictions of the pair-wise Lennard-Jones potential.

Modeling transition metals thus requires a model that goes beyond the simple pair-wise potentials, yet retaining enough computational simplicity to handle systems containing thousands of atoms. In this spirit, different semi-empirical energy models were developed. Miedema[23] performed some pioneering, mainly empirical work, leading to the development of the Macroscopic Atom Model (MAM). The first attempt to describe the metallic bond starting from quantum mechanical considerations was the Embedded Atom Method (EAM), later extended to account for directional bonding in the Modified Embedded Atom Method (MEAM). The equations governing these semi-empirical energy models will be given in the next paragraphs. An excellent review of the EAM and its applications, on which the following paragraph is mainly based, is given by Daw et al.[24]

4.1 The Embedded Atom Method

The Embedded Atom Method[22,24,25] (EAM) defines the energy E_i of an atom i in a lattice as the sum of the energy due to *electrostatic interactions* with its neighbors and the energy needed to *embed* this atom in the local electron density as generated by the other atoms in the system. This embedding energy depends only on the nature of atom i and can hence be derived from material constants of pure i. The energy of an atomic configuration x is then calculated as the sum of the energies of the atoms i within x

$$E_x = \sum_i E_i = \sum_i \left[F_i(\rho_{h,i}) + \frac{1}{2} \sum_j \phi_{ij}(R_{ij}) \right] \tag{44}$$

where F_i is the embedding energy function of atom i, $\rho_{h,i}$ is the host electron density around atom i due to the other atoms, and ϕ_{ij} is the electrostatic interaction energy between atoms i and j separated by a distance R_{ij}.

This separation into a pairwise interaction energy and a non-linear embedding energy can be derived starting from the expression for the cohesive energy of a solid in the density functional theory,[24,26] and by introducing two basic assumptions. The first assumption is that the embedding energy can be written as a function of the local electron density and its lower derivatives. Secondly, the electron density is approximated as a linear superposition of the densities of the individual atoms. The first assumption is justified by studies of the response theory of the nearly uniform electron gas, while the second is reasonable provided that covalent bonding effects in the metal are negligible. This naturally limits the range of applicability to simple metals and late and early transition metals.

The host electron density $\rho_{h,i}$ is obtained by summing the contributions of all the neighbors j of atom i

$$\rho_{h,i} = \sum_j \rho_j^a(R_{ij}) \tag{45}$$

Only the distance between atom i and the other atoms is taken into account and angular characteristics of directional bonds are thus neglected.

For practical calculations, the ρ_j^a-functions and the ϕ_{ij}-functions are represented by parameterized analytical expressions. The most commonly used parameterizations are the ones by Foiles et al.[25] and by Voter et al.[27] Besides these parameterizations, a lot of other formulations were developed, starting from the basic idea of the EAM.[28-30] Here, the parameterization of Foiles et al. will be further elaborated.

The electron density ρ_j of atom j is computed from the Hartree-Fock wave functions.[31,32] To account for possible electronic rearrangements upon alloying, the fixed total number of valence electrons N_j of element j is allowed to redistribute over the s and d sublevels

$$\rho_j = n_{s,j}\rho_{s,j} + n_{d,j}\rho_{d,j} = n_{s,j}\rho_{s,j} + (N_j - n_{s,j})\rho_{d,j} \tag{46}$$

where $n_{s,j}$ and $n_{d,j}$ denote the number of outer s and d electrons and N_j the total number of valence electrons. $\rho_{s,j}$ and $\rho_{d,j}$ are the densities associated with the s and d ground state wave functions, respectively.

The pair interaction term ϕ_{ij} is calculated as the electrostatic Coulomb repulsion between the screened, effective nuclear charges[25]

$$\phi_{ij}(R_{ij}) = \frac{1}{4\pi\varepsilon_0}\frac{Z_i(R_{ij})Z_i(R_{ij})}{R_{ij}} \tag{47}$$

with Z_i the effective charge of atom i and ε_0 the permittivity of vacuum

$$Z_i(R_{ij}) = Z_{0,i}(1+\beta_i R_{ij}^{\nu_i})e^{-\alpha_i R_{ij}} \tag{48}$$

where $Z_{0,i}$ denotes the number of outer electrons of atom i, α_i and β_i are constants that depend on the nature of atom i . ν_i is a parameter that is fixed empirically.

Once the atomic electron densities ρ_j and the pair interactions ϕ_{ij} are known, the embedding energy function F_i is derived by considering the pure element i. At equilibrium (lattice constant a_0), the energy of this pure metal i is given by the negative of the sublimation energy $-E_i^0$. In isotropically expanded or compressed form (lattice constant a) the energy of this configuration is given by Rose's universal equation of state[33]

$$E_i(a) = -E_i^0(1+a^*)e^{-a^*} \tag{49}$$

where a^* denotes the relative deviation from the equilibrium lattice constant a_0

$$a^* = \frac{a/a_0 - 1}{\sqrt{E_i^0/9B_i\Omega_i}} = \alpha\left(\frac{a}{a_0} - 1\right) \tag{50}$$

with B_i the bulk modulus of pure i and Ω_i the equilibrium volume per atom i. The embedding function F_i is now readily obtained by setting Eq. 44 equal to Eq. 49. This embedding function is assumed to be universally valid, which means that it depends only on the type of atom embedded in the background electron density and not on the origin of this electron density. The same embedding function can thus be used to calculate the energy of an atom i in an alloy.

Foiles globally optimized the EAM parameters towards the group of the six elements Cu, Ag, Au, Ni, Pd and Pt. This was done by setting up a least square optimization to solve an overdetermined set of equations for the elastic constants and vacancy formation energy of each of the six elements as well as for the heats of solution in the dilute limits of all binary systems. The sum over j in Eq. 44 in principle extends over the whole lattice. In practice, however, it is restricted to second nearest neighbors. Deurinck and Creemers[34,35] demonstrated that EAM often more reliably reproduces the driving forces for segregation if specific EAM parameters are calculated, specifically optimized for the binary alloy system under study.

4.2 The Modified Embedded Atom Method

The EAM has been successfully applied to the face cubic centered (fcc) transition metals with a nearly filled d-band. *In an empirical way and not justified by strong physical arguments* Baskes[36] extended the main idea behind the EAM in order to make it applicable to materials with directional bonding as well.

Similarly to the EAM, this modified embedded atom method (MEAM) defines the configurational energy E_i of an atom i as the sum of energy contributions due to electrostatic interactions with its neighbors and the energy release upon embedding this atom in the local electron density generated by the other atoms of the system

$$E_x = \sum_i E_i = \sum_i \left[F_i\left(\frac{\overline{\rho}_{h,i}}{Z_i}\right) + \frac{1}{2}\sum_j \phi_{ij}(R_{ij}) \right] \qquad (51)$$

where $\overline{\rho}_{h,i}$ is the host electron density at the site of atom i due to the remaining atoms of the system renormalized by the number of nearest neighbors Z_i in the reference structure of a type-i atom, $F_i(\rho)$ is the embedding energy of atom i into the background electron density ρ, and $\phi_{ij}(R_{ij})$ is the core-core pair repulsion energy between atoms i and j separated by the distance R_{ij}. In the original formulation by Baskes, the sum over j is restricted to nearest neighbors only, by an angular dependent many-body screening function (see later). Baskes' MEAM further follows the EAM concept, except for the improved approximation of the electron density. While the EAM uses a linear superposition of spherically averaged electron densities, in the MEAM angular characteristics of the electron densities are included. Directional bonds are taken into account in an empirical way by adding gradient and higher-order corrections to the simple EAM formulation.

The host electron density of atom i $\overline{\rho}_{h,i}$ is composed of the spherically symmetric partial electron density $\rho_i^{(0)}$ and angular dependent contributions $\rho_i^{(1)}$, $\rho_i^{(2)}$ and $\rho_i^{(3)}$. These partial electron densities have the following form:

$$\begin{aligned}
\rho_i^{(0)} &= \sum_{j\neq i} \rho_j^{a(0)}(R_{ij}) \\
(\rho_i^{(1)})^2 &= \sum_{\alpha}\left(\sum_{j\neq i} x_{ij}^{\alpha}\rho_j^{a(1)}(R_{ij})\right)^2 \\
(\rho_i^{(2)})^2 &= \sum_{\alpha,\beta}\left(\sum_{j\neq i} x_{ij}^{\alpha}x_{ij}^{\beta}\rho_j^{a(2)}(R_{ij})\right)^2 - \frac{1}{3}\left(\sum_{j\neq i}\rho_j^{a(2)}(R_{ij})\right)^2 \\
(\rho_i^{(3)})^2 &= \sum_{\alpha,\beta,\gamma}\left(\sum_{j\neq i} x_{ij}^{\alpha}x_{ij}^{\beta}x_{ij}^{\gamma}\rho_j^{a(3)}(R_{ij})\right)^2 - \frac{3}{5}\sum_{\alpha,\beta,\gamma}\left(\sum_{j\neq i} x_{ij}^{\alpha}\rho_j^{a(3)}(R_{ij})\right)^2
\end{aligned} \tag{52}$$

with $x_{ij}^{\alpha} = R_{ij}^{\alpha}/R_{ij}$ and R_{ij}^{α} the α component of the separation vector between atoms i and j. The second term in $\rho_i^{(3)}$ is a recent modification, and did not appear in the original formulation. In the present form, the partial electron density functions are orthogonal Legendre functions.[37,38] Old parameter sets can be transformed to new ones, just by replacing $t_i^{(1)}$ (see further) by $t_i^{(1)} + 3/5 \cdot t_i^{(3)}$.[38]

The atomic electron densities $\rho_j^{a(l)}$ are devised as exponential functions

$$\rho_i^{a(l)} = e^{-\beta_i^{(l)}(R_{ij}/R_{ij}^0 - 1)} \tag{53}$$

with the $\beta_i^{(l)}$ adjustable parameters and R_{ij}^0 the equilibrium distance between atom i and atom j in the reference structure.

The partial electron densities are combined into the total background density via

$$\overline{\rho}_{h,i} = \rho_i^{(0)}\sqrt{1+\Gamma_i} \tag{54}$$

and

$$\Gamma_i = \sum_{l=1}^{3} t_i^{l}\left(\frac{\rho_i^{(l)}}{\rho_i^{(0)}}\right)^2 \tag{55}$$

with $t_i^{(l)}$ adjustable weighing factors. Beside Eq. 55, some alternative formulations exist for the combination of partial electron densities into the total background density. Among them are the following expressions[39]

$$\bar{\rho}_{h,i} = \rho_i^{(0)} \frac{2}{1+\exp(-\Gamma_i)} \tag{56}$$

and

$$\bar{\rho}_{h,i} = \rho_i^{(0)} \exp\left(\frac{\Gamma_i}{2}\right) \tag{57}$$

The most commonly used formula is however Eq 55.

When all weighing factors are set equal to zero, all angular dependent terms vanish and the MEAM reduces to the EAM.

The embedding functions F_i are devised as functions of the electron density

$$F_i\left(\frac{\bar{\rho}_{h,i}}{Z_i}\right) = A_i E_i^0 \frac{\bar{\rho}_{h,i}}{Z_i} \ln\left(\frac{\bar{\rho}_{h,i}}{Z_i}\right) \tag{58}$$

with A_i an adjustable parameter and E_i^0 the ground state energy of atom i in the reference state.

Having now devised an analytical expression for the embedding energy, the electrostatic pair interaction energy must be calculated from the equation of state for a reference structure. For pure metals, the most stable phase of the specified atomic species is generally taken as the reference state. The energy as a function of interatomic distance is again described by Rose's equation of state, so that

$$\phi_{ii}(R) = \frac{2}{Z_i}\left(-E_i^0\left(1+a^*\right)e^{-a^*} - F_i\left(\frac{\bar{\rho}_0}{Z_i}\right)\right) \tag{59}$$

with $\bar{\rho}_0$ the background electron density for the reference structure.

For the calculation of the electrostatic interaction between atoms of a different type, a Rose-like equation for the reference state of the alloy is constructed. Equiatomic binary compounds, where each i atom is surrounded by only j neighboring atoms, show only pairwise interactions between unlike species and are thus ideal reference structures. Simple examples of such structures are B1 (simple cubic, NaCl) or B2 (bcc, CsCl). $\phi_{ij}(R)$ can then simply be derived from

$$E_{ij} = \frac{1}{2}\left(E_i + E_j\right) = \frac{1}{2}\left[F_i\left(\frac{Z_{ij}\overline{\rho}_j^{a(0)}}{Z_i}\right) + F_j\left(\frac{Z_{ij}\overline{\rho}_i^{a(0)}}{Z_j}\right) + Z_{ij}\phi_{ij}(R)\right] \tag{60}$$

with Z_{ij} the number of nearest neighbors in the equiatomic reference structure. The energy E_{ij} can be estimated by the equation of Rose, but with an 'average' sublimation energy $E_{ij}^0 = (E_i^0 + E_j^0)/2 - E_{ij}^{\text{mix}}$, with E_{ij}^{mix} the energy of mixing, with an 'average' bulk modulus $\alpha_{ij} = (\alpha_i + \alpha_j)/2$ and with an 'average' lattice parameter calculated from the assumed equilibrium intermetallic atomic volume $\Omega_{ij} = (\Omega_i + \Omega_j)/2$. From these two equations, $\phi_{ij}(R)$ is then readily obtained once the host electron densities and the embedding functions are known. A common problem arises when no experimental or theoretical values are known for the properties of the binary reference structure. In that case, other reference structures can be considered, e.g. the $L1_2$ (fcc, Cu_3Au) structure, resulting in somewhat more complicated formulas for the electrostatic repulsion energy.[40] The MEAM contains 8 adjustable parameters, which must be fitted to experimental properties. Apart from these 8 parameters, the MEAM contains 3 additional parameters, incorporated into Rose's equation of state, to ensure that the sublimation energy, bulk modulus and equilibrium volume of the reference structure are exactly reproduced. Due to the lack of consistent, accessible and reliable experimental results for a large class of materials, Baskes introduced some arbitrariness in the parameterization of the MEAM. The input set of experimental data consists of two elastic shear constants, two structural energy differences and the vacancy formation energy.

However, recent increased computational power allows for first principles calculations on real and hypothetical model systems. These data can then complement the experimental results to a complete input data set that is necessary for a unique and well defined parameterized potential.[41]

In the MEAM, the range of interactions is governed by a many-body radial screening function, proposed by Baskes,[42] using a simple elliptical construction. The screening function S_{ik} between atoms i and k depends on all other atoms between them, thus

$$S_{ik} = \prod_{j \neq i,k} S_{ijk}$$
$$S_{ijk} = f_c\left(\frac{C - C_{\min}}{C_{\max} - C_{\min}}\right) \tag{61}$$

where $f_c(x)$ is a function of the material dependent parameters C_{max} and C_{min}. The parameter C is determined using the following equation:

$$C = \frac{2\left(X_{ij} + X_{jk}\right) - \left(X_{ij} - X_{jk}\right)^2 - 1}{1 - \left(X_{ij} - X_{jk}\right)^2} \tag{62}$$

where $X_{ij} = (r_{ij} / r_{ik})^2$ and $X_{jk} = (r_{jk} / r_{ik})^2$. The r_{ij}, r_{jk} and r_{ik} are the distances between the corresponding atoms. The smooth cut-off function assures that no discontinuities occur in the forces on atoms and is given by

$$\begin{aligned} &\text{if } x \geq 1, && f_c(x) = 1 \\ &\text{if } 0 < x < 1, && f_c(x) = \left[1 - (1 - x)^4\right]^2 \\ &\text{if } x \leq 0, && f_c(x) = 0 \end{aligned} \tag{63}$$

The question whether to include other than nearest neighbors is addressed by Baskes et al.[43] for the particular case of Al. It appears that, at least for Al, the question of whether a short-range angular many-body potential is a more appropriate model than a long-range central many-body potential has not yet been resolved. Lee et al. introduced a MEAM formalism, that explicitly incorporates second nearest-neighbors, for bcc metals first and later for fcc metals.[44,45]

The original framework of the (M)EAM, namely the separation of the energy into a pair-wise electrostatic energy (entirely repulsive) and an embedding energy into the local electron density (attractive), is somewhat obsolete. Firstly, the nonuniqueness between ϕ and F allows more general forms for ϕ.[46,47] Furthermore, especially in the MEAM, the purely repulsive character of ϕ is completely abandoned, as F becomes zero and ϕ is negative for the pure metal in the equilibrium reference structure. Another viewpoint to the embedding function F was indicated by Van Beurden et al.[41] They point out that F is rather a coordination function, instead of an attractive embedding energy. To illustrate this, one can write the energy of any monoatomic cubic structure with a center of symmetry

$$E_i(R) = \frac{Z'}{Z}\left(-E_i^0\left(1 + a^*\right)e^{-a^*}\right) + \left(F_i\left(\frac{Z'\bar{\rho}_i^{a(0)}}{Z_i}\right) - \frac{Z'}{Z}F_i\left(\bar{\rho}_i^{a(0)}\right)\right) \tag{64}$$

The first term accounts for a strictly linear behavior of the energy with respect to the number of nearest neighbors Z', while the second term takes additional (non-linear) effects into account. This viewpoint was confirmed by DFT calculations for Rh in different metastable cubic structures. A more accurate name for F would therefore be *coordination function.*

4.3 Evaluation

EAM is a semi-empirical energy model. As explained above, it is firmly based on the many-electron Schrödinger equation but relies on parameterized expressions. The parameters are (in)directly related to material properties that can be determined experimentally. In this way a sound compromise arises: the method has a solid theoretical basis but is also anchored to accurately known material properties. This also results in a good equilibrium between accuracy and computational complexity. Its main asset is that it accounts for many-body interactions: the bond energies depend not only on the distance between atoms i and j, but also on the number and identity of their respective neighbors. EAM has been applied successfully to a variety of bulk and interface problems.[24]

MEAM surpasses EAM by considering the angular configuration between the different atoms. The angular dependence is however rather assumed than derived from explicit quantum mechanical considerations. Nevertheless, MEAM has proven to produce (more) accurate results for a number of alloys.[40,48,49] The extrapolation of the semi-empirical potentials, which are fitted to bulk properties only, to a wide range of surface phenomena confirms that they indeed contain the necessary ingredients to describe well various subtleties of the metallic bonding.

5. CASE STUDIES

In order to verify the simulation approach, developed in the previous sections, five typical case studies are next presented, binary alloys of catalytic importance: $Au_{75}Pd_{25}$, $Cu_{75}Pd_{25}$, $Pt_{50}Ni_{50}$, $Pt_{80}Fe_{20}$ and $Pt_{75}Sn_{25}$. In each case, the aim is to characterize the segregation behavior at the surface and subsurface, together with the order-disorder transition temperature in the bulk, surface relaxations, and possible reconstructions. As will be shown, each case is different and illustrates some specific considerations. Following the synergetic approach philosophy, the phase diagram is first presented for each selected system, together with an overview of the relevant material properties. Next, available experimental results on surface segregation results are briefly reported.

Table 3. The global EAM parameters for Cu, Ag, Au, Ni, Pd, Pt as obtained by Foiles et al.[25] Parameters α and β define the effective charges for the pair interactions and the parameters n_s and ν are used for the calculation of the atomic electron density. The last row specifies the atomic configuration used for the calculation of the atomic electron density.

	Cu	Ag	Au	Ni	Pd	Pt
α	1.7227	2.1395	1.4475	1.8633	1.2950	1.2663
β	0.1609	1.3529	0.1269	0.8957	0.0595	0.1305
n_s	1.000	1.6760	1.0809	1.5166	0.8478	1.0571
N	11	11	11	10	10	10
ν	2	2	2	1	1	1
Electronic configuration	$3d^{10}4s^1$	$4d^9 5s^2$	$5d^{10}6s^1$	$3d^8 4s^2$	$4d^9 5s^1$	$5d^9 6s^1$

Table 4. The specific EAM parameters for $Cu_{75}Pd_{25}$, $Au_{75}Pd_{25}$ and $Ni_{50}Pt_{50}$ using the scheme of Foiles et al.[25] but optimized for the alloy under study. Parameters α and β define the effective charges for the pair interactions and the parameters n_s and ν are used for the calculation of the atomic electron density. The last row specifies the atomic configuration used for the calculation of the atomic electron density.

	$Au_{75}Pd_{25}$		$Cu_{75}Pd_{25}$		$Pt_{50}Ni_{50}$	
	Au	Pd	Cu	Pd	Pt	Ni
α	1.4476	1.2953	1.7224	1.2723	1.3829	2.0040
β	0.1267	0.0595	0.1644	0.0695	0.2670	1.2567
n_s	1.0810	0.8478	1	0.9079	1.0571	1.5529
N	11	10	11	10	10	10
ν	2	1	2	1	1	1
Electronic configuration	$5d^{10}6s^1$	$4d^9 5s^1$	$3d^{10}4s^1$	$4d^9 5s^1$	$5d^9 6s^1$	$3d^8 4s^2$

As already pointed out in the previous section, Foiles et al.[25] optimized the EAM parameters for Ni, Cu, Pd, Ag, Pt and Au simultaneously. These parameters are tabulated in Table 3. Deurinck et al. however, demonstrated that very often the EAM yields better results when the parameters are optimized for the specific alloy system under study. The improved EAM parameters for $Au_{75}Pd_{25}$, $Cu_{75}Pd_{25}$ and $Pt_{50}Ni_{50}$ are given in Table 4. For the $Pt_{80}Fe_{20}$ alloy, the MEAM is used to account for the directional bonding due to incomplete *d*-band filling. The MEAM parameters can be found in Ref. 36. Finally, the $Pt_{75}Sn_{25}$ is studied with Miedema's Macroscopic Atom Model. For further details about this MAM model and its parameters, we refer to Ref. 23.

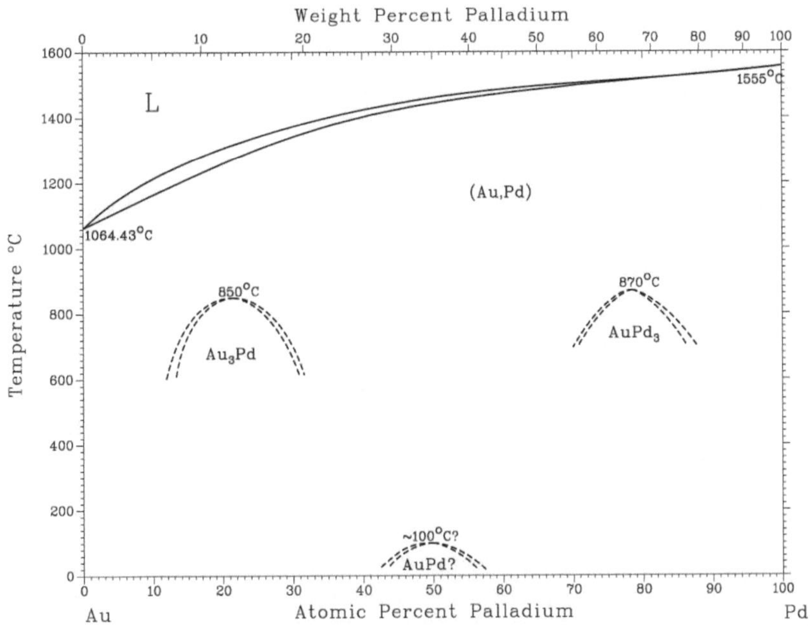

Figure 10. Phase diagram for the Au-Pd system.[53]

5.1 Surface structure and segregation profile of the alloy $Au_{75}Pd_{25}(110)$

Au-Pd alloys find increasing application in pollution control, e.g. for CO oxidation and decomposition of N_2O and formic acid.[51-52] The aim of this section is to investigate the structure and composition of the $Au_{75}Pd_{25}(110)$ surface.

The phase diagram for the Au-Pd system is presented in Figure 10.[53,58] The melting region has a slight upward curvature, which is an indication for moderate attractive chemical interactions. This is confirmed by the presence of ordered stoichiometric compounds at low temperatures. Ordering has indeed been experimentally confirmed for compositions near Au_3Pd and $AuPd_3$. For AuPd, an ordered structure may also exist at low temperatures (below 100° C).[53] Other relevant material properties with respect to surface segregation are summarized in Table 5.

Kuntze et al.[59] experimentally investigated this alloy surface using quantitative Low Energy Electron Diffraction (LEED) and Low Energy Ion Scattering Spectroscopy (LEIS). The experimental results give evidence of a disordered solid solution and of segregation of Au, resulting in pure Au in

Table 5. Relevant material properties with respect to surface segregation in the Au-Pd system.

	Au	Pd
Heat of sublimation (kJ/mole)	366.1[54]	378.2[54]
Surface energy (J/m²)	1.500[23]	2.050[23]
	1.333[55]	2.000[55]
Crystal structure	fcc[56,57]	fcc[56,57]
Lattice parameter (Å)	4.08[56]	3.8908[56]
Atomic radius (Å)	1.34[57]	1.376[57]

the two topmost layers. The surface features a (1×2) missing row reconstruction, similarly to Au (110), with a significant contraction of the first interlayer spacing and a buckling in the third layer. These experimental data will be compared with simulations in order to gain a better insight into the segregation behavior at the $Au_{75}Pd_{25}$(110) surface.

Specific EAM parameters are derived for the alloy system Au-Pd (Table 4) but, as can be noticed, these values are almost identical to the global ones in Table 3. Bulk simulations are first performed to derive the degree of bulk order, which determines the way $\Delta\mu$ is derived. Even when starting from a perfectly chemically ordered initial configuration, the simulations yield a disordered bulk at 300 K. This is in complete agreement with and the X-ray Diffraction (XRD) data obtained by Kuntze et al.[59] For disordered alloys, $\Delta\mu$ is obtained by performing Grand Canonical Ensemble bulk simulations, with an iteratively improved value of $\Delta\mu$, and with the atoms on a fixed lattice (atomic vibrations are not included), until the bulk concentration matches the desired bulk concentration (75 at.% Au).

The results of the surface simulations are shown in Figure 11.[59] The MC/EAM simulations completely confirm the experimental results: segregation of Au to the first two layers, a (1×2) missing row reconstructed surface and a significant contraction of the first interlayer spacing (–9% versus –13% found in the LEED experiments). Near the surface, the material shows a monotonic depth profile (region 1) that evolves to an oscillating one, closer to the bulk (region 2). This damped oscillatory concentration profile that is observed from the third layer on is the normal behavior that is expected for an alloy with a negative heat of mixing. Heat is evolved when a single Pd(Au) atom is embedded in pure Au(Pd),[25] which means both elements feature attractive interactions. Ultimately, this can lead to the formation of chemically ordered intermetallic compounds. In less exothermic alloys such as Au-Pd, the Au-Pd attractions are weaker and generate these composition oscillations only at the surface. The monotonic Au enrichment on top of this oscillating profile originates from the large difference in surface energy: with EAM one calculates 0.98 J/m^2 for Au

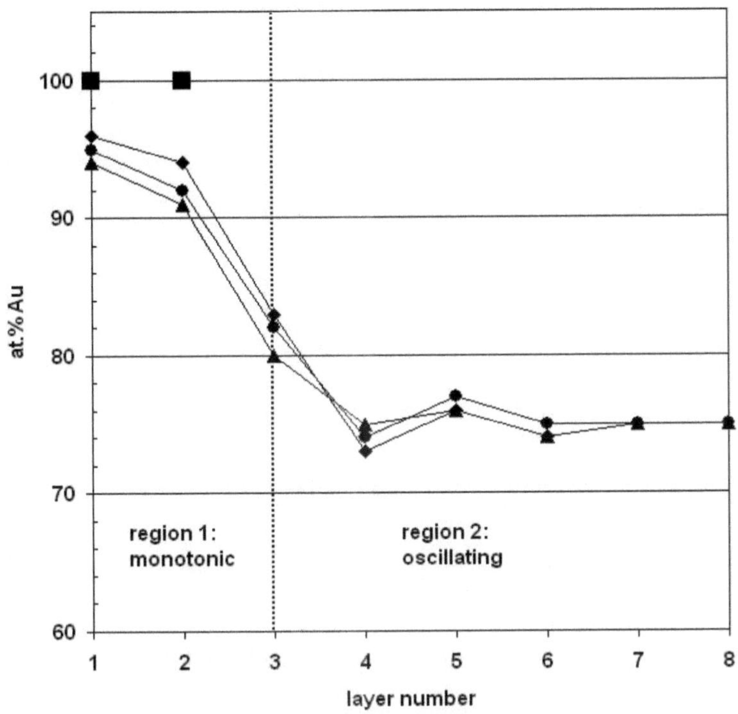

Figure 11. $Au_{75}Pd_{25}(110)$: Au-concentration as a function of depth at different temperatures (◆, 300 K; ●, 600 K; ▲, 800 K) according to MC/EAM simulations including the (2×1) missing row reconstructed surface, surface relaxations and atomic vibrations. The closed squares (■) indicate the experimental LEIS/LEED results for comparison.[59]

compared to 1.49 J/m^2 for Pd. The surface causes a strong Au enrichment and the presence of the surface is felt up to the third atomic layer, two layers deeper than would be estimated from only pairwise interactions between the nearest-neighbors in a close-packed (111) surface.[2] On the one hand, the second layer of a (110) surface is still part of the surface, because its atoms are also incompletely coordinated. On the other hand, our EAM-treatment also takes next-nearest-neighbor interactions into account, which again extends the influence of the surface one layer deeper than with nearest-neighbor interactions only. The normal oscillatory profile is therefore delayed until the third layer.

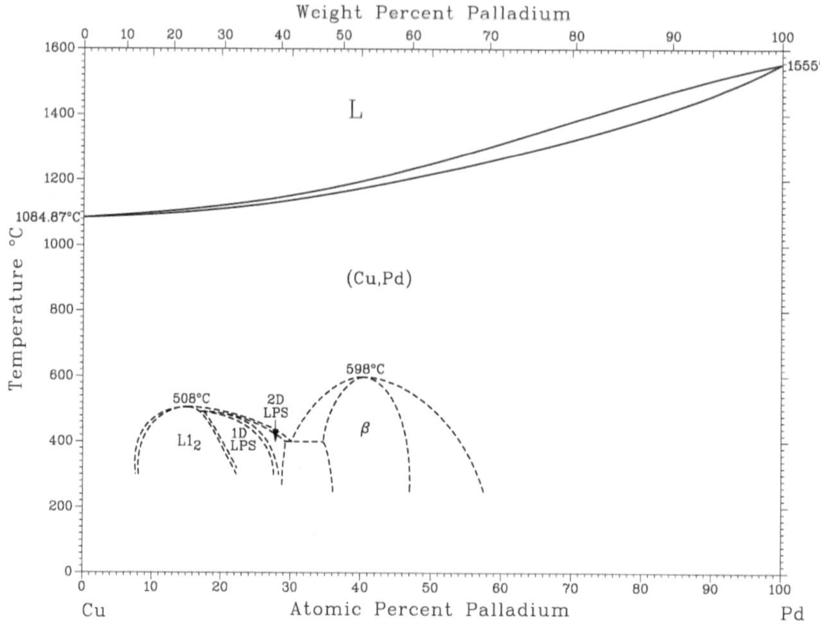

Figure 12. Phase diagram for the Cu-Pd system[53]

5.2 Cu segregation and ordering at the (110) surface of $Cu_{75}Pd_{25}$

Detailed information about the surface composition and structure of Cu-Pd alloys plays an important role in the interpretation of their catalytic behavior. The addition of Pd is responsible for a change in the reaction parameters as compared to pure Cu(110).[60,61] For example, the dehydrogenation rate of methoxy and formate species on the alloy surface is significantly increased compared to the pure metal.

The phase diagram for the Cu-Pd system is given in Figure 12.[53] At low temperatures, two ordered compounds Cu_3Pd and CuPd exist. The ordered compound Cu_3Pd has the LI_2 structure over the entire composition range from 10 to 25 at% Pd, with a maximum transition temperature at 500 °C and at the off-stoichiometric composition of 15 at% Pd. Other relevant material properties with respect to surface segregation are summarized in Table 6.

In order to characterize the surface behavior, Bergmans et al.[62] performed a detailed LEIS and LEED analysis of the $Cu_{85}Pd_{15}(110)$ surface at room temperature after annealing and equilibrating at 600 K. The composition and the structure of the outermost atom layers were determined: an oscillating concentration profile with a slight Cu enrichment in the first layer (89 ±

Table 6. Relevant material properties with respect to surface segregation in the Cu-Pd system

	Cu	Pd
Heat of sublimation (kJ/mole)	337.4[54]	378.2[54]
Surface energy (J/m²)	1.825[23]	2.050[23]
	1.566[55]	2.000[55]
Crystal structure	fcc[56,57]	fcc[56,57]
Lattice parameter (Å)	3.6075[56]	3.89[56]
Atomic radius (Å)	1.278[57]	1.376[57]

2 at.% Cu) and a strong Cu depletion in the second layer (60 ± 8 at.% Cu). In accordance with other work,[60,63] a (2×1) LEED reconstruction was suggested, caused by the ordering in the second layer as evidenced by the detailed LEIS analysis. In this section, these experimentally observed changes in the surface layers are explained in terms of the physical effects and driving forces governing segregation and (dis)ordering, as they emerge from simulations.

For the simulations, we have chosen to concentrate on the $Cu_{75}Pd_{25}$ alloy rather than on the $Cu_{85}Pd_{15}$ alloy as in the experiments for the following reason.[34] As explained in Section 3.4, the surface simulations are most accurately performed using the grand canonical ensemble. This simulation scheme requires the determination of the difference in chemical potential between the two species at the required temperatures. At 600 K, $Cu_{75}Pd_{25}$ is stoichiometrically ordered and $\Delta\mu$ can be calculated analytically. At this temperature, the off-stoichiometric $Cu_{85}Pd_{15}$ is also ordered, but the calculation of $\Delta\mu$ according to our alternative analytical approach for off-stoichiometric ordered alloys[20] is only reliable for small deviations from perfect stoichiometry. For this reason, we decided to proceed with the $Cu_{75}Pd_{25}$ alloy.

The MC/EAM simulation results with the global parameters (Table 3) determined by Foiles et al. are shown in Figure 13. In contradiction with the experimental facts, they predict a pronounced Pd-segregation and a disordered bulk, which is also in contradiction with the phase diagram.[53] In an attempt to obtain better agreement with the experiments, simulations were performed with newly derived EAM parameter values specific for the Cu-Pd alloy (Table 4). At 600 K, an ordered bulk is now obtained. An order-disorder transition temperature of ~ 770 K was found, in excellent agreement with the transition temperature of ~ 750 K in the Cu-Pd phase diagram. At temperatures well below 770 K, one can then safely assume complete long-range order in the bulk. In this temperature region, $\Delta\mu$ can be estimated analytically[64] and iteratively refined by MC simulations. For temperatures above T_c, $\Delta\mu$ can be estimated by assuming an ideal alloy as a first approximation, but again reliable values can only be obtained by iterative Monte Carlo simulations.

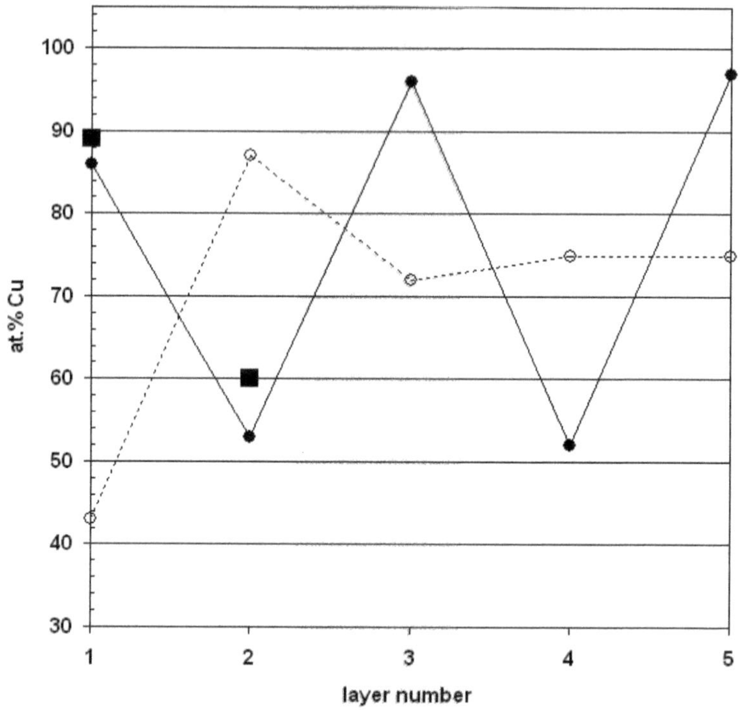

Figure 13. Concentration profiles for the $Cu_{75}Pd_{25}(110)$ alloy obtained at 600 K. The open circles (○) with the dashed line show the simulation results with the global parameters by Foiles et al. (Table 3) while the closed circles (●) with the full line are those for the simulations with the specific parameters (Table 4). The closed squares (■) pertain to the experimental LEIS/LEED data[62] for comparison. Direct comparison is complicated by the 10% higher Cu content of the sample used in the experiments.

When repeating the simulations with the new, specific EAM parameters (Figure 13), an excellent agreement between the simulation results and the experimental data is now observed. The first layer is enriched in Cu and followed by bulk-like alternating Cu_{100} and $Cu_{50}Pd_{50}$ layers. In agreement with the LEIS/LEED experiments, we also found a second atom layer of nearly 50 at.% Cu, showing a substantial degree of order that explains the experimentally observed (2×1) LEED pattern.

One of the most important driving forces for segregation of Cu to the surface is the difference in surface energy. The calculated relative difference in surface energy between pure Cu and Pd is nearly correct: ~11.5% as obtained by EAM calculations compared to 10.5% as an experimental value for the "average" surfaces of the pure metals.[55] This difference in driving force can be understood when looking at the behavior of the embedding functions. The second derivative of the embedding function with respect to the electron density is indeed directly related to the surface energy: the larger

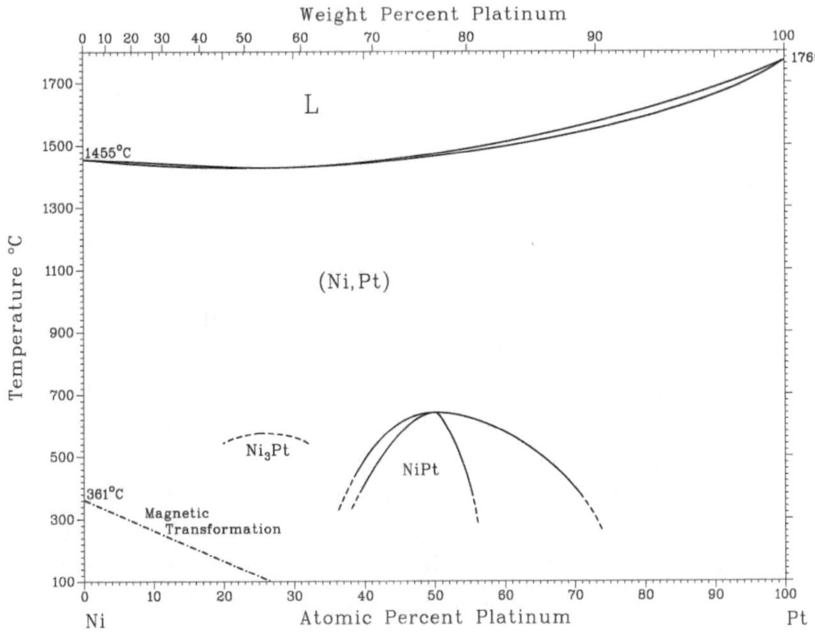

Figure 14. Phase diagram for the Pt-Ni system.[53]

the curvature the smaller the surface energy. The surface energy is the difference between the energy of an atom in the bulk and the same atom at the surface and in a first approximation equals half the energy in the missing bonds. However, due to the many-body effect, the strength of a bond depends on its environment. More in particular, the strength of a bond increases with decreasing coordination. A surface atom is thus subject to less, but stronger bonds than a bulk atom, leading to a contraction between the first and the second atomic layer. This effect in turn tends to lower the surface energy. The many-body interactions are accounted for in EAM because $F_i(\rho_{h,i})$ increases non-linearly with $\rho_{h,i}$. This relation features a positive curvature and the larger the curvature, the stronger the many-body effect and the smaller the surface energy.

5.3 Face-related segregation reversal at $Pt_{50}Ni_{50}$ surfaces

Pt-Ni alloys present a specific challenge because Pt and Ni are known to have catalytic properties for oxidation reactions and hydrogenation reactions respectively. In a Pt-Ni alloy, these properties can be enhanced. For instance, the (111) surface of these alloys can hydrogenate butadiene into butenes significantly better than does the pure Pt(111) surface.

Table 7. Relevant material properties with respect to surface segregation in the Pt-Ni system

	Pt	Ni
Heat of sublimation (kJ/mole)	565.3[54]	429.7[54]
Surface energy (J/m²)	2.475[23]	2.450[23]
	2.203[55]	2.380[55]
Crystal structure	fcc[56,57]	fcc[56,57]
Lattice parameter (Å)	3.924[56]	3.5238[56]
Atomic radius (Å)	1.38[57]	1.246[57]

Segregation in Pt-Ni alloys is somewhat special in that for the (111) and the (100) planes, a pronounced Pt segregation is observed, while the (110) surface is strongly enriched in Ni. This section concentrates on the surface configuration at the (110) surface of the $Pt_{50}Ni_{50}$ alloy that was not fully explained for a long time. The phase diagram for the Pt-Ni system is given in Figure 14.[53] Again, this system is characterized by moderately exothermic interactions, resulting in the formation of two ordered NiPt and Ni_3Pt compounds at lower temperatures. The Ni_3Pt compound has the well-known LI_2 structure, while NiPt exhibits the LI_0 structure. They are characterized by order-disorder transition temperatures of 645° C and 480° C respectively.[53] The other relevant material properties with respect to surface segregation are tabulated in Table 7.

To characterize and understand the catalytic behavior of Pt-Ni alloys, Pt-Ni surfaces have been extensively studied using X-ray Photoelectron Spectroscopy (XPS),[65,66] LEIS,[65,67-71] LEED,[72-77] Auger Electron Spectroscopy (AES)[65,68,69] and Incidence Dependent Excitation for Auger Spectroscopy (IDEAS).[78] These studies reveal a damped oscillating depth profile that is heavily dependent on both the bulk concentration and the surface orientation. As mentioned, at the (111) surface and the (100) surface a Pt enrichment is observed for a wide variety of bulk compositions.[65-69,71-75,78] For the (110) surface of $Pt_{10}Ni_{90}$ and $Pt_{50}Ni_{50}$, on the contrary, Ni segregates to the surface.[68,70,71,76-78] This section aims precisely at explaining the face related segregation behavior at the $Pt_{50}Ni_{50}$ (100), (110) and (111) surfaces. These surfaces have already been the subject of several attempts of modeling[79-81] but none of these former simulations could satisfactorily explain the Ni segregation to the (110) surface.

We first concentrate on the $Pt_{50}Ni_{50}(110)$ alloy at 1200 K. MC simulations with the Canonical Ensemble indicate a disordered bulk in agreement with the phase diagram.[53] It is again observed that only the MC/EAM simulations[35] with specific EAM parameters (Table 4) yield a correct characterization of the $Pt_{50}Ni_{50}(110)$ alloy. The transition temperature, $T_{c,exp}$ = 900 K, is much more accurately reproduced with the specific parameters ($T_{c,sp}$ = 875 K) than with the global parameterization

Table 8. The relative relaxations (in %) for the three $Pt_{50}Ni_{50}$ low-index surfaces from the MC/EAM simulations compared with the experimental values from LEED.[74,76,83]

	(100)		(110)		(111)	
	Simulation	Experiment	Simulation	Experiment	Simulation	Experiment
Δd_{12}	2 ± 0.5	4.6 ± 3	-14 ± 1	-19 ± 0.6	-1 ± 0.5	-2 ± 1
Δd_{23}	-3.5 ± 0.5	-9 ± 3	8 ± 1	10 ± 1	-0.8 ± 0.5	-2 ± 1

($T_{c,gl}$ = 720 K). The iterative procedure for disordered alloys yields a value of $\Delta\mu = 3.62 \cdot 10^{-19}$ J/atom at 1200 K. Figure 15 shows the results of the simulations for the (110) surface. The results based on the specific parameterization agree very well with the experiments: a damped oscillatory composition profile with Ni at the (110) surface, a contraction between the first and the second layer and an expansion between the second and third layer.

It is further shown that allowing a very large number of very small displacements of the individual atoms in the MC/EAM simulations can account for the effects of relaxations and atomic vibrations. This procedure yields results that agree well with a more classical treatment of vibrational entropy.[82] In the $Pt_{50}Ni_{50}(110)$ alloy, these vibrations, rather than relaxations between the atomic planes, have an attenuating influence on the segregation to the (110) surface. For the other low index surfaces, the effect of atomic vibrations is negligible. The resulting depth profiles for the (100), (110) and (111) surfaces are shown in Figure 16. The face related segregation reversal is clearly observed: whereas Pt segregates to the (100) and (111) surfaces, the (110) surface is enriched in Ni. All profiles are damped oscillatory, reflecting the preference for Pt-Ni bonds in this exothermic alloy. From the sixth layer on, the composition becomes constant at 50 at.% Ni. The simulation results are in rather good agreement with the experiments. The results for the relative interlayer relaxations are collected in Table 8 together with the experimental values from the LEED measurements.[74,76,83] These relaxations appear to be the most important for the (110) surface.

The face related reversal of the segregation observed in the experiments can be explained as follows. Ni has a surface energy that is only slightly less than that of Pt. Since the surface energy effect is relatively small, this effect is in close competition to the strain relaxation effect (size effect) promoting Pt-segregation. At the rougher (110) surface the difference in surface energy is somewhat more important[35,58] than at the other low index surfaces and causes Ni segregation. At the (100) and (111) surface this effect is overpowered by the size effect, promoting Pt segregation.

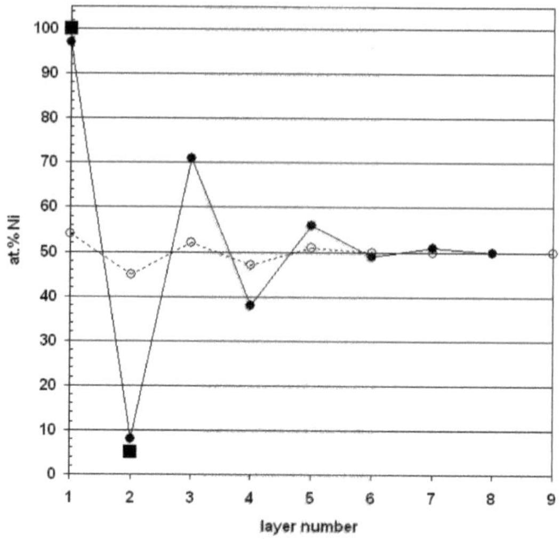

Figure 15. Concentration profiles for the $Pt_{50}Ni_{50}(110)$ alloy obtained at 1200 K. The open circles (○) with the dashed line show the simulation results with the global parameters by Foiles et al. (Table 3) while the closed circles (●) with the full line reflect the simulations with the specific parameters (Table 4). The closed squares (■) pertain to the experimental LEIS/LEED data[76] for comparison.

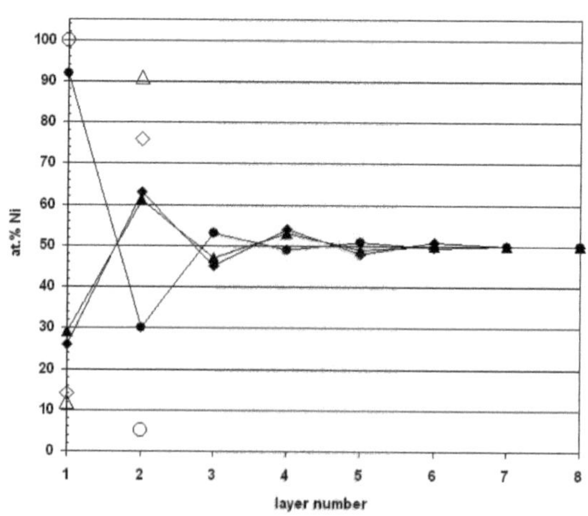

Figure 16. Concentration profiles for the $Pt_{50}Ni_{50}(100)$, (110) and (111) alloys obtained at 1200 K (◇,◆) (100); (○,●) (110); (△,▲) (111)). The closed symbols (◆,●,▲) with the full line show the simulation results with the specific parameters (Table 4). The open (◇,○,△) symbols reproduce the experimental LEIS/LEED data[76] for comparison.

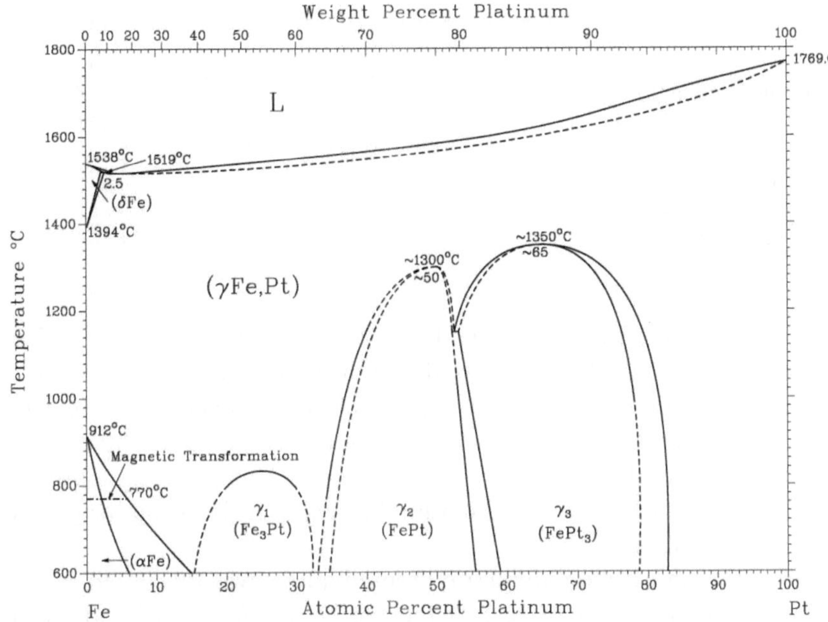

Figure 17. Phase diagram for the Pt-Fe system.[53]

5.4 Pt segregation to the (111) surface of ordered $Pt_{80}Fe_{20}$

The Pt-Fe alloy system exhibits pronounced catalytic activity in a narrow composition range around 80 at.% Pt.[84,85] In finely dispersed (4-5 nm) Pt-Fe/C form, this catalyst is used to speed up the hydrogenation of cinnamaldehyde.[84,86] Single crystals of the same composition have also proven to be catalytically active for the hydrogenation of crotonaldehyde and methylcrotonaldehyde.[87]

The phase diagram of the Pt-Fe system is given in Figure 17.[53] This system shows a variety of phases, among them Pt_3Fe and $PtFe_3$ with the $L1_2$ structure and PtFe with the $L1_0$ structure. The Pt_3Fe phase has a limited solubility for excess Pt atoms. The order-disorder transition temperature is maximum at 65% Pt and reaches a value of 1350 °C. Other relevant material properties are summarized in Table 9.

An experimental characterization of the $Pt_{80}Fe_{20}(111)$ alloy was performed by Beccat et al.[84] with LEED I/V at room temperature after annealing and equilibrating at high temperature up to 1200-1350 K. We studied the surface composition of the $Pt_{80}Fe_{20}(111)$ surface as a function of

Table 9. Relevant material properties with respect to surface segregation in the Pt-Fe system.

	Pt	Fe
Heat of sublimation (kJ/mole)	565.3[54]	398.5[54]
Surface energy (J/m²)	2.475[23]	2.475[23]
	2.203[55]	2.123[55]
Crystal structure	fcc[56,57]	bcc[56,57]
Lattice parameter (Å)	3.924[56]	2.861[56]
Atomic radius (Å)	1.38[57]	1.165[57]

temperature with LEIS.[12] From these experiments, it follows that Pt segregates up to 100 at.%, the second and third layer are also enriched in Pt and the segregation decreases with increasing temperature (exothermic segregation). The observed (2×2) LEED pattern is explained by a buckling of the surface caused by an ordered second layer. This buckling gives rise to two different Pt surface sites, which might very well be at the origin of the enhanced catalytic activity.[85]

Segregation in off-stoichiometric ordered alloys was discussed in section 2.2.2. It was pointed out there that the preference for A-B bonds leads to segregation of the majority component, in this case, Pt. For over-stoichiometric alloys, as is the case in $Pt_{80}Fe_{20}$, the excess Pt is accommodated on the Fe sublattice and this excess tends to segregate up to 100% Pt at the surface in an exothermic process. This is further enhanced by the fact that Pt features the lower surface energy, which again points at Pt as the segregating species. Finally, Pt atoms are larger than Fe, and the segregation of the 'compressed' excess Pt on the Fe sublattice allows for the relaxation of local elastic strain. In view of these arguments, the pronounced Pt segregation to 100%, only slightly decreasing at higher temperatures, is expected and perfectly normal. More refined and reliable results can be expected from the MEAM for the description of the energy of the alloy.[12] The MEAM has to be used since the *d* subshell of Fe is only half-filled so that EAM is no longer adequate to describe the $Pt_{80}Fe_{20}$ alloy.

Starting from either a disordered or an ordered bulk, bulk simulations show an ordered bulk at low temperature. The degree of disorder of course increases slightly with temperature. At higher temperatures, one grain boundary develops in the simulation slab, splitting it into two ordered domains. The ordered alloy under study is off-stoichiometric, where the sites of the Pt sublattice are occupied by Pt atoms, and the remaining excess Pt atoms are randomly distributed together with the Fe atoms over the sites of the Fe sublattice. MC/MEAM simulations (parameters from Ref. 36), taking into account surface relaxations, yield the composition and degree of ordering in the first three atomic layers near to the surface (Table 10): the

Table 10. Surface equilibrium for the $Pt_{80}Fe_{20}(111)$ ordered alloy as a function of temperature obtained by MC/MEAM simulations. The third and fourth columns show the Pt concentration and the amount of Pt that has been exchanged between the two sublattices in the first three layers, which is a measure for the long range order.

Temperature (K)	Pt concentration (at.%)		Disorder (%)
	Layer 1	100 ± 0.8	-
700	Layer 2	77.5 ± 0.8	5.6 ± 0.8
	Layer 3	82.1 ± 0.8	2.3 ± 0.8
	Layer 1	99.2 ± 0.8	0.4 ± 0.8
800	Layer 2	76.5 ± 0.8	6.3 ± 0.8
	Layer 3	80.3 ± 0.8	2.8 ± 0.8
	Layer 1	98.1 ± 0.8	0.8 ± 0.8
900	Layer 2	76.4 ± 0.8	7.6 ± 0.8
	Layer 3	80.6 ± 0.8	2.9 ± 0.8

first atomic layer is almost pure Pt and the depth profile is weakly oscillating, with a third layer concentration almost equal to the bulk concentration. The simulations yield a slightly decreasing Pt surface concentration with increasing temperature, as expected for a slightly exothermic process. The three-layer profile predicted by MEAM agrees with the LEED I/V conclusions within the experimental error ($C_1 = 96 \pm 4$ at.% Pt, $C_2 = 84 \pm 7$ at.% Pt and $C_3 = 85 \pm 15$ at.% Pt). Furthermore, this model predicts surface relaxation effects that are also in good agreement with the experimental LEED I/V evidence. It can be concluded that MEAM correctly accounts for all aspects of the surface segregation and relaxation in this alloy system.

5.5 Sn-segregation behavior and ordering at the alloy $Pt_{75}Sn_{25}(111)$

As a last example, the segregation behavior and ordering at various temperatures at the surface of the $Pt_{75}Sn_{25}(111)$ alloy is studied. Applications of Pt-Sn alloys are found in the catalysis of hydrocarbon reforming reactions[88-90] and the electrocatalysis of the direct oxidation of methanol in fuel cells.[91-97]

The phase diagram for the Pt-Sn system is presented in Figure 18.[53] This system is characterized by very strong attractive interactions, resulting in a wide variety of ordered compounds with no solubility for excess atoms. The ordered compounds exist up to the melting temperature. Other relevant properties with respect to surface segregation are tabulated in Table 11.

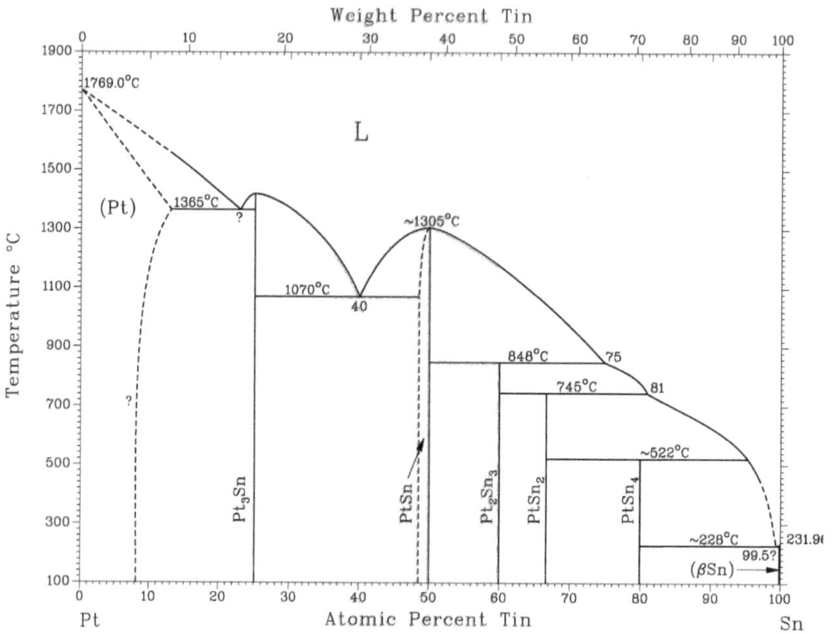

Figure 18. Phase diagram for the Pt-Sn system.[53]

The experimental starting point for the simulations is the work by Ceelen et al.[98] The composition, morphology and domain structure of the $Pt_{25}Sn_{75}(111)$ surface is studied as a function of temperature with LEIS, AES and Spot Profile Analysis (SPA) LEED. The experiments clearly show Sn depletion in the surface region after room temperature (RT) sputtering due to bombardment-induced segregation (BIS) and preferential sputtering. From 500 K on, Sn segregation is observed up to a maximum of 32 at.% in the outermost layer around 750 K and a $(\sqrt{3}\times\sqrt{3})R30°$ reconstruction develops. Along with these phenomena, the second layer is strongly Sn-depleted (< 2 at.%). A further increase up to 1200 K leads to the Sn replenishment of the subsurface up to the bulk value and to an outermost layer where the Sn percentage declines. The LEED pattern transforms into a (2×2) structure, corresponding to a truncated bulk situation at the surface.[99] Annealing beyond 1200 K, shows a slight reversible increase in Sn concentration with increasing temperature. During the cooling trajectory the Sn content in the outermost layer decreases to about 27 at.% ((2×2) periodicity) at 900 K and remains constant below this temperature. The simulation slab (18 layers of 288 atoms each) is chosen sufficiently large to minimize unrealistic changes in bulk composition. The simulations are carried out starting from two totally different initial configurations. The first one (further referred to as

Table 11. Relevant material properties for surface segregation in the Pt-Sn system.

	Pt	Sn
Heat of sublimation (kJ/mole)	565.3[54]	301.2[54]
Surface energy (J/m²)	2.475[23]	0.675[23]
	2.203[55]	0.661[55]
Crystal structure	fcc[56,57]	tetragonal[56,57]
Lattice parameter (Å)	3.924[56]	a = 5.8316[56]
		b = 3.1813[56]
Atomic radius (Å)	1.38[57]	1.405[57]

configuration A) consists of a surface layer with (√3×√3)R30° order (33 at.% Sn), a second layer of pure Pt and underneath the layers 3-18 with bulk composition and (2×2) order, which reproduces the experimental observed situation after sample preparation. The second configuration (B) starts from 18 unperturbed ordered bulk layers. The simulations are carried out starting from initial configuration A and swapping between the two outermost layers only, or using starting point B and swapping between all layers. The results (Figure 19) show that the starting situation A is preserved at temperatures around 750 K: the first layer contains about 33 at.% Sn and features a (√3×√3)R30° superstructure, while the second layer consists of nearly pure Pt. Starting from the second situation, there is a small Sn segregation to around 29 at.% Sn and a small increase at higher temperatures. Independent from temperature, a (2×2) order is observed. The behavior of the (√3×√3)R30° structure in the outermost layer at 750 K is investigated in more detail. Starting from different compositional combinations of the first and second layer in a random configuration, local equilibrium in the two topmost layers is established and the degree of (√3×√3)R30° order is measured. The simulations (Figure 20) clearly show a narrow region where the (√3×√3)R30° periodicity prevails. The long range order in excess of 99% is found around a straight line where the total initial Sn content in the first two layers amounts to 33 at.%. Even for small deviations from this region, the order already drops significantly. This proves the strong tendency to form a (√3×√3)R30° superstructure on top of an almost pure Pt second layer, even when starting from different and randomly chosen starting situations.

Following the temperature trajectory as indicated Figure 19, we can distinguish five regions, each with their own physical interpretation. At temperatures below 500 K (region 1), the atoms are not mobile. The surface and subsurface are Sn-depleted due to preferential sputtering and bombardment induced segregation of Sn. For temperatures between 600 K and 750 K (region 2), the atoms start to become mobile, first at the surface and this leads to a local 2-layer equilibrium. A complete Sn-exchange from the second to the surface layer creates a surface concentration of 33 at.% Sn on top of an almost pure Pt second layer. The atoms in the first layer are arranged in a (√3×√3)R30° superstructure and, together with the pure Pt second layer, this is a situation with a maximum number of Pt-Sn bonds, favored by the negative heat of solution. We also demonstrated that this configuration is very stable but well confined to a small region. Therefore, we conclude that the (√3×√3)R30° formation is an artifact rather than an intrinsic property of the Pt_3Sn catalyst. It is a consequence of the sample preparation (preferential sputtering of Sn and bombardment induced segregation) and the limited atomic mobility that, at moderate temperatures, is restricted to the first two layers. In region 3, at temperatures between 800 K en 1000 K, the local two-layer equilibrium evolves to a full equilibrium between bulk and surface. The additional Sn supply from the bulk to the second layer of pure Pt disturbs the surface ordering and consequently the optimum number of Pt-Sn bonds between the first two layers. As a result, the

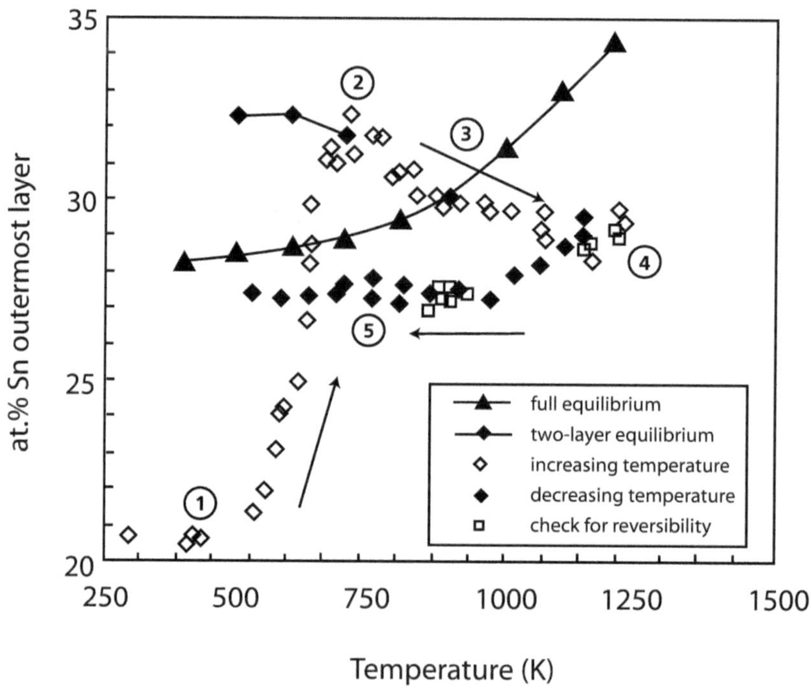

Figure 19. Comparison of the simulations with the experimentally[98] observed Sn concentration in the outermost layer of the $Pt_{75}Sn_{25}(111)$ alloy as a function of the anneal temperature and history. The closed circles show the results for the increasing temperature trajectory while the open circles are those for the decreasing trajectory. The open squares at 900 K and 1100 K pertain to the measurements performed to check the reversibility of the Sn segregation at high temperatures. Results of the calculations for the sputter-depleted surface where swaps are allowed between the two outermost layers only are indicated with a triangle, results for the full equilibrium with a stoichiometric bulk are shown as diamonds.

Sn content in layer 1 drops and the (√3×√3)R30° order is disrupted and is finally transformed into a (2×2) order. In such an ordered alloy, segregation is endothermic and this is confirmed by the tendency towards increasing Sn surface concentrations at the highest temperatures (region 4). This is the only temperature region where a fully reversible variation of the Sn surface concentration with the temperature was observed. For decreasing temperatures below 1000 K (region 5), the Sn concentration remains constant, while the (2×2) LEED pattern is conserved. In this region the equilibrium situation at 1000 K is frozen up due to insufficient atomic mobility.

The MC/MAM simulations are shown to reproduce the experimental results very well. Once again, the confrontation between simulations and experiment proves extremely useful for the interpretation of the experimental data and helps to understand the physical phenomena behind the experiments.

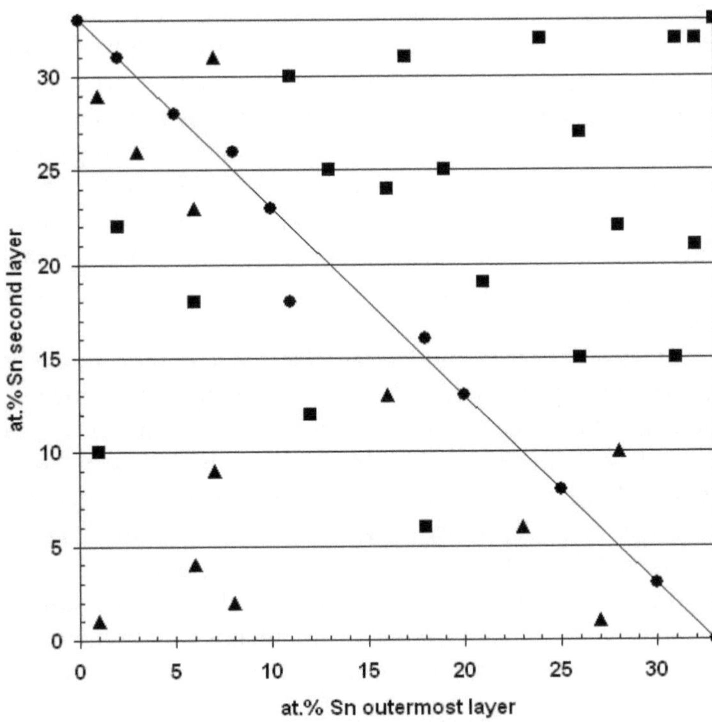

Figure 20. $(\sqrt{3}\times\sqrt{3})R30°$ order in the outermost layer at 750 K as a function of the composition in the outermost layer and the second one. A degree of $(\sqrt{3}\times\sqrt{3})R30°$ order of more than 99% is indicated by circles, $(\sqrt{3}\times\sqrt{3})R30°$ order between 80% and 99% is shown as triangles and a $(\sqrt{3}\times\sqrt{3})R30°$ order of less than 80% is indicated by squares.

6. CONCLUSIONS

This chapter tried to emphasize the synergism between surface science experiments and model calculations on alloys who are both governed by the same material properties, which in turn are also reflected in the bulk phase diagram. A few basic thermodynamical principles show how both alloy formation and surface segregation are basically determined by the same interatomic interactions.

Often valid conclusions can already be drawn from reasoning in terms of these fundamental interactions. Four distinct driving forces for segregation are recognized. When assuming constant bond energies a simple quasi-chemical approximation allows to formulate the resulting equilibrium in terms of quasi-chemical reactions and the matching equilibrium constant equations.

If the interatomic interactions lead to rather delicate balances, this simple approach is clearly insufficient. More elaborate energy models are then necessary that describe the interatomic interactions in a more subtle way, taking into account the dependence of bond strengths with the distance to the (second-) nearest-neighbors and with the detailed atomic surroundings of the atoms in question. Precisely the non-linearity aspects in this energy formulation allow for a correct description of surface segregation, surface relaxations and surface reconstruction processes. Several semi-empirical energy models are mentioned: EAM, MEAM and BFS that are all firmly based on quantum mechanical principles, yet are also soundly anchored to well-established values of material properties, obtained from experiments or DFT calculations.

These 'advanced' energy models are far too complex to be combined with a simple analytical energy equation as mentioned above for CBE. Furthermore, and far more important, more subtle, 'shaded', 'gray-scale' situations in the configurational entropy must be allowed for. This is precisely what Monte Carlo-simulations make possible: the detailed atomic stacking emerges naturally from these simulations and the configurational entropy is rendered correctly, with all intermediate possibilities in long- and short-range order between, on the one hand, ideally random solid solutions and on the other hand, perfectly ordered intermetallic compounds. The body of this chapter therefore presents an overall simulation procedure for Monte Carlo simulations combined with the (M)EAM, either in the Canonical Ensemble or in the Grand Canonical Ensemble. In the latter case, the difference in chemical potential $\Delta\mu(T)$ between the components is used to determine the constant bulk composition.

The power of MC/(M)EAM simulations is next illustrated by applying this operating procedure to five different alloy systems: $Au_{75}Pd_{25}(110)$,

$Cu_{75}Pd_{25}(110)$, $Pt_{50}Ni_{50}(100)$, (110) and (111), $Pt_{80}Fe_{20}(111)$ and $Pt_{75}Sn_{25}(111)$, each with a different aspect or special facet of (M)EAM. In each case study, the experimentally observed equilibrium surface configurations are compared to the modeled equilibrium situations of minimal Gibbs free energy. In all the investigated cases, and provided the parameters are determined with sufficient accuracy (based on known material properties), the simulations predict and describe how the surface jointly deploys all the possible compositional and structural rearrangements in order to achieve the equilibrium state of minimal Gibbs free energy. In this way it becomes possible to interpret these results in depth and to gain an improved insight into the sometimes delicate balances between the different driving forces for the investigated surface phenomena.

REFERENCES

1. R.A. Swalin, *Thermodynamics of solids*, 2nd edition, Wiley (1972).
2. F.L. Williams and D. Nason, *Surf. Sci.* **45**, 377 (1974).
3. R.A. Van Santen and W.M.H. Sachtler, *J. Catal.* **33**, 202 (1974).
4. W.M.H. Sachtler and R.A. Van Santen, *Appl. Surf. Sci.* **3**, 121 (1979).
5. M. Methfessel, D. Hennig, M. Scheffler, *Phys. Rev. B* **46**, 4816 (1992).
6. Brejnak M., Modrak P., *Surf. Sci.* **247**, 215 (1991).
7. D. McLean, *Grain boundaries in metals*, Clarendon, Oxford, 1957.
8. N. Sano, T. Sakurai, *J. Vac. Sci. Technol. A* **8**, 3421 (1990).
9. B. Sundman and J. Ågren, *J. Phys. Chem. Solids* **42**, 297 (1981).
10. B. Sundman and J. Ågren, *Mat. Res. Soc. Symp. Proc.* **19**, 115 (1983).
11. B. Sundman, *Anal. Fis. B* **86**, 69 (1990).
12. C. Creemers and P. Deurinck, *Surf. Interface Anal.* **25**, 177 (1997).
13. A.V. Ruban, *Phys. Rev. B* **65**, 174201 (2002).
14. M.E.J. Newman and G.T. Barkema, *Monte Carlo Methods in Statistical Physics*, Oxford University Press (1998).
15. F.C. Nix and W. Schockley, *Rev. Mod. Phys.* **10**, 2 (1938).
16. J. M. Cowley, *Phys. Rev.* **77**, 669 (1950).
17. G. Bozzolo, J. Ferrante, R.D. Noebe, B. Good, F.S. Honecy and P. Abel, *Comp. Mat. Sci.* **15**, 169 (1999).
18. N. Metropolis, A.W. Rosenbluth, M.N. Rosenbluth. A.H. Teller and E. Teller, *J. Chem. Phys.* **21**, 1087 (1953).
19. K. Binder, in: K. Binder (Ed.), *Topics in Current Physics*, vol.7, Springer-Verlag, Berlin, 1979, 1.
20. C. Creemers, P. Deurinck, S. Helfensteyn and J. Luyten, *Appl. Surf. Sci.* **219**, 11 (2003).
21. F. Ercolossi, M. Parrinello and E. Tosatti, *Phil. Mag. A* **58**, 213 (1988).
22. M.S. Daw and M.I. Baskes, *Phys. Rev. B* **29**, 6443 (1984).
23. F.R. de Boer, R. Boom, W.C.M. Mattens, A.R. Miedema and A.K. Niessen, F.R. de Boer and D.G. Pettifor (Eds.), *Cohesion and structure* Vol. 1, North-Holland, Amsterdam, 1988, 1.
24. M.S. Daw, S.M. Foiles and M.I. Baskes, *Mat. Sci. Rep.* **9**, 251 (1999).
25. S.M. Foiles, M.I. Baskes and M.S. Daw, *Phys. Rev. B* **33**, 7983 (1986).

26. P. Hohenberg and W. Kohn, *Phys. Rev. B* **136**, 864 (1964).
27. A.F. Voter, S.P. Chen, R.C. Albert, A.M. Boring and P.J. Hay, V. Vitek and D.J. Srolovitz (Eds.), *Atomistic Simulation of Materials: Beyond Pair Potentials*, Plenum, New York, 1989, p. 223.
28. R.A. Johnson, *Phys. Rev. B* **37**, 3924 (1988).
29. M. W. Finnis and J. E. Sinclair, *Phil. Mag. A* **50**, 45 (1984).
30. F. Ercolessi, E. Tosatti and M. Parrinello, *Phys. Rev. Lett.* **57**, 719 (1986).
31. E. Clementi and C. Roetti, At. *Data Nucl. Data Tables* **14**, 177 (1974).
32. A.D. McLean and R.S. McLean, At. *Data Nucl. Data Tables* **26**, 197 (1981).
33. J.H. Rose, J.R. Smith, F. Guinea and J. Ferrante, *Phys. Rev. B* **29**, 2963 (1984).
34. P. Deurinck and C. Creemers, *Surf. Sci.* **419**, 62 (1998).
35. P. Deurinck and C. Creemers, *Surf. Sci.* **441**, 493 (1999).
36. M. I. Baskes, *Phys. Rev. B* **46**, 2727 (1992).
37. M. I. Baskes, *Mater. Sci. Eng. A* **261**, 165 (1999).
38. G. Wang, M. A. Van Hove, P.N. Ross and M. I. Baskes, *J. Chem. Phys.* **121**, 5410 (2004).
39. M.I. Baskes, *Mat. Chem. Phys.* **50**, 152 (1997).
40. M.I. Baskes, J.E. Angelo and C.L. Bisson, *Modell. Sim. Mater. Sci. Eng.* **2**, 505 (1994).
41. P. van Beurden and G.J. Kramer, *Phys. Rev. B* **63**, 165106 (2001).
42. M.I. Baskes, *Mater. Chem. Phys.* **50**, 152 (1997).
43. M.I. Baskes, M. Asta and S.G. Srinivasan, *Phil. Mag. A* **81**, 991 (2001).
44. B.-J. Lee, M.I. Baskes, H. Kim and Y.K. Cho, *Phys. Rev. B* **64**, 184102 (2001).
45. B.-J. Lee, J.H. Shim and M.I. Baskes, *Phys. Rev. B* **68**, 144112 (2003).
46. A.F. Voter and S.P. Chen, in *Characterization of defects in materials*, MRS Symposia Proceedings No. 82, edited by R.W. Siegal, J.R. Weertman, and R. Sinclair, Materials Research Society, Pittsburgh, 1987.
47. M.S. Daw, *Phys. Rev. B* **39**, 7441 (1989).
48. M.I. Baskes, J.S. Nelson, and A.F. Wright, *Phys. Rev. B* **40**, 6085 (1989).
49. M.I. Baskes and R.A. Johnson, *Mod. Simul. Mater. Sci. Eng.* **2**, 147 (1994).
50. R.L. Moss and L. Whalley, *Advances in Catalysis*, Vol. 22, Academic Press, New York, 1983.
51. J.H. Sinfelt, *Bimetallic Catalysts: Discoveries, Concepts and Applications* (Wiley, New York, 1983).
52. F. Besenbacher, I. Chorkendorff, B.S. Clauesn, B. Hammer, A.M. Molenbroek, J.K. Nørskov and I. Stengaard, *Science* **279**, 1913 (1998).
53. T.B. Massalski, H. Okamoto, P.R. Subramanian and L. Kacprzak, *Binary Alloy Phase Diagrams*, American Society for Metals, Ohio, 1990.
54. D.R. Lide (Ed.), *Handbook of Chemistry and Physics*, CRC Press, Boca Raton, 1992.
55. W.R. Tyson and W.A. Miller, *Surf. Sci.* **62**, 267 (1977).
56. W.B. Pearson, *A Handbook of Lattice Spacings and Structure of Metals and Alloys*, Pergamon, New York, 1958.
57. J. Emsley, *The Elements*, Clarendon Press, Oxford, 1991.
58. R. Hultgren, P.A. Desai, D.T. Hawkins, M. Gleiser and K.K. Kelly, *Selected Values of the Thermodynamic Properties of Binary Alloys*, American Society for Metals, Metals Park, OH, 1973.
59. J. Kuntze, S. Speller, W. Heiland, P. Deurinck, C. Creemers, A. Atrei and U. Bardi, *Phys. Rev. B* **60**, 1 (1999).
60. M.A. Newton, S.M. Francis, Y. Li, D. Law and M. Bowker, *Surf. Sci.* **259**, 45 (1991).
61. A. Newton and M. Bowker, *Surf. Sci.* **307-309**, 445 (1994).

62. R.H. Bergmans, M. van de Grift, A.W. Denier van der Gon and H.H. Brongersma, *Surf. Sci.* **345**, 303 (1996).
63. D.J. Holmes, D.A. King and C.J. Barnes, *Surf. Sci.* **227**, 179 (1990).
64. S.M. Foiles and M.S. Daw, *J. Mater. Res.* **2**, 5 (1987).
65. Bertolini, J. Massardier, P. Delichere, B. Tardy and B. Imelik, *Surf. Sci.* **119**, 95 (1982).
66. A. Pantforder, J. Skonieczny, E. Janssen, G. Meister, A. Goldmann and P. Varga, *Surf. Sci.* **331**, 824 (1995).
67. E. van de Riet, S. Deckers, F.H.P.M. Habraken and A. Niehaus, *Surf. Sci.* **243**, 49 (1991).
68. P. Weigand, B. Jelinek, W. Hofer and P. Varga, *Surf. Sci.* **301**, 306 (1994).
69. P. Weigand, W. Hofer and P. Varga, *Surf. Sci.* **287**, 350 (1993).
70. Weigand, B. Jelinek, W. Hofer and P. Varga, *Surf. Sci.* **295**, 57 (1993).
71. P. Weigand, P. Novacek, G. van Husen, T. Neidhart and P. Varga, *Surf. Sci.* **269**, 1129 (1992).
72. Y. Gauthier, R. Baudoing, Y. Joly and J. Rundgren, *Surf. Sci.* **162**, 342 (1985).
73. Y. Gauthier, W. Hoffmann and M. Wuttig, *Surf. Sci.* **233**, 239 (1990).
74. Y. Gauthier, Y. Joly, R. Baudoing and J. Rundgren, *Phys. Rev. B* **31**, 6216 (1985).
75. R. Baudoing, Y. Gauthier, M. Lundgren J. Rundgren, *Solid State Phys.* **19**, 2825 (1986).
76. Y. Gauthier, R. Baudoing, M. Lundgren and J. Lundgren, *Phys. Rev. B* **35**, 7867 (1987).
77. Y. Gauthier, R. Baudoing and J. Jupille, *Phys. Rev. B* **40**, 1500 (1989).
78. D. Dufayard, R. Baudoing and Y. Gauthier, *Surf. Sci.* **233**, 223 (1990).
79. H. Stadler, W. Hofer, M. Schmid and P. Varga, *Surf. Sci.* **287**, 366 (1993).
80. M. Lundgren, *Phys. Rev. B* **36**, 4692 (1987).
81. G. Tréglia, B. Legrand, F. Ducastelle, *Phys. Rev. B* **41**, 4422 (1990).
82. R. Najafabadi and D.J. Srolovitz, *Surf. Sci.* **286**, 104 (1996).
83. S.M. Foiles, in: P.A. Dowben and A. Miller (Eds.), *Surface Segregation Phenonomena*, CRC Press, Boca Raton, FL, 1990.
84. P. Beccat, Y. Gauthier, R. Baudoing-Savois and J.C. Bertolini, *Surf. Sci.* **238**, 105 (1990).
85. R. Baudoing-Savois, Y. Gauthier and W. Moritz, *Phys. Rev. B* **44**, 12977 (1991).
86. D. Goupil, P. Fouilloux and R. Maurel, *React. Kinet. Catal. Lett.* **35**, 185 (1987).
87. P. Beccat, J.C. Bertolini, Y. Gauthier, J. Massardier, P. Ruiz, *J. Catal.* **126**, 451 (1990).
88. B.H. Davis, *J. Catal.* **46**, 348 (1977).
89. F.M. Dautzenberg, J.N. Helle, P. Biloen and W.M.H. Sachtler, *J. Catal.* **63**, 119 (1980).
90. Z. Karpinski and J.K.A. Clarke, *J. Chem. Soc. Faraday Trans.* **75**, 893 (1975).
91. K.J. Cathro, *J. Electrochem. Soc.* **116**, 1608 (1969).
92. M.M.P. Janssen and J. Moolhuysen, *J. Catalysis* **46**, 289 (1977).
93. B. Beden, F. Kardigan, C. Lamy and J.M. Leger, *J. Electroanal. Chem.* **127**, 75 (1981).
94. A.N. Haner and P.N. Ross, *J. Phys. Chem.* **95**, 3740 (1991).
95. S.A. Campbell and R. Parsons, *J. Chem. Soc. Faraday Trans.* **88**, 833 (1992).
96. K. Wang, H.A. Gasteiger, N.M. Markovic and P.N. Ross Jr, *Electrochim. Acta* **41**, 2587 (1996).
97. C. Panja, N. Saliba and B.E. Koel, *Surf. Sci.* **395**, 248 (1998).
98. W.C.A.N. Ceelen, A.W. Denier van der Gon, M.A. Reijme, H.H. Brongersma, I. Spolveri, A. Atrei and U. Bardi, *Surf. Sci.* **406**, 264 (1998).
99. A. Atrei, U. Bardi, M. Torrini, E. Zanazzi, G. Rovida, H. Kasamura and M. Kudo, *J. Phys. Condens. Matter* **5**, L207 (1993).

Chapter 6

INTEGRATION OF FIRST-PRINCIPLES CALCULATIONS, CALPHAD MODELING, AND PHASE-FIELD SIMULATIONS

Zi-Kui Liu and Long-Qing Chen
The Pennsylvania State University, University Park, PA 16802, USA

Abstract: Computational methods are playing an increasingly important role in materials science and engineering. In this chapter, our recent activities are presented in relation to an integrated framework for multi-scale materials simulation and design. We focus on microstructure evolutions of Ni-base superalloys starting from the generation of materials data needed for such simulations and their integration in terms of information flow and data processing. The first-principles calculations of free energy at finite temperatures, free energy and lattice parameters of solution phases, and interfacial energy are presented. The CALPHAD modeling, using both experimental and first-principles calculations, is discussed for free energy, atomic mobility, and molar volume. Using the thermodynamic and kinetic data thus generated, results from 2D and 3D phase-field simulations are reviewed. It is demonstrated that the integration of first-principles calculations, CALPHAD modeling, and phase-field simulations provides a powerful approach in materials research and development.

Keywords: First-principles calculations, CALPHAD modeling, phase-field simulations, Ni-base superalloys

1. INTRODUCTION

Traditionally, the field of materials science and engineering is predominantly focused on establishing the relationships among processing, microstructure, properties, and performance through large-scale experimental investigations. This traditional and often highly empirical approach is increasingly shifting towards the design of materials to achieve optimal functionality, driven largely by advances in information technology and computational materials science. Today, we are witnessing a paradigm shift for materials research and development from experimental based investigations to integrated computational-prediction and experimental-validation approaches. In materials research and development the bottleneck in terms of cost and time is the prototype design and test. The controllable variables are materials chemistry and processing route and parameters. There are two options for improvement: one is to reduce the number of prototype designs by optimal selection of materials chemistry and processing parameters, and the other is to lower the cost and shorten the time of the prototype test, or even better, the combination of both options.

Based on advances in information technology and computational tools, these two options can be broadly categorized as material design and materials simulations and can be significantly enhanced simultaneously with the new paradigm through computational based approaches as materials simulations can provide a more complete knowledge base needed for materials design, hopefully with lower cost and shorter time than experimental investigations alone.

Recently, we developed an integrated framework for multi-scale materials simulation and design,[1] named MatCASE (Materials Computation and Simulation Environments), which involves four major computational steps: (1) Atomic-scale first-principles calculations to predict thermodynamic properties, lattice parameters, and kinetic data of unary, binary and ternary compounds and solutions phases; (2) CALPHAD data optimization approach to model thermodynamic properties, lattice parameters, and kinetic data of multicomponent systems; (3) Multicomponent phase-field approach to predict the evolution of microstructures in one to three dimensions; and (4) Finite element analysis to generate the mechanical response from simulated microstructures. These four stages are integrated with advanced discretization and parallel algorithms and a software architecture for distributed computing systems. In the present chapter, we focus on materials simulations as a function of controllable variables. Specifically, we will address microstructure evolutions of Ni-base superalloys starting from the generation of materials data needed for such simulations and their integration in terms of information flow and data processing. The following four sections are constructed: Section 2. Phase-field simulation principles; Section 3. CALPHAD modeling of materials properties; Section 4. First-principles calculations of materials properties, and Section 5. Application to Ni-Al and Ni-Al-Mo systems.

2. PHASE-FIELD SIMULATION PRINCIPLES

In a phase-field model, a microstructure is described using one or more physical and/or artificial field variables.[2] These variables are uniform inside a phase or domain away from the interfaces. The same phase or the same types of domains have the same values for the field variables. Different values of the field variables, for example, 0 and 1, distinguish different phases or domains. Across the interfaces between different phases or domains, the field variables vary continuously from one uniform value corresponding to one type of phase or domain to another corresponding to another phase or domain. Therefore, the interfaces in a phase-field model are diffuse and possess a certain thickness.

Field variables can be either conserved or non-conserved, depending on if they satisfy the local conservation law, $\partial\phi/\partial t = -\nabla \bullet \boldsymbol{J}$ where ϕ is a field variable and $\boldsymbol{J}$ is the corresponding flux. For example, composition and temperature fields are both conserved while long-range order parameter fields describing ordered domain structures are non-conserved. It is easy to understand that the artificial phase-field in solidification modeling of a single-component liquid is non-conserved since its value can go from 0 to 1 for the whole system.

Artificial fields are introduced for the sole purpose of avoiding tracking the interfaces. Essentially all phase-field models of solidification employ an artificial field called the "phase field". The interfacial width described by artificial fields has no directional relationship to the physical width of a real interface. The thermodynamic and kinetic coefficients in the phase-field equations are chosen to match the corresponding parameters in the conventional sharp-interface equations through sharp- or thin-interface analyses.[3-7] Physical fields refer to well-defined order parameter fields which can be experimentally measured. The interfacial width described by a physical field is also expected to reflect the actual interfacial width. In phenomenological theories of phase transformations, order parameter fields are used to characterize the nature and the critical temperatures of a phase transformation that produces the microstructure. A well known example is the long-range order parameter for order-disorder transformations. The corresponding order parameter field can be employed to describe the antiphase domain structure resulted from ordering. Another example is a composition field which describes the morphological evolution during phase separation either through nucleation and growth or spinodal decomposition or during precipitate coarsening.

In the phase-field approach, the thermodynamics of an inhomogeneous microstructure is described by the diffuse-interface theory of Cahn and Hilliard.[8] For a more general inhomogeneous system described by a set of conserved field c_i and a set of non-conserved field η_i, the total free energy can be written as

$$F = \int\left[f(c_1,c_2,...,c_n,\eta_1,\eta_2,...\eta_n) + \sum_{i=1}^{n}\alpha_i(\nabla c_i)^2 + \sum_{i=1}^{3}\sum_{j=1}^{3}\sum_{k=1}^{p}\beta_{ij}\nabla_i\eta_k\nabla_j\eta_k \right]d^3r \qquad (1)$$
$$+ \iint G(\boldsymbol{r}-\boldsymbol{r}')d^3rd^3r'$$

where f is the usual local free energy density as a function of all the field variables, a_i and β_{ij} are the gradient energy coefficients. The second double-integral represents the contributions from pair-wise, non-local, long-range interactions ($G(\boldsymbol{r}-\boldsymbol{r}')$) such as elastic, electrostatic, and magnetic interactions between volume elements d^3r and d^3r'.

The main difference among different phase-field models lies in the construction of f as a function of field variables. A simple and familiar system is a two-phase binary system described by a physical composition field. For the case of isothermal solidification of a binary liquid, an artificial field, ϕ, is introduced, in addition to the physical field, composition c. The artificial field, also called the phase field, is used to distinguish solid and liquid phases and to automatically take into account the boundary conditions at the interfaces. The local free energy as a function of composition c and ϕ can be expressed as,[9]

$$f(c,\phi) = h(\phi)f^S(c) + [1-h(\phi)]f^L(c) + wg(\phi) \qquad (2)$$

where $f^S(c)$ and $f^L(c)$ are the free energy densities of solid and liquid as a function of composition at a given temperature, respectively. $g(\phi)$ is a double-well potential which is only a function of the artificial field. A possible functional form for g is

$$g(\phi) = 16w\phi^2(\phi-1)^2 \qquad (3)$$

which has minina at $\phi = 0$ and $\phi=1$. w in Eq. 3 represents the depth of the two wells at $\phi=0$ and $\phi=1$ with respect to the local maximum at $\phi=½$. The function $h(\phi)$ in Eq. 2 is required to have the following properties:

$$h(0) = 0, h(1) = 1, \left. dh/d\phi\right|_{\phi=0} = \left. dh/d\phi\right|_{\phi=1} = 0 \qquad (4)$$

These properties ensure that the equilibrium values, 0 and 1, for the phase field in the double-well potential are not affected by the chemical free energies f^S and f^L. An example which satisfies the conditions given in Eq. 4 is

$$h(\phi)=\phi^3\left(6\phi^2-15\phi+10\right) \tag{5}$$

Therefore, in this model, ϕ = 0 represents the liquid phase since $h(0)$=0 and $f(c,0)=f^L(c)$. Similarly, ϕ = 1 describes the solid phase with $h(1)$=1 and $f(c,1)=f^S(c)$. Across the interface, the local free energy has contributions from both the liquid and solid chemical free energies as well as from the double-well potential.

Recently Kim et al.[5] proposed a variation of the above model (which they refer to as the WBM model), based on an earlier work by Steinbach et al.[10] They considered the interfacial region to be a mixture of solid and liquid phases with compositions c_S and c_L and with the same chemical potential, i.e. c_S and c_L satisfy the following set of equations for a binary system,

$$c=[1-h(\phi)]c_L+h(\phi)c_S \tag{6}$$

$$\frac{\partial f^L(c_L)}{\partial c_L}=\frac{\partial f^S(c_S)}{\partial c_S} \tag{7}$$

The main advantage of Kim's approach (called KKS model hereafter) compared to the WBM model is the fact that for an interface at equilibrium, there is no contribution of the actual chemical free energy, f^S and f^L, to the total interfacial energy, i.e. Δf is eliminated at equilibrium because of conditions (6) and (7). The interfacial energy and interfacial width are entirely determined by the double-well potential and the gradient energy coefficient in the artificial phase field. As a result, a larger interfacial width may be employed to fit the same interfacial energy, thus increasing the length scale of a phase-field simulation even with the usual numerical methods using uniform grids. However, it should be cautioned that the implementation of this model requires the numerical solution to the above coupled Eqs. 6 and 7 for c_L and c_S for each set of c and ϕ at each time step, and this process can be computationally very expensive. Furthermore, the depth of the double-well potential (w) cannot be made too small compared to the actual chemical driving force (described by f^S and f^L) for phase transformations during a phase-field simulation, and thus the interfacial width that one can use is also limited. Otherwise numerically instability may develop in a phase-field simulation, leading to incorrect path for the microstructure evolution.

For many solid-state phase transformations, the local free energy function can be expressed as a polynomial of order parameters using a conventional Landau-type of expansion since the field variables correspond to well-defined physical order parameters. All the terms in the expansion are required to be invariant with respect to the symmetry operations of the high-temperature phase. For example, for precipitation of an ordered phase ($L1_2$) from a disordered face-centered-cubic (fcc) matrix in a binary alloy, with expansion terms up to the fourth order, the local free energy function is given by[11-14]

$$f(c,\eta_1,\eta_2,\eta_3)=f_d(c,T)+\frac{1}{2}A_2(c,T)\left(\eta_1^2+\eta_2^2+\eta_3^2\right)+\frac{1}{3}A_3(c,T)\eta_1\eta_2\eta_3$$

$$+\frac{1}{4}A_{41}(c,T)\left(\eta_1^4+\eta_2^4+\eta_3^4\right)+\frac{1}{4}A_{42}(c,T)\left(\eta_1^2\eta_2^2+\eta_2^2\eta_3^2+\eta_1^2\eta_3^2\right) \tag{8}$$

where $f_d(c,T)$ is the free energy of the disordered phase at η_i=0, A_2, A_3, A_{41}, and A_{42} are the expansion coefficients which are functions of temperature and composition. The free energy function (Eq. 8) has four degenerate minima with respect to the order parameters. If $A_3(c,T)<0$, the free energy minima are located at

$$(\eta_0,\eta_0,\eta_0),(\eta_0,-\eta_0,-\eta_0),(-\eta_0,\eta_0,-\eta_0),(-\eta_0,-\eta_0,\eta_0), \tag{9}$$

where η_o is the equilibrium value for the long-range parameter at a given composition and temperature. The four sets of long-range order parameters given in Eq. 9 describe the four energetically equivalent antiphase domains of the $L1_2$ ordered phase related by a primitive lattice translation of the parent disordered fcc phase.

As mentioned above, in the diffuse-interface description, the free energy of an inhomogeneous system, such as a microstructure, also depends on the gradients of the field variables. The gradient energy coefficients which characterize the energy penalty due to the field inhomogeneity at the interfaces, i.e. the interfacial energy contribution to the total free energy. For a given free energy model and a given set of gradient energy coefficients, the specific interfacial energy (interfacial energy per unit area) can be calculated for an equilibrium interface. Analytical expressions for the interfacial energy in terms of free energy parameters and the gradient energy coefficients are only available for very simple cases that an analytical solution for the equilibrium profile of field variable across the interface can be derived. For more general cases, the interfacial energy has to be computed numerically.

The second integral in Eq. 1 represents the nonlocal contributions to the total free energy from long-range interactions such as elastic interactions, electric dipole-dipole interactions, and electrostatic interactions. These long-range interactions are usually obtained by solving the corresponding mechanical and electrostatic equilibrium equations for a given microstructure. For example, for the long-range elastic interactions, the following mechanical equilibrium equation has to be solved under given mechanical boundary conditions,

$$\frac{\partial \sigma_{ij}}{\partial r_j} = 0 \quad \text{with} \quad \sigma_{ij}(r) = \lambda_{ijkl}(r)\left[\varepsilon_{kl}(r) - \varepsilon_{kl}^{o}(c, \phi, \eta_i, \ldots)\right] \tag{10}$$

where σ_{ij} is the local elastic stress, r_j the jth component of the position vector, $\boldsymbol{r}$, $\lambda_{ijkl}(r)$ the elastic stiffness tensor which varies with space, $\varepsilon_{jk}(r)$ the total strain state at a given position in a microstructure, and ε_{kl}^{o} the local stress-free strain or transformation strain or eigenstrain which is also a function of position through its dependence on field variables. The resulted elastic energy is a function of phase-field variables and thus the microstructure.[15] Various approximations and different approaches have been proposed to solve the elasticity Eq. 10 with arbitrary distribution of eigenstrains, i.e. microstructure. For the case of homogeneous approximation and periodic boundary conditions, it was shown by Khachaturyan[16] that an analytical solution for the displacements, strains, and thus the strain energy could be obtained in the Fourier space. Therefore, in the case of homogeneous approximation, the elastic energy calculation does not incur any significant computation. For systems with small elastic homogeneity, first order approximations may be employed.[17-19] For large elastic inhomogeneities, first-order approximations are not sufficient and it is numerically more expensive to compute the elastic energy contributions. However, recently a number of approaches have been proposed for obtaining elastic solutions in systems with large elastic inhomogeneity.[20-23]

In all phase-field models, the temporal and spatial evolution of the field variables follows the same set of kinetic equations. All conserved fields, c_i, evolve with time according to the Cahn-Hilliard equation,[24] or simply the diffusion equation in the case that no gradient energy is introduced for the conserved variable, whereas the non-conserved fields, η_p, are governed by the Allen-Cahn equation,[25] i.e.

$$\frac{\partial c_i(\mathbf{r},t)}{\partial t} = \nabla\left[M_{ij}\nabla\frac{\delta F}{\delta c_j(\mathbf{r},t)}\right] \tag{11}$$

$$\frac{\partial \eta_p(\mathbf{r},t)}{\partial t} = -L_{pq} \frac{\delta F}{\delta \eta_q(\mathbf{r},t)} \tag{12}$$

where M_{ij} and L_{pq} are related to atom or interface mobility. F is the total free energy of a system which is a functional of all the relevant conserved and non-conserved fields given by Eq. 1.

In order to relate the phase-field parameters to the experimentally measurable thermodynamic and kinetic properties, one has to examine the phase-field equations in the sharp- and/or thin-interface limit. This is particularly true for phase-field models with artificial field variables for which the corresponding kinetic parameters are not directly related to the measurable physical properties. A sharp-interface analysis[3] matches the phase-field parameters at the limit of zero interfacial thickness to experimentally measured thermodynamic and kinetic properties while a thin-interface analysis[4,5,7] allows the variation of the phase-field variable over a certain thickness for the interface. It is shown by Karma[4] that a phase-field simulation using the thin-interface asymptotics permits one to use a larger interface width and thus larger grid size.

With all the input parameters that characterize the thermodynamics of a microstructure and the kinetic coefficients that enter to the evolution equations, the microstructure evolution of a system can be obtained by numerically solving the systems of Cahn-Hilliard diffusion equations and the Allen-Cahn relaxation equations subject to appropriate initial and boundary conditions. Most of the phase-field simulations employ the second-order finite-difference discretization in space using uniform grids and the forward Euler method for time stepping to solve the phase-field equations for simplicity. It is well known that in such an explicit scheme, the time step has to be small to keep the numerical solutions stable. Dramatic savings in computation time and improvement in numerical accuracy can be achieved by using more advanced numerical approaches such as the semi-implicit Fourier Spectral method,[26-27] and adaptive-grid finite-element method.[28-31]

3. CALPHAD MODELING OF MATERIALS PROPERTIES

The CALPHAD (CALculation of PHAse Diagrams) approach was originally developed for modeling of thermodynamic properties and has been extended to modeling of atomic mobility and molar volume, which are discussed in the present section. We are currently developing models for elastic constants in multicomponent systems.

3.1 CALPHAD modeling of thermodynamics

The history of the CALPHAD approach development up to 1998 was reviewed by Saunders and Miodownik.[32] This approach was pioneered in the middle of the 20th century by Kaufman who also coined the name of CALPHAD.[33] It is based on mathematically formulated models describing the thermodynamic properties of individual phases. The model parameters are evaluated from thermochemical data of the individual phases and phase equilibrium data between phases, as phase equilibria are a manifestation of the thermodynamic properties of the phases involved. More specifically, under typical experimental conditions of constant temperature and pressure, phase equilibrium is obtained by minimization of the Gibbs energy of a closed system.

The CALPHAD approach is particularly valuable in materials science and engineering in comparison with physics and chemistry due to more complicated systems studied such as multicomponent solution phases. It has produced reliable phase diagrams and stability maps for complicated multicomponent commercial alloys.[34,35] The CALPHAD approach begins with the evaluation of the thermodynamic descriptions of unary and binary systems. By combining the constitutive binary systems with ternary data, ternary interactions and Gibbs energy of ternary phases are obtained. The modeling of the Gibbs energy of individual phases and the coupling of phase diagram and thermochemistry are the keys to developing unambiguous thermodynamic descriptions of multi-component materials because these two sets of independently measured data are deduced from the Gibbs energy of individual phases under given constraints.

Models for the Gibbs energy are based on the crystal structures of the phases. For pure elements and stoichiometric compounds, the most commonly used model is the one suggested by the Scientific Group Thermodata Europe (SGTE)[36] and has the following form (for simplicity, the pressure dependence and the magnetic contribution are not shown here),

$$G_m - H_m^{SER} = a + bT + cT\ln T + \sum d_i T^i \tag{13}$$

The left hand side of the equation is defined as the Gibbs energy relative to a stable element reference state (SER) where H_m^{SER} is the enthalpy of the element in its stable state at 298.15 K. Coefficients, a, b, c, and d_i, are the model parameters. The SGTE data for pure elements have been compiled by Dinsdale.[37]

For multicomponent solution phases, the molar Gibbs energy has the following general formula,

$$G_m = {}^{ref}G_m + {}^{ideal}G_m + {}^{xs}G_m \tag{14}$$

where ${}^{ref}G_m$ represents the reference Gibbs energy of the phase, ${}^{ideal}G_m$ the ideal mixing contribution, and ${}^{xs}G_m$ the excess Gibbs energy of mixing due to non-ideal interactions. A variety of mathematical models have been developed for these three terms.[32] For metallic solutions, sublattice models developed by Hillert and co-workers have been widely used.[38,39] The phenomenological model can account for a broad range of phases by defining internal parameter relationships to reflect different crystal structures including order-disorder transformations. For example, in an *A-B* binary phase with two types of lattice sites and with both components entering two sublattices, the sublattice model is written as $(A,B)_a(A,B)_b$, where subscripts *a* and *b* denote the number of sites of each sublattice, respectively. The three terms in Eq. 14 are written as,

$${}^{ref}G_m = y_A^I y_A^{II}\,{}^0G_{A:A} + y_A^I y_B^{II}\,{}^0G_{A:B} + y_B^I y_A^{II}\,{}^0G_{B:A} + y_B^I y_B^{II}\,{}^0G_{B:B} \tag{15}$$

$${}^{ideal}G_m = aRT\left(y_A^I \ln y_A^I + y_B^I \ln y_B^I\right) + bRT\left(y_A^{II} \ln y_A^{II} + y_B^{II} \ln y_B^{II}\right) \tag{16}$$

$$\begin{aligned} {}^{xs}G_m = {} & y_A^I y_B^I\left(y_A^{II} \sum_{k=0} {}^kL_{A,B:A}(y_A^I - y_B^I)^k + y_B^{II} \sum_{k=0} {}^kL_{A,B:B}(y_A^I - y_B^I)^k \right) \\ & + y_A^{II} y_B^{II}\left(y_A^I \sum_{k=0} {}^kL_{A:A,B}(y_A^{II} - y_B^{II})^k + y_B^I \sum_{k=0} {}^kL_{B:A,B}(y_A^{II} - y_B^{II})^k \right) \\ & + y_A^I y_B^I y_A^{II} y_B^{II} \; L_{A,B:A,B} \end{aligned} \tag{17}$$

where y^I and y^{II} are the mole fractions of *A* or *B* in the first and second sublattices, respectively, commonly called site fractions. ${}^0G_{I:J}$ are the Gibbs energy of the compound I_aJ_b in the formula represented by Eq. 13. ${}^kL_{A,B:*}$ (${}^kL_{*:A,B}$) is the interaction parameter between component *A* and *B* in the first (second) sublattice, and $L_{A,B:A,B}$ the cross-interaction parameter. In this notation, a colon separates components occupying different sublattices, and a comma separates interacting components in the same sublattice. These equations can be generalized for phases with multicomponents and multisublattices,[39] and they reduce to a random substitutional model when all sublattices are the same. The site fractions are related to the regular mole fraction through the following equations:

$$x_A = \frac{ay_A^I + by_A^{II}}{a+b} \qquad x_B = \frac{ay_B^I + by_B^{II}}{a+b} \tag{18}$$

In principle, nearest neighbor bonds should be placed between sublattices. If this is not the case, such as the two-sublattice model for the $L1_2$ phase, there will be constraints among interaction parameters as discussed by Andersson et al.[39] and Huang and Chang[40] based on bond energies and by Ansara et al.[41,42] based on the constraint that the Gibbs energy should have an extreme at the disordered state. For the $L1_2$ phase, $(A,B)_3(A,B)_1$, those constraints for the regular and subregular interactions were derived by Ansara et al.[41,42] as

$$\begin{aligned}
&{}^oG_{A:B} = u_1 && {}^oG_{B:A} = u_2 \\
&{}^0L_{A,B:A} = 3u_1 + \frac{1}{2}u_2 + 3u_3 && {}^1L_{A,B:A} = 3u_4 \\
&{}^0L_{A,B:B} = \frac{1}{2}u_1 + 3u_2 + 3u_3 && {}^1L_{A,B:B} = 3u_5 \\
&{}^0L_{A:A,B} = \frac{1}{2}u_2 + u_3 && {}^1L_{A:A,B} = u_4 \\
&{}^0L_{B:A,B} = \frac{1}{2}u_1 + u_3 && {}^1L_{B:A,B} = u_5 \\
&{}^0L_{A,B:A,B} = 4u_4 - 4u_5
\end{aligned} \tag{19}$$

To separate the ordering contribution from the Gibbs energy of a disordered phase so the disordered phase can be modeled independently, Ansara et al.[42] splitted the Gibbs energy into three terms as follows:

$$G_m = {}^{dis}G_m(x_i) + {}^{ord}G_m(y_i^I, y_i^{II}) - {}^{ord}G_m(x_i) \tag{20}$$

with ${}^{dis}G_m(x_i)$ representing the Gibbs energy of the disordered phase equal to $\sum x_i {}^0G_i + RT\sum x_i \ln x_i + x_A x_B \sum_{k=0} {}^kL_{A,B}(x_A - x_B)^k$ and ${}^{ord}G_m(y_i^I, y_i^{II})$ the Gibbs energy of the ordered state as described by the sublattice model, and ${}^{ord}G_m(x_i)$ the contribution from the disordered state to the ordered phase. The last two terms cancel each other when the phase is disordered, i.e., the site fractions are equal in all sublattices and equal to mole fractions.

Considering subregular interactions in the disordered state, i.e., ${}^{dis-xs}G_m = x_A x_B[{}^0L + {}^1L(x_A - x_B)]$, the Gibbs energy of the ordered state becomes:

$$ {}^{o}G_{A:B} = u_1 + \frac{3}{16}\,{}^{0}L + \frac{6}{64}\,{}^{1}L \qquad {}^{o}G_{B:A} = u_2 + \frac{3}{16}\,{}^{0}L - \frac{6}{64}\,{}^{1}L $$

$$ {}^{0}L_{A,B:A} = 3u_1 + \frac{1}{2}u_2 + 3u_3 + \frac{9}{16}\,{}^{0}L + \frac{27}{64}\,{}^{1}L \qquad {}^{1}L_{A,B:A} = 3u_4 + \frac{27}{64}\,{}^{1}L $$

$$ {}^{0}L_{A,B:B} = \frac{1}{2}u_1 + 3u_2 + 3u_3 + \frac{9}{16}\,{}^{0}L - \frac{27}{64}\,{}^{1}L \qquad {}^{1}L_{A,B:B} = 3u_5 + \frac{27}{64}\,{}^{1}L $$

$$ {}^{0}L_{A:A,B} = \frac{1}{2}u_2 + u_3 + \frac{1}{16}\,{}^{0}L + \frac{9}{64}\,{}^{1}L \qquad {}^{1}L_{A:A,B} = u_4 + \frac{1}{64}\,{}^{1}L $$

$$ {}^{0}L_{B:A,B} = \frac{1}{2}u_1 + u_3 + \frac{1}{16}\,{}^{0}L - \frac{9}{64}\,{}^{1}L \qquad {}^{1}L_{B:A,B} = u_5 + \frac{1}{64}\,{}^{1}L $$

$$ {}^{0}L_{A,B:A,B} = 4u_4 - 4u_5 \tag{21} $$

The contributions due to the ${}^{0}L$ and ${}^{1}L$ of the disordered state in Eq. 21 have been directly programmed in Thermo-Calc.[43] It should be mentioned that when evaluating model parameters, it is often very difficult to find a set of model parameters for all individual phases to represent all the information in the system with high accuracy, particularly for systems with several solution phases. For practical applications phase stability represented by phase equilibria is usually more important than thermochemical data of individual phases.

3.2 CALPHAD modeling of atomic mobility

Diffusion coefficients are needed for simulating materials processing. The most basic diffusion coefficients are the tracer diffusion coefficients as the intrinsic and chemical diffusion coefficients can be calculated from tracer diffusivity when the thermodynamic model of the phase is available. The tracer diffusion coefficient (D_i^*) is related to the atomic mobility (M_i) by the Einstein equation

$$ D_i^* = RTM_i \tag{22} $$

with D_i^* expressed by $D_i^o = D_i^o \exp(-Q_i / RT)$, where D_i^o and Q_i are the prefactor and activation energy. The atomic mobility can be written as

$$M_i = \frac{D_i^*}{RT} = \frac{D_i^o}{RT}\exp(-\frac{Q_i}{RT}) = \frac{1}{RT}\exp\left[-\frac{1}{RT}(Q_i - RT\ln D_i^o)\right]$$

$$= \frac{1}{RT}\exp\left[-\frac{\Delta\Phi_i}{RT}\right] \tag{23}$$

with $\Delta\Phi_i = Q_i - RT\ln D_i^o$. Similar to the CALPHAD modeling of thermodynamics, Andersson and Ågren[44] developed the following model of atomic mobility as a function of compositions for simple phases, $(A,B,...)_a(C,Va\ ...)_b$ with the first and second sublattices being substitutional and interstitial, respectively,

$$\Delta\Phi_i = \sum_j\sum_m y_j^I y_m^{II}\Delta\Phi_i^{jm} + \sum_j\sum_{k>j}\sum_m y_j^I y_k^I y_m^{II}\Delta\Phi_i^{j,k:m} +$$
$$+ \sum_j\sum_n\sum_{m>n} y_j^I y_n^{II} y_m^{II}\Delta\Phi_i^{j:n,m} \tag{24}$$

$\Delta\Phi_i^{j:m}$ is the model parameter for species i when the first and second sublattices are occupied by species j and m, $\Delta\Phi_i^{j,k:m}$ the interaction parameter between j and k in the first sublattice when the second sublattice is occupied by m, $\Delta\Phi_i^{j:n,m}$ the interaction parameter between n and m in the second sublattice when the first sublattice is occupied by j. The atomic mobility can be used to calculate the intrinsic diffusivity in the lattice-fixed frame of reference and the interdiffusion coefficients in the volume-fixed frame of reference.[44]

To describe the effect of chemical ordering in a two-sublattice model discussed in the previous section, Helander and Ågren[45] suggested the following expression

$$\Delta\Phi_i = \Delta^{dis}\Phi_i + \Delta^{ord}\Phi_i \tag{25}$$

where $\Delta^{dis}\Phi_i$ and $\Delta^{ord}\Phi_i$ are the contributions from the disordered state and the chemical ordering, respectively, with the former defined by Eq. 24 and the latter by

$$\Delta^{ord}\Phi_i = \sum_j\sum_{k\neq j}\Delta^{ord}\Phi_i^{j:k}(y_j^I y_k^{II} - x_j x_k) \tag{26}$$

where $\Delta^{ord}\Phi_i^{j:k}$ represents the effect of chemical ordering of $j-k$ atoms on atomic mobility of element *i*.

3.3 CALPHAD modeling of molar volume

The molar volume of a phase can be calculated from its molar Gibbs energy as

$$V_m = \left(\frac{\partial G_m}{\partial P}\right)_{T,x_i \text{ or } y_i} \tag{27}$$

From Eq. (14), one can obtain

$$V_m = \frac{\partial^{ref} G_m}{\partial P} + \frac{\partial^{xs} G_m}{\partial P} \tag{28}$$

For the same sublattice model used in Section 3.1, i.e., $(A,B)_a(A,B)_b$, the equation can be written as

$$\begin{aligned} V_m = &\sum_i \sum_j y_i^I y_j^{II} {}^0V_{i:j} + y_A^I y_B^I \sum_i y_i^{II} \sum_{k=0} {}^kV_{A,B:i}(y_A^I - y_B^I)^k \\ &+ y_A^{II} y_B^{II} \sum_i y_i^I \sum_{k=0} {}^kV_{i:A,B}(y_A^{II} - y_B^{II})^k + y_A^I y_B^I y_A^{II} y_B^{II} \cdot V_{A,B:A,B} \end{aligned} \tag{29}$$

where ${}^0V_{i:j}$ is the molar volume of the compound $(i)_A(j)_B$, and ${}^kV_{A,B:i}$, ${}^kV_{i:A,B}$ and $V_{A,B:A,B}$ are interaction parameters in the first, second, and both sublattices, respectively, all of which could be temperature and pressure dependent.

Under ambient pressure the contribution to the Gibbs energy due to the volume change is very small and often neglected. Consequently, molar volume models can be developed separately from the Gibbs energy, and only the temperature dependence needs to be considered for ${}^oV_{i:j}$ and the interaction parameters in Eq. 29. For the $L1_2$ phase modeled with two sublattices $(A,B)_3(A,B)$, constraints on interaction parameters would be inherited from the Gibbs energy model presented in Section 3.1.

$$
\begin{aligned}
&V_{A:B} = v_1 \qquad V_{B:A} = v_2 \\
&{}^0V_{A,B:A} = 3v_1 + \frac{1}{2}v_2 + 3v_3 \qquad {}^1V_{A,B:A} = 3v_4 \\
&{}^0V_{A,B:B} = \frac{1}{2}v_1 + 3v_2 + 3v_3 \qquad {}^1V_{A,B:B} = 3v_5 \\
&{}^0V_{A:A,B} = \frac{1}{2}v_2 + v_3 \qquad {}^1V_{A:A,B} = v_4 \\
&{}^0V_{B:A,B} = \frac{1}{2}v_1 + v_3 \qquad {}^1V_{B:A,B} = v_5 \\
&{}^0V_{A,B:A,B} = 4v_4 - 4v_5
\end{aligned}
\tag{30}
$$

It is evident that the splitting of Gibbs energy into the disordered and ordered contributions can be directly applied for the molar volume through the derivative of Eq. 20 with respect to pressure, i.e., $V_m = {}^{dis}V_m(x_i) + {}^{ord}V_m(y_i^I, y_i^{II}) - {}^{ord}V_m(x_i)$. Similarly, with subregular interactions in the disordered state, i.e., ${}^{dis-xs}V_m = x_A x_B[{}^0V + {}^1V(x_A - x_B)]$, the molar volume of the ordered state becomes:

$$
\begin{aligned}
&V_{A:B} = v_1 + \frac{3}{16}{}^0V + \frac{6}{64}{}^1V \qquad V_{B:A} = v_2 + \frac{3}{16}{}^0V - \frac{6}{64}{}^1V \\
&{}^0V_{A,B:A} = 3v_1 + \frac{1}{2}v_2 + 3v_3 + \frac{9}{16}{}^0V + \frac{27}{64}{}^1V \qquad {}^1V_{A,B:A} = 3v_4 + \frac{27}{64}{}^1V \\
&{}^0V_{A,B:B} = \frac{1}{2}v_1 + 3v_2 + 3v_3 + \frac{9}{16}{}^0V - \frac{27}{64}{}^1V \qquad {}^1V_{A,B:B} = 3v_5 + \frac{27}{64}{}^1V \\
&{}^0V_{A:A,B} = \frac{1}{2}v_2 + v_3 + \frac{1}{16}{}^0V + \frac{9}{64}{}^1V \qquad {}^1V_{A:A,B} = v_4 + \frac{1}{64}{}^1V \\
&{}^0V_{B:A,B} = \frac{1}{2}v_1 + v_3 + \frac{1}{16}{}^0V - \frac{9}{64}{}^1V \qquad {}^1V_{B:A,B} = v_5 + \frac{1}{64}{}^1V
\end{aligned}
\tag{31}
$$

The lattice parameter of a crystal phase, used to calculate the lattice mismatch between phases, is directly related to its molar volume with the relation depending on its crystal structure. For cubic crystals, it is simply $V_m = a^3N/4$ where a is the lattice parameter and N the Avogadro number. Alternatively, one can model the lattice parameter directly using the similar phenomenological approach as recently carried out by Wang et al.[46] Since the change of molar volume and lattice parameter is relatively small with respect to temperature and compositions, both approaches would give similar results.

4. FIRST-PRINCIPLES CALCULATIONS OF MATERIALS PROPERTIES

First-principles calculations, based on density functional theory, require only knowledge of the atomic species and crystal structure, and hence, are predictive in nature. The widely used first-principles methods include the full-potential linearized augmented plane wave (FLAPW) method (the "benchmark" for accuracy in density-functional-based methods), and the highly efficient Vienna *ab initio* Simulation Package (VASP).[47-49] These methods yield quantities related to the electronic structure and total energy of a given system and can be used to accurately predict phase stabilities of compounds at 0 K. Coupling with frozen phonon or linear response techniques opens the possibility for exploring finite-temperature vibrational effects. Furthermore, these approaches are applicable to any phases of a given alloy system, not only the equilibrium ones. Hence, first-principles techniques can provide a method to obtain properties of metastable phases, which are often crucial to mechanical properties (e.g., strengthening precipitates) but can be difficult to isolate and study experimentally. In addition to thermodynamic properties, first-principles calculations can also provide thermal expansion, diffusivity,[50] and interfacial energies.[51,52] In this section, we will briefly summarize first-principles calculations for finite temperatures, solutions phases, and interfacial energies.

4.1 First-principles calculations for finite temperatures

For thermodynamic properties of alloys, one currently needs to calculate three additive contributions to the free energy. The first contribution is the cold energy or the 0 K total energy. In this case, the atoms are kept fixed at their static lattice positions. Secondly, for thermodynamic properties at finite temperatures, the contribution of lattice thermal vibration needs to be taken into account. Theoretically, the commonly accepted method is the lattice dynamics or phonon approach. When temperature is increased, especially for cases when the electronic density of state at the Fermi level is high, the third contribution to be included is the thermal electronic contribution (TEC).

Therefore, in a system with an atomic volume V at temperature T, the Helmholtz free-energy $F(V,T)$ can be approximated as

$$F(V,T) = E_c(V) + F_{ph}(V,T) + F_{el}(V,T) \tag{32}$$

where E_c is the 0 K total energy, F_{ph} the vibrational free energy of the lattice ions given by[53]

$$F_{ph}(V,T) = k_B T \sum_{\mathbf{q}} \sum_{j} \ln\left\{2\sinh\left[\frac{\hbar\omega_j(\mathbf{q},V)}{2k_B T}\right]\right\} \tag{33}$$

where $\omega_j(\mathbf{q},V)$ represents the frequency of the jth phonon mode at wave vector $\mathbf{q}$. When the magnetic contribution and the electron-phonon interactions are neglected, TEC to the free energy F_{el} is obtained from the energy and entropy contributions, i.e. E_{el}–TS_{el}. The bare electronic entropy S_{el} takes the form

$$S_{el}(V,T) = -k_B \int n(\varepsilon,V)[f \ln f + (1-f)\ln(1-f)]d\varepsilon, \tag{34}$$

where $n(\varepsilon, V)$ is the electronic density of states (DOS) with f being the Fermi distribution. The energy E_{el} which is due to the electron excitations can be expressed as

$$E_{el}(V,T) = \int n(\varepsilon,V) f \varepsilon d\varepsilon - \int^{\varepsilon_F} n(\varepsilon,V)\varepsilon d\varepsilon, \tag{35}$$

where ε_F is the Fermi energy.

The current first-principles implementations of phonon theory are divided into two categories:[53-56] the linear-response method and the supercell method. In the linear-response method, the normal frequencies (i.e. phonon frequencies) associated with microscopic displacements of atoms in a crystal are calculated by means of the dynamical matrix which is the reciprocal-space expression (Fourier transform) of the interatomic force constant matrix. Utilizing the electronic linear response upon the undistorted crystals, the evaluations of the dynamical matrix are then carried out through the density-functional perturbation theory without the approximation of the cutoff in neighboring interactions. Compared with the linear-response method, the supercell method is conceptually simple and computationally straightforward. The supercell method adopted the frozen phonon approximation through which the changes in total energy or forces are calculated in the real space by displacing the atoms from their equilibrium positions. If one is only interested in the gross phonon frequencies, the results of supercell calculation and linear-response calculation are very similar.[57,58] The Alloy-Theoretic Automated Toolkit (ATAT) code developed by van de Walle et al.[56] serves as the interface in calling the VASP code for phonon calculations.

4.2 First-principles calculations of solution phases

While the calculation of the electronic structure of perfectly ordered periodic structures with first-principles methods is well established, this is not the case for phases with homogeneity ranges, which is still under development. A couple of representative methods used to obtain the properties of random solid solutions from theoretical calculations are listed below:

i) One strategy to tackle this problem is through the use of large supercells. In this case, the sites of the supercell can be randomly occupied by either *A* or *B* atoms to yield the desired $A_{1-x}B_x$ composition. In order to reproduce the statistics corresponding to a random alloy, such supercells must necessarily be very large. Since most of first-principles codes are usually limited to calculations involving up to about 100 atoms, the supercell approach is computationally prohibitive.

ii) The Coherent Potential Approximation, CPA, relies on the average occupation of *A* and *B* atoms corresponding to the $A_{1-x}B_x$ alloy. With this mean-field method it is possible to analyze the energetics of random alloys throughout the composition range, however, local relaxations cannot be considered explicitly—at least in the most common version of this technique—and therefore the effects of alloying on the distribution of local environments within the solution phase cannot be fully taken into account. The calculation of disordered hcp phase in the Mo-Ru system showed that the local relaxation can be considered with the CPA approach with additional steps.[59]

iii) Another approach is to apply the Cluster Expansion.[60,61] In this case, a generalized Ising model is used and the spin variables can be related to the occupation of either atom *A* or *B* in the parent lattice. In order to obtain an expression for the configurational energy of the solid phase, the energies of multiple configurations (typically in the order of a few dozens) based on the parent lattice must be calculated to obtain the parameters that describe the energy of any given $A_{1-x}B_x$ composition in the alloy. This approach yield valuable information regarding short range ordering, relaxation energies and obviously the energetics of random alloys but it is computationally expensive, and the choice of clusters is critical.

iv) The Special Quasirandom Structures (SQS) mimic a random solution phase through creating a small (4-48 atoms) periodic structure that best satisfies the pair and multi-site correlation functions corresponding to a random solid solution, up to a certain coordination shell. For a binary $A_{1-x}B_x$ substitutional alloy, many properties are dependent on the *configuration*, or the substitutional arrangement of A and B atoms on the lattice. These configurationally-dependent properties (such as the energy) can be characterized very efficiently by a "lattice algebra".[62] Pseudo-spin variables are assigned to each site, S_i=-1 (+1) if an A (B) atom sits at site i. One can further define geometric *figures*, f, symmetry-related groupings of lattice sites, e.g., single site, nearest-neighbor pair, three-body figures, etc. These figures, f=(k, m), can have k vertices and span a maximum distance of m (m=1, 2, 3... are the first, second and third-nearest neighbors, etc.). By taking the product of the spin variable over all sites of a figure and averaging over all symmetry-equivalent figures of the lattice, we obtain the correlation functions $\overline{\Pi}_{k,m}$. For the perfectly random alloys, there is no correlation in the occupation between various sites, and therefore $\overline{\Pi}_{k,m}$ simply becomes the product of the lattice-averaged site variable. The SQS approach amounts to finding small-unit-cell ordered structures that possess $\left(\overline{\Pi}_{k,m}\right)_{SQS} \cong \left\langle\overline{\Pi}_{k,m}\right\rangle_R$ for as many figures as possible.

It is true that describing random alloys by small unit-cell periodically-repeated structures will surely introduce erroneous correlations beyond a certain distance. However, since interactions between nearest neighbors are generally more important than interactions between more distant neighbors, we can construct SQS's that exactly reproduce the correlation functions of a random alloy between the first few nearest neighbors, deferring errors due to periodicity to more distant neighbors. Furthermore, one single calculation on SQS can give various properties of random alloys, and only the SQS's allow direct calculations of real-space quantities (e.g. lattice parameters and bond lengths) of alloys in comparison with the other approaches. The SQS's for fcc alloys alloys have been generated by Zunger et al.[62] We have developed the SQS's for bcc,[63] B2,[64] Laves phases,[65] halite ,[66] hcp,[67] and $L1_2$,[68] using the ATAT code.[56]

4.3 First-principles calculations of interfacial energy

In first-principles calculations of interfacial energy, the following steps were taken[51,52]

i) Define the interfacial supercells considering the usual degrees of freedom of interfaces such as where to "cut" the crystals of two phases, how to "join" them – interfacial orientation, termination, alignment, and atomic configurations,

ii) Calculate the total energy of the supercell with full atomic relaxations, E_{tot},
iii) Calculate the total energies for each phase using the same supercell size and fixed lattice parameters in the interface and allowing the third lattice parameter to be relaxed. Their total energies are denoted by E_1 and E_2, respectively.

The interfacial energy is then evaluated by the following equation

$$\sigma = \left\{ E_{tot} - \frac{1}{2}\left[E_1 + E_2\right] \right\} / 2S \tag{36}$$

where S is the interface area.

5. APPLICATIONS TO Ni-Al

Ni-base superalloys consist of ordered intermetallic γ'-$L1_2$ precipitates embedded in a disordered face-centered cubic γ-fcc matrix. The γ' precipitate morphology and spatial distribution are known to depend on a number of factors including the alloy compositions, temperature, presence of dislocations and applied stress direction. Therefore, extensive efforts have been devoted to understanding the evolution of the γ' precipitate microstructure, both experimentally and theoretically, over the past fifty years.[69] In this section, we present our recent work on Ni-Al through combining first-principles calculations, CALPHAD modeling, and phase-field simulations to determine the relationships between chemistry and microstructure at high temperatures. The ternary Ni-Al-Mo system is one of the basic systems for Ni-base superalloys,[70] and our research results on Ni-Al-Mo have been submitted for publication.[71]

5.1 First-principles calculations

Unless specified otherwise, we use the Vienna *ab initio* simulation package (VASP)[48] with the ultrasoft pseudopotentials and the generalized gradient approximation (GGA).[72] Spin polarized configurations are used for checking the magnetism of Ni and Ni-rich compounds.

5.1.1 Interfacial energy between γ and γ' [73,74]

In the calculations of interfacial energies, the energy cutoff was determined by the choice of "high accuracy" in VASP. Two types of

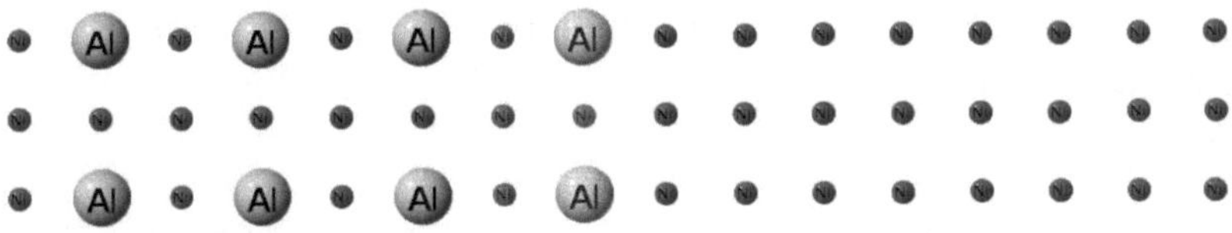

Figure 1. Supercell for the (100) interfacial boundary between γ' and γ.

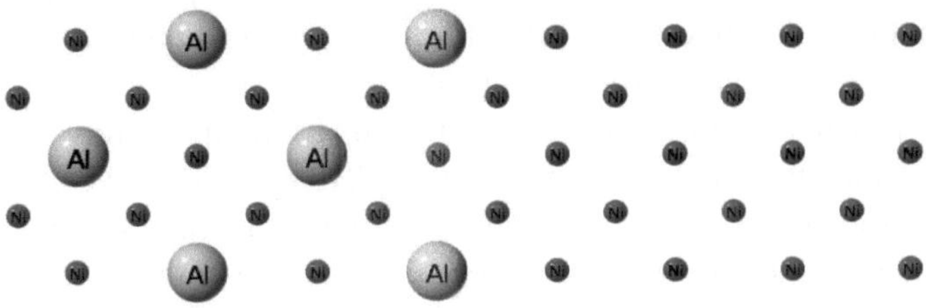

Figure 2. Supercell for the (110) interfacial boundary between γ' and γ.

interfaces were considered: (001) with 4 cubic $L1_2$ γ'-Ni_3Al unit cells and 4 cubic fcc γ-Ni cell (Figure 1), and (011) with 4 $L1_2$ γ'-Ni_3Al unit cells and 4 fcc γ-Ni cell cut in the [011] direction (Figure 2). The 12×12×6 and 8×12×8 Monkhost *k* points are used for (001) and (011) interfaces, respectively, to provide a higher accuracy in the [001] direction. The interfacial energies thus obtained are 39.6 and 63.8 mJ/m^2 for (100) and (110), respectively.

5.1.2 Structural stability of Ni-Mo compounds[75]

In the accepted Ni-Mo binary system, there are three stable compounds. They are designated as β-Ni_4Mo ($D1_a$), γ-Ni_3Mo ($D0_a$), and δ-$Ni_{24}[Ni_{4x}Mo_{4(5-x)}]Mo_{12}$ ($P2_12_12_1$) (hereafter referred as δ-NiMo). There are experimental evidences that suggest Ni_2Mo and Ni_8Mo might be stable at low temperatures. Their stability is investigated by first-principles calculations. In each calculation, with the cell shape and the internal atomic coordinates being fully relaxed, the total energies are calculated at seven fixed volume points near the experimental volume. The lowest energy is found by fitting the total energy curve with the Morse function. Energy cutoff of "high accuracy" in the VASP, 15x15x15 Monkhost *k* points for the pure elements, 6x6x6 for δ-NiMo, and 11x11x11 for the other compounds are employed. The primitive cell of δ-$Ni_{24}[Ni_{4x}Mo_{4(5-x)}]Mo_{12}$ has 56 atomic sites divided into three types.[76] The 20 sites in the square bracket can be occupied either by Ni or Mo. Therefore, the various configurations for δ-$Ni_{24}[Ni_{4x}Mo_{4(5-x)}]Mo_{12}$ are considered in the calculation. It is found that the δ-NiMo phase is meta-stable at 0 K, and Ni_2Mo and Ni_8Mo are stable in the system.

5.1.3 Thermodynamic properties of Al, Ni, NiAl and Ni_3Al[57,58]

The thermodynamic properties of Al, Ni, NiAl and Ni_3Al are systematically calculated by means of the linear-response theory (LRT)[57] and the supercell method (SC) (also known as frozen phonon method[58]). In this section, we present the results for the $L1_2$ Ni_3Al.

In the linear-response approach, the phonon frequency $\omega_j(\mathbf{q},V)$ was calculated using the Plane-Wave Self-Consistent Field (PWSCF) method.[55] The electronic DOS $n(\varepsilon, V)$ were calculated using VASP. The 0 K calculations were performed in a wide volume range with a step of 0.1 a.u. in the lattice parameter. At a given temperature T, the F_{ph} in Eq. 33 was calculated in the same lattice parameter step as the 0 K calculations. E_{el} and S_{el} were evaluated using one-dimensional integrations as shown in Eqs. 34 and 35. In the supercell method, first-principles calculations were carried out using Projector Augmented-Wave (PAW) pseudo-potentials. An energy cutoff of 350 eV for all the calculations was set and an isotropic mesh of 5000 k-points per unit cell was used. For the quasi-harmonic approximation, 5 volumes were generated (with the maximum total strain being equal to the lattice constant near the melting point of each of the structures).

Figure 3 shows the phonon dispersion curves for the Ni_3Al $L1_2$ structure. In general, the results for the three supercell sizes of 8, 16, and 32 atoms and those from the LRT calculations agree well with each other and with the experimental results[77] except along the ΓR direction which is also observed when comparing the LRT calculations for this phase. The discrepancy at this point of the phonon dispersion curve is somewhat reduced for the 32-atom supercell calculations. Despite the discrepancies, the convergence of the calculated phonon DOS is remarkable and it is expected that the calculated thermodynamic properties for this structure will be almost independent of the size of the supercell used.

In order to compare the different approximations to the free energy of the structures, the thermodynamic properties of the systems were calculated using three approximations, harmonic approximation (har) at the theoretical volume calculated at 0K, the quasi-harmonic approximation (q-h) with five different volumes, and the electronic DOS using the quasi-harmonic approximation (q-h,el). Figure 4 shows the enthalpy of $L1_2$ Ni_3Al.

In this case, it is to be expected that the TEC is considerable and in fact it can be seen in the figure that using only the q-h calculations is not enough to reproduce the suggested values.

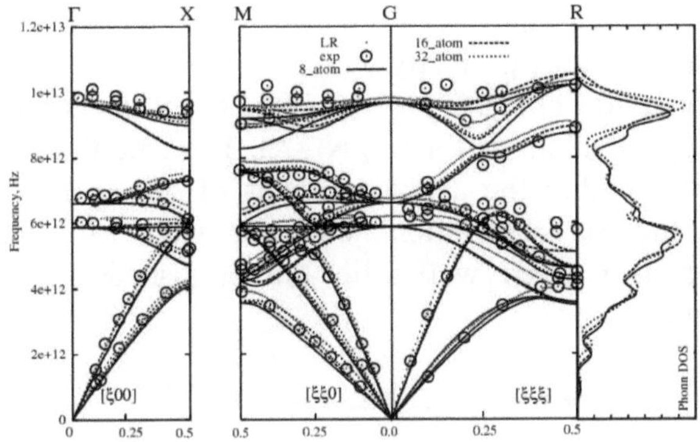

Figure 3. Phonon dispersion calculations for L1$_2$ Ni_3Al. The solid and dashed lines correspond to the SC calculations.[58]

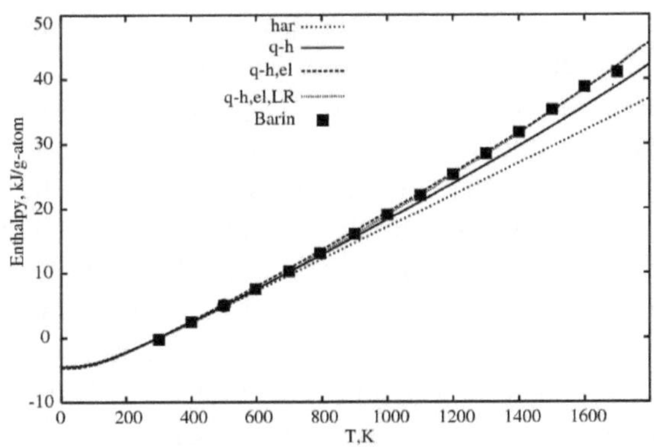

Figure 4. Enthalpy of L1$_2$ Ni_3Al.[58]

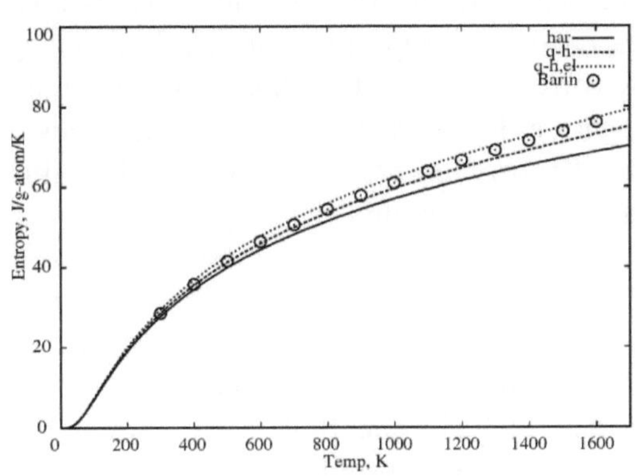

Figure 5. Entropy of L1$_2$ Ni_3Al.[58]

With TEC, the present q-h, el and LRT calculations are virtually indistinguishable and agree perfectly with the tabulated data by Barin.[78] The TEC effect accounts for about 8% of the enthalpy of this structure. Figure 5 shows the entropy calculations for $L1_2$ Ni_3Al. The data from Ref. 78 lies, at high temperatures, within the approximations of q-h and q-h,el. It can be concluded that the SC calculations compare well with linear response calculations and also agree well with the published phonon dispersion experimental data.

5.1.4 SQS calculations of bcc, B2, and $L1_2$[63,64,68]

For the perfectly random $A_{1-x}B_x$ bcc alloys, there is no correlation in the occupation between various sites and the pair and multisite correlation functions $\overline{\Pi}_{k,m}$ are given simply as $\langle \overline{\Pi}_{k,m} \rangle_R = (2x-1)^k$. Various SQS-$N$ structures (with N=2, 4, 8 and 16 atoms per unit cell) for the random bcc alloys at composition x=0.50 and 0.75 were generated using the *gensqs* code in the (ATAT).[56] For each composition x, our procedure can be described as follows: (1) Using *gensqs,* we exhaustively generate all structures based on the bcc lattice with N atoms per unit cell and composition x. (2) We then construct the pair and multisite correlation functions $\overline{\Pi}_{k,m}$ for each structure. (3) Finally, we search for the structure(s) that best match the correlation functions of random alloys over a specified set of pair and multisite figures. First-principles calculations were performed using the plane wave method with Vanderbilt ultrasoft pseudopotentials. The k-point meshes for Brillouin zone sampling were constructed using the Monkhorst–Pack scheme and the total number of k-points times the total number of atoms per unit cell was at least 6000 for all systems.

As an example, the Mo-Nb system was studied with a plane wave cutoff energy of 233.1 eV. Mo and Nb form a continuous bcc solid solution. No intermediate phases have been reported. The equilibrium lattice parameters of Mo-Nb bcc alloys obtained from the relaxed SQS's are plotted in Figure 6 showing a small negative deviation from the Vegard's law. In Figure 7, the predicted enthalpies of formation of random Mo-Nb bcc alloys are compared with the experimental measurements by Singhal and Worrell[79] at 1200 K, and the calculated data by Sigli et al.[80] using the TB-CPA-GPM approach.

For the random pseudobinary $A_{1-x}B_xC$ B2 phase, we generated various SQS-N structures (with N=4 and 16 simple cubic sites per unit cell, or a total of $2N$ atoms per unit cell including the common C sublattice) at composition x=0.5 and 0.25, respectively.[64] The SQS-N structures for x=0.75 can be simply obtained by switching the A and B atoms in SQS-N for x=0.25.

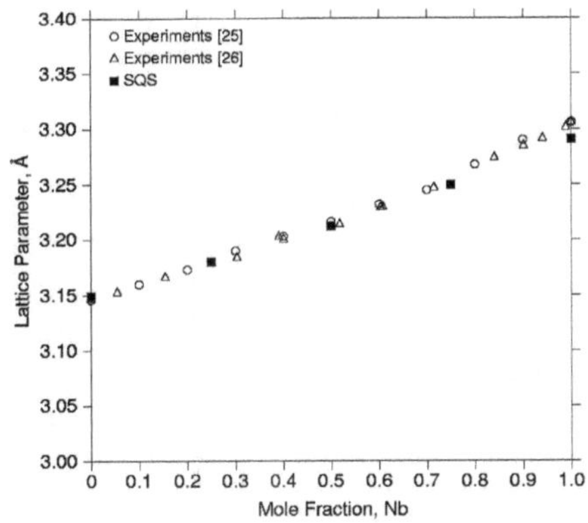

Figure 6. Equilibrium lattice parameters of Mo-Nb bcc alloys,[63] O,[81] Δ.[82]

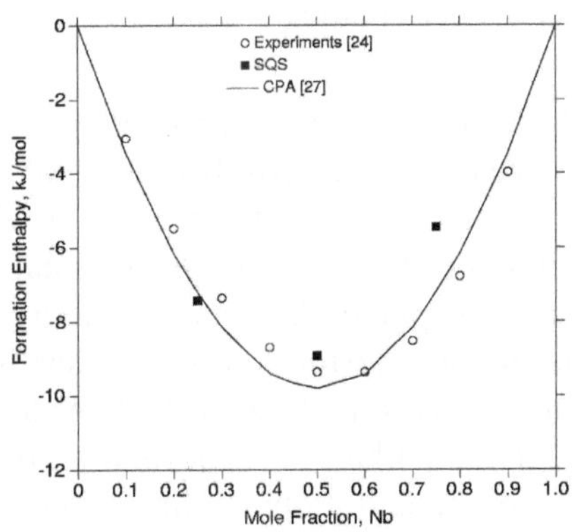

Figure 7. Formation enthalpies of Mo-Nb bcc alloys by SQS[63] in comparison with experiment[79] and CPA.[80]

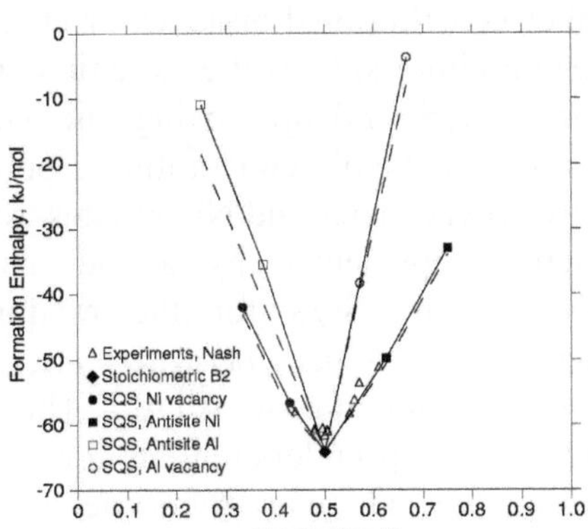

Figure 8. First-principles calculated and experimentally observed[83] enthalpy of formation of B2 NiAl. The solid and dashed lines for volume relaxations only and complete relaxations, respectively.[63]

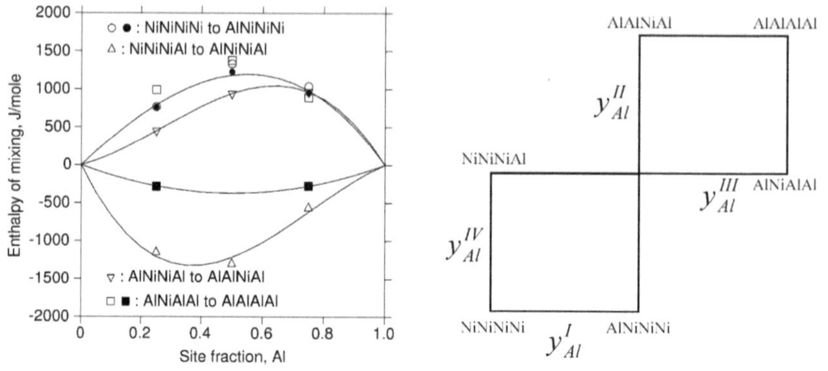

Figure 9. Enthalpies of mixing of $L1_2$ Ni_3Al by SQS calculations, open symbols from volume relaxations only and solid symbols from complete relaxations (with symmetry preserved) (a) with the corresponding composition square (b).[68]

First-principles calculations are performed using the all-electron Blöchl's projector augmented wave (PAW) approach for B_2 NiAl. The *k*-point meshes for Brillouin zone sampling are constructed using the Monkhorst–Pack scheme, and the total number of *k*-points times the total number of atoms per unit cell is at least 10000. A plane wave cutoff energy of 459.9 eV is used. Spin-polarized calculations are performed to account for the ferromagnetic nature of Ni though our calculations showed B2 NiAl as non-magnetic.

In Figure 8, the predicted enthalpies of formation of B2 NiAl containing each of the four types of constitutional point defects are plotted as four branches, respectively. We consider a canonical ensemble containing a total one mole of Al and Ni atoms and the total number of lattice sites may vary when vacancies are present. Since deviation from stoichiometry can be accommodated by either constitutional antisites or constitutional vacancies, there are two branches on either side of the stoichiometric composition, the branch with lower formation enthalpy being the stable one. Figure 8 unambiguously shows that the stable constitutional point defects in Al-rich and Ni-rich B2 NiAl are Ni vacancies and Ni antisites, respectively.

For the $L1_2$ structure represented by a four sublattice model, i.e. *(Al,Ni)(Al,Ni)(Al,Ni)(Al,Ni)*, the SQS for the random mixing in each sublattice are equivalent, i.e. *(Al,Ni)(i)(j)(k)* where *i*, *j*, and *k* are either *Al* or *Ni*. A 64 atom SQS[68] was developed using ATAT. The calculated enthalpies of formation using the ultrasoft pseudopotentials for four combinations of *i*, *j*, and *k* are shown in Figure 9. There are the same amount of combinations where Al substitutions take place in different sublattices though not shown here. It is observed that the atomic environments have significant impacts on enthalpy of mixing.

5.1.5 Lattice distortion and lattice parameters[71]

We employ 108-atom supercells with one solute atom in each supercell to study the lattice distortions caused by solute atoms adopting ultrasoft pseudopotentials. Ten commonly used alloying elements in Ni-base alloys were chosen, namely, Al, Co, Cr, Hf, Mo, Nb, Re, Ru, Ta, Ti and W. The set of k points is adapted to the size of the primitive cell, and a 4x4x4 k-point mesh is selected for the supercell used in the present calculations. The energy cutoff is determined by the choice of "high accuracy" in VASP. The lattice distortions due to solute atoms can be separated into two categories, the local lattice distortion and the macroscopic lattice distortion with the latter represented by the overall lattice parameter change represented by $\Delta a = a_{sol} - a_{pure}$ with a_{pure} and a_{sol} being the lattice parameters of pure solvent and the solution containing the solute atoms. For dilute solutions, one can define a linear regression coefficient defined as $k_i = N_s \Delta a$ with N_s being the number of atoms in the supercell.

The linear regression coefficients and local relaxations for the above ten elements evaluated from first-principles calculations can be found elsewhere.[71] It is found that the calculated data agree with the experimental results within the experimental uncertainties. Among them, Co exhibits the difference between the macroscopic lattice parameter change and the local lattice distortion. Macroscopically, Co atoms expand the fcc Ni while locally they decrease the nearest-neighbor distances. Lattice parameters of several Ni-Al-Cr-Co-Mo-Nb-Re-Ta-Ti-W alloys are calculated using the above linear regression coefficients and compared with the experimental data in with a good agreement.[71]

5.2 CALPHAD modeling

5.2.1 Thermodynamic modeling of Ni-Mo[84]

Thermodynamic modeling of the Ni-Mo system was first carried out by Frisk[85] in which six phases were considered. Cui and Jin[86] modified the thermodynamic description of the Ni-Mo system by modeling the Ni_3Mo phase with a two-sublattice model, $(Mo,Ni)_3(Mo,Ni)_1$, to account for its homogeneity range. The calculated phase diagram by Cui and Jin[86] revealed unrealistic behavior of the δ-NiMo phase at low temperature as shown by dotted lines in Figure 10. The calculated enthalpies of formation are plotted in Figure 11. The isostructure enthalpies of mixing of bcc and fcc from SQS calculations are plotted in Figure 12.

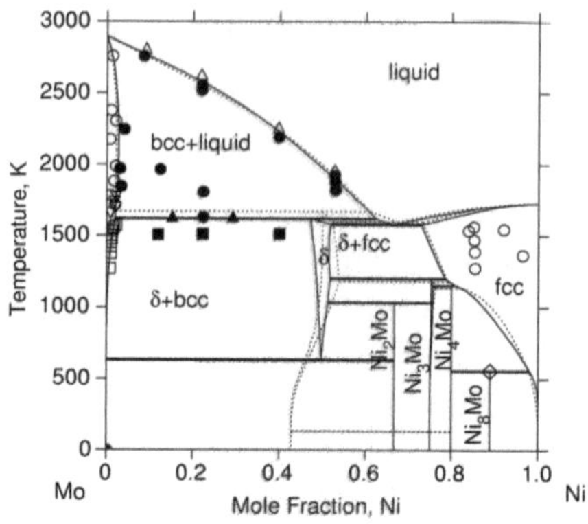

Figure 10. Calculated phase diagrams with the experimental data.[84]

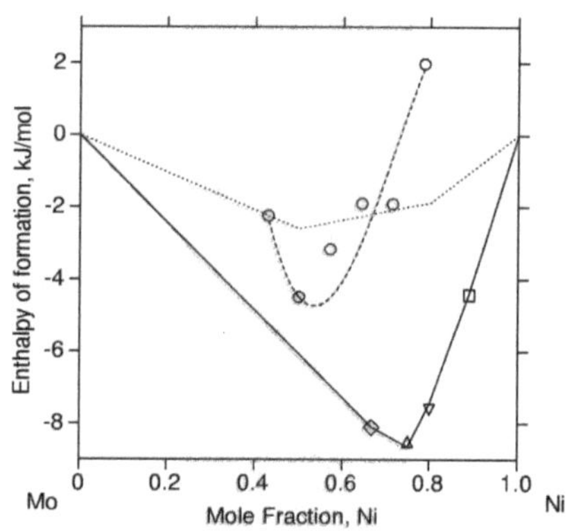

Figure 11. Calculated enthalpy of mixing (---) of the compound δ-NiMo at 0 K and enthalpy of formation (—) (details see Ref. 84).

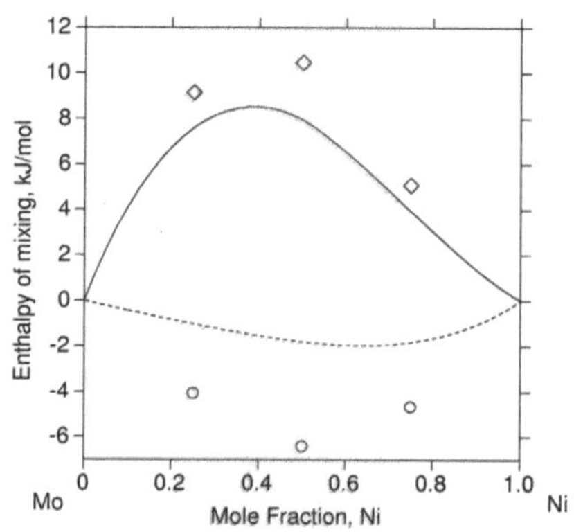

Figure 12. The calculated isostructure enthalpy of mixing of bcc (—) referred to bcc-Ni and bcc-Mo and fcc (---) referred to fcc-Ni and fcc-Mo with the first-principles calculation data ◇ for bcc and O for fcc.[84]

Based on the newly obtained enthalpy of formation of compounds by first-principles calculations, the enthalpy of mixing in fcc and bcc by SQS, and experimental data in the literature, the thermodynamic modeling of the Ni-Mo system was carried out, and a new Ni-Mo phase diagram was obtained (see Figure 10).

5.2.2 Thermodynamic modeling of Ni-Al-Mo[87]

Thermodynamic modeling of the ternary Ni-Al-Mo system was first performed by Kaufman and Nesor[88] and re-evaluated by Lu et al.[89] without considering the two ternary N and X phases. In the present modeling of the ternary system, the Gibbs energy functions of individual phases in the binary Ni-Al,[90] Ni-Mo[84] and Al-Mo[91] were used. The parameters of the non-stoichiometric phases from three binary systems were evaluated by considering the solubility of the third component. The first-principles calculated enthalpy of formation for Al_8NiMo_3 (the N phase with the stoichiometric composition) with the slightly distorted $D0_{22}$ structure is –38.17 kJ/mol at 0 K, while the enthalpy of mixing of the Al_8Mo, $AlMo_3$ and B2 NiAl phases is –41.49 kJ/mol and –40.62kJ/mol calculated by CALPHAD and the first-principles approaches, respectively. This indicates that Al_8NiMo_3 is not stable at 0 K.

According to the experimental data,[92-94] the ternary compound N was described using a three-sublattice model. The calculated enthalpies of mixing using ATAT combined with VASP were used to evaluate the model parameters of the N phase. Figure 13 shows the calculated enthalpy of mixing of the N phase with the ATAT calculation data. Due to lack of experimental data and structure information, X phases were treated as stoichiometric compounds. The calculated ternary phase diagrams of the Ni-Al-Mo system at 1474 K are shown in Figure 14.

5.2.3 Atomic mobility modeling in Ni-Al and Ni-Al-Mo[95]

The atomic mobility in γ Ni-Al was modeled by Engstrom and Agren.[96] The atomic mobility in γ' of Ni-Al is needed along with the atomic mobility in γ and γ' of Ni-Mo, Al-Mo, and Ni-Al-Mo. There are two main challenges, i.e., Mo is bcc and has very limited solubility in Ni and Al, thus there are very few experimental data available. The first-principles calculations of Mo fcc remains an unresolved issue.[97] Based on available experimental information, the atomic mobility for Al-Mo and Ni-Mo were evaluated.[95]

For atomic mobility in γ', there are experimental data on tracer diffusion of Ni,[98-102] inter-diffusions of Ni-Al alloys in the temperature range from 1073 K to 1473 K by Watanabe et al.[103-105] and between 1423 K and 1523 K

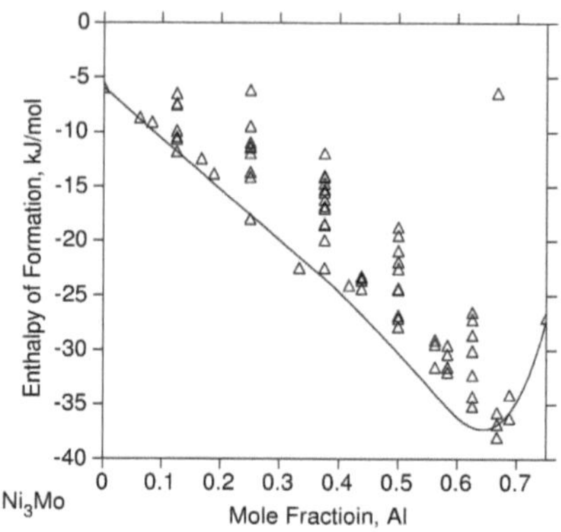

Figure 13. Calculated enthalpy of formation of the N phase for x_{Mo}=0.25.[87]

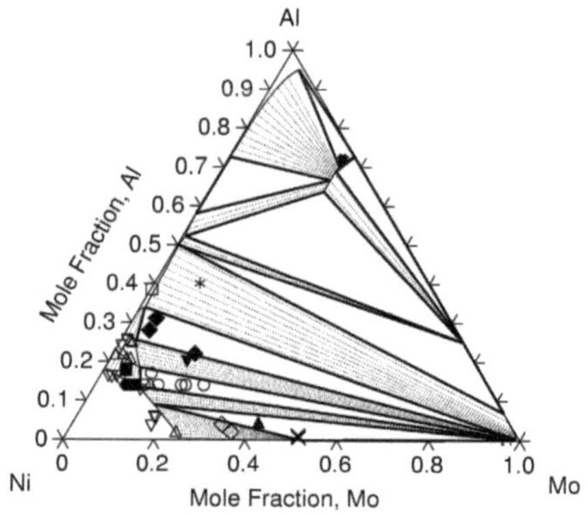

Figure 14. Calculated phase diagram at 1473K (details see Ref. 87).

by Fujiwara and Horita.[106] To extract the ordering effect from the experimental data, both the kinetic description for the related disordered phase and the related thermodynamic description are needed. In this work, the thermodynamic database for Ni-Al from Dupin et al.[90] and the mobility database for Al-Ni alloys from Engstrom and Agren[96] are used to evaluate the model parameters $\Delta\Phi_{ijk}^{ord}$. The parameter evaluation was carried out using the PARROT module of DICTRA.[43]

Since Mo prefers to occupy Al-sites in γ', the effect of Al-Mo ordering is ignored, and the atomic mobility of Mo in γ'-Ni_3Al is assumed to be the same of that of Al. The atomic mobility in the hypothetical γ'-Ni_3Mo phase is assumed to be the same of that in γ'-Ni_3Al.[95]

5.2.4 Lattice parameter modeling in Ni-Al and Ni-Al-Mo[46,71]

The lattice parameter of γ in Ni-Al can be modeled by the following equation:

$$a = x_{Al}{}^{0}a_{Al} + x_{Ni}{}^{0}a_{Ni} + x_{Al}x_{Ni}(A_{Al,Ni} + B_{Al,Ni}T) \tag{37}$$

where ${}^{0}a_{Al}$ and ${}^{0}a_{Ni}$ are the lattice parameter of Al and Ni, and $A_{Al,Ni}$ and $B_{Al,Ni}$ the model parameters to be evaluated. The γ and γ′ phases are described with one single Gibbs energy by two sub-lattices using the formula $(Al,Ni)_3(Al,Ni)_1$. Its lattice parameter a can be expressed by the following equation, similar to Eqs. 14-17,

$$\begin{aligned} a = &\sum_i \sum_j y_i^I y_j^{II}\, {}^{0}a_{i:j} + \sum_i \sum_{j>i} y_i^I y_j^I \sum_k y_k^{II} I_{i,j:k} \\ &+ \sum_i \sum_{j>i} y_i^{II} y_j^{II} \sum_k y_k^I I_{k:i,j} + \sum_i \sum_{j>i} \sum_k \sum_{l>k} y_i^I y_j^I y_k^{II} y_l^{II} I_{i,j:k,l} \end{aligned} \tag{38}$$

The lattice parameter of the end member $i_p j_q$, ${}^{0}a_{i:j}$, can be written as

$${}^{0}a_{Al:Al} = {}^{0}a_{Al} \qquad\qquad {}^{0}a_{Ni:Ni} = {}^{0}a_{Ni}$$

$${}^{0}a_{Ni:Al} = 0.25\,{}^{0}a_{Al} + 0.75\,{}^{0}a_{Ni} + C_{Ni:Al} + D_{Ni:Al}T$$

$${}^{0}a_{Al:Ni} = 0.75\,{}^{0}a_{Al} + 0.25\,{}^{0}a_{Ni} + C_{Al:Ni} + D_{Al:Ni}T \tag{39}$$

where $C_{i:j}$ and $D_{i:j}$ are model parameters. All model parameters for the lattice parameters of γ and γ′ in Ni-Al are listed in Ref. 46.

Using the lattice parameters of γ and γ′, the lattice mismatch between them was predicted and plotted in Figure 15. The solid curve shows the mismatch between the two equilibrium phases. As the experimental holding time is not long enough, the γ and γ′ phases in the samples would maintain their compositions at the aging temperature (973 K), and their lattice parameters should thus be calculated using the corresponding equilibrium compositions at the aging temperature. The corresponding mismatches thus calculated are shown by the dashed line in Figure 15 for the aging temperature of 973 K, in agreement with experimental data.

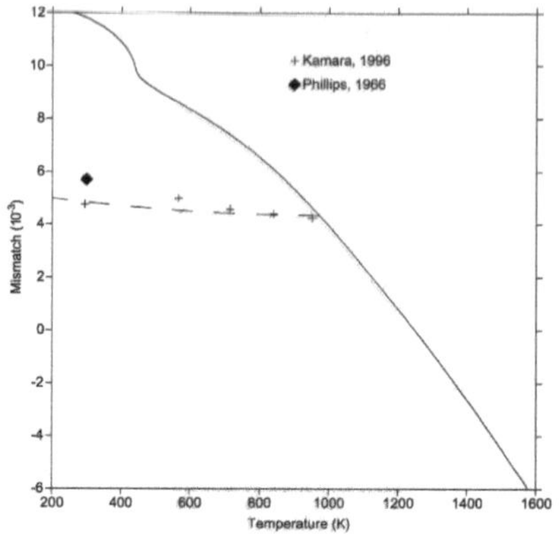

Figure 15. Calculated mismatch[46] between γ and γ′ (Kamara et al.[107] and Philips[108]).

The effect of Mo on the lattice parameters of γ and γ′ was studied through first-principles calculations in the γ′ phase, its substitution in either sublattice was considered separately and modeled through the lattice change $\Delta a_C = \sum\sum k_i^s y_i^s$ where s indicates different sublattices and i denotes solute or anti-site elements, y_i^s is the atomic fraction of i in sublattice s, and k_i^s is the related linear coefficient. Combining with the lattice parameter of γ′-Ni_3Al, the lattice mismatch between γ and γ′ is calculated.[71,95]

5.3 Phase-field simulations

Despite the remarkable success of phase-field simulations in providing fundamental understanding of underlying thermodynamic and kinetic mechanisms leading to various morphological evolution of γ′ precipitates, existing phase-field simulation results are largely qualitative due to the simplifications in thermodynamic and kinetic properties and the existing numerical approaches and computer power which limit most simulations to two-dimensions and small sizes. In this section, we present our recent approaches in quantitative predictions of the coarsening kinetics of γ′ in Ni-Al and Ni-Al-Mo alloys.

5.3.1 Interface models[109]

Two different types of phase-field models have been employed:[109] a model based on the physical order parameters which is called the physical model and the KKS model[5] for describing the γ/γ′ interfaces and microstructures. The kinetics of interface motion from the two models was

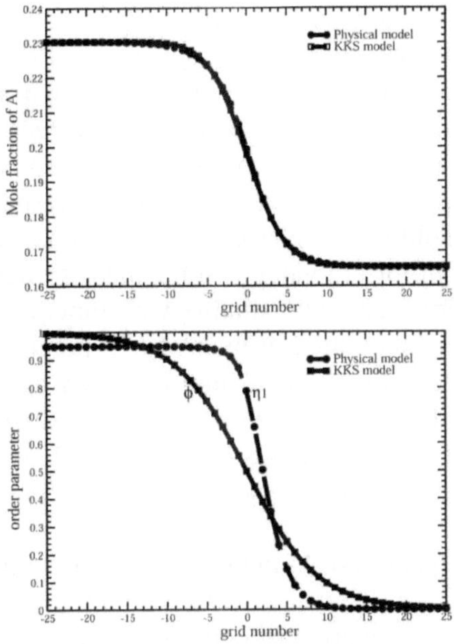

Figure 16. Composition (a) and order parameter (b) across the interface.[109]

compared to that obtained from the DICTRA code based on a sharp-interface description.[43] In order to compare with the results by DICTRA simulations which do not include the elastic energy calculation, a Ni-Al system at a high temperature was considered, in which the lattice mismatch between precipitate and matrix is small enough that the elastic energy contribution can be ignored. The diffusion mobility M is related to the atomic mobility through $M = c_{Al}c_{Ni}[c_{Al}M_{Ni} + c_{Ni}M_{Al}]$ where M_{Ni} and M_{Al} are obtained from the atomic mobility database.

In the KKS model, since thermodynamic databases use a non-linear description of the Gibbs free energy, numerical approximation has to be employed. For simplicity, the free energy of both phases can be approximated as parabolic functions whose first and second derivatives are imported from the thermodynamic database. Another solution is to tabulate the introduced equilibrium compositions in terms of composition and order parameter by using the Gibbs free energy in the thermodynamic database. It was found that no significant difference exists for the equilibrium compositions between the parabola approximation and tabulation approach.

Figure 16 shows the composition profile and order parameter profile across an equilibrium γ/γ' interface obtained using the physical and the KKS models at 1300 K. In Figure 16.a a smooth change of the equilibrium composition value from $c_{\gamma'} = 0.2287$ in γ' to $c_{\gamma} = 0.1595$ in γ was observed.

The equilibrium composition profiles obtained by both models match each other very well. However, there is a significant difference in the order parameter profiles between the results obtained by two models as shown in Fig. 16.b. In the physical model, the equilibrium values of the composition and order parameters, predetermined by the minimization of the free energy with respect to the composition and order parameters, are automatically obtained through the solution of phase-field equations. In the KKS model, the order parameter equilibrium value is 0 for the disordered phase and 1 for the ordered phase determined by the double-well function $g(\phi)$.

Figure 17 shows the simulated growth of a γ' precipitate by plotting the composition profiles as a function of time obtained by different models at 1300 K and $c_0 = 0.2$. The system size is $2\mu m$. The diffuse-interface model simulations agree well with the results from DICTRA as seen by the dashed line (the physical model) and dot-dashed line (the KKS model).

5.3.2 3D simulations of Ni-Al using the physical model[110]

A single Gibbs energy function of composition and temperature for γ and γ' with a four-sublattice model is used.[42] The mole fraction of Ni and Al in each sublattice can then be represented by the three-component order parameter η_i. Figure 18 plots the chemical free energy as a function of composition at 1300 K separated into two curves for γ and γ', respectively. The maximum driving force Δg_{max} at 1300 K is approximately 125 J/mol. The total free energy G is the sum of the incoherent free energy G_c and the coherent elastic energy E_{el}, i.e. $G = G_c + E_{el}$. The simulations were performed on a cubic domain with periodic boundary conditions.[110] Figure 19 shows an example of microstructural evolution in a lattice with 128x128x128 grid points at an equilibrium particle volume fraction ~35%. Initially the system is in a high-temperature homogeneous state where the composition deviation from the average value is only caused by fluctuation. The nucleation, growth and coarsening of γ' particles are evident in Fig. 19, driven by the decrease of the total free energy. During the nucleation and growth periods where the particle size is small, the particles are spherical (Figures 19.a and 19.b). The average domain size increases at later stages accompanying with the particle shape changing to cuboidal. Particle coarsening by the coalescence of neighboring domains are observed.

5.3.3 3D simulations of Ni-Al and Ni-Al-Mo using the KKS model[95,111]

In these simulations, nucleation of γ' in the supersaturated γ phase was explicitly introduced into the system at the beginning of a phase-field simulation using an approach similar to that described by Simmons et al.[112]

Figure 20 shows the 3D morphological evolution of γ' in a Ni-13.8 at. % Al alloy at a temperature of 1023 K. The simulation was performed using a 64 x 256 x 256 grid with a unit grid size of 2.5 nm.

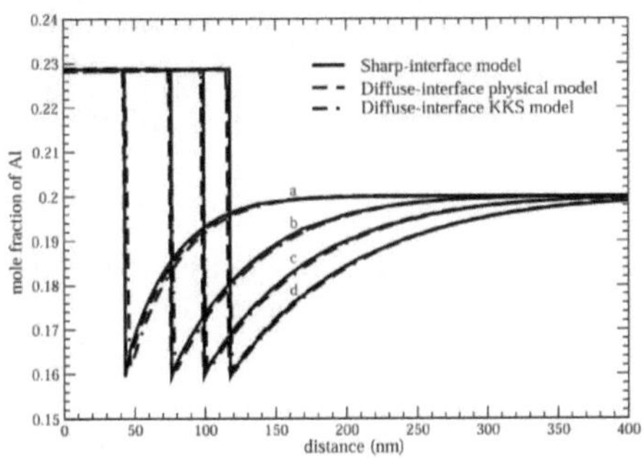

Figure 17. Temporal evolution of composition profiles (a)-(d) 0.1, 0.3, 0.5, 0.7 (s).[109]

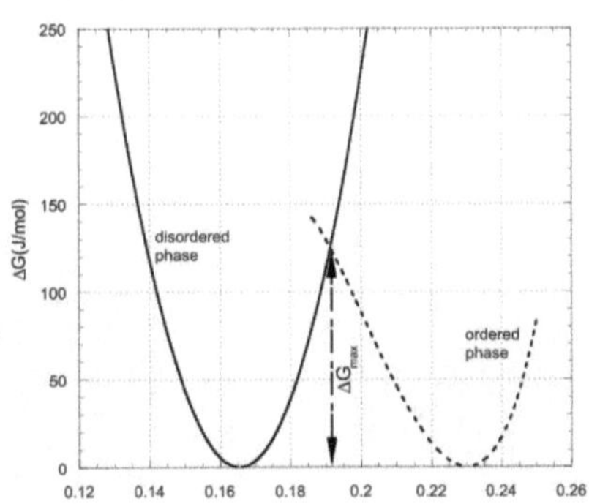

Figure 18. Calculated chemical free energy as a function of composition for both the disordered and ordered phases at 1300 K.[110]

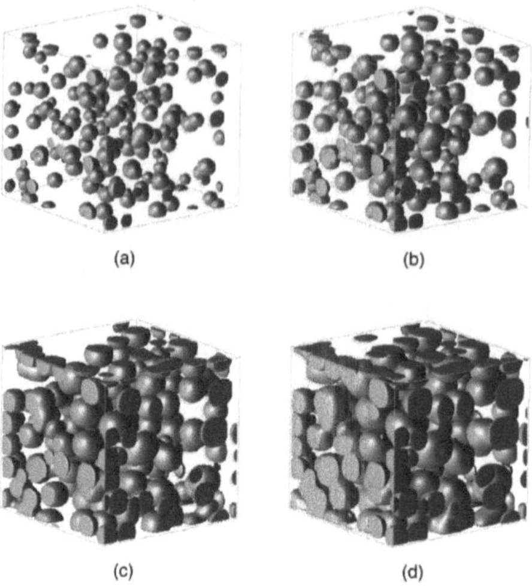

Figure 19. Morphological evolution from 3D simulations showing the nucleation, growth and coarsening of γ′ precipitates at 1300 K, (a)-(d) t^*= 4, 8, 12, 24.[110]

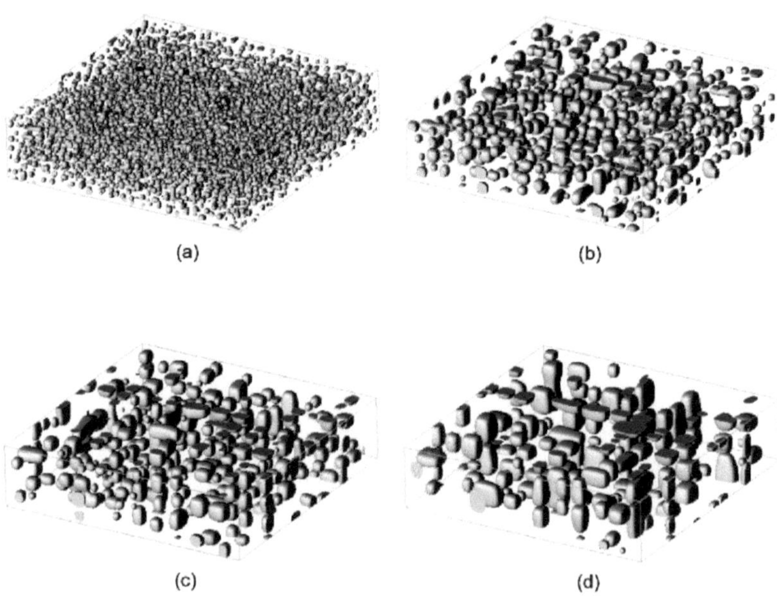

Figure 20. Morphological evolution of γ′ from 3D simulations in a Ni-13.8 at% Al alloy aged at 1023 K (a) 15 min; (b) 2 hr; (c) 4 hr; (d) 8 hr.[111]

In other words, the system is 160 x 640 x 640 nm in its physical size. The initial state was a homogeneous solution with small composition fluctuations around the average composition. Spherical nuclei with an average radius of 7.5 nm were randomly introduced to the computational domain at $t > 0$ with the mass balance maintained at all times. Approximately 6000 particles were formed during the nucleation stage. According to experimental observations,[113] precipitation occurs very rapidly in this system. Thus we switched off the nucleation event at an early stage of aging ($t = 5$ min).

The volume fraction of γ' at the four different aging times in Figure 20 is approximately 9-12%, which is very close to the equilibrium volume fraction of 10%. Precipitate coarsening is clearly seen where the number of particles in the computational domain decreases from about 2500 at $t = 15$ min to 110 at $t = 8$ hr. Shrinking and growth of precipitates, as well as the coalescence of two neighboring domains, are observed. Driven by the reduction of the interfacial energy and elastic energy, the average precipitate size grows, accompanied by a change of particle shape from spherical to cuboidal, rod-like, or plate-like at later stages.

The shape change from spheres to cubes occurs when the average particle size is around 20-25 nm at an aging time about 2-4 hr, which is in good agreement with experimental observations.[113]

The simulated 3D microstructures were projected onto a 2D plane for comparison with experimental transmission electron microscope (TEM) images. The thin foils used in TEM measurements were typically from 20 to 120 nm thick.[113] As shown in the left column of Figure 21, we obtained 2D images projected from part of the 3D system, but with a thickness comparable to that of the experimental TEM samples. Experimental dark-field electron micrographs at the same aging time are also shown in the right column for comparison. All the morphologies shown are contained in a physical area measuring 640 x 640 nm. The general morphological patterns obtained in the simulations agree well with those determined experimentally, particularly the precipitate shapes, alignment along the <100> crystallographic directions at the later stages, and the particle densities.

For cube-shaped particles, the relationship between the average cube edge length $\bar{a}$ and aging time can be expressed as $\bar{a}^3 - \bar{a}_0^3 = Kt$. The mean particle edge length was obtained by following the same procedures used in the experiments.[113] Figure 22 shows the simulated and experimental results plotted as $\bar{a}^3 - \bar{a}_0^3$ vs. t on a logarithmic scale. The solid and dashed lines are linear fits to the data from the simulations and experiments, respectively.

While the cubic growth law provided reasonably good fits observed in both cases, the data from the simulation seems to fit better than the experimental data. The rate constant K and the initial particle size $\bar{a}_0$ are 4.60 x 10^3 nm^3/hr and -2.99 nm from experimental data,[113] and 5.57 x 10^3 nm^3/hr and -4.07 nm from simulation data, respectively, with the rate constant K being approximately 20% higher than the experimental value.

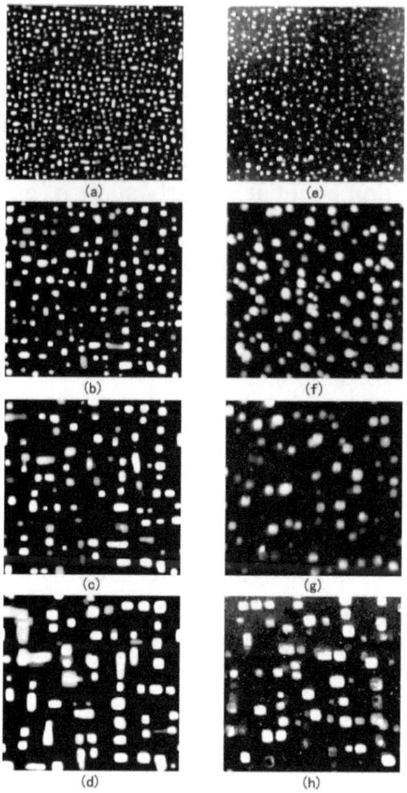

Figure 21. Comparison of precipitate morphologies, projected in 2D (640 x 640 nm), obtained by simulations (left column) and experiments (right column) in a Ni-13.8 at% Al alloy aged at 1023 K (a)(e) 15 min; (b)(f) 2 hr; (c)(g) 4 hr; (d)(h) 8 hr. Sample thickness in experiments is from 20 to 120 nm, and in simulation is (a) 20 nm; (b) 40 nm; (c) 50 nm; (d) 80nm.[111]

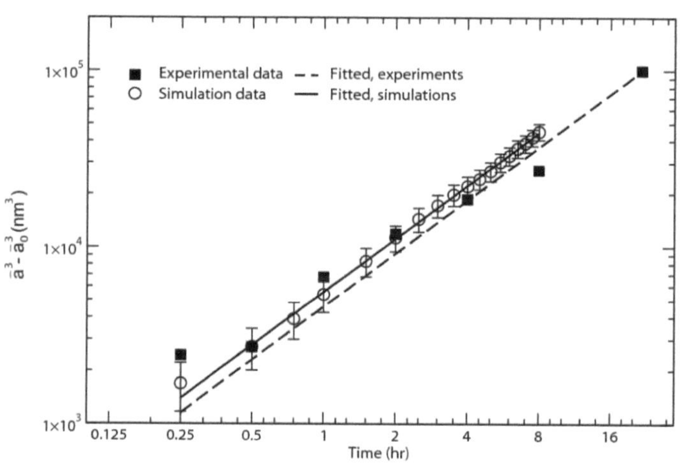

Figure 22. Average length scale evolution in a Ni-13.8 at% Al alloy aged at 1023 K.[111]

Considering the faster coalescence of neighboring particles in the simulations than in reality, due to the use of wider interface and the uncertainty in experiments and simulations, the agreement between the rate constant obtained from the simulations and the experimentally measured value is excellent.

It can be concluded that by artificially increasing the interfacial width, while keeping the interfacial energy constant, it is possible to simulate microstructure evolution in physical systems with length and time scales comparable to experimental results. The simulated morphological pattern agrees well with experimental observations in terms of the particle density, morphology and spatial correlations. Good quantitative agreement was achieved between the simulations and experiments regarding the growth law, the coarsening rate constant, and the particle size distributions.

Recently, phase-field simulations of the morphological evolution and coarsening kinetics in Ni-Al-Mo have also been carried out.[95] In the Ni-Al-Mo ternary system, two composition variables $c_i(\mathbf{r},t)$ $(i = Al, Mo)$ are needed in addition to the four artificial order parameters $\eta_j(\mathbf{r},t)$ (j=1,2,3,4). The diffusion mobility of i with respect to the concentration gradient of element m, M_{im}, is related to atomic mobility of elements as the follows

$$M_{im} = \sum_{j=1}^{n}[\delta_{mj} - c_m][\delta_{ji} - c_i]c_j M_j \tag{40}$$

where M_j is the atomic mobility of element j.

To compare with experimental results,[114] a number of phase-field simulations at a temperature of 1048 K were performed with the γ' volume fraction close to 0.15. The effects of volume fractions of precipitates were studied by selecting more alloys along the tie-lines of above alloys. The lattice misfit between γ and γ' is evaluated by combining first-principles calculations, existing experimental data and phenomenological modeling.[71] The detailed 2D and 3D simulation results are presented in Ref. 95. It is observed that the morphologies of γ' in the simulated and experimental microstructures agree well with each other. With increasing the Mo concentration from A1 to A3, the lattice mismatch decreases, resulting in the γ' morphology changing from rectangular to circular. For all the alloy compositions, the coalescence is observed between neighboring domains with the same order parameter, similar to the previous phase-field simulations in Ni-Al alloys.[111] The coarsening rate constants as a function of Mo contents and γ' volume fractions have been obtained and compared with available experimental data in the literature.[95]

6. CONCLUSIONS

In this chapter, we presented a hierarchical approach integrating first-principles calculations, CALPHAD modeling, and phase-field simulations and their applications to Ni-base alloys. It is demonstrated that first-principles calculations can provide many important, fundamental data of individual phases with only the input of atomic positions and atomic numbers. The interfacial energy can also be calculated when interface structures are defined. These data together with experimental information can be cast into the CALPHAD formalism to create a continuum description of properties of individual phases as a function of composition and temperature for multicomponent systems. Through proper modeling of phase interfaces, the phase-field simulations can predict microstructure evolutions that are in quantitative agreement with experimental observations in terms of both the length and time scales for binary and ternary systems.

ACKNOWLEDGEMENTS

The authors are grateful for the financial supports from NASA under grant no. NCC3-920 and National Science Foundation under the grants DMR-9983532, DMR-0122638, DMR-0205232 and DMR-0510180. Many postdoctoral fellows and graduate students in the authors' research labs at the Pennsylvania State University contributed to the research activities, particularly Dr. Raymundo Arroyave, Dr. Chao Jiang, Dr. Yi Wang, Dr. Shihuai Zhou, Dr. Jingzhi Zhu, and Dr. Tao Wang. The authors greatly appreciate the collaborations with Dr. Chris Wolverton at Ford Company and Dr. Jorge Sofo at the Pennsylvania State University.

REFERENCES

1. Z.-K. Liu, L.-Q. Chen, P. Raghavan, Q. Du, J.O. Sofo, S.A. Langer and C. Wolverton, *J. Comput-Aided Mater. Des.* **11**, 183 (2004).
2. L.Q. Chen, *Ann. Rev. Mater. Res.* **32,** 113 (2002).
3. G. Caginalp and W. Xie, *Phys. Rev. E* **48,** 1897 (1993).
4. A. Karma and W.J. Rappel, *Phys. Rev. E* **53,** R3107 (1996).
5. S.G. Kim, W.T. Kim and T. Suzuki, *Phys. Rev. E* **60,** 7186 (1999).
6. K.R. Elder, M. Grant, N. Provatas and J.M. Kosterlitz, *Phys. Rev. E* **6402,** 1604 (2001).
7. A. Karma, *Phys. Rev. Lett.,* **8711,** art. no.-115701 (2001).
8. J.W. Cahn and J.E. Hilliard, *J. Chem. Phys.,* **28,** 258 (1958).
9. A.A. Wheeler, W.J. Boettinger and G.B. McFadden, *Phys. Rev. A* **45,** 7424 (1992).

10. I. Steinbach, F. Pezzolla, B. Nestler, M. Seesselberg, R. Prieler, G.J. Schmitz and J.L.L. Rezende, *Physica D* **94**, 135 (1996).
11. Z.W. Lai, *Phys. Rev. B* **41**, 9239 (1990).
12. R.J. Braun, J.W. Cahn, G.B. McFadden, H.E. Rushmeier and A.A. Wheeler, *Acta Mater.* **46**, 1(1997).
13. D.Y. Li and L.Q. Chen, *Acta Mater.* **46**, 2573 (1998).
14. Y. Wang, D. Banerjee, C.C. Su and A.G. Khachaturyan, *Acta Mater.* **46**, 2983 (1998).
15. A.G. Khachaturyan, *Theory of Structural Transformations in Solids* (John Wiley & Sons, New York, 1983).
16. A.G. Khachaturyan and G. A. Shatalov, *Sov. Phys. Solid State* **11**, 118 (1969).
17. A. Onuki, *J. Phy. Soc. Jap.* **58**, 3069 (1989).
18. H. Nishimori and A. Onuki, *Phys. Rev. B* **42**, 980 (1990).
19. C. Sagui, D. Orlikowski, A. Somoza and C. Roland, *Phys. Rev. E* **58**, 569 (1998).
20. P.H. Leo, J.S. Lowengrub and H. J. Hou, *Acta Mater.* **61**, 2113 (1998).
21. S.Y. Hu and L.Q. Chen, *Acta Mater.* **49**, 1879 (2001).
22. J.Z. Zhu, L.Q. Chen and J. Shen, *Model. Simul. Mater. Sci. Eng.* **9**, 499 (2001).
23. Y.U. Wang, Y.M. Jin and A.G. Khachaturyan, *Appl. Phys. Lett.* **80**, 4513 (2002).
24. J.W. Cahn, *Acta Metall.* **9**, 795 (1961).
25. S.M. Allen and J.W. Cahn, *J. de Physique* **C7**, 51 (1977).
26. L.Q. Chen and J. Shen, *Comput. Phys. Commun.* **108**, 147 (1998).
27. J.Z. Zhu, L.Q. Chen, J. Shen and V. Tikare, *Phys. Rev. E* **60**, 3564 (1999).
28. N. Provatas, N. Goldenfeld and J. Dantzig, *Phys. Rev. Lett.* **80**, 3308 (1998).
29. J.H. Jeong, N. Goldenfeld and J.A. Dantzig, *Phys. Rev. E* **6404**, art. no.-041602 (2001).
30. J. H. Jeong, J. A. Dantzig and N. Goldenfeld, *Metall. Mater. Trans. A* **34**, 459 (2003).
31. C.W. Lan and Y.C. Chang, *J. Cryst. Growth* **250**, 525 (2003).
32. N. Saunders and A.P. Miodownik, *CALPHAD (Calculation of Phase Diagrams): A Comprehensive Guide* (Pergamon, Oxford, New York, 1998).
33. L. Kaufman and H. Bernstein, *Computer Calculation of Phase Diagram* (Academic Press Inc., New York, 1970).
34. U.R. Kattner, et al., *CALPHAD* **24**, 55 (2000).
35. L. Kaufman, *CALPHAD* **25**, 141 (2001).
36. I. Ansara and B. Sundman, The Scientific Group Thermodata Europe, P. S. Glaeser, Ed., Computer Handling and Dissemination of Data (Elsevier Science Pub. Co, 1987), pp. 154-158.
37. A.T. Dinsdale, *CALPHAD* **15**, 317 (1991).
38. B. Sundman and J. Agren, *J. Phy. Chem Solids* **42**, 297 (1981).
39. J.O. Andersson, A.F. Guillermet, M. Hillert, B. Jansson and B. Sundman, *Acta Metall.*, **34**, 437 (1986).
40. W. Huang and Y.A. Chang, *Intermetallics* **6**, 487 (1998).
41. I. Ansara, B. Sundman and P. Willemin, *Acta Metall.* **36**, 977 (1988).
42. I. Ansara, N. Dupin, H.L. Lukas and B. Sundman, *J. Alloy. Compd.* **247**, 20 (1997).
43. J.O. Andersson, T. Helander, L.H. Hoglund, P.F. Shi and B. Sundman, *CALPHAD* **26**, 273 (2002).
44. J.O. Andersson and J. Agren, *J Appl Phys* **72**, 1350 (1992).
45. T. Helander and J. Agren, *Acta Mater.* **47**, 1141 (1999).
46. T. Wang, J.Z. Zhu, R.A. Mackay, L.Q. Chen and Z.K. Liu, *Metall. Mater. Trans. A* **35**, 2313 (2004).
47. G. Kresse and J. Furthmüller, *Comp. Mater. Sci.* **6**, 15 (1996).

48. G. Kresse, *Vienna ab initio simulation package (VASP)*, http://cms.mpi.univie.ac.at/vasp/vasp/vasp.html, 2003.
49. G. Kresse and J. Furthmuller, *Phys. Rev. B* **54**, 11169 (1996).
50. A. Van der Ven and G. Ceder, *Phys. Rev. Lett.* **94**, 045901 (2004).
51. M. Sluiter and Y. Kawazoe, *Phys. Rev. B* **54**, 10381 (1996).
52. V. Vaithyanathan, C. Wolverton and L. Q. Chen, *Phys. Rev. Lett.* **88**, 125503 (2002).
53. A. van de Walle and G. Ceder, *Rev. Mod. Phys.* **74**, 11 (2002).
54. S. Baroni, S. de Gironcoli, A. Dal Corso and P. Giannozzi, *Rev. Mod. Phys.* **73**, 515 (2001).
55. S. Baroni, A. Dal Corso, S. de Gironcoli and P. Giannozzi, *PWscf PseudoPotential*, http://www.pwscf.org/pseudo.htm, 2005.
56. A. van de Walle, M. Asta and G. Ceder, *CALPHAD* **26**, 539 (2002).
57. Y. Wang, Z.K. Liu and L. Q. Chen, *Acta Mater.* **52**, 2665 (2004).
58. R. Arroyave, D. Shin and Z.K. Liu, *Acta Mater.* **53**, 1809 (2005).
59. A. E. Kissavos, S. Shallcross, V. Meded, L. Kaufman and I.A. Abrikosov, *CALPHAD* **29**, 17 (2005).
60. C. Wolverton and A. Zunger, *Phys. Rev. Lett.* **75**, 3162 (1995).
61. V. Ozolins, C. Wolverton and A.A. Zunger, *Phys. Rev. B* **57**, 6427 (1998).
62. A. Zunger, S.H. Wei, L. G. Ferreira and J.E. Bernard, *Phys. Rev. Lett.* **65**, 353 (1990).
63. C. Jiang, C. Wolverton, J. Sofo, L.Q. Chen and Z.K. Liu, *Phys. Rev. B* **69**, 214202 (2004).
64. C. Jiang, L.Q. Chen and Z.K. Liu, *Acta Mater.* **53**, 2643 (2005).
65. Y. Zhong, A.A. Luo, J.O. Sofo and Z.K. Liu, *Mater. Sci. Forum* **488-489**, 169 (2005).
66. M. Yang and Z.-K. Liu, *unpublished research,* (2005)
67. D. Shin, R. Arroyave and Z.-K. Liu, *Phys. Rev. B, submitted,* (2005).
68. T. Wang, L. Q. Chen and Z.-K. Liu, *unpublished research,* (2005).
69. A. Baldan, *J. Mater. Res.* **37**, 2379 (2002).
70. V. Biss and D.L. Sponseller, *Metall. Trans., Aug.* **4**, 1953 (1973).
71. T. Wang, L.-Q. Chen and Z.-K. Liu, *Acta Mater., submitted,* (2005).
72. J.P. Perdew and Y. Wang, *Phys. Rev. B* **45**, 13244 (1992).
73. C. Woodward, A. van de Walle and M. Asta, *unpublished research,* (2005).
74. Y. Wang, C. Woodward, L.Q. Chen and Z.K. Liu, *unpublished research,* (2005).
75. Y. Wang, C. Woodward, S.H. Zhou, Z.K. Liu and L.Q. Chen, *Scr. Mater.* **52**, 17 (2005).
76. C.B. Shoemaker and D.P. Shoemaker, *Acta Cryst.* **16**, 997 (1963).
77. C. Stassis, F. X. Kayser, C.-K. Loong and D. Arch, *Phys. Rev. B* **24**, 3048 (1981).
78. I. Barin, *Thermochemical Data of Pure Substances* (Weinheim, New York, 1995).
79. S.C. Singhal and W.L. Worrell, *Metall. Trans.* **4**, 1125 (1973).
80. C. Sigli, M. Kosugi and J.M. Sanchez, *Phys. Rev. Lett.* **57**, 253 (1986).
81. H.J. Goldschmidt and J.A. Brand, *J. Less-Common Met.* **3**, 44 (1961).
82. J. A. Catterall and S. M. Barker, *Plansee Proc.* 577 (1964).
83. P. Nash and O. Kleppa, *J. Alloy. Compd.* **321**, 228 (2001).
84. S. H. Zhou, Y. Wang, C. Jiang, J. Z. Zhu, L. Q. Chen and Z. K. Liu, *Mater. Sci. Eng. A* **397**, 288 (2005).
85. K. Frisk, *CALPHAD,* **14**, 311-320 (1990).
86. Y. Cui, Z. Jin and X. Lu, *Met. Mater. Trans. A* **30**, 2735 (1999).
87. S.H. Zhou, Y. Wang, J.Z. Zhu, T. Wang, L.Q. Chen, R.A. MacKay and Z.-K. Liu, Computational Tools for Designing Ni-Base Superalloys, K.A. Green, et al., Eds., Superalloy 2004 (TMS, 2004), pp. 969-975.
88. L. Kaufman and H. Nesor, *Metall. Trans.* **5**, 1623 (1974).

89. X. Lu, Y. Cui and Z. Jin, *Metall. Mater. Trans. A* **30**, 2735 (1999).
90. N. Dupin, I. Ansara and B. Sundman, *CALPHAD* **25**, 279 (2001).
91. N. Saunders, *J. Phase Equilib.,* **18**, 370-378 (1997).
92. K. Schubert, A. Raman and W. Rossteutscher, *Naturwissenschaften* **51**, 506 (1964).
93. A. Raman and K. Schubert, *Z. Metallkd.* **56**, 99 (1965).
94. V.Y. Markiv, V.V. Burnashova, L. I. Pryakhina and K. P. Myasnikova, *Izv. Akad Nauk SSSR Metall.* **5**, 180 (1969).
95. T. Wang, PhD dissertation, The Pennsylvania State University (2006).
96. A. Engstrom and J. Agren, *Z. Metallkd.* **87**, 92-97 (1996).
97. Y. Wang, S. Curtarolo, C. Jiang, R. Arroyave, T. Wang, G. Ceder, L.Q. Chen and Z. K. Liu, *CALPHAD* **28**, 79 (2004).
98. G.F. Hancock, *Physica Status Solidi (a)* **7**, 535 (1971).
99. M.B. Bronfin, G.S. Bulatov and I.A. Drugova, *Fiz. Metal. Metalloved.* **40**, 363 (1975).
100. K. Hoshino, S.J. Rothman and R. S. Averback, *Acta Metall.* **36**, 1271 (1988).
101. Y. Shi, G. Frohberg and H. Wever, *Phys Status Solidi A* **152**, 361 (1995).
102. S. Frank, U. Sodervall and C. Herzig, *Phys Status Solidi B* **191**, 45 (1995).
103. M. Watanabe, Z. Horita, D.J. Smith, M.R. Mccartney, T. Sano and M. Nemoto, *Acta Metall. Mater.* **42**, 3381 (1994).
104. M. Watanabe, Z. Horita and M. Nemoto, *Defect Diffus Forum* **143**, 345 (1997).
105. M. Watanabe, Z. Horita and M. Nemoto, *Defect Diffus Forum* **143**, 637 (1997).
106. K. Fujiwara and Z. Horita, *Acta Mater.* **50**, 1571 (2002).
107. A. B. Kamara, A. J. Ardell and C. N. J. Wagner, *Metall. Mater. Trans. A* **27**, 2888 (1996).
108. V. A. Phillips, *Acta Metall.* **14**, 1533 (1966).
109. J.Z. Zhu, T. Wang, S.H. Zhou, Z.K. Liu and L.Q. Chen, *Acta Mater.* **52**, 833 (2004).
110. J.Z. Zhu, Z.K. Liu, V. Vaithyanathan and L.Q. Chen, *Scr. Mater.* **46**, 401 (2002).
111. J.Z. Zhu, T. Wang, A.J. Ardell, S. H. Zhou, Z.K. Liu and L.Q. Chen, *Acta Mater.* **52**, 2837 (2004).
112. J. P. Simmons, C. Shen and Y. Wang, *Scr. Mater.* **43**, 935 (2000).
113. A.J. Ardell and Nicholso.Rb, *J. Phys. Chem. Solids* **27**, 1793 (1966).
114. M. Fahrmann, P. Fratzl, O. Paris, E. Fahrmann and W.C. Johnson, *Acta Metall. Mater.* **43**, 1007 (1995).

Chapter 7

QUANTUM APPROXIMATE METHODS FOR THE ATOMISTIC MODELING OF MULTICOMPONENT ALLOYS

Guillermo Bozzolo[1,2], Jorge Garcés[3], Hugo Mosca[4], Pablo Gargano[4], Ronald D. Noebe[2] and Phillip Abel[2]

[1]Ohio Aerospace Institute, Cleveland OH 44142, USA; [2]NASA Glenn Research Center, Cleveland, OH 44135, USA; [3]Centro Atomico Bariloche, 8400 Bariloche, Argentina; [4]Comisión Nacional de Energía Atómica, CAC-U.A.Materiales, Av. Gral Paz 1499, (B1650KNA) San Martín, Buenos Aires, Argentina

Abstract: This chapter describes the role of quantum approximate methods in the understanding of complex multicomponent alloys at the atomic level. The need to accelerate materials design programs based on economical and efficient modeling techniques provides the framework for the introduction of approximations and simplifications in otherwise rigorous theoretical schemes. As a promising example of the role that such approximate methods might have in the development of complex systems, the BFS method for alloys is presented and applied to Ru-rich Ni-base superalloys and also to the NiAl(Ti,Cu) system, highlighting the benefits that can be obtained from introducing simple modeling techniques to the investigation of such complex systems.

Keywords: Quantum approximate methods, superalloys, high temperature ordered intermetallics, defect structure, site substitution schemes, Monte Carlo simulations

1. INTRODUCTION

Due to its applications, it is accepted that Ni-based superalloys are among the most advanced engineering materials today. These materials have evolved over the years to the point that they can contain a dozen or more intentional elements, each with a specific purpose. Certain elements are added to promote the precipitation of the γ' fcc ordered phase, others to solid solution strengthen either the matrix phase or the γ' phase, still other elements are used to enhance interfacial or grain boundary strength, and other additions are intended to promote superior environmental resistance against oxidation, sulfidation, or hot corrosion. In addition, there are elements that have a major effect on rupture strength, creep properties, or fatigue.

It becomes apparent that for a full understanding of these systems, besides the individual role that each addition could have in the original system, it is equally necessary to isolate and understand the interactions between them and how they affect, change, and sometimes invalidate, their original purpose. However, given the complex nature of these alloys, further optimization and design can be quite tedious because of the multidimensional space that is defined by the current compositions and the number of experimental iterations that would be needed to understand the effect of even one new alloying addition on the properties of the base alloy. Consequently, further experimental design of these advanced systems has become quite expensive and time consuming, while providing significantly less potential benefit as the alloy system matures.

It must be recognized, however, that given the current research environment, it is unlikely that any new alloy system will be allowed to evolve over such a long time frame. Various high-temperature ordered intermetallic alloys, once heralded as potential replacements for the superalloys, did not receive the sustained interest and funding needed for their full development, as superalloys did. Of the thousands of potential systems that could have been studied, only a few simple formulations were examined in any detail. A frequent obstacle is that as the number of alloying additions grows, it is inevitable that complex interactions may arise besides the individual role intended for each addition. The most common consequence is that multiple alloying additions may interact more strongly with each other than with base alloy resulting, for example, in precipitation of a second phase that was never intended. This was the case during the evolution of Ni-base superalloys, where refractory elements meant as solid-solution strengthening agents of both matrix and γ' precipitates, when added at high levels, resulted in the precipitation of refractory-metal-rich topologically closed-packed phases, deleterious to the mechanical properties of these materials. Unfortunately, overcoming these problems is a lengthy and expensive process, with no guarantee of success. With increasing demand for materials with specific properties, something needs to be done in order to develop new alloys in a more efficient manner. Changes have been forthcoming in the last few years, where advances in computing power have enabled the growth of computational modeling as an ever more realistic way to supplement, or in some cases replace, experiment. It is also possible that full implementation of modeling as an integral part of the alloy design process could yield unexpected and viable new alternatives. In any case, it is essential to develop tools that could help researchers make educated decisions when designing new experiments, diminishing the serendipitous character of materials design by providing necessary guidance and criteria during the research process.

However, while the virtual design of new materials through computer simulations is gaining momentum, its long term success will depend, among other factors, on the availability of a single approach that provides the same level of simplicity and accuracy for almost any possible system and works equally well for surface or bulk analysis. Although *ab initio* or first-principles (FP) approaches provide the most accurate framework for such studies and to the problem of alloy phase stability, their substantial computational requirements still impose limitations that prevent these techniques from becoming economical predictive tools for systems as complex as most commercial structural or functional alloys. Alternative approaches exist, such as atomistic quantum approximate method (qam), that can handle simulations of more complex systems, in a more qualitative fashion than FP techniques. In broad terms, two different modeling and simulation approaches can be recognized. First, one where advances in computing power favor the development of faithful descriptions of the system at hand. Second, one where simple calculational schemes on idealized systems are implemented which, besides the obvious gains in computational efficiency and data analysis, lead to simple and basic interpretations of the phenomenon under study. In both cases, the goal is to test the response of a given system and, through this knowledge, gain better understanding. An advantage of the second approach is that it is not strictly necessary to adhere to known concepts or definitions, as long as the alternative virtual concepts or processes created to describe the system meet all the necessary requirements for consistency and predictability. The 'system' itself does not have to be a detailed replica of the real system. If the defined concepts and tools find a correlation with the real system, it is straightforward to create a path where the modeled results translate into alternative descriptions which can be associated to mainstream concepts that characterize more realistic approaches.

The purpose of this chapter is to investigate one such approach which, from its very basic formulation and implementation, relies on interpretations and definitions that are different from those used in other methods. Due to the simplified reality that it creates, the method allows for simple and straightforward descriptions of complex systems that would otherwise be difficult to understand by traditional methods. As mentioned above, this is a necessary requirement when dealing with one of the basic questions in materials design, where basic simple systems need to be modified by means of alloying additions in order to meet strict specific properties for the application for which they are intended.

In order to demonstrate the depth and breadth to which these computational modeling techniques can be used to study the behavior of complex alloys, this chapter concentrates on two applications to high-temperature ordered intermetallic systems, one based on RuAl and the other on NiAl. In the first example, we examine a system which has recently become a promising line of research in the development of new superalloys due to the rather unexpected finding that the presence of Ru could lead to the formation of precipitates which meet the necessary requirements for which these materials are intended. In this case, the interaction between alloying additions adds to their individual role. In the second example, however, this is not the case: additions (Ti, Cu) to NiAl alloys change their individual behavior due to the interactions between them. One feature is common to both examples: the need to fully understand not just the individual behavior of each alloying element but also the system as a whole, which could only arise from a modeled reality where each element, whether in the majority or not, is described on an equal basis.

To accomplish this, we first describe the Bozzolo-Ferrante-Smith (BFS) method for alloys,[1] its definitions and concepts, in order to provide the appropriate framework in which the analyses are made, and to explain how we can deal with such complex systems in a straightforward manner. Three main aspects of the method are shown: 1) the definition of virtual processes and their consequences, which allows for a simplified description of what is clearly a complex problem, 2) the relationship of the method to *ab initio* approaches, establishing the necessary link with formal descriptions and the credibility that ensues from such link, and 3) computational and operational schemes appropriate for the problem at hand, defining efficient and clear tools for what is, ultimately, this interpretation of modeling.

2. THE BFS METHOD

Materials development programs could benefit from two different theoretical approaches: FP, or qam, depending on the theoretical input needed in such program. At the most fundamental level, the solution of Schrödinger's equation contains all the information on the properties of the system. It is well known, however, that such solution only exists for very simple systems. This limitation has been circumvented with the use of two different approaches: one, dealing with the search of alternative, simplified, solutions which hopefully reproduce the main features of the (unknown) exact, true, solution. This is, in essence, the strategy followed by the Hartree-

Fock (HF) method. A second approach, as implemented in Density Functional Theory (DFT), consists of replacing Schrödinger's equation with one that is easier to solve, once again assuming that it is selected in such a way that its solutions are a good representation of the exact answer.[2] In spite of these simplifications, the corresponding equations are still hard to solve in either case, requiring a great deal of computer power and fast and efficient algorithms. Still, substantial progress has been realized in recent years, allowing for proper computational treatment of realistic problems.

Both approaches have one thing in common: either by limiting the search of possible solutions (as in HF), or by altering the equation that properly describes the system (as in DFT), no adjustable parameters appear in these methods. In other words, simplifications are introduced without any particular reference to the system under study. Underlying this potential limitation is the fact that when implementing the simplifications that characterize either approach, the real system is replaced by a virtual one, and it is assumed that the essential features of the real system are faithfully captured.[2] The payoff warrants making these necessary changes, as complex systems can be systematically described by known and manageable algorithms with a proven record of accuracy.

In essence, approximate methods follow the same path: substantial simplifications are made for the sake of efficiency, by replacing the real system or the real process by a virtual one, but one that allows for proper tuning and optimized performance within their limited framework by means of a minimal number of adjustable parameters.

This chapter concentrates on one member of the growing family of quantum approximate methods, the BFS method for alloys,[1] which fulfills several requirements for applicability in terms of simplicity, accuracy, and range of application. It has no limitations in its formulation on the number and type of elements present in a given alloy, thus showing promise in describing diverse problems, from alloy design to the study of surface alloys.[1,3-9] The BFS method relies on approximations, where the exact process of alloy formation is replaced with virtual processes whose results are, or are expected to be, a good description of the result of the real process. In terms of validation, the same way that DFT requires that the virtual system describes the real electron density with a high level of accuracy, BFS is expected to reproduce the essential features of the equation of state of the solid at zero temperature and, in particular, around equilibrium. Unlike *ab initio* methods that provide a full description of the system at hand (including band structure, density of states, charge density, etc.), BFS is limited to structural information that is, ultimately, contained in the binding energy curve describing the solid under study at zero temperature. This trade-off, i.e. greatly increased computational efficiency and a minimum number of universal parameters at the expense of detailed electronic structure information, allows for a full description of several aspects of interest. Starting from the same initial state, BFS tries to provide an alternative, virtual path leading to the final state with a minimum number of parameters to guide its way. While more flexibility can be gained by letting these parameters vary according to the specific problem at hand, no such degree of freedom is added to the method. The parameters remain fixed and fully transferable for any case dealing with the same elements, regardless of their number, type, or structural properties. This restriction implies that in order for the method to be equally valid in a number of diverse situations, as DFT is, the parameters must contain all the necessary information to warrant the accuracy of the virtual path chosen for describing the process of alloy formation. If this description is correct, then the method should accurately reproduce the most critical properties of the solid in its final state, including the cohesive energy per atom, compressibility, and equilibrium Wigner-Seitz radius. Their accuracy is essential for a proper description of atomic defects such as the site preference behavior of alloying additions in multicomponent systems.[10,11]

It is also worth noting that the parameterization of the BFS method implies a somewhat different approach for the interaction between different atoms. In general, most approaches introduce some sort of interaction potential, with the parameters describing each constituent remaining unchanged. In BFS, it is precisely the set of parameters describing the pure element that is perturbed in order to account for the distortion introduced by the presence of a different element or defect nearby. The additive nature of the perturbative theory means that only information from binary systems is needed within the BFS framework. Multicomponent systems are thus studied via only binary perturbations.

As in FP-based methods, BFS does not require experimental information and only the quantified chemical element descriptions, the crystal structure, and the binary perturbation factors are needed as input. The method then provides a simple algorithm for the calculation of the energy of formation ΔH of an arbitrary alloy (the difference between the energy of the alloy and that of its individual constituents), written as the superposition of elemental contributions ε_i of all the atoms in the alloy, where ε_i denotes the difference in energy between a given atom in the equilibrium alloy and in an equilibrium single crystal of species i,

$$\Delta H = \sum_i \varepsilon_i \qquad (1)$$

For each atom, we partition the energy into two parts: a strain energy, ε_i^S, and a chemical energy, ε_i^C. By definition, the BFS strain and chemical energy contributions deal with geometry and composition as isolated effects. A coupling function, g_i, restores the relationship between the two terms by considering the asymptotic behavior of the chemical energy, where chemical effects are negligible for large separations between dissimilar atoms. A reference chemical energy, ε_i^{Co}, is also included to insure a complete decoupling of structural and chemical features. Summarizing, the contribution to the energy of formation of atom *i* is then

$$\varepsilon_i = \varepsilon_i^S + g_i\left(e_i^C - e_i^{C_0}\right) \tag{2}$$

The BFS strain energy, ε_i^S, differs from the commonly defined strain energy in that the actual chemical environment is replaced by that of a monoatomic crystal. Its calculation is then straightforward, even amenable to FP methods. In this work, however, we use for its computation the Equivalent Crystal Theory (ECT),[12] due to its ability to provide accurate and computationally economical answers to most general situations. The BFS strain energy contribution, ε_i^S, is obtained by solving the ECT perturbation equation

$$NR_1^{p_i} e^{-\alpha_i R_1} + MR_2^{p_i} e^{-(\alpha_i + 1/\lambda_i)R_2} = \sum_j r_j^{p_i} e^{-\left(\alpha_i + S(r_j)\right) r_j} \tag{3}$$

where N and M are the number of nearest-neighbors (NN) and next-nearest neighbors (NNN) at distances R_1 and R_2 (in the equivalent crystal), respectively, and where p, l, α and λ are ECT parameters[12] that describe element *i* in the real crystal, r denotes the distance between the reference atom and its NN and NNN, and $S(r)$ describes a screening function for NNN. This equation is used for the calculation of the lattice parameter a_i^S of a perfect crystal where the reference atom *i* has the same energy as it has in the geometrical environment of the alloy under study. Once the lattice parameter of the (strain) equivalent crystal, a_i^S, is determined, ε_i^S is computed using the universal binding energy relation (UBER) of Rose et al.,[13] which contains all the relevant information concerning a single-component system:

$$\varepsilon_i^S = E_{C,i}\left\{1 - \left(1 + a_i^{S*}\right) e^{-a_i^{S*}}\right\} \tag{4}$$

where $E_{C,i}$ is the cohesive energy of atom *i* and where the scaled lattice parameter a_i^{S*} is given by

$$a_i^{S*} = q\frac{\left(a_i^S - a_{e,i}\right)}{l_i} \tag{5}$$

Table 1. Results for the lattice parameter, cohesive energy, and bulk modulus for the bcc phases of a) Ru, Al, Ta, Ni, W, Co and Re, obatined with LAPW, and b) Ni, Al, Ti., Cu, obtained with LMTO. Also listed are the resulting ECT parameters p, α, *l* and λ and the BFS perturbative parameters Δ_{AB} and Δ_{BA}.

(a)

	Lattice parameter (Å)	Cohesive energy (eV)	Bulk modulus (GPa)	ECT parameters p	α(Å^{-1})	*l*(Å)	λ(Å)
Ru	3.0484	6.5514	294.52	8	3.5974	0.2508	0.7048
Al	3.2400	3.4225	69.37	4	1.7609	0.3623	1.0180
Ta	3.3241	9.4396	194.32	10	4.3467	0.3549	0.9973
Ni	2.7985	5.6001	198.35	6	3.0597	0.2949	0.8288
W	3.1858	8.9000	302.16	10	4.3672	0.2823	0.7933
Co	2.7591	5.2842	250.32	6	3.0317	0.2568	0.7216
Re	3.1137	7.7270	358/24	10	4.3762	0.2444	0.6866

BFS parameters Δ_{AB} and Δ_{BA} (in Å^{-1}) for the bcc elements used in Sec. 4.

A\B	Ru	Al	Ta	Ni	W	Co	Re
Ru		-0.041861	-0.100202	0.285035	0.013210	0.105500	-0.048171
Al	-0.024831		-0.047743	0.039742	-0.051695	0.051118	-0.046416
Ta	0.334780	0.068865		0.780169	0.075347	-0.054963	0.166979
Ni	-0.043618	-0.040894	-0.097999		-0.053251	-0.031848	-0.042922
W	-0.020386	0.204440	-0.066284	0.293955		-0.051474	0.134775
Co	-0.037791	-0.041359	0.027972	0.074819	0.054312		-0.042531
Re	0.205503	0.155822	-0.089358	0.250198	-0.014582	0.161530	

(b)

	Lattice parameter (Å)	Cohesive energy (eV)	Bulk modulus (GPa)	ECT parameters p	α(Å^{-1})	*l*(Å)	λ(Å)
Ni	2.752	5.869	249.21	6	3.1486	0.2716	0.7632
Al	3.192	3.942	77.97	4	1.8917	0.3695	1.0383
Ti	3.213	6.270	121.03	6	2.8211	0.3728	1.0476
Cu	2.822	4.438	184.55	6	3.0492	0.2710	0.7615

BFS parameters Δ_{AB} and Δ_{BA} (in Å^{-1}) for the bcc elements used in Sec. 5.

A\B	Ni	Al	Ti	Cu
Ni		-0.06078	-0.09062	0.01914
Al	0.0916		-0.08649	0.05438
Ti	0.4958	0.23399		0.20565
Cu	-0.01708	-0.04993	-0.07356	

where q is the ratio between the equilibrium Wigner-Seitz radius and the equilibrium lattice parameter, $a_{e,i}$.

The BFS chemical energy, e_i^C, is obtained by a similar procedure. As opposed to the strain energy term, the surrounding atoms retain their chemical identity, but are forced to be in equilibrium lattice sites of an equilibrium crystal of atom *i*. The BFS equation for e_i^C is given by

$$NR_1^{p_i}e^{-\alpha_i R_1} + MR_2^{p_i}e^{-(\alpha_i+1/\lambda_i)R_2} =$$

$$= \sum_j \{N_{ij}r_q^{p_i}e^{-\alpha_{ij}r_1} + M_{ij}r_2^{p_i}e^{-(\alpha_{ij}+1/\lambda_i)r_2}\} \tag{6}$$

where N_{ij} and M_{ij} are the number of NN and NNN of species *j* around atom *i*. The chemical environment surrounding atom *i*, reflected in the parameter α_{ij}, is given by $\alpha_{ij} = \alpha_i + \Delta_{ji}$, where the BFS parameter Δ_{ji} (a perturbation on the single-element ECT parameter α_i) describes the changes of the wave function in the overlap region between atoms *i* and *j*. Once Eq. 6 is solved for the equivalent chemical lattice parameter, a_i^c, the BFS chemical energy is then

$$e_i^C = \gamma_i E_{C,i}\left\{1-(1+a_i^{C*})e^{-a_i^{C*}}\right\} \tag{7}$$

where $\gamma_i = 1$ if $a_i^{C*} > 0$ and $\gamma_i = -1$ if $a_i^{C*} < 0$, and the scaled chemical lattice parameter a_i^{C*} is given by

$$a_i^{C*} = q\frac{\left(a_i^C - a_{e,i}\right)}{l_i} \tag{8}$$

The reference chemical energy term, $e_i^{C_o}$, is computed following the same procedure outlined for the chemical energy, but setting $\Delta_{k,i} = 0$ in Eq. 6. The chemical energy, e_i^C, is then free of strain and structural effects. Finally, as mentioned above, the BFS strain and chemical energy contributions are linked by a coupling function g_i, which describes the influence of the geometrical distribution of the surrounding atoms in relation to the chemical effects and is given by

$$g_i = e^{-a_i^{C*}} \tag{9}$$

The computation of ε_i^S and e_i^C, using ECT,[12] involves three pure element properties for atoms of species *i*: cohesive energy, lattice parameter and bulk modulus. The chemical energy, e_i^C, includes two BFS perturbative parameters (Δ_{ki} and Δ_{ik}, with *i*, *k* including all possible binary combinations of the alloy constituents). Table 1 lists all the parameters for the pure elements used in the different examples in this chapter. Two different FP methods are used in order to show that the parameterization scheme and its effect on the performance of the method is independent of the particular FP method selected for the calculation of the parameters: the linerarized augmented plane wave (LAPW) method[14] for Ru, Al, Ta, Ni, W, Co and Re, in the first example, and the linear muffin tin orbital (LMTO) method[15] for Ni, Al Ti and Cu in the second example.

3. RELATIONSHIP BETWEEN BFS AND AB INITIO METHODS

In this section we elaborate further on the concepts introduced in Sec. 2 and examine the performance of the BFS method with respect to FP calculations. Comparable in simplicity but different in its formulation from other qam, the BFS method for alloys is based on a novel way of interpreting the alloy formation process. Any given system, regardless of its composition and structure, is always modeled in terms of two independent virtual processes which, when properly coupled, are meant to result in the final state that is being studied. Both processes are based on the concept of ideal equivalent crystals, assuring accuracy when describing bulk or surface problems, as both will be mapped onto isotropically deformed equivalent crystals and a unique set of parameters. However, the correspondence between the virtual processes and real situations, or their interpretation as components of the process of alloy formation, does not necessarily guarantee that the results will be accurate or comparable to those obtained with FP methods. Once the choice is made to depart from a straight description of the real process and virtual processes are chosen to replace them, the freedom in the features describing each virtual process is constrained by the fact that the coupling between them has to be such that the final state coincides with the actual alloy or system under study. The coupling function g links the structural and chemical information of the system in a straightforward and computationally economical way (no additional calculations are required to determine the value of g for every single atom in the computational cell). As defined in Eq. 9, g ensures the correct asymptotic behavior of the chemical energy, emphasizing its effect in compressed systems and diminishing it in expansion.

This simple way of coupling the two processes can be shown to yield results that are comparable to those obtained from FP calculations. To this end, we consider a multicomponent compound of X atoms in the unit cell. The energy of formation of an alloy, ΔH, is the difference between the total energy of the alloy and the sum of the equilibrium energy of each one of its components evaluated at the equilibrium lattice parameter of the alloy,

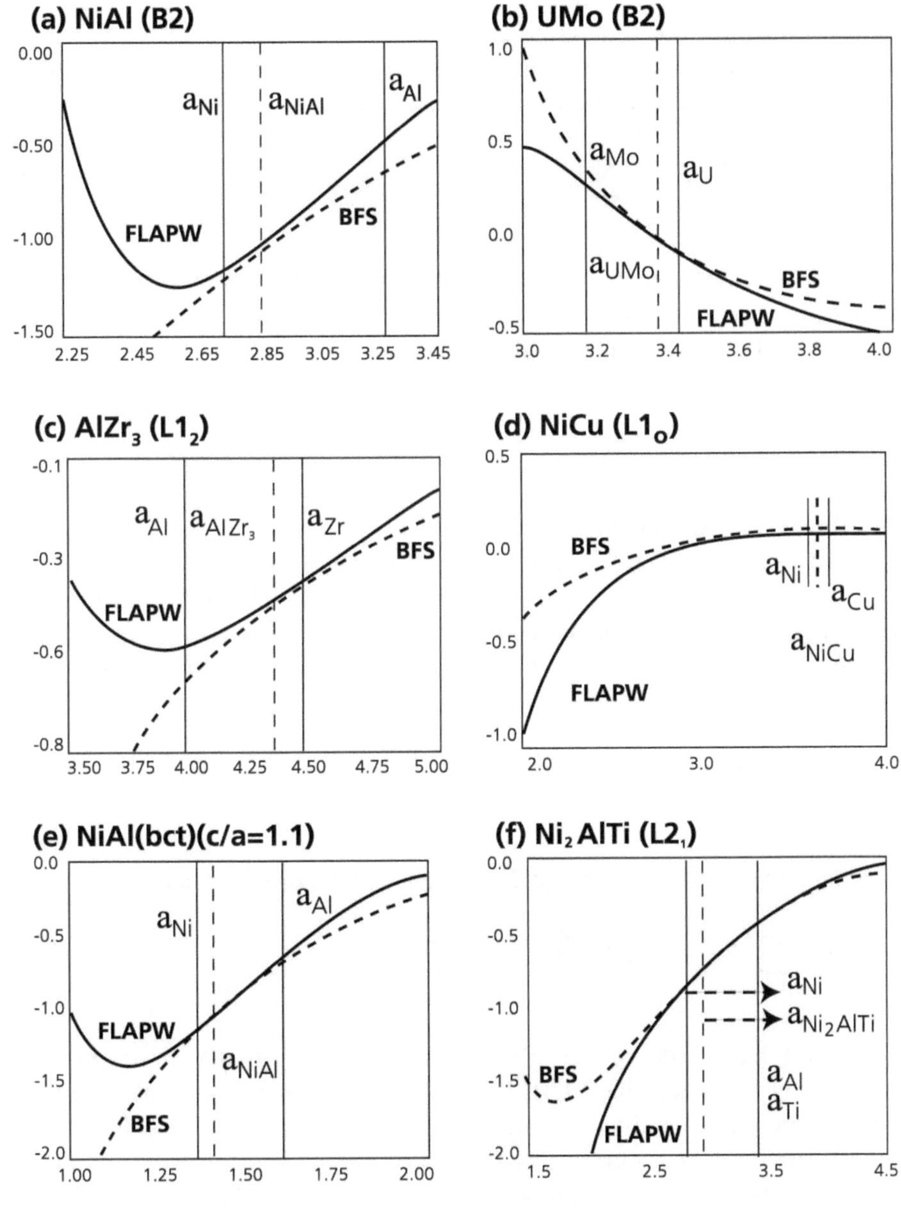

Figure 1. Comparison of FP results (solid curves, in eV/atom) and BFS results (dashed curves), as described by Eq. 13, and as a function of lattice parameter (in Å), for (a) NiAl (B2), (b) UMo (B2), (c) $AlZr_3$ ($L1_2$), (d) NiCu ($L1_0$), (e) body-centered tetragonal NiAl (c/a = 1.1), and (f) Ni_2AlTi ($L2_1$). In this last case, the Wigner-Seitz radius is used instead of the lattice parameter. The vertical lines denote the equilibrium lattice parameters of the individual elements (in the symmetry of the alloy, solid lines) and the lattice parameter of each ordered structure (dashed line), as predicted by FP methods.

computed from the minimum of the UBER (or any other equation of state at T = 0 K). In turn, the total energy of the compound can also be written in terms of the energy of each of its components, $E_{X,i}$, and an 'alloying' term, $E_{residual}$. Combining these two definitions, we obtain

$$\Delta H = E - \sum_{X} E_X^0 = \sum_{X,i} \left(E_{X,i} - E_X^0\right) + E_{residual} \tag{10}$$

This expression can now be compared to the BFS expression for ΔH in terms of strain and chemical components, reformatted to single out the energy contributions from monoatomic crystals,

$$\Delta H = \sum_{X,i} \varepsilon_{X,i}^S + \varepsilon_{residual} \tag{11}$$

A direct comparison of Eqs. (10) and (11) shows that BFS extracts the maximum amount of information of a given compound from the single element UBER, with $\varepsilon_{residual}$ responsible for any additional information regarding the mixing process. In BFS, this quantity is written as a linear combination of the coupling functions $g_{X,j}$ assigned to each atom j of element X,

$$\varepsilon_{residual} = \sum_{X,j} \mu_{X,j} g_{X,j} \tag{12}$$

where $\mu_{X,j}$ are the chemical energies of atom j of species X. Finally,

$$\Delta H - \sum_{X,j} \left(E_{X,j} - E_X^0\right) = \sum_{X,j} \mu_{X,j} g_{X,j} \tag{13}$$

The left hand side of Eq. 13 denotes quantities that can be properly described by FP-determined UBERs, whereas the right hand side denotes a quantity that is exclusively computed within the context of BFS. If the method provides an accurate description of the mixing process and if the parameterization of the elements is properly included, then the validity of BFS is warranted when the identity between both terms is satisfied within the range of validity of the description of the system as provided by the UBER (i.e., in the vicinity of equilibrium). To illustrate this point, Fig. 1 displays results for a variety of binary systems. Fig. 1.a and 1.b show the results for two ordered B2 (bcc) compounds, NiAl and UMo, highlighting the fact that BFS is equally accurate regardless of the type of element

Table 2. Calculated (LAPW) properties compared to measured and other theoretical values for NiAl and RuAl. The first theoretical value corresponds to LAPW results (* this work).

		a_0(Å)	B (GPa)	ΔH (ev/atom)
NiAl	Experiment	2.882[39], 2.881[40]	158-166[40], 189[42]	0.61[52], 0.6831[53]
	Theory	2.89*,2.84[44], 2.881[40], 2.86[41], 2.88[48], 2.84[45]	185*,200[41], 170[43], 184[45]	0.79*,0.697[45], 0.68[46], 0.61[44] 0.733[47]
RuAl	Experiment	2.95-3.03[39],2.95[49]	208[41]	0.64[50]
	Theory	3.005*, 2.967[45], 3.02[47]	203*, 223[45], 220[47]	0.99*, 0.61[45], 0.776[47], 0.81[51]

considered. Figs. 1.c and 1.d refer to fcc compounds, $AlZr_3$ ($L1_2$) and NiCu ($L1_0$), respectively, indicating that the approach is equally applicable regardless of the lattice mismatch between the constituents or the type of lattice (bcc or fcc) or the specific geometry of the cell, as shown in Fig. 1.e for the case of body-centered-tetragonal NiAl (with $c/a = 1.1$).

Fig. 1.f shows results for a ternary case, an $L1_2$ Ni_2AlTi Heusler alloy, addressing the extended validity of the method for systems with more non-equivalent atoms and number of components. Results by Légaré et al.[16] indicate that similar conclusions can be drawn for hcp systems. Similarly, the methodology does not impose restrictions on the number of elements in the system at hand, and no loss in accuracy should be expected for higher-order systems. Recent applications to systems with up to 12 elements[17] indicate that this correspondence between FP and BFS results could be expected for any number of components.

The remaining sections in this chapter are examples of the application of BFS as used to investigate the properties of ternary and higher order alloys, providing a modeling scheme that would make the analysis of complex systems equally simple, regardless of the number of components.

4. MODELING OF RuAlX ALLOYS

In comparison with nickel or cobalt aluminides, B2 RuAl has appreciable room temperature toughness and plasticity and maintains considerable strength at high temperatures.[18,19] These properties, in combination with excellent oxidation resistance,[20] make this alloy a potential candidate for the challenging environments encountered in aerospace applications though cost and high density are a significant concern. In an effort to drive down both cost and weight and improve upon its other properties, several studies[21-24]

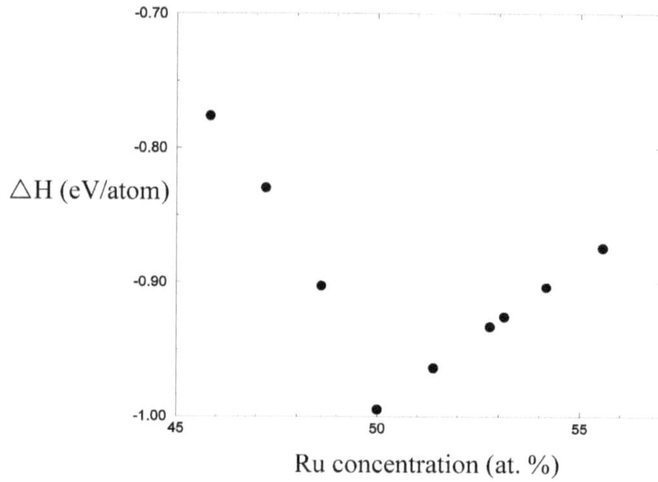

Figure 2. Energy of formation of B2 Ru-Al (in eV/atom) vs. Ru concentration. The data points shown correspond to the lowest energy configuration found at each composition.

have looked at alloying schemes for replacing Ru or Al with other elements that generally form an isostructural B2 phase such as Co and Fe for Ru and Ti for Al. But by far, the most widely studied ternary alloying addition has been Ni.[25-31] Even then, there is disagreement as to the structure of ternary Ni-Ru-Al alloys that exist between the NiAl and RuAl B2-phase fields. Some studies[27,28,32] have reported a miscibility gap centered between the two binary phases resulting in a region consisting of two distinct B2 compounds. In apparent agreement, Sabariz and Taylor[29] have also observed a two-phase alloy at the composition $Ni_{25}Ru_{25}Al_{50}$. While Horner et al.[31] found results similar to Chakravorty and West[28] in that many, but not all, of the ternary compounds seemed to exhibit two distinct components, the evidence seemed to suggest coring as opposed to actual formation of two distinct B2 phases. Furthermore, a sample within the miscibility gap claimed by Chakravorty and West was heavily milled and annealed so that diffusion distances would be much smaller and a better opportunity for obtaining a near equilibrium structure would exist. In this case, only a single B2 phase was observed. Furthermore, Liu et al.[30] observed only a single B2 phase across the (NiRu)Al system in mechanically alloyed samples when Ni was varied between 10 and 25 at.%, indicating complete mutual solubility between NiAl and RuAl.

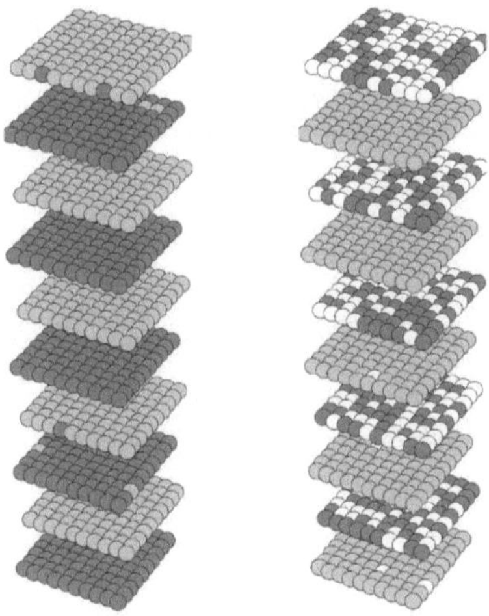

Figure 3. Final state (room temperature) of Monte Carlo NN simulations for (a) RuAl and (b) $Ni_{25}Ru_{25}Al_{50}$. Ni, Ru, and Al atoms are denoted with white, dark, and light grey spheres, respectively.

For Ta additions, Feng et al.[33,34] demonstrated the existence of a Ru_2AlTa Heusler (H) phase, which is also an equilibrium component in certain Ni-base superalloys containing high levels of Ru. Interest in Ru-containing alloys is growing since Ru additions seem to improve the high-temperature properties of Ni-base superalloys.[35,36] A similar case can be made about the beneficial strengthening effects of fine Heusler precipitates.[33-35,37] However, besides the recently reported H phase[33] and a single phase B2-ordered $Ru_{55}Al_{32}Ta_{13}$ alloy,[34] very little is known about the ternary Ru-Al-Ta phase diagram. The existence of these two ordered structures, in addition to the fact that the binary RuAl phase diagram does not show Ru-rich ordered phases, suggests that Ta plays an important role in stabilizing highly ordered phases in the Ru-Al-Ta system. A similar case could be made for Nb additions, due to the existence of a Ru_2AlNb H phase.[38]

BFS can be used to determine a number of properties for Ru-Al-X alloys, as a guide to help understand the structure of such systems. For example, the lattice parameter and energy of formation of B2 RuAl and $(Ru_{50-x}Ni_x)Al_{50}$ alloys can be determined as a function of Ni concentration. Also, Monte Carlo simulations for compositions close to $Ru_{25}Ni_{25}Al_{50}$ can be used to determine the structure of alloys where a miscibility gap was once reported. For Ru-Al-Ta alloys the modeling leads to a better understanding of the recently discovered $L2_1$ Ru_2AlTa Heusler phase, as well as other ordered structures in the Ru-Al-Ta system.

Table 3. NN (NNN) coordination matrices for the Ru-Ni-Al simulation shown in Fig. 3.b. The matrix element in row *i* and column *j*, denotes the probability that an atom of species *i* has a NN (NNN) of species *j*, where *i* (or *j*) = 1, 2, 3 corresponds to Ru, Al, Ni, respectively.

NN (*NNN*)	Ru	Al	Ni
Ru	0.000	0.992	0.008
	(0.480)	(0.008)	(0.512)
Al	0.496	0.016	0.488
	(0.004)	(0.985)	(0.011)
Ni	0.008	0.976	0.016
	(0.512)	(0.023)	(0.465)

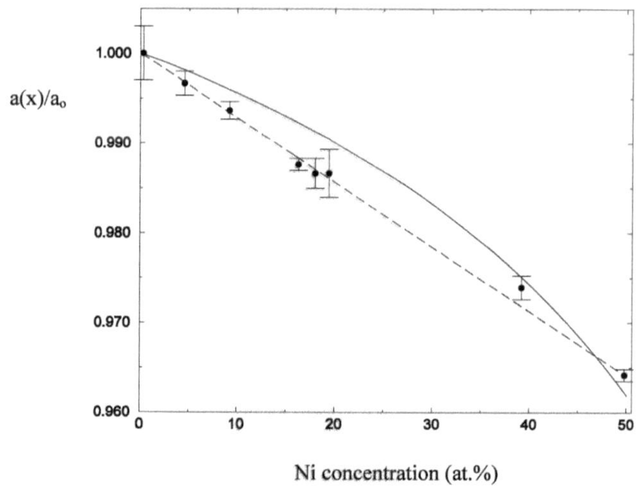

Figure 4. Comparison of experimental results and BFS predictions (solid curve) for the lattice parameter of $(Ru_{50-x}Ni_x)Al_{50}$ alloys as a function of Ni concentration. The dashed line indicates the average values.

4.1 The Ru-Al system

Starting with the basic binary system, RuAl, a general comparison between the LAPW values and other calculated and experimental values of the lattice parameter, cohesive energy, and bulk modulus for NiAl and RuAl is shown in Table 2. This comparison raises confidence in the parameterization used, which remains the same when the methodology is applied to higher order systems.

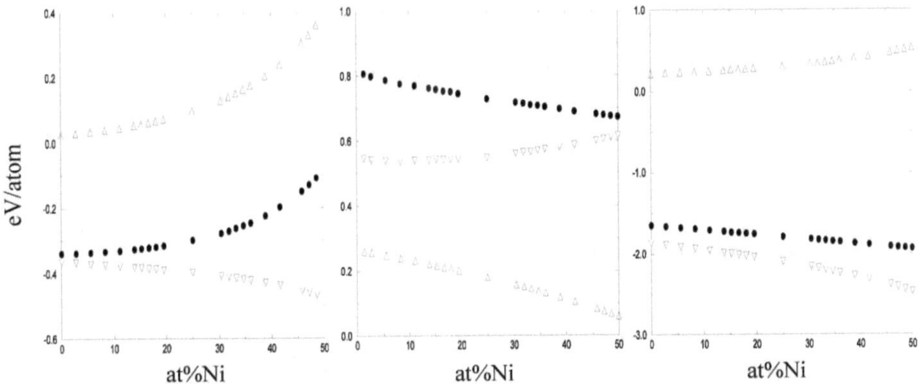

Figure 5. Individual average BFS contributions for a) Ru, b) Ni and c) Al atoms in a $(Ru_{50-x}Ni_x)Al_{50}$ alloy, as a function of Ni concentration. In each case, the strain (triangle up), chemical (triangle down) and total (solid disk) energies are shown.

We first study the B2 RuAl phase field, by considering a large number of possible computational cells defining various defect structures for off-stoichiometric compositions. Regardless of whether structural vacancies are included or not in this set of possible configurations, it is seen that the RuAl phase is unstable with respect to changes in stoichiometry, in the sense that there is a noticeable change in energy of formation off-stoichiometry. Furthermore, this change is more pronounced for Al-rich alloys, as can be seen in Fig. 2. For Ru-rich alloys, the lowest energy configurations correspond to substitutional alloys characterized by clustering of excess Ru atoms (as opposed to alternative ordered patterns). This can be explained in terms of the individual BFS contributions e_i (i = Ru, Al) to the energy of formation of the computational cell. For B2 RuAl, e_{Ru} = -0.34 eV/ atom and e_{Al} = -1.65 eV/atom, indicating that both atoms contribute to the formation of the alloy (negative contributions to ΔH). The 'weak' contribution of Ru atoms (compared to that of Al atoms) and the similarity between the atomic volume per atom in the B2 RuAl alloy and the atomic volume in a pure Ru crystal, lead to the phase separation observed for Ru-rich alloys.[23] Monte Carlo simulations using the BANN approximation[54] show that the B2 RuAl structure is highly ordered and stable up to 2300 K.

4.2 The Ru-Al-Ni system

Given the possibility of reducing the cost and weight of RuAl alloys through substitution of Ru with Ni, the potential to use RuAl as a reinforcing phase in other intermetallic systems, and given the controversy surrounding the phase structure of ternary Ru-Ni-Al alloys in the region between the two binary B2 phases NiAl and RuAl,[27-31] Ru-Ni-Al is an ideal system for further theoretical analysis.

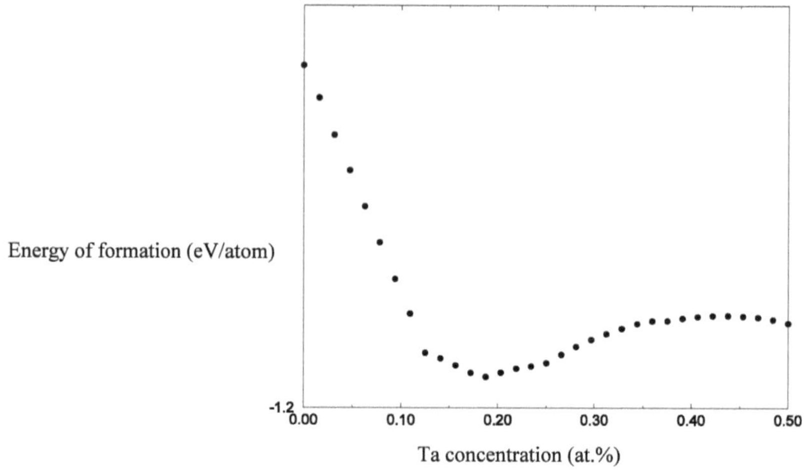

Figure 6. Energy of formation (in eV/atom) of $Ru_{50}Al_{50-x}Ta_x$ (0 < x < 50 at%) alloys as a function of Ta concentration. The data points shown correspond to the lowest energy configuration found at each composition by means of Monte Carlo simulations.

For the ternary case, we first determine the preferred site occupancy for dilute additions of Ni in RuAl. For convenience, we introduce a simple notation to denote single or double substitutional defects where A(B) denotes an A atom in the B sublattice and A(B)B(C) denotes an A atom in a B site, next to the displaced B atom in a C site. For $(Ru_{50-x}Ni_x)Al_{50}$ alloys, Ni(Ru) = -0.98 eV/atom and Ni(Al)Al(Ru) = -0.88 eV/atom, indicating that Ni prefers available Ru sites. However, for $Ru_{50}(Al_{50-x}Ni_x)$ alloys Ni(Al) = -0.96 eV/atom and Ni(Ru)Ru(Al) = -0.95 eV/atom, indicating a slight preference for Al sites, although the difference in energy is so small that it is difficult to conclude any obvious preference for either site under these conditions. The strong preference of Ni for Ru sites when Ni+Ru < 50 at.% is not surprising, as it allows for the formation of strong Ni-Al bonds.

Additional information can be obtained from the NN (or NNN) coordination matrices b_{ij} (or c_{ij}), shown in Table 3, which denote the probability that an atom *i* has an atom *j* as a NN (or NNN). In terms of the phase structure of the ternary alloys bridging the two binary B2 phases NiAl and RuAl, there are reports that suggest either a miscibility gap and phase separation for compositions near $Ni_{25}Ru_{25}Al_{50}$[41,43,48] or a single ternary B2 phase throughout.[30,31] While the simulations show the formation of the B2 RuAl phase, as can be seen in Fig. 3.a, the results are just as clear for $Ni_{25}Ru_{25}Al_{50}$, as shown in Fig. 3.b. Both Ru and Ni share the same sublattice. If there is phase separation, it would be expected that atoms in the Ru/Ni sublattice will have, mostly, atoms of their same species as NNN. If there was a ternary ordered phase other than B2, the NNN coordination of Ru or Ni atoms would reflect this fact with an increasing number of NNN of the other species. For example, the NNN coordination matrix for a Heusler $L2_1$ ternary phase would maximize the value of b_{NiRu} ($= b_{RuNi} = 1$). Phase separation, on the other hand, would lead to a very small value of b_{NiRu} (and b_{RuNi}), maximizing in turn, b_{NiNi} (and b_{RuRu}). Instead, the results of the simulations reflect the existence of a (Ru,Ni)Al B2 phase where Ru and Ni are randomly located throughout their own sublattice, in agreement with Liu et al.[30] and Horner et al.[31]

The BFS values of the lattice parameter for $(Ru_{50-x}Ni_x)Al_{50}$ as a function of Ni concentration (in at%), $a(x)$, show a slight deviation with respect to the average values, as shown in Fig. 4. Reported measurements[29,30] of $a(x)$ are also considered as having a linear behavior. The linear fit of the experimental values[29] (normalized to the equilibrium B2 value a_0) is $a(x)/a_0 = 1 - 0.0006949x$. In spite of the slight positive deviation from linearity, the best linear fit of the BFS prediction is almost identical, $a(x)/a_0 = 1 - 0.0007266x$.

To further understanding of the behavior of the ternary alloys, it is possible to define within the framework of BFS, the role that each atom plays in $(Ru_{50-x}Ni_x)Al_{50}$ as a function of Ni concentration, by separately computing the BFS energy contributions. Fig. 5 shows the strain, chemical and total energy contributions of Ru, Ni, and Al atoms to ternary Ru-Ni-Al alloys, computed as the average over all atoms of similar species. Ru atoms provide a favorable (i.e., $\varepsilon^C < 0$, favoring alloying) chemical contribution to the total energy of formation, diminished by increasing strain as the Ni concentration increases. As a result, the role of Ru in favoring compound formation is diminished with increasing Ni content (Fig. 5.a). Ni atoms display the opposite behavior: as the concentration of Ni increases, the strain energy becomes substantially smaller as the average volume per atom in the alloy becomes closer to that of Ni. The chemical contribution, which does not favor alloying, increases, but at a slower rate than the decrease in strain, resulting in a total contribution that does not favor the stability of the alloy (Fig. 5.b). Al atoms display a mild increase in strain with increasing Ni content, easily compensated by the growing (much more favorable than Ru) chemical contribution, leading to a net decrease of the total energy (Fig. 5.c), which along with the chemical contribution from Ru ultimately favors the formation of ternary B2 compounds.

Table 4. Individual BFS contributions to the energy of formation (in eV/atom) of each non-equivalent atom for various substitutional defects, relative to a perfect B2 RuAl cell. The first three columns show the total change in energy of the substitutional atom, the surrounding 8 NN and 6 NNN. The last column displays the net change in energy relative to B2 RuAl.

Defect	Subst. atom	(8x) NN	(6x) NNN	ΔE(eV/atom)
Ta(Al)	-4.78	2.99	0.35	-1.44
	Ta(Al)	(Ru)	(Al)	
Ta(Ru)	0.88	10.57	0.06	11.50
	Ta(Ru)	(Al)	(Ru)	
Ru(Al)	1.68	0.81	-0.26	2.23
	Ru(Al)	(Ru)	(Al)	
Al(Ru)	0.56	6.13	-0.01	6.67
	Al(Ru)	(Al)	(Ru)	

Table 5. Individual BFS contributions to the energy of formation (in eV/atom) of each non-equivalent atom in (B2) RuAl, ($L2_1$) Ru_2AlTa and (B2) RuTa alloys.

Alloy	Atom	E(strain) (eV/atom)	Coupling Function	E(chemical) (eV/atom)	E(total) (eV/atom)
RuAl	Ru	0.03	1.09	-0.33	-0.34
	Al	0.23	1.38	-1.36	-1.65
Ru_2AlTa	Ru	0.01	0.95	1.41	1.35
	Al	0.10	1.25	-1.13	-1.31
	Ta	0.70	1.41	-4.93	-6.25
RuTa	Ru	0.11	0.82	3.89	3.31
	Ta	0.34	1.28	-4.71	-5.69

4.3 The Ru-Al-Ta system

The site preference of Ta in $Ru_{50}(Al,Ta)_{50}$ alloys can be determined from two types of substitutions: a) Ta occupying available Al sites, Ta(Al), ('direct' substitutions) or b) Ta occupying Ru sites, leading to the creation of additional antisite defects: Ta(Ru)Ru(Al). Direct substitutions lower the energy of the system, indicating an absolute site preference of Ta for Al sites.

Configurations containing Ru(Al) antistructure atoms are much higher in energy due to the energy cost of creating antisite defects. This is detailed in Table 4, which displays the energy contributions of each substitutional atom and its immediate local environment, to the total energy of formation. Relative to a homogeneous B2 RuAl cell, Ta substitution in an Al site lowers

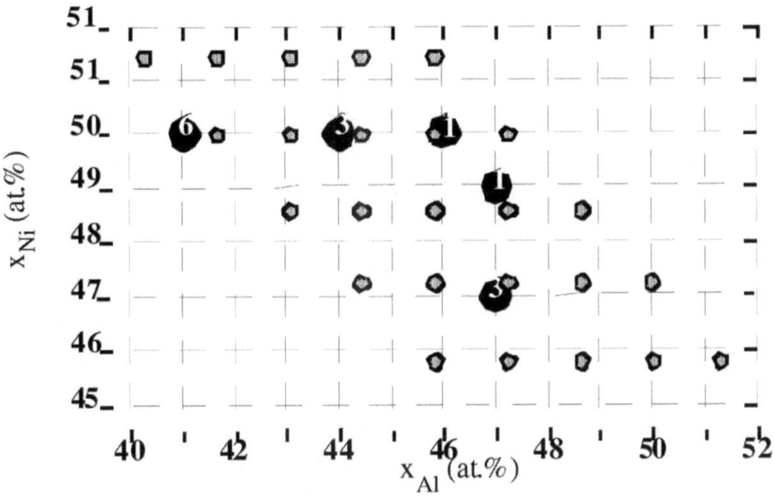

Figure 7. Composition of the NiAl-Ti-Cu alloys modeled. Those for which experimental data exist are denoted with filled circles. The numbers inside the circles indicate the concentration of Cu (in at.%). The horizontal and vertical axes indicate the concentration of Al and Ni, respectively. The Ti concentration was held constant at 3%. The grey solid circles denote the alloys studied analytically with BFS.

the energy of the system by 1.44 eV/atom, while any other substitution raises the energy. Within the framework of BFS, the results could be understood as ranging from a very favorable situation for Ta in Al sites (which lowers the energy by 4.78 eV relative to an Al atom in that site), to a very unfavorable situation for the Al nearest-neighbors of Ta in a Ru site (which in the presence of Ta(Ru), collectively raise the energy by 10.56 eV relative to the case when a Ru atom occupies that site). Similar calculations for increasing number of Ta atoms show that the absolute Ta preference for Al sites in $Ru_{50}(Al,Ta)_{50}$ alloys is independent of concentration.

The role of Ta in stabilizing the ternary Heusler phase can be seen in the evolution of the energy of formation (per atom) of $Ru_{50}Ta_{50-x}Al_x$ alloys as a function of Ta concentration, as shown in Fig. 6, for the range $0 < x_{Ta} < 50$ at%. Two distinct regimes can be seen: a low x_{Ta} regime, leading to the formation of the Heusler Ru_2AlTa structure (in the range $12.5 < x_{Ta} < 27.5$ at%), in agreement with experiment,[34] and a high x_{Ta} regime where the energetics of the metastable B2 RuTa structure dominate.[55]

Table 6. Percentage of Ti and Cu on Al sites.[57]

Alloy	Ti	Cu
$Ni_{50}Al_{46}Ti_3Cu_1$	93.1±5.2	77.1±6.4
$Ni_{50}Al_{45}Ti_3Cu_2$	90.1±10.0	91.8±17.9
$Ni_{50}Al_{44}Ti_3Cu_3$	95.5±8.0	88.1±13.1
$Ni_{49}Al_{47}Ti_3Cu_1$	85.9±6.4	29.6±8.3
$Ni_{48}Al_{47}Ti_3Cu_3$	87.3±15.9	6.8±18.7

The energy of formation (per atom) as a function of x_{Ta} shows increasingly large deviations from the average values for low x_{Ta}. For higher values, however, the 'average' behavior is restored in a relatively linear fashion. The source for this behavior can be traced to the BFS contributions to the total energy of formation from each type of atom in RuAl, Ru_2AlTa, and RuTa, as displayed in Table 5.

As mentioned above, the stability of the B2 RuAl structure (within the framework of BFS) is due to the fact that Ru and Al atoms, have negative chemical energies, leading to the formation of strong bonds. Added to the low strain of Ru atoms in the equilibrium B2 cell, the contributions of Ru and Al atoms to the total energy of formation are negative (-0.34 and -1.65 eV, respectively), meaning that both favor alloying. The substitution of Al for Ta changes the balance, but not the end result: Ta plays a similar role as that of Al, in that Ta atoms provide strong negative chemical energy contributions. Increasing amounts of Ta, however, lead Ru to change its role. While the strain energy contribution of Ru atoms in Ru(Al,Ta) alloys continues to be small, the net contribution of each Ru atom to the energy of formation is now positive (1.35 and 3.31 eV in Ru_2AlTa and RuTa, respectively). For $Ru_2(Al,Ta)$ alloys, in spite of Ru atoms 'rejection' of Ta atoms as nearest-neighbors, the large negative contribution of Ta atoms is strong enough to offset increasing positive contributions from Ru atoms (unfavorable for alloying), leading to a highly stable $L2_1$ structure, as observed experimentally.[33] In contrast, the $Ru_{55}Al_{32}Ta_{13}$ alloy, was experimentally found to have a B2 structure with Ru-rich precipitates.[34] The system does not undergo further ordering to a Heusler-type structure, retaining B2 $Ru_{50}(Al,Ta)_{50}$ order at any temperature.

Table 7. NN coordination matrices. All alloys contain 3 at% Ti.

(a) $Ni_{50}Al_{41}Ti_3Cu_6$

i\j	Ni	Al	Ti	Cu
Ni	4.3	80.2	6.0	6.6
Al	97.7	0.	0.0	2.3
Ti	98.4	0.0	0.0	1.6
Cu	80.3	15.6	0.8	3.3

(b) $Ni_{50}Al_{44}Ti_3Cu_3$

Ni	2.7	86.6	6.0	4.7
Al	98.5	0.0	0.0	1.5
Ti	99.2	0.0	0.0	0.8
Cu	77.0	21.4	0.8	0.8

(c) $Ni_{50}Al_{46}Ti_3Cu_1$

Ni	1.2	91.4	6.0	1.3
Al	99.4	0.0	0.0	0.6
Ti	99.6	0.0	0.0	0.4
Cu	68.8	27.5	1.2	2.5

(d) $Ni_{49}Al_{47}Ti_3Cu_1$

Ni	0.0	93.7	6.1	0.2
Al	97.8	0.3	0.0	1.8
Ti	98.4	0.8	0.0	0.08
Cu	10.0	87.5	2.5	0.0

(e) $Ni_{47}Al_{47}Ti_3Cu_3$

Ni	0.0	93.8	6.2	0.0
Al	93.8	0.0	0.0	6.2
Ti	96.4	0.0	0.0	3.6
Cu	0.0	96.4	3.6	0.0

4.4 The Ru-Al-Ta-Ni-W-Co-Re system

Finally, simulations can be extended to any alloy system regardless of the number of components. For example, Monte Carlo simulations were performed on a $Ru_{44.96}Al_{31.77}Ta_{12.92}Ni_{7.21}W_{2.01}Co_{1.01}Re_{0.11}$ alloy,[33] equilibrated at different temperatures. Heusler ordering is observed below 1500 K.

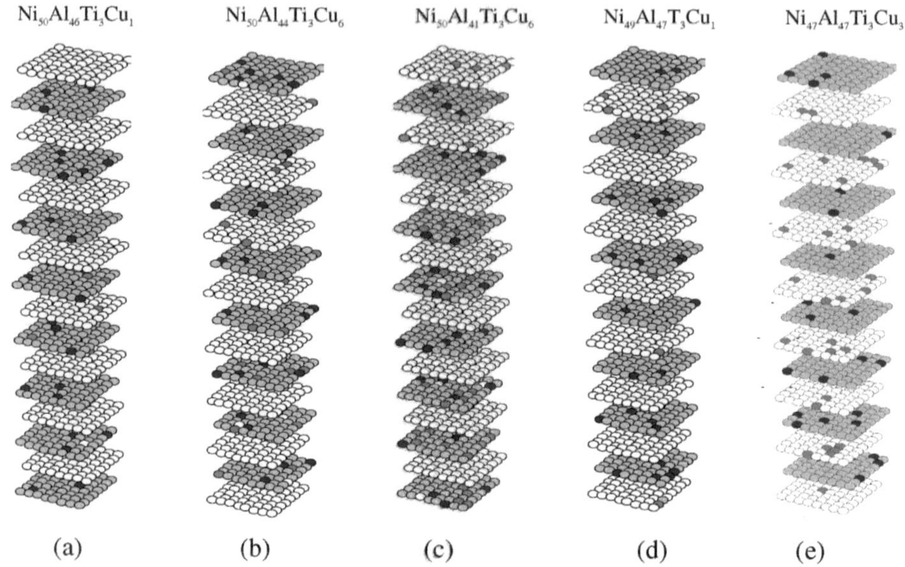

Figure 8. (a-e) Final structures of the Monte Carlo/Metropolis/BFS simulations of the five alloys studied in Ref. 57. White, light grey, dark grey and black spheres denote Ni, Al, Cu and Ti atoms, respectively.

At lower temperatures, Ni atoms mostly occupy Ru sites, while Co, W and Re atoms reside primarily on Ta sites. The lattice parameter of the T = 300 K cell is 6.062 Å, just 0.5% smaller than the experimental value[33] of 6.089 Å. At extremely low temperatures (below 100 K), however, the formation of a Ni_4W precipitate is seen, indicating a low solubility for W in the Heusler structure. The formation of a metastable 'Ni_4W' phase at low temperatures is, to a certain extent, open to interpretation. The simulations are performed in rigid bcc-based lattices and all atoms are constrained to seek order only within that particular geometry. Besides experimental verification, certainty about the formation of such phase can only arise from considering alternative symmetries. However, the results indicate that solubility for W in the alloy is low and would lead to precipitation of some kind of W-rich compound.

5. NiAlTiCu MODELING

In this section we move one step forward, both theoretically and experimentally, in understanding the role of simultaneous alloying additions by studying a quaternary system, a NiAl based alloy with Ti and Cu alloying additions, and use modeling to determine the detailed interactions between all the component elements.

5.1 Site Occupancy of Ti and Cu (experiment)

As the number of elements increases, the amount of available experimental evidence of any kind for a particular system decreases, assigning modeling techniques a growing role in filling the gaps in our knowledge. The purpose of this section is to highlight this point, by presenting recent computational (BFS) and experimental work making use of Atom Location by Channelling Enhanced Microanalysis (ALCHEMI)[56] to study NiAl(Ti,Cu) alloys.[57]

Five alloys were created as variations of the base composition $Ni_{50}Al_{47}Ti_3$ (at%). In Alloys #1, 2, and 3, the amount of Al replaced by Cu was 1, 3, and 6 at.%, respectively. Alloys #4 and 5 contain 1 and 3 at.% Cu, respectively, added to replace Ni. Transmission Electron Microscope (TEM) and Energy-Dispersive X-ray Spectroscopy (EDS) analysis on all five alloys resulted in the site occupancies for Ti and Cu listed in Table 6, along with the calculated uncertainties. The results demonstrate that the Cu content on Al sites is strongly dependent upon the stoichiometry of the alloy, whereas the Ti strongly prefers Al sites for all alloys. The calculation indicates that Cu substitutes for either Ni or Al sites, depending on stoichiometry.

5.2 Site Occupancy of Ti and Cu (BFS and Monte Carlo Simulations)

The first step in the modeling effort consists of testing the ability of the theoretical method to reproduce the experimental results for the five concentrations studied via ALCHEMI, as shown in Fig. 7. Clearly, no single theoretical technique can reproduce all the subtleties of an experiment, much less simultaneously deal with all the length and time scales involved. For the purpose of this work, however, it suffices to implement a ground state search where a large collection of atoms is allowed to evolve to its lowest energy state following a predetermined temperature dependent process. This is best achieved by means of Monte Carlo-Metropolis simulations where only the lattice parameter of the alloy (i.e., computational cell) is varied until the energy is minimized. The results for the five alloys are shown in Fig. 8. Table 7 displays the NN coordination matrices for the five alloy concentrations studied. As long as the cell follows the basic B2 ordering (in this case, Ni and Al atoms occupying their own sublattices) then b_{mNi} and b_{mAl} can be taken as an *approximate* measure of the likelihood that an atom m occupies a site in the Al or Ni sublattice, respectively. If $P[m(n)]$ denotes the probability of an atom m occupying a site in the n sublattice, then the previous statement can be written as $b_{mNi} \sim P[m(Al)]$ and

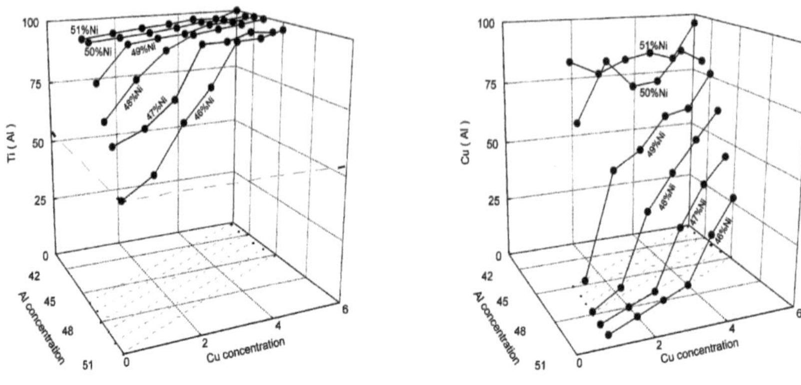

Figure 9. Matrix elements (a) a_{TiNi} and (b) a_{CuNi}, taken as a measure of P[*Ti(Al)*] and P[*Cu(Al)*], respectively, as a function of Al and Cu concentration, for different values of x_{Ni}. The projection of these curves onto the horizontal plane (dashed lines), highlights the fact that Cu(Al) goes to zero for high Al concentration and low Ni and Cu concentration.

$b_{mAl} \sim$ P[*m(Ni)*]. Large values of the diagonal elements in each matrix can indicate either antiphase boundaries or the presence of precipitates (i.e., a large value of b_{NiNi} indicates that many Ni atoms are at NN distance, which would be highly unlikely in a NiAl alloy where, if perfectly ordered, b_{NiNi} = 0). If the diagonal elements are small, then the off-diagonal elements can be taken as a good approximation of the site preference. In other words, if b_{NiNi} and b_{AlAl} are small, the closer the cell is to a highly ordered state, which translates into b_{TiNi} being a true measure of the likelihood of finding Ti in an Al site. In agreement with experimental results, the simulations show that Cu does change site preference in NiAl, depending on the ratio of Ni to Al, as seen in the variations of b_{CuNi} in Table 7. For the five alloys shown, P[*Ti(Al)*] > P[*Ti(Ni)*], whereas P[*Cu(Al)*] > P[*Cu(Ni)*] for Ni-rich alloys and P[*Cu(Ni)*] > P[*Cu(Al)*] otherwise. A comparison of the computed site occupation probabilities P[*Ti(Al)*] and P[*Cu(Al)*] with those determined experimentally,[43] clearly show, the agreement between experiment and theory, providing a great degree of confidence in the BFS simulations. The possibility of a general trend in site occupancy as a function of Ni and Al concentration warrants additional simulations for a range of concentrations that includes the five studied experimentally. Fig. 9 shows (a) P[*Ti(Al)*] and (b) P[*Cu(Al)*] as a function of Al, Ni and Cu concentration.

As expected, P[*Ti(Al)*] is always significantly greater than zero. The minimum value obtained is 54% for $Ni_{46}Al_{51}Ti_3$. In contrast, Cu(Al) decreases much more rapidly than Ti(Al) for increasing Al concentration and decreasing Ni and Cu concentration, eventually becoming zero. This behavior indicates a switch in site preference exclusively favoring Ni sites

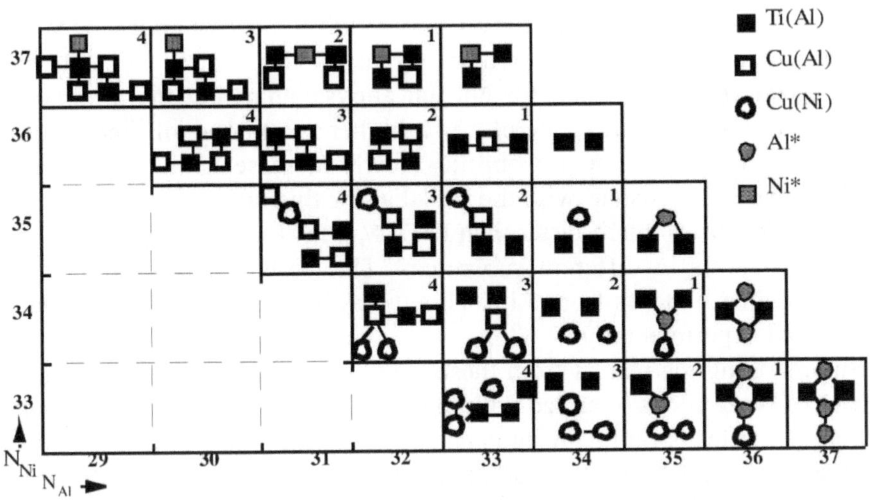

Figure 10. Basic schemes of the ground state for each alloy modeled (see Fig. 7). Squares (circles) denote atoms in the Al (Ni) sublattice. Connected squares (circles) indicate NN bonds and isolated squares (circles) denote atoms that are separated by distances greater than NN distance. The ground state structures range from patterned Ti(Al) and Cu(Al) substitutions (upper left corner), to Ti(Al) and Cu(Ni) substitutions (lower right). Ni and Al content is indicated along the vertical and horizontal axes, respectively. Every configuration includes two Ti atoms and Cu is balance, as indicated by the number in the upper right corner of each box. Asterisks denote antistructure atoms.

(b_{CuNi} = 0). This can also be seen in Fig. 9.b, in the shape of a vaguely defined 'boundary'. This transitional regime, loosely defined by alloys with a 1:1 Ni:Al ratio, is a region where P[*Cu(Al)*] (or the matrix element b_{CuNi}) displays a sharp drop for decreasing Ni:Al ratios, and where the corresponding Ti site preference for Al sites (P[*Ti(Al)*], or b_{TiNi}) in Fig. 9.a decreases at a much slower rate and is mostly independent of Cu concentration. In this sense, mapping of the simulated results indicate much greater sensitivity of P[*Cu(Al)*] with concentration than that seen for P[*Ti(Al)*]. While the experimental results suggest a sudden reversal in site preference for Cu at a specific stoichiometry, the theoretical results indicate that this is a more gradual change beginning for alloys with 1:1 Ni:Al ratio (i.e, the 'surface' that could be built from the data displayed in Fig. 9 for either P[*Ti(Al)*] or P[*Cu(Al)*] would resemble a 'cascade', smoothly evolving from one regime to another, as opposed to a sharp 'step' where there is no transitional region). General conclusions can be drawn regarding the interaction between Ti and Cu atoms in the NiAl matrix. While b_{TiTi} = 0 in every single alloy studied (indicating a strong repulsion between Ti atoms), b_{CuCu} is (in most cases) finite, allowing the possibility of clustering of a small fraction of Cu atoms, particularly for Ni-rich alloys (for example, b_{CuCu} = 3.0

for $Ni_{50}Al_{43}Ti_3Cu_4$, indicating that a Cu atom has a 3% probability of having another Cu atom as a NN). Moreover, the interaction matrix elements b_{TiCu} and b_{CuTi} are generally small for Ni-rich alloys, indicating the presence of few Ti and Cu NN pairs, therefore, a higher likelihood of finding both elements in Al sites. These probabilities slowly increase as the concentration of Ni decreases, consistent with the fact that for this range of concentrations P[*Cu(Al)*] decreases much faster than P[*Ti(Al)*], thus favoring the location of Ti and Cu atoms in different sublattices. This results in a switch of site preference for Cu atoms from Al to Ni sites with a change in stoichiometry, as observed experimentally.

Summarizing the results from these simulations, the strong ordering tendencies of NiAl alloys, coupled with the strong preference of Ti for Al sites, as well as the less prominent interaction between Cu and Ti atoms and the small energy difference for Cu atoms in Ni versus Al sites, all contribute to make Cu atoms the ones most likely to fill in for any deficiency on either side of stoichiometry. This translates into a smooth transition from Al to Ni site occupancy as the change in composition becomes Ni-poor.

5.3 Atom-by-atom Analysis of the Ground State

Detailed analytical calculations can provide information on the structure of the low-lying energy states (i.e., most likely to appear) and the energetics of individual atoms or groups of atoms. A reasonably large set of different atomic configurations in a 72-atom cell (sufficiently large to represent the most relevant distributions) is defined and its energy of formation computed using BFS. In what follows, we will use the notation $Ni_{[i]}Al_{[j]}Ti_{[k]}Cu_{[l]}$ to denote the concentration of the alloy in terms of the number of atoms in the 72-atom cell (e.g., [i]= 36 corresponds to 50 at% Ni).

In order to match the compositions studied either experimentally or with Monte Carlo simulations as closely as possible, we define 25 compositions that properly cover the whole range of Ni, Al and Cu concentrations studied before. These states, denoted with grey solid circles in Fig. 7, correspond to alloys $Ni_{[A]}Al_{[B]}Ti_{[2]}Cu_{[C]}$, where the subindex indicates the number of atoms of each species in the 72-atom cell (A+B+C = 70). For each concentration a catalogue of configurations is built so that, if large enough, it will contain every possible arrangement of atoms that is likely to occur in the real alloy. We then compute the energy of formation of each cell and plot the results in the form of an energy level spectrum, where, for each concentration, we can understand what atomic configurations are energetically favored and which ones are not.

5.4 Ground state structure versus Cu concentration

First, we concentrate on the structure of the ground states as a function of Cu concentration. Figure 10 summarizes the ground state results for all the alloys defined in Fig. 7. The strong site preference of Ti for Al sites guides this choice, as it allows for a clear understanding of the behavior of Cu as it increases its role in the system.

A series of $Ni_{[38-x]}Al_{[32+x]}Ti_{[2]}$, (x=1,...,5) alloys helps establish the behavior of Ti, when it is the only alloying addition. The behavior is dictated by the strong site preference of Ti for Al sites, regardless of the ratio between Ni and Al atoms. This preference for Al sites is absolute, in that Al antistructure atoms are created, if necessary, to accommodate all the Ti atoms in the Al sublattice (Ti(Al)). Ti(Al) atoms, however, are somewhat sensitive to the presence of antisite defects, whether these are Ni atoms in the Al sublattice (which attract other Ti(Al) atoms at NNN distance) or Al atoms in the Ni sublattice (which attract Ti(Al) atoms at NN distance). The number of antistructure Al atoms is reflected in the rapidly increasing energy of formation for this set, with an average increase of 0.055 eV/atom per antisite defect.

The second set of alloys, $Ni_{[38-x]}Al_{[31+x]}Ti_{[2]}Cu_{[1]}$, (x=1,...,5) shows the first indication of the interaction between Ti and Cu. Once again, Ti(Al) substitutions dominate, leaving for Cu second choice for available Al sites. Ti(Al) preference is strong enough as to induce the creation of Al antistructure atoms. Moreover, Ti(Al) seems to attract available Cu(Al) atoms along the [100] direction, thus inducing an ordered pattern where Ti and Cu atoms occupy alternating sites in the Al sublattice. However, the interaction between Ti(Al) and Cu atoms seems to be restricted to those cases where Cu goes to Al sites only. If Cu atoms are forced to go to Ni sites, they ignore the presence of Ti(Al) linking themselves only to Al antistructure atoms. These results highlight dominant (Ti(Al) -> antistructure atoms) and secondary (Cu(Al) linked to Ti(Al)) features characterizing this group of alloys. So far, the main features that characterize the N_{Cu}=0 and N_{Cu}=1 cases are: 1) absolute preference of Ti for Al sites, 2) first choice of Ti for Al sites, overriding the preference of Cu for Al sites, 3) creation of antistructure Al atoms when the number of Ti atoms exceeds the number of available Al sites (i.e., for N_{Al} >34), 4) clustering of Ti(Al) around Al(Ni) antisite defects, 5) clustering of Cu(Al) around Ti(Al) (at NNN distance) and 6) decoupling between Ti(Al) and Cu atoms when Cu goes to available Ni sites.

The N_{Cu}=2 ($Ni_{[38-x]}Al_{[30+x]}Ti_{[2]}Cu_{[2]}$, (x=1,...,5)) set highlights the interaction between Cu atoms and the enhanced competition between Ti and Cu additions with increasing Cu concentration. A common feature to the alloys in this set continues to be the absolute preference of Ti for Al sites, and the ensuing creation of antistructure Al atoms when Al sites become unavailable. Ti(Al) clusters form around the Ni antistructure atom. If Al sites are available, Cu(Al) atoms couple to Ti(Al) atoms, but as more Cu moves to the Ni sublattice, the coupling with Ti(Al) is replaced by either complete decoupling (as in $Ni_{[34]}Al_{[34]}Ti_{[2]}Cu_{[2]}$) or attraction with other Cu atoms (as in $Ni_{[33]}Al_{[35]}Ti_{[2]}Cu_{[2]}$).

The N_{Cu}=3 group ($Ni_{[38-x]}Al_{[29+x]}Ti_{[2]}Cu_{[3]}$, for x= 1,...,5), underscores the behavior of Cu, as its percentage now exceeds that of Ti. The only additional feature observed in this group, a consequence of the increased Cu content, consists of the apparent clustering of Cu atoms (either in the Ni or Al sublattice), decoupled from any Ti(Al) atoms. This highlights the competition between Ti-Cu ordering in the Al sublattice and the clustering of Cu atoms in either sublattice.

The N_{Cu}=4 group ($Ni_{[38-x]}Al_{[28+x]}Ti_{[2]}Cu_{[4]}$, for x= 1,...,5), displays trends already apparent from the previous cases. Although no new features are observed, the larger number of Ti and Cu atoms helps understand the main characteristic of the Ni:Al 1:1 alloys where most additions are in solution and the clustering tendencies clearly observed in other systems are much less pronounced (i.e., coupling of Ti(Al) and Cu(Al) atoms or clustering of Cu atoms in either sublattice). In all cases, as expected, the minimum energy occurs for $Ni_{[a]}Al_{[b]}Ti_{[2]}Cu_{[c]}$ alloys, where $b+c+2 = a$, for which no antistructure atoms exist. The increase in Al concentration is the only source of changes in lattice parameter, due to the small difference in size between Ni and Cu atoms.

Visual examination of Fig. 10 provides an indication of the characteristic features of each region in the range of concentrations studied. The Ti-Cu ordering in the Al sublattice that characterizes alloys where $N_{Ni}>N_{Al}$ transitions to an intermediate regime where no specific pattern dominates ($N_{Ni} \sim N_{Al}$), to be replaced later by another ordering pattern in the Ni sublattice (Al-rich alloys) governed by Cu-Cu interactions.

5.5 Local environment analysis of atomic coupling

The picture that emerges from Fig. 10, while retaining the essential features already observed in Monte Carlo simulations describes the results of the interaction between different alloying additions, Ti and Cu, as a function of composition. A complementary analysis would be to perform an atom-by-atom energy analysis in order to identify the trends and reasons for the observed behavior of the different elements in the NiAlTiCu alloy. Based on the energetics of small groups of atoms surrounding specific atoms in the cell, the idea consists of analyzing the resulting site substitution behavior not

Table 8. BFS energy of formation of single and double local environments (in eV/atom), relative to a 72-atom B2 cell. For the double-centered LE, the center atoms are located at NN ($[\ ]_1$), NNN ($[\ ]_2$) or greater than NNN distance ($[\ ]_f$). The subindex H ($[\ ]_H$) indicates that the center atoms (Ti and Cu) locate themselves following a Heusler ($L2_1$) pattern in the Al-sublattice (i.e., sharing an Al atom as a NNN).

LE	Δe_T	LE	Δe_T
<Ni(Ni)>	0.000	<[Cu(Al)+Ni(Al)]$_f$>	0.020
<Ni(Al)>	0.011	<[Cu(Al)+Ni(Al)]$_2$>	0.021
<Al(Ni)>	0.094	<[Al(Ni)+Cu(Al)]$_f$>	0.103
<Ti(Al)>	0.008	<[Al(Ni)+Cu(Al)]$_1$>	0.094
<Ti(Ni)>	0.163	<[Ti(Al)+Ti(Al)]$_f$>	0.015
<Cu(Ni)>	0.016	<[Ti(Al)+Ti(Al)]$_2$>	0.016
<Cu(Al)>	0.009	<[Ti(Al)+Ti(Al)]$_H$>	0.016
<[Ti(Al)+Al(Ni)]$_f$>	0.102	<[Cu(Al)+Cu(Al)]$_f$>	0.018
<[Ti(Al)+Al(Ni)]$_1$>	0.087	<[Cu(Al)+Cu(Ni)]$_1$>	0.025
<[Ti(Al)+Cu(Al)]$_2$>	0.015	<[Cu(Al)+Cu(Ni)]$_1$>	0.024
<[Ti(Al)+Cu(Al)]$_H$>	0.016	<[Ti(Ni)+Ni(Al)]$_1$>	0.151
<[Ti(Al)+Cu(Al)]$_f$>	0.017	<[Ti(Ni)+Ni(Al)]$_f$>	0.174
<[Ni(Al)+Ti(Al)]$_f$>	0.019	<[Ni(Al)+Cu(Ni)]$_1$>	0.026
<[Ni(Al)+Ti(Al)]$_2$>	0.017	<[Ni(Al)+Cu(Ni)]$_f$>	0.027

just in terms of individual interactions between individual atoms, but through the effect that these bonds might have in the immediate vicinity of such atoms.

In order to implement this approach, we define atomic 'local environments' (LE), denoted as <X(B)>, consisting of a given central X atom in the B-sublattice and its eight NN and six NNN (in a bcc lattice), under the assumption that it is this group of atoms that will be most affected by the presence of a central substitutional defect atom, and that any change in the energy of formation of the cell will arise mostly from changes in energy within this environment. It is then sufficient to examine the energetics of this limited group of atoms in order to understand the full effect of a given point defect (i.e., substitution of one atom by another, creation of a single vacancy, etc.). The formation of an antisite defect, X(A)A(B), could then be seen as the superposition of the LE surrounding two individual point defects: one centered around the X atom, <X(A)>, and the other centered around the displaced A atom, <A(B)>. These could be 'non-interacting', <[X(A)A(B)]$_f$> (where the subindex f denotes that the two point defects are separated by more than one lattice parameter distance), or 'overlapping', <[X(A)A(B)]$_1$>, LE's, where the two center atoms (X and A) are NN thus sharing a large number of NN and NNN atoms.

To make a fair comparison of the different single and extended local environments, it is necessary to embed each LE in a B2 cell so that all bonds affected by the presence of the defect are accounted for. To do so, we locate the LE at the center of a 72-atom equilibrium B2 NiAl cell and refer all energies to the pure AB version of that cell. Because all LE's, and the reference B2 cell are evaluated at the same lattice parameter, the energy difference between the cell with the defect and the reference cell, Δe, indicates the energy cost (in terms of chemical energy differences) of performing the specific substitutions that characterize the defect (as it contains only the contributions from the atoms in the LE, and not the rest of the cell).

Table 8 lists the energy of formation (relative to an ideal B2 NiAl cell) of different local environments: <X(B)> and <X(A)> (for X = Ti, Cu and A, B = Ni or Al), <Ni(Al)> and <Al(Ni)>. The corresponding values for the extended local environments, including two substitutions are also listed. It is reasonable to expect that the site preference is mainly determined by the relative energy of formation $\Delta e\{LE\}$ of these local environments. For example, if $\Delta e\{<X(B)>\} < \Delta e\{<[X(A)A(B)]_f>\}$ or $\Delta e\{<X(B)>\} < \Delta e\{<[X(A)A(B)]_1>\}$, then X chooses a site in the B sublattice, as the formation of a single <X(B)> entails a lower energy cost than that required by the combination of a substitutional defect and the creation of an antistructure A atom, regardless of the relative position of these defects. For consistency, this analysis has to be performed in unrelaxed atomic positions, under the assumption that relaxation effects, while influential, are not ultimately responsible for the observed site preference behavior.

5.6 Local environment analysis of the ternary system

We start with the LE study of ternary additions (Ti or Cu) to NiAl, followed by the quaternary case (Ti and Cu interactions in NiAl). In the ternary cases, where the goal is to identify the reasons that explain the different site substitution options, it is not necessary to consider the relative locations of the substitutional and antistructure atoms. While important, the gain or loss of energy due to different relative positions are not comparable in magnitude with the corresponding gain or losses due to substitutions themselves.

5.6.1 Ti site preference in NiAl

The relevant results listed in Table 8 are shown in Figs. 11.a-b, which schematically represent the energy cost (in terms of the energy of formation of unrelaxed LE's) in creating substitutional defects and antistructure atoms

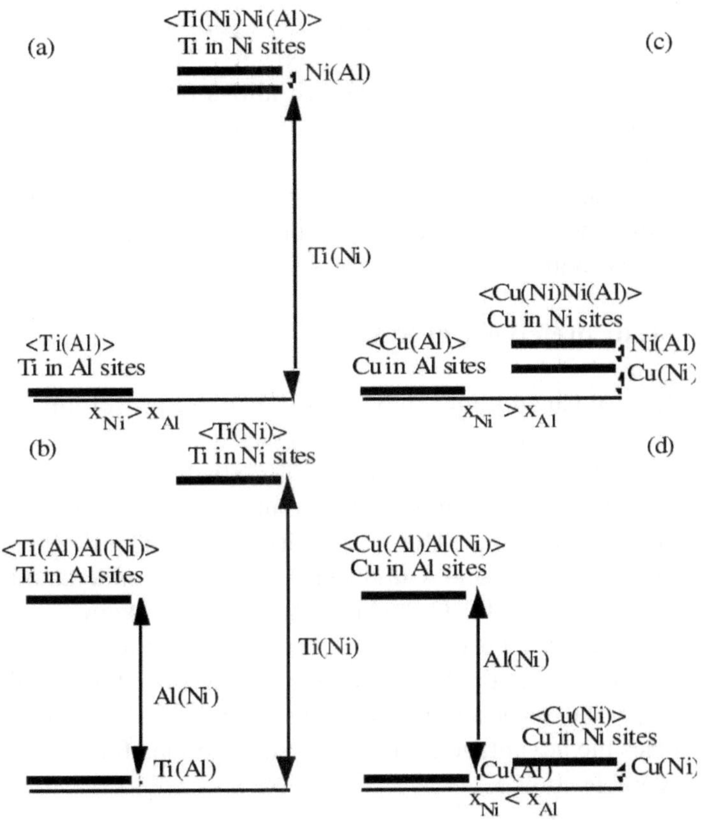

Figure 11. Energy level diagram for the energies of formation of different local environments needed to describe the site preference behavior of (a) Ti in Ni-rich, (b) Ti in Al-rich, (c) Cu in Ni-rich and (d) Cu in Al-rich NiAl alloys. Diagrams (a) and (c) indicate the possible substitutions in Ni-rich alloys, indicating a clear preference for Al sites for Ti and Cu. Diagrams (b) and (d) correspond to Al-rich alloys, still showing a clear (but less marked) Ti preference for Al sites and Cu preference for Ni sites. The energies are referenced to a pure B2 NiAl alloy.

for (a) Ni-rich and (b) Al-rich alloys. For Ni-rich alloys, which contain available Al sites by direct substitution or Ni sites through antisite defects, the competition between Ti(Al) and Ti(Ni)Ni(Al) substitutions is explained by the high energy cost of the Ti(Ni) substitution (<Ti(Ni)>) which, compounded with the small energy cost of creating a Ni antistructure atom (<Ni(Al)>) results in an almost insurmountable energy gap greatly favoring Ti(Al) substitutions: $\Delta e\{<\text{Ti(Ni)Ni(Al)}>_f\} - \Delta e\{<\text{Ti(Al)}>\} = 0.167$ eV/atom. In Al-rich alloys (Ni available sites by direct substitution or Al sites by antisite defects), it is the high energy cost of creating an antistructure Al atom that dominates the site preference behavior, as seen in Fig. 11.b. However, in spite of being large ($\Delta e\{<\text{Al(Ni)}>\} = 0.094$ eV/atom), the combination of a Ti(Al) and a Al(Ni) substitution still does not match the high energy cost of locating a Ti atom in a Ni site. The energy gap between Ti(Ni) and Ti(Al)Al(Ni) defects in Al-rich alloys 0.061 eV/atom) is not as large as the one between Ti(Ni)Ni(Al) and Ti(Al) defects in Ni-rich alloys, indicating that while the preference of Ti for Al sites is shared by Ni-rich and Al-rich alloys, it is weaker for the latter, consistent with the results of the MC simulations where there was an increasing proportion of Ti atoms finding their way to the Ni sublattice for Al-rich alloys.

5.6.2 Cu site preference in NiAl

The analysis of Cu additions to NiAl is shown in Figs. 11.c-d. For Ni-rich alloys (Fig. 11.c), the energy cost of locating a Cu atom in a Ni site, added to a comparable energy cost for creating a Ni antisite defect, result in a net preference for Al sites although not of the same magnitude as that found for Ti in a similar situation. The proximity between the energy levels corresponding to Cu(Al) and Cu(Ni)Ni(Al)$_f$ can be understood as more options become available for Cu, which can be easily excited to reside in a Ni site with low energy cost. For Al-rich alloys, in spite of the fact of the direct preference of Cu for Al sites (Δe{<Cu(Al)>} = 0.009 eV/atom vs. Δe{<Cu(Ni)>} = 0.016 eV/atom), the high energy cost of creating an antisite Al defect erases the advantage of direct Cu substitutions for Al sites, making Ni sites a favorable choice.

5.6.3 Ti and Cu additions and interaction between point defects

Before continuing with a detailed analysis of the quaternary case, some conclusions can be extracted from the ternary cases, as described in Fig. 11. It is interesting to compare the Cu energy level diagrams with those obtained for Ti in spite of the fact that both elements display a 'direct' preference for Al sites regardless of composition (i.e., either Ti or Cu are energetically more stable in an Al site than in a Ni site). Comparing Figs. 11.a-b with Figs. 11.c-d and the corresponding LE formation energies in Table 8, the energy difference between Cu(Al) and Cu(Ni)Ni(Al) substitutions is only 0.019 eV/atom, an order of magnitude smaller than that found for Ti. This fact, coupled to the smaller energy cost in locating Ti atoms in Al sites in Ni-rich alloys than the much larger one required for substitutions in Ni sites, justifies the observed prevalence of Ti in using the available sites in the Al sublattice. Conversely, for Al-rich alloys, the resulting energy difference between Cu(Al)Al(Ni) and Cu(Ni) substitutions (0.087 eV/atom), explains the ease with which Cu favors sites in the Ni sublattice instead of competing

a. $Ni_{[37]}Al_{[33]}Ti_{[2]}$	Ni* — Ti | Ti	Ni* — Ti + Ti┌ ┐ Ti└ ┘ Ti coupling to Ni* Ti preference for Heusler sites
b. $Ni_{[37]}Al_{[32]}Ti_{[2]}Cu_{[1]}$	Ni* — Ti | | Ti — Cu	Ni* — Ti + Ti┌ ┐ Ti└ ┘ + Ti — Cu Cu coupling to Ti
c. $Ni_{[37]}Al_{[31]}Ti_{[2]}Cu_{[2]}$	Ti — Ni* — Ti | | Cu Cu	Ni* — Ti + Ti — Cu Cu coupling to Ti takes over Ti's preference for H sites
d. $Ni_{[37]}Al_{[30]}Ti_{[2]}Cu_{[3]}$	Ni* | Ti — Cu | | Cu — Ti — Cu	Cu | Ni* — Ti + Ti — Cu
e. $Ni_{[37]}Al_{[29]}Ti_{[2]}Cu_{[4]}$	Ni* | Cu — Ti — Cu | | Cu — Ti — Cu	Abundance of Cu favors Ti/Cu coupling over Ti coupling to Ni

Figure 12. Coupling series Cu ↔ Ti ↔ Ni, as manifested in Ni-rich alloys with increasing Cu content. The last column describes the individual elements, in order of importance, leading to the final state (center column).

with Ti for Al sites. It could then be assumed that if this behavior continues in the case of quaternary systems, then Ti will always choose Al sites while Cu will move from Al sites to Ni sites depending on the Ni:Al ratio. This assumption, based solely on extrapolating features characterizing the ternary systems, ignores effects that could arise from the interaction between the two alloying additions. The atomic distributions shown in Fig. 10, clearly indicate that the ternary site substitution behavior alone does not completely explain all the salient characteristics of each ground state, as several other features, besides the specified site preference for Ti and Cu, can be identified. These results indicate that the characteristic energies of the single element site preference behavior (Ti or Cu in NiAl) are of the order of 0.1 eV/atom. There are clearly secondary effects, at a lower energy scale, that also intervene in determining the final atomic distribution. Such effects arise from the coupling between the different alloying additions and should not be ignored if a proper description of the ground state is desired. Several of these effects have already been identified, although not explained, by inspecting the structure of the ground states shown in Fig. 10. In what follows, we describe and quantify the nature of these effects, based on the analysis of non-interacting and overlapping LE's.

5.6.4 Ti and Cu interaction with antisite defects

One common feature of all the ground states (Fig. 10) consists of the preference of Ti for Al sites. It is interesting to discuss how Ti(Al) atoms interact with antistructure atoms, when present, or Cu alloying additions. The top row in Fig. 10 contains alloys for which it has been established that Ti and Cu additions occupy sites in the Al sublattice. Moreover, for $x_{Ni} > 50$ at.%, there are also Ni antistructure atoms present. With these three types of atoms in the Al sublattice, we now investigate the possible coupling schemes, i.e. the interaction between different types of atoms that result in the formation of NN or, when appropriate, NNN bonds that might appear in such systems.

Among the three possible couplings, [Ti(Al)+Ni(Al)], [Cu(Al)+Ti(Al)] and [Cu(Al)+Ni(Al)], only the first two involve a gain in energy when a NNN bond is formed (i.e., $\Delta e\{[]_2\} < \Delta e\{[]_f\}$). The creation of a $[Ti(Ni)+Ni(Al)]_2$ bond introduces a gain of 0.002 eV/atom over $[Ti(Al)+Ni(Al)]_f$. This explains the observed Ti(Al) ↔ Ni(Al) coupling that characterizes the ground states of the Ni-rich quaternary alloys in the top row in Fig. 10. A comparable gain is realized if a $[Cu(Al)+Ti(Al)]_2$ bond is formed. No such gain exists for $[Cu(Al)+Ni(Al)]_2$ bonds, thus establishing a coupling scheme Cu(Al) ↔ Ti(Al) ↔ Ni(Al). Table 8 also shows that the local environment energy of Ti(Al)+Cu(Al) atoms varies little depending on the relative location of the two substitutional atoms (0.015 eV/atom when they are located at NNN distance, 0.017 eV/atom when they are separated by greater distances, and 0.016 eV/atom when they are located at the opposite corners of a cube in the Al sublattice). The proximity of these three levels explains the ordering pattern observed, for example, in the ground state for the $Ni_{[37]}Al_{[29]}Ti_{[2]}Cu_{[4]}$ alloy.

The coupling scheme (Cu(Al) ↔ Ti(Al) ↔ Ni(Al)) results in the characteristic feature of other Ni-rich alloys, as shown in Fig. 12. Some additional patterns with respect to that coupling hierarchy are also apparent. With the coupling Ti(Al) ↔ Ni(Al) firmly established, we now consider its consequences when several Ti(Al) atoms are present. Of the three cases listed in Table 8 for the relative location of two substitutional Ti(Al) atoms ($[Ti(Al)+Ti(Al)]_f$, $[Ti(Al)+Ti(Al)]_2$ and $[Ti(Al)+Ti(Al)]_H$), the NNN bond is the less energetically favored, followed by the 'Heusler' arrangement and finally, with the lowest energy cost, the two atoms far apart. As mentioned above, Ti(Al) ↔ Ni(Al) coupling will rule out this last option, thus favoring sites at third-neighbor distance (Heusler-like sites), as seen in Fig. 12, where the number of favorable Ti-Al bonds is maximized. This effect can be understood as the role of Ni(Al) promoting the attraction of Ti atoms into an ordered pattern. While the link Cu(Al) ↔ Ti(Al) can be interpreted as responsible for 'closing' the square formed by Ni(Al), Ti(Al), Cu(Al) and Ti(Al), the 'repulsion' between Cu(Al) and Ni(Al) results in a competition between the number of Ti(Al)-Ni(Al) and Cu(Al)-Ti(Al) bonds which explains the loss of one Ti(Al)-Ni(Al) bond at the expense of extra Cu(Al)-Ti(Al) bonds. However, the alternating pattern of Cu-Ti-Cu-Ti atoms persists, suggesting that in alloys with higher Ti concentration than the ones studied in this work, Cu additions might partition to the $L2_1$ (Heusler) phase.

The coupling of Ti(Al) atoms with antisite defects persists in Al-rich alloys. For these cases, the Ti(Al)-Al(Ni) coupling introduces comparable energy gains (0.015 eV/atom) with respect to the non-interacting case. Every alloy with Al antistructure atoms shows Ti(Al)-Al(Ni) coupling, regardless of the Cu contents. Moreover, Ti(Al)-Al(Ni) coupling overrides the tendency of Ti(Al) atoms to locate themselves in Heusler sites. Overall, this coupling seems to be the leading effect relating the defects in each sublattice in these alloys: Ti(Al) substitutional atoms in the Al sublattice and Cu atoms in the Ni sublattice both bond to the available antistructure Al atoms.

5.6.5 Ti and Cu interactions

As the number of available Al sites decreases ($N_{Ni} < 36$ in Fig. 10), the leading role of Ti(Al) substitutions forces Cu atoms to occupy the remaining Al sites as well as the increasing number of Ni available sites, thus coexisting in both sublattices. Table 8 indicates that the interaction between Cu(Ni) and Cu(Al) atoms involves changes in energy of similar or smaller magnitude than the ones described in the previous subsection. As a consequence, clustering of Cu atoms is observed. The NN bond $[Cu(Ni)+Cu(Al)]_1$ introduces a 0.002 eV/atom energy gain over the two atoms in solution, translating into Cu clustering in Ni-rich alloys with high Cu concentration. Finally, the smallest gains are realized by Ti(Al) atoms locating themselves at NNN distance (0.001 eV/atom). Although small, it suffices to explain the observed trends in all alloys with nearly the same number of Ni and Al sites where equal partitioning of Ti atoms in solution in one sublattice (Al) and Cu atoms in the other (Ni), without interaction between them, is seen.

Having discussed individual and collective behavior, it is useful to summarize the main features by a quick review of Figs. 10 and 12. While the top row of Fig. 10 highlights the role of the coupling series Cu(Al) $\leftrightarrow$ Ti(Al) $\leftrightarrow$ Ni(Al) in $Ni_{[37]}Al_{[34-x]}Ti_{[2]}Cu_{[x]}$ ($x = 1,...,5$) alloys, the next lines in Fig. 10 are examples of the same coupling feature but without the Ni antistructure atom, indicating that it is precisely the presence of antisite defects and other substitutional atoms (like Cu or Ti) that ties substitutional Ti atoms together. Their absence, as in alloys with $N_{Al} = 33$ and $N_{Al} = 34$, result in Ti(Al) atoms remaining in solution. The set with $N_{Ni} = 35$ marks the transition between the Cu(Al) ↔ Ti(Al) ↔ Ni(Al) regime, characteristic of Ni-rich alloys, to the migration of Cu atoms to the Ni-sublattice that is dominant in Al-rich alloys.

For example, $Ni_{[35]}Al_{[34]}Ti_{[2]}Cu_{[1]}$ represents the reduced importance of Cu-Ti coupling once Cu occupies Ni sites, while alloys with the same number of Ni atoms but increasing Cu content highlight the dual role of Cu linking Ti atoms when residing in Al sites and leading to Cu clustering when occupying sites in either sublattice. The NN bonds Cu(Ni)-Cu(Al) compete with Cu(Al) ↔ Ti(Al), thus explaining the small energy advantage of $Ni_{[35]}Al_{[33]}Ti_{[2]}Cu_{[2]}$ over a similar distribution where the isolated Ti atom is linked to both Cu(Al) and Cu(Ni), or, similarly, the advantage of $Ni_{[35]}Al_{[32]}Ti_{[2]}Cu_{[3]}$ over a similar configuration where the isolated Ti atom closes the square of NNN bonds between Ti(Al) and the two Cu(Al) atoms. Finally, alloys with N_{Ni}= 33 and 34 transition to a regime where dominance of coupling between Cu atoms is the main feature.

6. CONCLUSIONS

In this chapter we have reviewed the use of a quantum approximate method as a viable option in modeling of complex systems, with simple and straightforward schemes to understand the role of alloying additions and their interactions. This is particularly important when dealing with multicomponent systems where, regardless of the inherent complexities of detailed electronic structure calculations, it is imperative to first develop some insight on the fundamental features of the system under study.

REFERENCES

1. G. Bozzolo and J. E. Garcés, in Atomistic modeling of surface alloys, Surface alloys and alloy surfaces, The Chemical Physics of Solid Surfaces, Vol. 10, Elsevier (2002).
2. S. Cottenier, When do we understand solids?, *Physicalia Magazine* **26**, 3 (2004).
3. D. Farías, M. A. Niño, J. J. de Miguel, R. Miranda, J. Morse and G. Bozzolo, Growth of Co and Fe on Cu(111): Experiment and BFS based calculations, *Appl. Surf. Sci.* 219 (2003) 80.
4. A. Canzian, H. Mosca and G. Bozzolo, Atomistic Modeling of Pt Deposition on Cu(111) and Cu deposition on Pt(111), *Surf. Rev. Lett.* **11**, 1 (2004).
5. G. Bozzolo, J. E. Garcés and G. Demarco, Atomistic modeling of Au deposition on a Cu substrate, *Surf. Sci.* **532/535**, 41 (2003).
6. J.E. Garcés, G. Bozzolo, P. Abel, and H. Mosca, Atomistic modeling of Pd/Cu(110) surface alloy formation, *Appl. Surf. Sci.* **167** 18 (2000).
7. H. Mosca, J.E. Garcés, and G. Bozzolo, Surface ternary alloys of (Cu,Au)/Ni(110), *Surf. Sci.* **454-456**, 707 (2000).

8. A. Canzian, H. Mosca and G. Bozzolo, Deposition of Fe on Nb(110) and Nb on Fe(110), *Surf. Sci.* **574**, 287 (2005).
9. G. Bozzolo, J. E. Garcés, R. D. Noebe, P. Abel and H. Mosca, Atomistic modeling of surface and bulk properties of Cu, Pd, and the Cu-Pd system, *Prog. Surf. Sci.* **73**, 79 (2003).
10. G. Bozzolo, R. D. Noebe and H. Mosca, Atomistic Modeling of Pd Site Preference in NiTi, *J. Alloys and Comp.* **386**, 125 (2005).
11. G. Bozzolo, R.D. Noebe, and C. Amador, Site occupancy of ternary additions to B2 alloys, *Intermetallics* **10**, 149 (2002).
12. J. R. Smith, T. Perry, A. Banerjea, J. Ferrante and G. Bozzolo, Equivalent crystal theory of metals and semiconductor surfaces and defects, *Phys. Rev. B* **44**, 6444 (1991).
13. James H. Rose, John R. Smith and John Ferrante, Universal features of bonding in metals, *Phys. Rev. B* **28**, 1835 (1983).
14. P. Blaha, K. Schwartz and J. Luitz, WIEN97, Vienna University of Technology. Improved and updated Unix version of the copyrighted WIEN Code, P. Blaha, K. Schwartz, P. Sorantin and S. B. Trickey, Comput. Phys. Commun. **59**, 399 (1990).
15. O. K. Andersen, Linear Methods in Band Theory, *Phys. Rev. B* **12**, 3060 (1975).
16. P. Légaré, G. F. Cabeza and N. J. Castellani, Platinum overlayers on Co(0001) and Ni(111): numerical simulation of surface alloying, *Surf. Sci.* **441**, 461 (1999).
17. P. Abel, and G. Bozzolo, Calculation of the thermal expansion coefficients of pure elements and their alloys, *Scripta Mater.* **46**, 557 (2002).
18. R. L. Fleischer, R. D. Field, C. L. Briant, Mechanical properties of high-temperature alloys of AlRu, *Metall. Trans. A* **22** 403 (1991).
19. R. L. Fleischer, Boron and off-stoichiometry effects on the strength and ductility of AlRu, *Met. Trans. A* **24** 227 (1993).
20. R. L. Fleischer, D. W. McKee, Mechanical and oxidation properties of AlRu-based high-temperature alloys, *Met. Trans. A* **24** 759 (1993).
21. R. Fleischer, Substitutional solutes in AlRu I. Effects of solute on moduli, lattice parameters and vacancy production, *Acta metal. mater.* **41** 863 (1993).
22. R. Fleischer, Substitutional solutes in AlRu II. Hardening and correlations with defect structure, *Acta metal mater.* **41** 119 (1993).
23. I. M. Wolff, G. Sauthoff, Role of an intergranular phase in RuAl with substitutional additions, *Acta mater.* **45** 2949 (1997).
24. S. Mi, S. Balanetskyy and B, Grushko, A study of the Al-rich part of the Al-Ru alloy system, *Intermetallics* **11** 643 (2003).
25. I. M. Wolff, G. Sauthoff, Mechanical properties of Ru-Ni-Al alloys, *Met. Mat. Trans. A* **27** 1395 (1996).
26. I. M. Wolff, G. Sauthoff, High-temperature behavior of precious metal base composites, *Met. Mat. Trans. A* **27** 2642 (1996).
27. S. Chakravorty, D. R. F. West, Phase equilibria between NiAl and RuAl in the Ni-Al-Ru system, *Scripta Metall.* **19** 1355 (1985).
28. S. Chakravorty, D. R. F. West, The constitution of the Ni-Al-Ru system, *J. Mater. Sci.* **21** 2721 (1986).
29. A. L. R. Sabariz, G. Taylor, Preparation, structure and mechanical properties of RuAl and (Ru,Ni)Al alloys, *Mat. Res. Soc. Symp. Proc.* **460** 611 (1997).
30. K. W. Liu, F. M. Mueckllich, W. Pitschke, R. Birringer, K. Wetzig, Formation of nanocrystalline B2-structured (Ru,Ni)Al in the ternary Ru-Al-Ni system by mechanical alloying and its thermal stability, *Mat. Sci. Eng. A* **313** 187 (2001).
31. I. J. Horner, N. Hall, L. A. Cornish, M. J. Witcomb, M. B. Cortie, T. D. Boniface, An investigation of the B2 phase between AlRu and AlNi in the Al-Ni-Ru ternary system, *J. Alloys and Compounds* **264** 173 (1998).
32. I. Vjunitsky, E. Schönfeld, T. Kaiser, W. Steurer and V. Shklover, Study of phase states and oxidation of B2-structure based Al-Ni-Ru-M alloys, *Intermetallics* **13** 35 (2005).

33. Q. Feng, T. K. Nandy, T. M. Pollock, Observation of a Ru-rich Heusler phase in a multicomponent Ni-base superalloy, *Scripta Mater.* **50**, 849 (2004).
34. Q. Feng, T. K. Nandy, B. Tryon, T. M. Pollock, Deformation of Ru-Al-Ta ternary alloys, *Intermetallics* **12**, 755 (2004).
35. Q. Feng, T. K. Nandy, T. M. Pollock, Solidification of high-refractory ruthenium-containing superalloys, *Acta Mater.* **51**, 269 (2003).
36. A. P. Ofori, C. J. Rossouw, C. J. Humphreys, Determining the site occupancy of Ru in the $L1_2$ phase of a Ni-base superalloy using ALCHEMI, *Acta Mater.* **53**, 97 (2005).
37. R. S. Polvani, W. S. Tzeng, P. R. Strutt, High temperature creep in a semi-coherent NiAl-Ni_2AlTi alloy, *Met. Trans. A* **7**, 33 (1976).
38. P. Cerba, M. Vilasi, B. Malaman, J. Steinmetz, Caractérisation des trois nouvelles phases ternaires, $Nb(Pd,Al)_2$ et $Nb(Ru,Al)_2$ de type $MgZn_2$, et $NbRu_2Al$ de type BiF_3, dans les systémes Nb-Pd-Al et Nb-Ru-Al, *J. Alloys and Compounds* **57**, 201 (1993).
39. P. Villars, L. D. Calvert, Pearson's Handbook of Crystallographic Data for Intermetallic Phases. Metals Park:American Society for Metals (1986).
40. M. H. Yoo, T. Takasuga, S. Hanada, O. Izumi, Slip modes in B2-type intermetallic alloys, *Mater. Trans. JIM* **31**, 435 (1990).
41. D. Hackenbracht, J. Kubler, Electronic, magnetic and cohesive properties of some nickel -aluminum compounds, *J. Phys. F* **10**, 427 (1980).
42. R. Fleischer, D. M. Dimiduk, H. A. Lipsitt, Intermetallic compounds for strong high-temperature materials: status and potential, *Annu. Rev. Mater. Sci.* **19**, 231 (1989).
43. V. L. Moruzzi, A. R. Williams, J. F. Janak, Electronic, magnetic and cohesive properties of some nickel-aluminium compounds, *Phys. Rev. B* **10**, 4856 (1974).
44. D. J. Singh, Electronic structure, magnetism, and stability of Co-doped NiAl, *Phys. Rev. B* **46**, 14392 (1992).
45. D. N. Manh, D. G. Pettifor, Electronic structure, phase stability and elastic moduli of AB transition metal aluminides, *Intermetallics* **7**, 1095 (1999).
46. I. Zou , C. L. Fu, Structural, electronic, and magnetic properties of 3d transition-metal aluminides with equiatomic composition, *Phys. Rev. B* **51**, 2115 (1995).
47. W. Lin , J. H. Xu, A. J. Freeman, Cohesive properties, electronic structure, and bonding characteristies of RuAl. A comparison to NiAl, *J. Mater. Res.* **7**, 592 (1992).
48. A. R. Williams, J. Kuebler, C. D. Gelatt Jr., Cohesive properties of metallic compounds: Augmented-spherical-wave calculations, *Phys. Rev. B* **19**, 6094 (1979).
49. L. E. Edshammar, An x-ray investigation of Ruthenium-Aluminium alloys, *Acta Chemica Scandinavica* **20**, 427 (1966).
50. W. G. Jung, O. J. Kleppa, Standard molar enthalpies of formation of MeAl (Me = Ru, Rh, Os, Ir), *Met. Trans. B* **23**, 53 (1992).
51. D. N. Manh, A. T. Paxton, D. G. Pettifor, A. Pasturel, On the phase stability of transition metal trialuminide compounds, *Intermetallics* **3**, 9 (1995).
52. R. Hultgren, P. D. Desai, D. T. Hawkins, M. Gleiser, K. K. Kelley. Selected Values of the Thermodynamic Properties of Binary Alloys, Metals Park:American Society of Metals; 1973; O. Kubaschewski, E. L. Evans, C. B. Alcock, Metallurgical Thermochemistry, 4th Ed. Oxford, Pergamon Press (1967).
53. K. Rzyman, Z. Moser, R. E. Watson, M. J. Weinert, Enthalpies of formation of AlNi: Experiment versus theory, *J. Phase Equil.* **19**, 106 (1998).
54. G. Bozzolo, J. Khalil, R. D. Noebe, Modeling of the site preference in ternary B2-ordered Ni-Al-Fe alloys, *Comp. Mat. Sci.* **24**, 257 (2002).
55. E. Raub, Die struktur der festen Tantal-Ruthenium-legierungen, *Z. Metallkd* **54**, 451 (1963).
56. I. M. Anderson, A. J. Duncan, and J. Bentley, Determination of site occupancies in aluminide intermetallics by alchemi, Mat. Res. Soc. Symp. Proc., **364**, 443 (1995).
57. A. Wilson and J. Howe, Statistical alchemi study of the site occupancies of Ti and Cu in NiAl, *Scripta Mater.* **41**, 327 (1999).

Chapter 8

MOLECULAR ORBITAL APPROACH TO ALLOY DESIGN

Masahiko Morinaga, Yoshinori Murata and Hiroshi Yukawa
Department of Materials Science and Engineering, Graduate School of Engineering, Nagoya University, Furo-Cho, Chikusa-Ku, Nagoya 464-8603, JAPAN

Abstract: A molecular orbital approach to alloy design has recently made great progress. This approach is based on the electronic structure calculations by the DV-Xα cluster method. New alloying parameters are obtained for the first time by the calculations and used for the prediction of phase stability of alloys and alloying properties as well. This approach is applicable not only to structural alloys, but also to functional alloys. For example, heat-resisting single crystal Ni-based superalloys for turbine blades and high Cr ferritic steels for turbine rotors have recently been developed following this approach. Light metal based alloys such as Ti alloys, Al alloys and Mg alloys have been developed with the aid of this approach. Crystal structure maps for intermetallic compounds have also been proposed. In addition, hydrogen storage alloys are also investigated. Through a series of molecular orbital calculations, a universal relation has been discovered between electron density minima and atomic (or ionic) radii in various materials.

Keywords: alloy design, DV-Xα cluster method, Ni-based superalloys, ferritic steels, Ti alloys, Al alloys, Mg alloys, intermetallic compounds, hydrogen storage alloys, electron density distributions

1. INTRODUCTION

A variety of approaches to materials design have been proposed to save both time and cost necessary for materials development. For example, a thermodynamic approach to the prediction of phase diagrams called CALPHAD[1] has been widely used for alloy design. The construction of materials system databases has been in progress for various materials (for example, INTERGLAD[2] for glasses). In addition, molecular dynamics and Monte Carlo simulations have recently made remarkable progress. The phase-field method[3] has been found to be effective in predicting the micro-structural evolution in alloys and oxides. Further, first-principles calculations of electronic structures have been performed over the world due to the progress in computer hardware as well as the software. Total energy calculations are now very common in every field of materials science.

Recently, a theoretical method for alloy design has been developed on the basis of the DV-Xα molecular orbital calculation of electronic structure of alloys. Employing this method, we can treat consistently a variety of materials (e.g., metals, semiconductors and ceramics) by standing on a common ground of electron theory. Any discrimination is no longer needed between ferrous and non-ferrous alloys.

In this chapter, our attention will be directed mainly toward this molecular orbital approach to alloy systems. New alloying parameters of elements in various metals are introduced and used for alloy design. It is stressed that such alloying parameters are determined for the first time in a long history of metallurgy and metal science. Along this approach, NEW PHACOMP (PHAse COMPutation) has been developed in order to predict the formation of topologically close-packed (TCP) phases (e.g., σ phase and μ phase) in nickel-based superalloys.[4,5] A *d*-electrons concept relevant to transition metal based alloys has also been developed with the knowledge of alloying parameters.[6] Even mechanical properties are predictable for both aluminum alloys and magnesium alloys with multiple components.[7-10] Furthermore, new structure maps have been proposed to predict the crystal structures of intermetallic compounds such as TiAl and $MoSi_2$.[11]

Following this molecular orbital approach, various alloys have been developed, for example, heat-resisting single crystal Ni-based superalloys,[12,13] high strength titanium alloys,[14,15] implant titanium alloys[16] and heat-resisting high Cr ferritic steels for power plants.[17] It is applicable not only to these structural materials, but also to functional materials such as hydrogen storage alloys.

In this chapter, recent progress in this method will be reviewed focusing mainly on the Ni-based superalloys, iron alloys, Ti alloys, Al alloys, Mg alloys and hydrogen storage alloys. A universal relation between electron density minima and atomic (or ionic) radii will be shown, as it has been discovered recently through a series of the DV-Xα molecular orbital calculations of materials.[18]

2. DV-Xα MOLECULAR ORBITAL METHOD

The DV-Xα cluster method is a molecular orbital calculation method assuming the Hartree-Fock-Slater approximation.[19,20] The exchange-correlation between electrons, V_{XC}, is expressed by using a Slater's Xα potential,[21]

$$V_{XC} = -3\alpha\left[\frac{3}{8\pi}\rho(r)\right]^{1/3} \tag{1}$$

where ρ(*r*) is the electron density at position *r*, the parameter α is fixed at 0.7 and the self-consistent charge approximation is used in the calculation. The matrix elements of the Hamiltonian and the overlap integrals are calculated by a random sampling method. The molecular orbitals are constructed by a linear combination of numerically generated atomic orbitals (LCAO).

Local electronic structures are calculated by this method even for a large size of the molecule. Depending on the crystal structure, appropriate cluster models are employed in this calculation. Here, "cluster" means a hypothetical molecule that represents a crystal structure. For example, a cluster model for Ni_3Al is shown in Figure 1. Ni_3Al is a strengthening phase (γ' phase) of the Ni-based superalloys. This cluster model is made on the basis of the $L1_2$ type structure of Ni_3Al shown in Fig. 1.a.[22] In pure Ni_3Al, a central Al atom is surrounded by twelve Ni atoms (first-nearest-neighbors) and by six Al atoms (second-nearest-neighbors), as shown in Fig. 1.b.

The density of states of electrons in pure Ni_3Al obtained from the cluster calculation[22] resembles the result of the band calculation by Flecher[23] and by Hackenbracht and Kubler[24] despite the use of a relatively small cluster in the calculation. Thus, the cluster approach is a good approximation to the electronic structure of Ni_3Al.

In order to get alloying parameters, a variety of alloying transition metals, M, are substituted for the Al atom at the center of the cluster, $[MNi_{12}Al_6]$, as shown in Fig. 1.b. This cluster model is also valid approximately for fcc Ni (γ phase) because the element which substitutes for Ni in the γ phase, is surrounded by twelve Ni atoms as first-nearest-neighbors, similarly to the atomic arrangement in the $[MNi_{12}Al_6]$ cluster. It is noticed that first-nearest-neighbor interactions are most predominant in fcc metals and alloys. The lattice parameter of Ni_3Al is similar to that of fcc Ni. As a result, the electronic structure of Ni_3Al resembles that of fcc Ni. Further explanation on the calculation method is given elsewhere.[25]

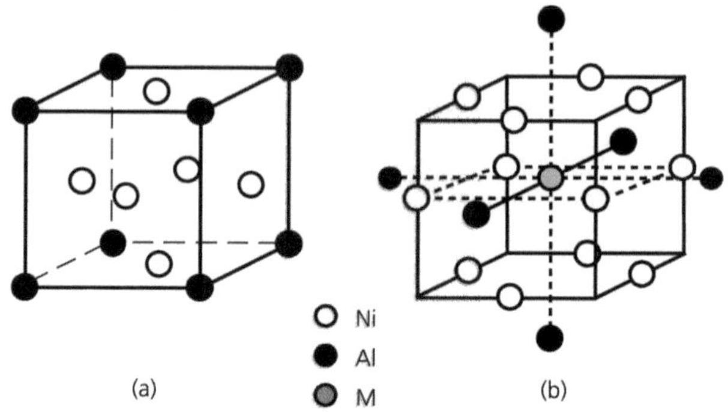

Figure 1. (a) Crystal structure of Ni_3Al, and (b) cluster model used in the calculation.

3. ALLOYING PARAMETERS

3.1 *d*-orbital energy level, Md

Two alloying parameters are obtained from the cluster calculation and used for alloy design.[22] One is the *d*-orbital energy level of alloying transition metal, M, in a base metal system. For example, the results of the level structure are shown in Fig. 2 for Ni_3Al alloyed with various 3*d* alloying elements. In this figure, the energy of the Fermi level (E_f) of Ni_3Al is set to be zero and used as a reference. In the case of a pure Ni_3Al cluster (i.e., M=Al) the levels of $13a_{1g}$ to $15e_g$ originate mainly from the Ni 3*d* orbital, and form the Ni 3*d* band where the Fermi energy level lies as is indicated by an arrow. In case of alloyed cluster, new energy levels due mainly to the *d* orbital of the alloying transition metal appear above the Fermi energy level. For instance, the $16e_g$ and $14t_{2g}$ levels (drawn as broken lines) correspond to these new levels. Their energy height changes systematically with the order of elements in the periodic table.

The M-d levels correlate with the electronegativity and the atomic radius of elements. In fact, it is well known that the eigen energy value determined by the Xα method itself is a minus value of the Mulliken's electronegativity. The M-d levels increase with decreasing electronegativity of alloying element, as shown in Fig. 3. In addition, as shown in Fig. 4, the M-d levels increase with increasing atomic radius of the element. The average energy of these two *d*-orbital levels is hereafter referred to as Md level. The Md values are listed in Table 1.

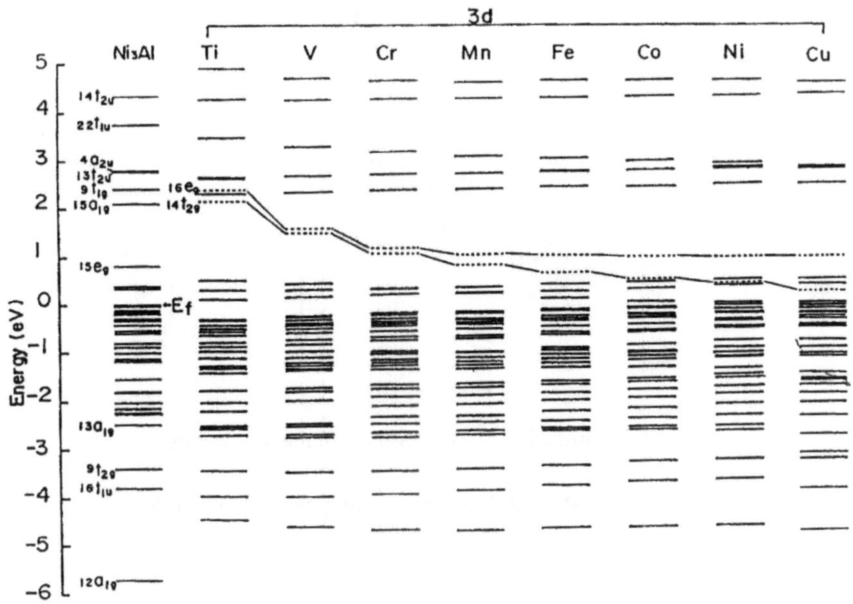

Figure 2. Energy level structures of pure and alloyed Ni_3Al with 3*d* transition elements.

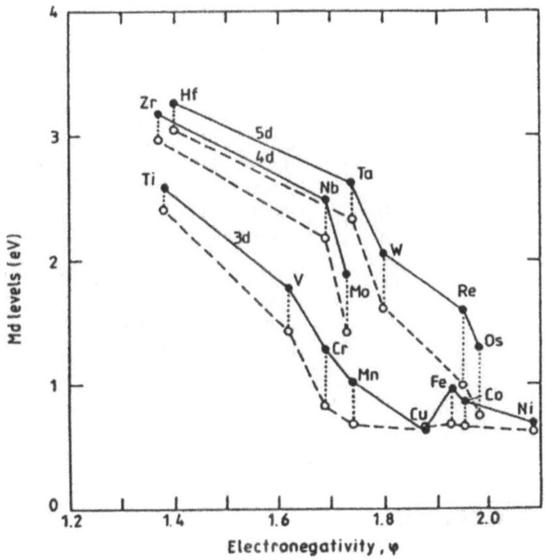

Figure 3. Correlation of M-d levels with electronegativity.

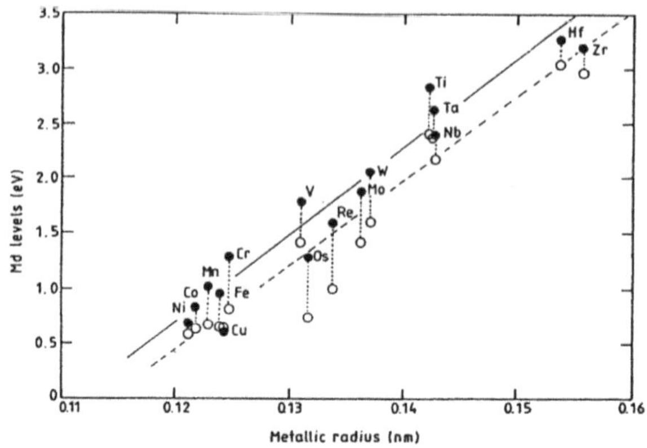

Figure 4. Correlation of M-d levels with metallic radius.

Table 1. List of Md and Bo values for alloying elements in Ni.

Element		Md(eV)	Bo
3d	Ti	2.271	1.098
	V	1.543	1.141
	Cr	1.142	1.278
	Mn	0.957	1.001
	Fe	0.858	0.857
	Co	0.777	0.697
	Ni	0.717	0.514
	Cu	0.615	0.272
4d	Zr	2.944	1.479
	Nb	2.117	1.594
	Mo	1.550	1.611
	Tc	1.191	1.535
	Ru	1.006	1.314
	Rh	0.898	1.068
	Pd	0.779	0.751
	Ag	0.659	0.391
5d	Hf	3.020	1.518
	Ta	2.224	1.670
	W	1.655	1.730
	Re	1.267	1.692
	Os	1.063	1.500
	Ir	0.907	1.256
	Pt	0.764	0.920
	Au	0.627	0.528
Others	Al	1.900	0.533
	Si	1.900	0,589

3.2 Bond order, Bo

The other alloying parameter is the bond order. Here, the overlap population, $Q_{\nu\nu'}$ of electrons between two atoms ν and ν' is defined as,

$$Q_{\nu\nu'} = \sum_{l}\sum_{ij} C_{il}^{\nu} C_{jl}^{\nu'} \int \Psi_i^{\nu} \Psi_j^{\nu'*} dV \tag{2}$$

Here, Ψ_i^{ν} and $\Psi_l^{\nu'}$ are the wavefunctions of the *i* and *j* orbitals of atoms ν and ν', respectively. C_{il}^{ν} and $C_{jl}^{\nu'}$ are the coefficients which show the magnitude of the linear combination of atomic orbitals in the *l*-th molecular orbital. In this chapter, $Q_{\nu\nu'}$ is called the bond order and it is used as a measure of the strength of the covalent bond between atoms.

For example, the values of the bond order (hereafter referred to as Bo) are listed in Table 1 for a variety of alloying elements in Ni_3Al or Ni.[26] This bond order (Bo) changes following the position of elements in the periodic table, as does the Md.

A number of alloying parameters have been proposed for describing alloying behavior (e.g., the electrons-per-atom-ratio (e/a), the atomic radius, the electronegativity and the electron vacancy number). However, most of them are assigned to pure metals, and inevitably no alloying effects are involved in them in a proper way. It is stressed that both the Md and the Bo are new alloying parameters obtained for the first time from the molecular orbital calculation of alloyed clusters. As mentioned earlier, these two parameters change following the order of elements in the two-dimensional periodic table. Both the Md and the Bo have been calculated for various metal systems, for example, Ni_3Al,[22,26] Cr,[27] Ti,[14] Zr,[28] Nb[29] and Mo.[29] Such calculations have recently been extended to heavy metals (e.g., W[30] and U[31]) by using a relativistic DV- Xα method.[25]

3.3 Average parameters for an alloy

For an alloy, the average values of Md and Bo are defined simply by taking the compositional average, $\overline{\text{Md}}$ and $\overline{\text{Bo}}$, denoted as,

$$\overline{\text{Md}} = \sum_{i=1}^{n} \text{X}_i \cdot (\text{Md})_i \,, \tag{3}$$

$$\overline{\text{Bo}} = \sum_{i=1}^{n} \text{X}_i \cdot (\text{Bo})_i \,. \tag{4}$$

Here, X_i is the atomic fraction of component *i* in the alloy, $(Md)_i$ and $(Bo)_i$ are the respective values for component *i*. The summation extends over the components, *i*=1, 2, ..., *n*. For simplicity, unless otherwise stated, the unit of the Md parameter, eV, is omitted.

4. NICKEL-BASED SUPERALLOYS

Ni-based superalloys are heat-resisting alloys widely used for the blades of jet engines and industrial gas turbines. As explained before, superalloys are strengthened by Ni_3Al (γ' phase) precipitates dispersed in fcc Ni (γ phase).

4.1 New PHACOMP

It is well known that the precipitation of topologically close-packed (TCP) phases (e.g., the σ phase) in the fcc (γ) matrix deteriorates the mechanical properties remarkably. The compositions of the alloys which are free from such precipitates have been predicted by the PHACOMP (PHAse COMPutation) method.[32,33] From a historical point of view, we have started using the words "alloy design" since this PHACOMP was developed in 1964.

In this method, the precipitation of the TCP phases in the alloy is predicted using the average electron vacancy number, Nv. Here, the electron vacancy number, N_v, is the number of electron vacancies or holes existing above the Fermi energy level in the *d* band. This parameter was introduced by Pauling to explain the magnetism of transition metals. It is expressed approximately as Nv=10.66-(e/a), where e/a is the electrons-per-atom ratio. For example, the Nv value for the 4A group elements, Cr, Mo and W, is 6.66. The compositional average of Nv is denoted as $\overline{Nv}$, and when the $\overline{Nv}$ value exceeds a certain value, the TCP phase will be formed in the γ matrix. This Nv-PHACOMP method has been employed widely for the design and the quality control of nickel-based superalloys. However, there are many difficulties and contradictions in this method.[5]

As explained earlier, the Md parameter is concerned with the electronegativity and the atomic radius of elements, both of which are classical parameters to be used for treating the solid solubility in alloys by Hume-Rothery[34] and Darken and Gurry.[35] Therefore, the Md parameter has a great possibility of dealing with the present solubility problem of alloys,

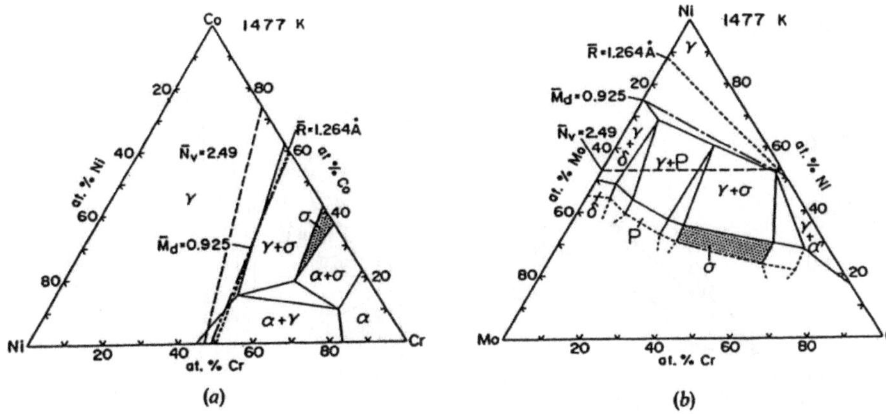

Figure 5. Phase diagram of (a) Ni-Co-Cr and (b) Ni-Cr-Mo. Here, $\overline{R}$ is the compositional average of atomic radii of each element in the alloy.

both solute and solvent being transition metals. So, instead of the Nv parameter, this Md parameter is applied to the prediction of the occurrence of the TCP phases in nickel alloys.[4,5,26] Two typical phase diagrams of Ni-Co-Cr and Ni-Cr-Mo are shown in Fig. 5. In each phase diagram, the iso-$\overline{\text{Md}}$ line of 0.925(eV) traces well the γ / $\gamma+\sigma$ phase boundary, whereas the iso-$\overline{\text{Nv}}$ line is far away from the boundary. The validity of this Md method has been confirmed through the examination of more than 25 ternary phase diagrams.[5] This Md method, called NEW PHACOMP, has been applied successfully to the design of commercially available alloys with multiple components.[36,37] It has also been applied to the prediction of solidification microstructures in the weld nickel alloys.[38,39]

In Table 1, recently calculated values of Md are given for several elements of interest (e.g., Ru, Ir, Y). The third generation single crystal superalloys contain Ru[40] and/or Ir,[41] because their addition reduces a tendency to form undesirable TCP phases[42] owing to the relatively low Md values of these elements.

4.2 *d*-electrons concept

A *d*-electron concept for alloy design has been constructed on the basis of NEW PHACOMP. Besides $\overline{\text{Md}}$, the $\overline{\text{Bo}}$ parameter is used in this concept.

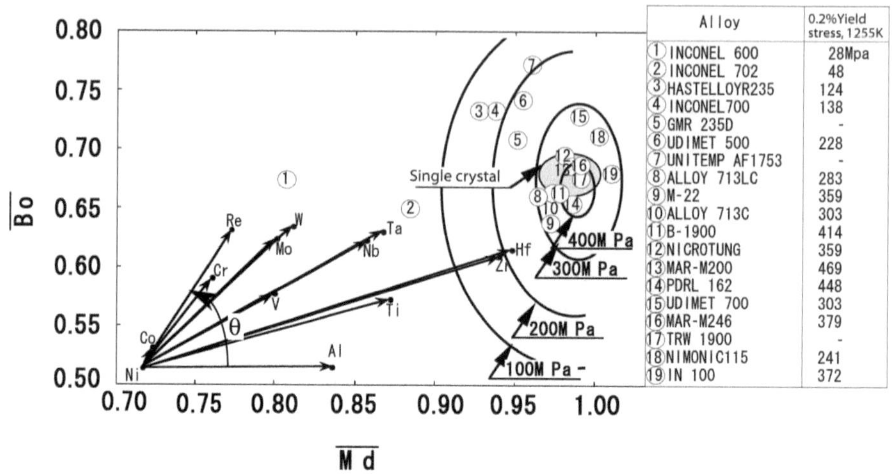

Figure 6. $\overline{\mathrm{Bo}}-\overline{\mathrm{Md}}$ diagram showing the locations of conventional cast Ni alloys.

4.2.1 Target region for alloy design

Nineteen conventional cast superalloys are plotted in the $\overline{\mathrm{Bo}}-\overline{\mathrm{Md}}$ diagram as shown in Fig. 6.[6] The contour lines showing the 0.2% yield strength level at 1255 K are also indicated by dotted lines. The 0.2% yield strength shows the maximum at the $\overline{\mathrm{Md}}$ value of about 0.98 and the $\overline{\mathrm{Bo}}$ value of about 0.67. The creep strength also shows the maximum around this position. Furthermore, all the single crystal superalloys are located near the maximum position as indicated by a shadow area in Fig. 6. Every alloy lying on such a maximum position contains a large volume fraction (about 60-65%) of the γ' phase without precipitating any TCP phases in it, resulting in the high strength of the alloy. Thus, a target region for alloy design can be specified concretely on the $\overline{\mathrm{Bo}}-\overline{\mathrm{Md}}$ diagram.

The $\overline{\mathrm{Md}}$ value, 0.98, in the target region is calculated using the alloy composition, so it is higher than the critical value, 0.925, for the γ phase shown in Fig. 5. However, assuming that the $\overline{\mathrm{Md}}$ value is 0.925 for the γ phase and 1.02 for the γ' phase, and also that the volume fraction of the γ' phase is 60%, then the $\overline{\mathrm{Md}}$ value for the whole alloy becomes about 0.98, in agreement with the $\overline{\mathrm{Md}}$ value of the target region.

4.2.2 Alloying vector

The concept of alloying vectors is also important for practical alloy design. As shown in Fig. 6, the vector starts from the position of pure Ni and ends at the position of Ni-10mol%M binary alloy in the $\overline{\mathrm{Bo}}-\overline{\mathrm{Md}}$ diagram, where M is an alloying element in Ni. The directions and the magnitudes of these vectors vary with alloying elements.[32] It is noticed that the vector directions are similar among the same group elements in the periodic table, for example, among Ti, Zr, Hf (4A group elements), V, Nb, Ta (5A elements) and Cr, Mo, W (6A elements). This is attributable to the fact that both the Bo and the Md parameters change following the order of elements in the periodic table, as mentioned earlier.

As explained before, any superalloy consists of the γ' (Ni_3Al) phase and the γ (fcc) matrix phase, and the γ' phase is a precipitation hardening phase in the alloy. The γ' stabilizing elements such as Al, Ti, V, Nb, Ta take relatively lower θ angles than the γ stabilizing elements such as Cr, Mo, W, Re as is shown in the figure.[43] So, the θ angle is an indication for the stability of the γ' phase with respect to the γ phase, and hence it is related closely to the γ' volume fraction of alloy.[43]

The concept of alloying vectors is, for example, useful for alloy modification. In order to keep the alloy position on the $\overline{Bo}-\overline{Md}$ diagram nearly unchanged with the modification, alloy compositions may be adjusted among those elements which have a similar vector direction. For example, Re addition may require to reduce the Cr content in the alloy, since their vectors point to the similar direction. Such a compositional change is actually seen in the modification from PWA 1480 (the first generation superalloy) to PWA 1484 (the second generation superalloy). Namely, the Cr content of PWA 1484 decreases by 6 mol%, whereas the Re content increases by 1 mol% compared to the respective values of PWA 1480. Further decrease in the Cr content and attendant increase in the Re content is seen in René N6 (the third generation superalloy).[44]

4.3 Design of nickel based single crystal superalloys

In advanced industrial gas-turbine systems, there has been a great demand for new single crystal superalloys with an excellent combination of high-temperature creep strength, hot-corrosion resistance and oxidation resistance. As mentioned above, there is a clear trend of lowering Cr and increasing Re content in the evolution from the 1st- to the 2nd- and to the 3rd-generation Ni-based single crystal (SC) superalloys.[37] However, it has been reported recently that high temperature oxidation resistance decreases largely with increasing Re content in the alloys.[45] Therefore, it is strongly needed to improve the high temperature oxidation resistance of the third generation superalloys in some ways. Standing on this background, we have designed ten nickel-based SC superalloys containing about 4-5 mass% Re with the aid of the *d*-electrons concept.[13]

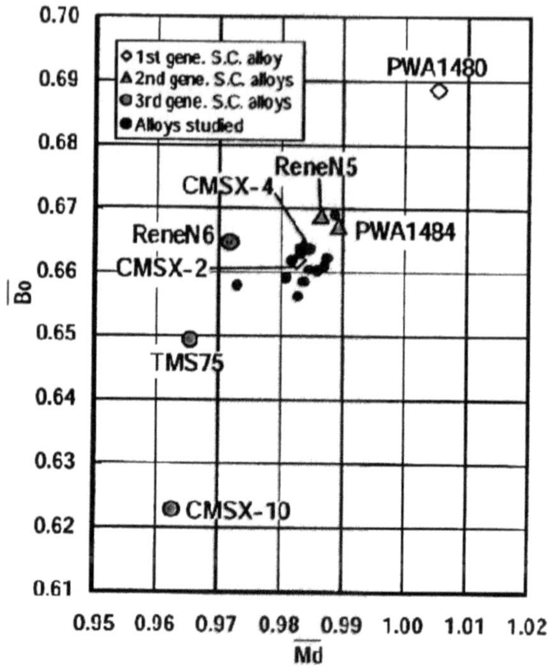

Figure 7. $\overline{Bo}-\overline{Md}$ diagram showing the location of the designed alloys and some reference alloys of the 1st, 2nd and 3rd generation superalloys.

Their chemical compositions are in the range of 1.2-1.5%Ti, 3.8-6.5%Cr, 11%Co, 0-1.4%Mo, 6.5-7.4%Ta, 5.0-6.0%W, 3.6-5.4%Re, 5.1-5.5%Al, 0.12-0.14%Hf and balanced Ni in mass% units. As shown in Fig. 7, the ten designed alloys are located nearer to the 2nd generation SC superalloys than the 3rd generation SC superalloys.

A series of experiments such as creep rupture tests, burner rig tests and cyclic oxidation tests is conducted with the heat-treated SC specimens of these alloys. Almost all the designed alloys are superior in creep rupture life to a 2nd generation superalloy, CMSX-4, currently used worldwide. In the hot-corrosion resistance estimated from the burner rig tests, any designed alloys are comparable or even superior to CMSX-4. The oxidation resistance is very different among the designed alloys, but some of them show better resistance than CMSX-4. Thus the SC alloys containing about 4-5 mass% Re have about 20 K higher temperature capability than the 2nd generation superalloy, CMSX-4, while exhibiting excellent hot-corrosion resistance and good oxidation resistance. Furthermore, it has good productivity, since a single-crystal blade with about 170 mm in length can be grown, as shown in Fig. 8.[13]

Figure 8. Single crystal blade with 170mm in length. The cuboidal γ' phases are arranged three-dimensionally as is shown in the figure.

5. IRON ALLOYS

The cluster models used for the calculation are shown in Fig. 9 for bcc Fe.[46] The alloying element, M, is located at the center of an MFe_{14} cluster, and $Fe^{(1)}$ and $Fe^{(2)}$ are the first and the second-nearest-neighbor Fe atoms from M, respectively. The electron density of states for pure bcc Fe is calculated using an Fe_{15} cluster (i.e., M = Fe in the cluster). The result of the cluster calculation shown in Fig. 10.b resembles the result of a band calculation shown in Fig. 10.a. The Fermi energy level, E_f, is located at the peak of 3*d* band in either case. Thus, the cluster approach is a good approximation to the electronic structure of iron.

5.1 Second-nearest-neighbor interactions in bcc Fe

The covalent bonds between the *d-d* electrons contribute largely to the total cohesive energy of transition metals like iron. Therefore, the bond order associated with the *d-d* electrons is calculated in Fig. 11.a. For instance, M-$Fe^{(1)}$ refers to the bond order between M-d and $Fe^{(1)}$-d, excluding the *s* and *p* contributions. In the figure, "Total" means the sum of the three components of M-$Fe^{(1)}$, M-$Fe^{(2)}$ and $Fe^{(1)}$-$Fe^{(2)}$.[46]

It should be noticed that the second-nearest-neighbor bond order, M-$Fe^{(2)}$, is smaller than the first-nearest-neighbor bond order, M-$Fe^{(1)}$, but is

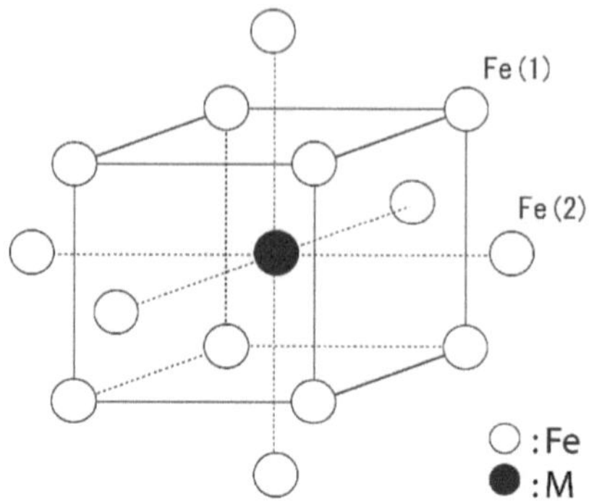

Figure 9. MFe_{14} cluster model employed in the calculation. Fe(1) and Fe(2) are the first- and second-nearest-neighbor iron atoms form M.

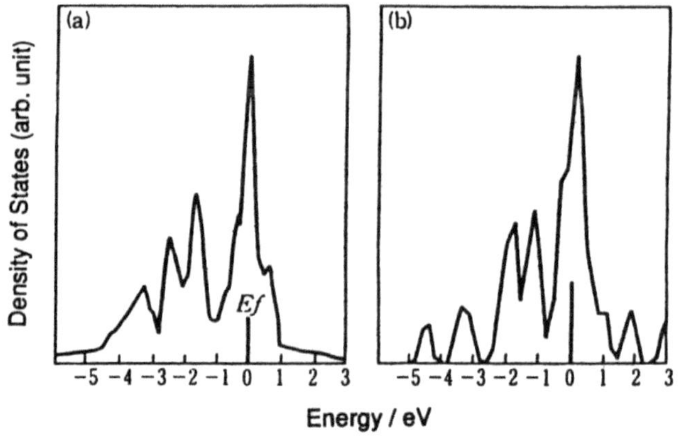

Figure 10. Comparison of the electron density of states for bcc Fe between (a) the band calculation and (b) the cluster calculation.

definitely present. The ratio of the bond orders, M-Fe$^{(2)}$/ M-Fe$^{(1)}$, is shown in Fig. 11.b. Of the 3*d* metals a large peak appears in Fe, indicating that the second-nearest-neighbor interactions are most predominant in pure bcc Fe, which is in agreement with the prediction by Pauling.[47] The ratio decreases rather drastically with the addition of M into bcc Fe.

The large magnitude of the second-nearest-neighbor interaction in bcc metals is interpreted primarily as due to the short second-nearest-neighbor distance that is only 15% larger than the first-nearest-neighbor distance in a bcc metal. In the case of fcc Fe, this distance is 41% larger, and hence the second-nearest-neighbor interaction is less significant. However, this is not necessarily true in every bcc metal. To see a general trend for the magnitude of the second-nearest-neighbor interaction, cluster calculations are

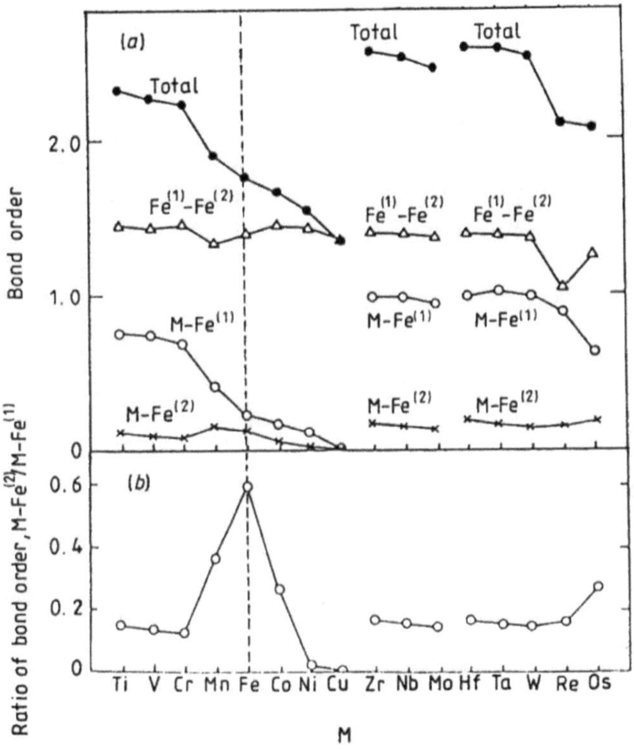

Figure 11. Change of (a) bond order and (b) ratio of the second-neighbor bond order to the first-neighbor bond order with alloying elements, M, in bcc Fe.

performed in pure bcc Cr, V, Ti, Zr, Nb and Mo using the same size of cluster shown in Fig. 9 (e.g., Cr_{15}). As shown in Table 2, the ratio of the second-nearest-neighbor bond order to the first-nearest-neighbor bond order is largest for Fe, while it is small for Cr, V, Nb and Mo. The ratio is intermediate for Ti and Zr.

Here, bcc Fe transforms to fcc Fe, and bcc Ti (or Zr) transforms to hcp Ti (or Zr), but bcc Cr ,V, Nb and Mo are stable, without any transformation. Thus, there is a trend in that the magnitudes of the second-nearest-neighbor interactions are large in those bcc metals which exhibit an allotropic transformation to the close-packed phase (fcc or hcp) at certain temperatures. In the close-packed phase an atom is surrounded by twelve first-nearest-neighbor atoms. In the bcc phase an atom is surrounded by fourteen atoms, if the second-nearest-neighbors are included, as shown in Fig. 9. So, the present calculation implies that the bcc metal, on transforming to the fcc (or hcp) phase, tends to hold the fcc (or hcp)-like atomic interaction even in the bcc phase. In other words, in such a bcc metal, the attractive atomic interactions operate over the first- and the second-neighbors, as if the metal has a higher coordination number such as in fcc or hcp. Thus, the second-nearest-neighbor interactions are significantly large in those bcc metals which show an allotropic transformation to the fcc or hcp phase. The bcc Fe (α-Fe) corresponds to this case.

It is known that such a second-nearest-neighbor interaction plays an important role in the ω-phase transformation in Fe, Ti and Zr alloys.[48]

Table 2. Comparison of the first- and second-nearest-neighbor bond orders among bcc metals.

	Bond order (first neighbours)	Bond order (second neighbours)	Ratio of bond order (second neighbours / first neighbours)
[Fe-Fe] in Fe_{15}	0.230	0.136	0.591
[Cr-Cr] in Cr_{15}	0.851	0.072	0.085
[V-V] in V_{15}	0.922	0.134	0.145
[Ti-Ti] in Fe_{15}	0.556	0.160	0.287
[Zr-Zr] in Zr_{15}	0.770	0.197	0.255
[Nb-Nb] in Nb_{15}	1.254	0.138	0.110
[Mo-Mo] in Mo_{15}	1.303	0.098	0.075

Table 3. List of Md and Bo values for alloying elements in bcc Fe.

Element		Md(eV)	Bo
3d	Ti	2.497	2.325
	V	1.610	2.268
	Cr	1.059	2.231
	Mn	0.854	1.902
	Fe	0.825	1.761
	Co	0.755	1.668
	Ni	0.661	1.551
	Cu	0.637	1.361
4d	Zr	3.074	2.551
	Nb	2.335	2.523
	Mo	1.663	2.451
5d	Hf	3.159	2.577
	Ta	2.486	2.570
	W	1.836	2.512
	Re	1.294	2.094
Others	C	-0.230	0
	N	-0.400	0
	Si	1.034	0

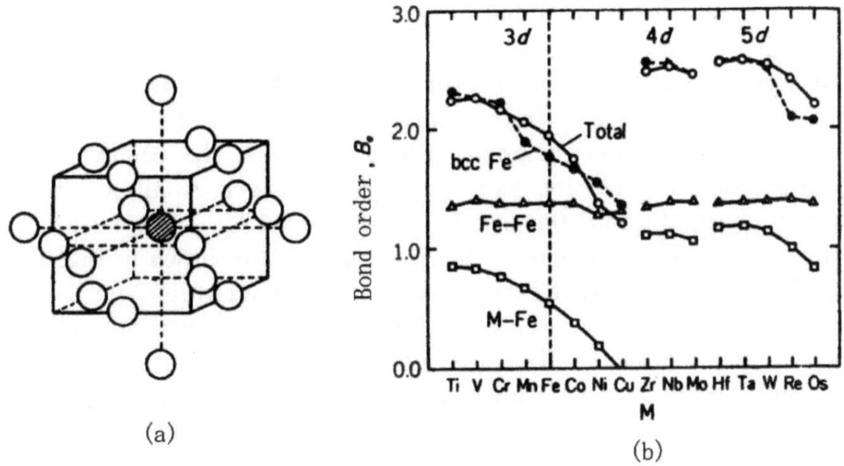

Figure 12. (a) Cluster model for fcc Fe and (b) change in the bond order with alloying elements, M in fcc Fe.

5.2 Alloying parameters in bcc Fe and fcc Fe

The calculated bond orders for alloying elements in bcc Fe are listed in Table 3 together with the Md values.[17] Also, the calculations are performed for alloying elements in fcc Fe, using a cluster model, MFe_{18}, shown in Fig. 12.a. The calculated results of bond orders are shown in Fig. 12.b. There is a resemblance in the variation of the total Bo with alloying elements between fcc Fe and bcc Fe.

Before explaining the alloy design, a local lattice-strain problem relevant to iron martensite will be shown, as the martensite is an important phase in the steel.

5.3 Local lattice strain induced by C and N in iron martensite

Iron martensite has usually a body-centered tetragonal (bct) lattice with its axial ratio, c/a, slightly larger than unity. However, this does not mean that the local strain around interstitial atoms is small. According to the experiments by Jack[49] and Bell and Owen,[50] when C or N atom occupies an interstitial (octahedral) site in bcc Fe, as shown in Fig. 13, the $Fe^{(1)}$-$Fe^{(1)}$ distance, l, expands by about 40% as compared to the non-distorted distance, l_0, (namely, $l/l_0 = 1.4$) and the $Fe^{(2)}$-$Fe^{(2)}$ distance, w, contracts by several percent as compared to the non-distorted distance, w_0. In other words, the first-nearest-neighbor iron atoms $Fe^{(1)}$, are displaced in the direction away from the central C or N atom, but instead the second-nearest-neighbor, $Fe^{(2)}$, move slightly toward the C or N atom.

To account for this local distortion, a cluster model, $Fe_{14}C$ or $Fe_{14}N$, shown in Fig. 13 is used.[51] In this cluster, $Fe^{(1)}$, $Fe^{(2)}$ and $Fe^{(3)}$ are the first-, second- and third-nearest-neighbor iron atoms from the central C or N interstitial atom, respectively. As the change in w is very small, w is fixed at w_0 and l is changed from 1.0 l_0 to 1.7 l_0 in steps of 0.1 l_0.

The calculated components of the bond orders are shown in Fig. 14 for $Fe_{14}C$. The C-Fe bond order is composed of the three terms, C-$Fe^{(1)}$, C-$Fe^{(2)}$ and C-$Fe^{(3)}$. Also, the Fe-Fe bond order is decomposed into $Fe^{(1)}$-$Fe^{(1)}$, $Fe^{(2)}$-$Fe^{(2)}$ and $Fe^{(1)}$-$Fe^{(3)}$. The magnitude of each component changes significantly with l/l_0. As shown in Fig. 14.b, the C-$Fe^{(1)}$ bond order decreases monotonously with increasing C-$Fe^{(1)}$ distance, while the C-$Fe^{(2)}$ bond order increases, probably due to a charge-transfer from C-$Fe^{(1)}$ bonds to C-$Fe^{(2)}$ bonds. The C-$Fe^{(3)}$ bond order scarcely changes with l/l_0. The sum of these three components makes a broad peak in the total C-Fe bond order curve at 1.4 for l/l_0, in agreement with the experimental result. Also, the increase in the C-$Fe^{(2)}$ bond order with l/l_0 implies that the C-$Fe^{(2)}$ distance tends to decrease, as is found experimentally.

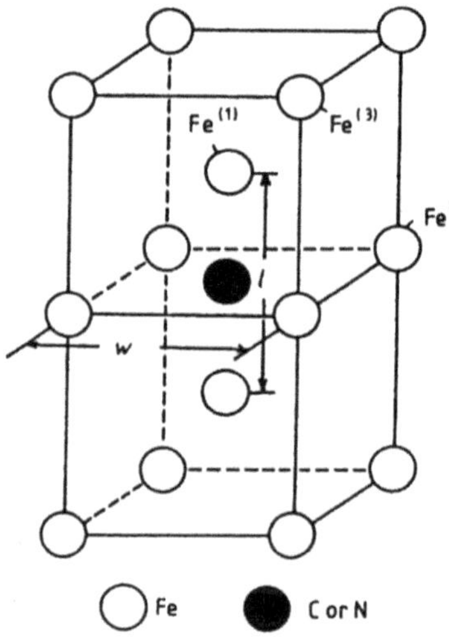

Figure 13. Cluster model, $Fe_{14}C$ or $Fe_{14}N$ employed in the calculation. $Fe^{(1)}$, $Fe^{(2)}$ and $Fe^{(3)}$ are the first-, second- and third-nearest-neighbor iron atoms from C or N.

For the Fe-Fe bonding, as shown in Fig. 14.a, the $Fe^{(2)}$-$Fe^{(3)}$ bond order stays almost constant against l/l_0. However, the $Fe^{(1)}$-$Fe^{(2)}$ bond order decreases as the $Fe^{(1)}$-$Fe^{(2)}$ distance increases with l/l_0 ,whereas the $Fe^{(1)}$-$Fe^{(3)}$ bond order increases as the $Fe^{(1)}$-$Fe^{(3)}$ distance decreases with l/l_0. The sum of them, namely the Fe-Fe (total) bond order increases with l/l_0, but it saturates near 1.4 for l/l_0. The saturated value is close to the bond order of pure Fe calculated by using a Fe_{14} cluster. Thus, both the C-Fe and the Fe-Fe bonds are optimized by local atomic displacements of Fe around C. Similar results are obtained in the Fe-N system.[51] Thus, the local strain around an interstitial C or N in iron martensite is understood by electronic structure calculations.

The electronic structures of interstitial C, N, and O in fcc Fe are also simulated in a similar way.[52] The magnitude of covalent interactions changes in the order, C-Fe > N-Fe > O-Fe.

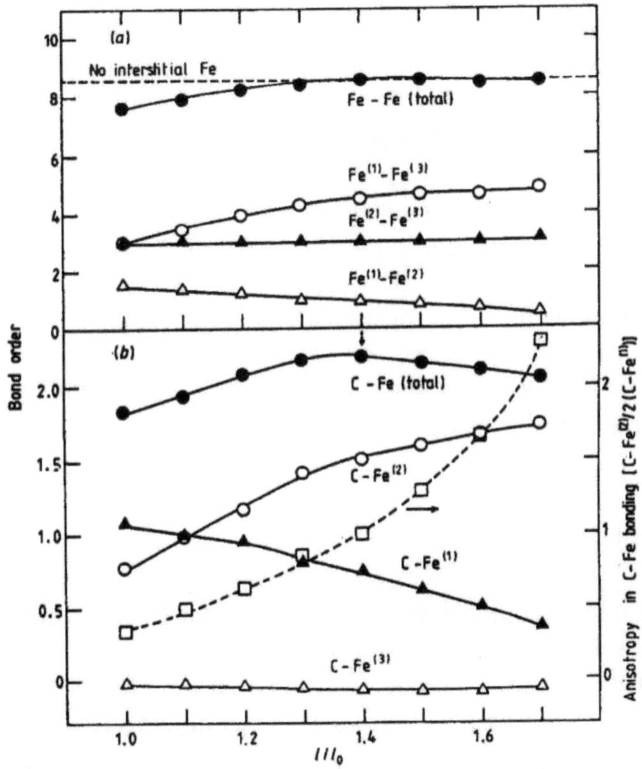

Figure 14. Components of (a) Fe-Fe bond order and (b) C-Fe bond order, and their changes with *l/l0* for $Fe_{14}C$.

5.4 Design of high Cr ferritic steels

Recently, there have been increasing demands for operating power generating plants under ultra super critical conditions.[53] For this reason, it has been strongly requested to develop advanced ferritic steels which are tolerable even in such severe operating conditions of power plants.

5.4.1 Alloying vector

In Fig. 15, alloying vectors starting from the position of pure bcc Fe and ending at the position of Fe-1 mol% M binary alloy are shown in the $\overline{Bo}-\overline{Md}$ diagram.[17] They are drawn using the Md and Bo values listed in Table 3. All the ferrite stabilizing elements (e.g., Cr and Mo) are located in the upper-right region except for Mn, whereas all the austenite stabilizing elements (e.g., Ni and Co) are located in the lower-left region in this diagram.

It is empirically known that the elements having high Bo and low Md are principal alloying elements in most structural alloys. The high Bo element will strengthen alloys by forming strong chemical bonds with the neighboring atoms, and the low Md element will increase the phase stability without forming any undesirable phases in the alloy. This is also true in the ferritic steels. In other words, principal alloying elements have the high $\overline{Bo}/\overline{Md}$ ratio, namely the large slope of alloying vectors shown in Fig. 15. The ratio changes in the order, Cr > Mo > W > Re > V > Nb > Ta > Zr > Hf> Ti. Being consistent in this order, Cr, Mo, W, V, Nb are indeed principal alloying elements in most ferritic steels. The only exception lies in Re, although the addition of Re has been found to improve creep strength significantly.[17] On the other hand, for austenite stabilizing elements, the alloying vector has the opposite direction and the $\overline{Bo}/\overline{Md}$ ratio changes in the order, Ni > Co > Cu.

5.4.2 δ ferrite formation

It is desirable to suppress the formation of the δ ferrite in ferritic steels because of the detrimental effect on the creep resistance and their fracture toughness. The measured volume fraction of the δ ferrite existing in the steels normalized at 1323 K, are plotted against the $\overline{Md}$ parameter as shown in Fig. 16. For Ni-free steels the δ ferrite starts forming as the $\overline{Md}$ value exceeds 0.852, and its volume fraction increases with increasing $\overline{Md}$ value. Also, as is evident from this figure, the existence of Ni in the steel increases the critical $\overline{Md}$ value for the δ ferrite formation. This is also the case of other austenite stabilizing elements, Co and Cu. So, the addition of Ni, Co or Cu into the steel gives room to increases in the amount of ferrite stabilizing elements such as Mo and W without the δ ferrite formation, resulting in the increase in creep properties at high temperatures. In fact, any advanced steels contain such austenite stabilizing elements to some extent

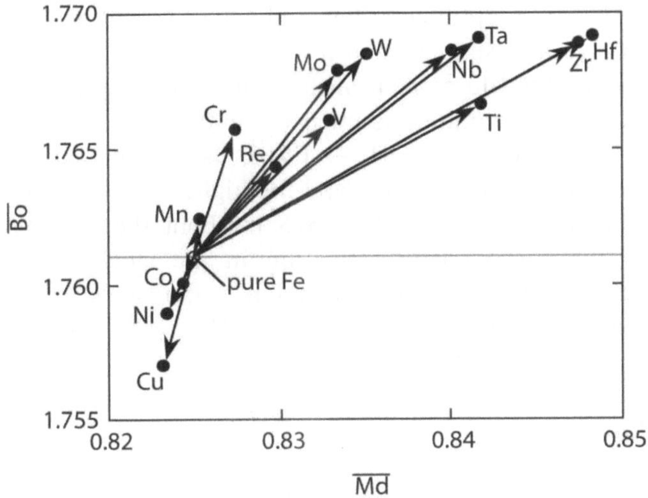

Figure 15. Alloying vectors of elements in bcc Fe.

(e.g., 3 mass% Co). In addition, it is stressed here that the δ ferrite formation is predicted more accurately by the $\overline{Md}$ parameter than the Cr equivalent, a parameter widely used in the steels.[54]

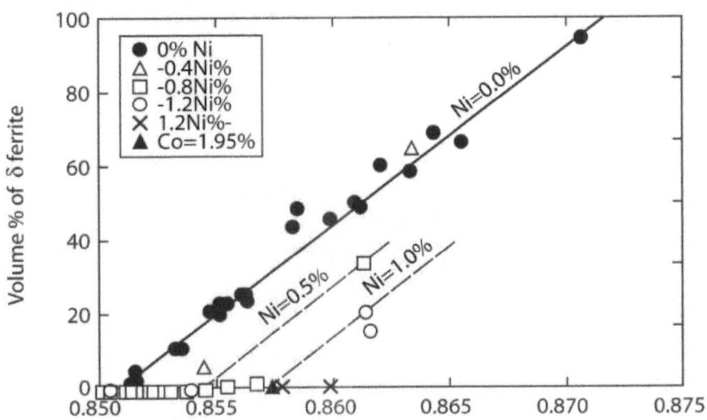

Figure 16. Correlation of the volume fraction of the δ ferrite with the $\overline{Md}$ parameter.

5.4.3 Trace of the evolution of ferritic steels

Masuyama[55] has shown the evolution process of ferritic steels for boiler applications. For example, 9% Cr steels have been developed in the sequence, T9 → (F9) → T91 → F616. Here, T91 (Mod. 9Cr-1Mo) is modified from T9 (9Cr-1Mo) by optimizing the contents of V and Nb, both of which form carbonitrides and strengthen the steel. Then NF616 is obtained from T91 by decreasing the Mo content, instead of increasing the W content. As shown in Fig. 17, this evolution is traceable in the $\overline{Bo}-\overline{Md}$ diagram.[17] The evolution of T9 → (F9) → T91 → F616, is interpreted as the alloy modification toward the higher $\overline{Md}$ and the higher $\overline{Bo}$ in the diagram. It is noted here that the $\overline{Md}$ value of NF616 is about 0.852, which is just a critical value for the δ ferrite formation in the steel without austenite stabilizing elements. Similarly, the evolution of the steels for turbine rotors can be traced in the $\overline{Bo}-\overline{Md}$ diagram.[17] Thus, this $\overline{Bo}-\overline{Md}$ diagram provides us a clue to alloy design.

5.4.4 Alloy design

A high Cr ferritic steel for steam turbine rotor has been designed with this approach.[17] A 10-tons turbine rotor has been made as shown in Fig. 18. The chemical composition of the steel is Fe-0.09%C-10%Cr-4%W-3.5% Co-0.2%Ni-0.1%Mo-0.15%V-0.06%Nb-0.08%Mn-0.08%Si-0.007%B-0.018 %N, in mass% units. The creep rupture life of this alloy is longer than 4000 hrs (14.4 Ms) in the creep test condition of the temperature, 923 K and the applied stress, 157 MPa. It is also found experimentally that the addition of 0.2% Re increases the creep rupture life considerably in high W-containing steels.

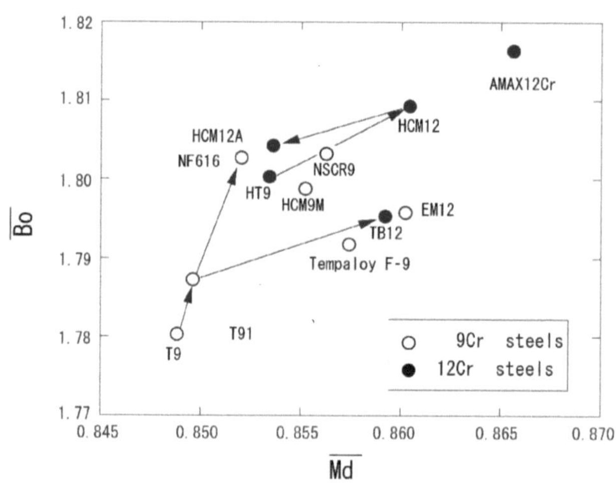

Figure 17. Location of 9-12% Cr steels in the $\overline{Bo}-\overline{Md}$ diagram.

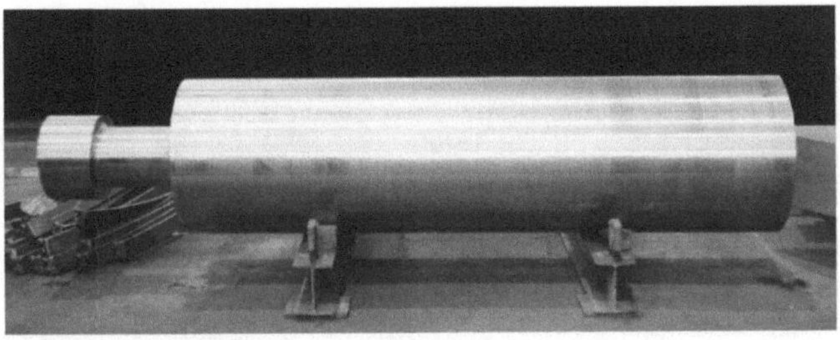

Figure 18. A 10-tons model steam turbine rotor made of designed steel.

6. TITANIUM ALLOYS

6.1 Alloying parameters in bcc Ti

There is an allotropic transformation between hcp Ti (α-Ti) and bcc Ti (β-Ti) around 1155 K. So, two alloying parameters, Bo and Md, are calculated using bcc and hcp cluster models.[14] As shown in Fig. 19, the tendency of the bond order (Bo) change with alloying elements resembles that between bcc Ti and hcp Ti. For convenience, the Bo and Md parameters are listed in Table 4 for various alloying elements in bcc Ti (β-Ti). Hereafter, unless otherwise mentioned, the parameters obtained for bcc Ti are used in the analysis.[14]

6.2 Classification of commercially available alloys into α, α+β and β-types

Titanium alloys are commonly classified into the α, α+β, and β types, according to the phases existing in the alloy. In Fig. 20, about fourty commercially available alloys are plotted in the $\overline{Bo}-\overline{Md}$ diagram, by calculating the $\overline{Bo}$ and $\overline{Md}$ values from the alloy compositions. It is apparent that the locations of three types of alloys are separated clearly in this diagram. The No.6 alloy, Ti-8Mn, is located in the β field in spite of the α+β type alloy. However, this is not a contradiction, because it is really a β type alloy, but heat-treated in the (α+β) temperature range to improve the mechanical properties, and incidentally grouped into the α+β type alloy.

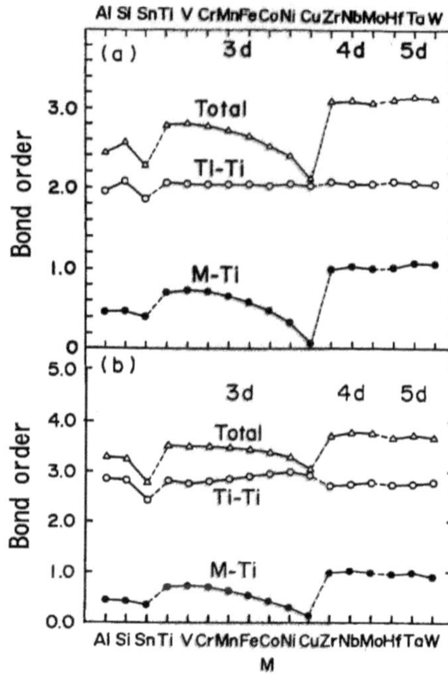

Figure 19. Changes in the bond order with alloying elements, M in (a) bcc Ti and (b) hcp Ti.

Utilizing this figure, we can predict readily the type of alloy if the alloy composition is given.

In order to understand the alloying effect of each element, M, in Ti, the $\overline{Bo} - \overline{Md}$ diagram is made for Ti-M binary alloys as shown in Fig. 21. When comparing Fig, 20 with Fig. 21, we may see that for instance in Ti-Al binary alloy, the alloying vector goes into the α–phase field with increasing Al content, indicating that Al acts as an α–stabilizing element. On the other hand, for instance, Mo, W, Nb and Ta are all β–stabilizing elements, since their alloying vectors are directed toward the β–phase field. These results agree with the well-known alloying behavior of elements in titanium.

Table 4. List of the Bo and Md values for various alloying elements in bcc Ti.

3d	*Bo*	*Md* (eV)	4d	*Bo*	*Md* (eV)	5d	*Bo*	*Md* (eV)	other	*Bo*	*Md* (eV)
Ti	2.790	2.447	Zr	3.086	2.934	Hf	3.110	2.975	Al	2.426	2.200
V	2.805	1.872	Nb	3.099	2.424	Ta	3.144	2.531	Si	2.561	2.200
Cr	2.779	1.478	Mo	3.063	1.961	W	3.125	2.072	Sn	2.283	2.100
Mn	2.723	1.194	Tc	3.026	1.294	Re	3.061	1.490			
Fe	2.651	0.969	Ru	2.704	0.859	Os	2.980	1.018			
Co	2.529	0.807	Rh	2.736	0.561	Ir	3.168	0.677			
Ni	2.412	0.724	Pd	2.208	0.347	Pt	2.252	0.146			
Cu	2.114	0.567	Ag	2.094	0.196	Au	1.953	0.258			

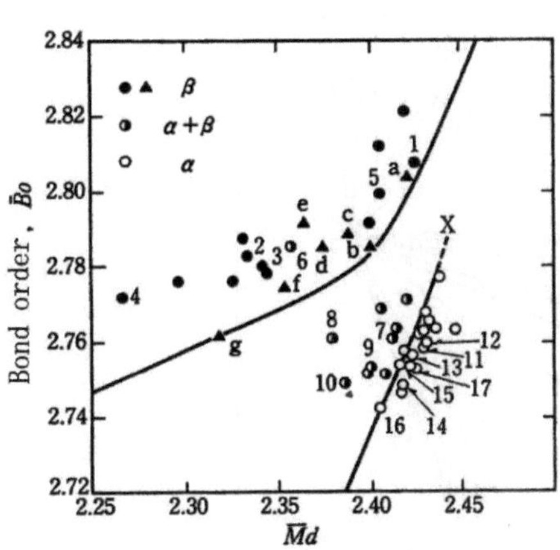

1. Ti-4.5 Sn-11.5 Mo-6 Zr (Beta III)
2. Ti-3 Al-8 V-4 Zr-4 Mo-6 Cr (Beta C)
3. Ti-3 Al-8 V-8 Mo-2 Fe (8-8-2-3)
4. Ti-3 Al-13 V-11 Cr (13-11-3)
5. Ti-15 Mo-5 Zr-3 Al
6. Ti-8 Mn (8 Mn)
7. Ti-6 Al-6 Mo-2 Sn-4 Zr (6-2-4-6)
8. Ti-5 Al-2 Sn-2 Zr-4 Mo-4 Cr (Ti-17)
9. Ti-6 Al-4 V (6-4)
10. Ti-6 Al-6 V-2 Sn (6-6-2)
11. Ti-2.25 Al-11 Sn-5 Zr-1 Mo-0.2 Si (IMI-679)
12. Ti-6 Al-0.5 Mo-5 Zr-0.2 Si (IMI-685)
13. Ti-6 Al-2 Mo-2 Sn-4 Zr (6-2-4-2)
14. Ti-5 Al-6 Sn-2 Zr-1 Mo-0.2 Si (5621 S)
15. Ti-5 Al-2.5 Sn (A-110)
16. Ti-8 Al-1 V-1 Mo (8-1-1)
17. Ti-5.5 Al-3.5 Sn-3 Zr-0.3 Mo-1 Nb-0.3 Si (IMI-829)

Ti-M (a) M=Mo, (b) Fe, (c) Cr, (d) Mn, (e) V, (f) Co, (g) Ni

Figure 20. Classification of the α-, α+β- and β-type alloys in the $\overline{Bo}-\overline{Md}$ diagram.

6.3 Design of β–type alloys

The β–type alloys possess high strength and good cold workability. They are deformed by either slip or twin mechanism, depending on the stability of the β–phase. Namely, the slip mechanism works dominantly when the stability increases with increasing content of β–stabilizing elements in the alloy. The boundary between the slip and twin mechanism is shown in Fig. 22, together with the region of martensite phase appearing upon quenching from high temperatures.[14] It is interesting to note that most practical alloys are located along this slip/twin boundary. Using this $\overline{Bo}-\overline{Md}$ diagram, a new alloy, Ti-13%V-3%Cr-2%Nb-(2-3)%Al has been designed.[15] It is superior in tensile strength and fracture toughness to a commercial alloy, Ti-15%V-3%Cr-3%Sn-3%Al (Ti-15-3-3-3). Recently, Niinomi et al[16] have developed a new β-type alloy for bio-implant applications, using the *d*-electrons concept. Its composition is Ti-29%Nb-13%Ta-4.6%Zr in mass% units, and its Young's modulus is very low. Saito et al[56] have reported the appearance of gum metals in the neighborhood of this bio-implant alloy, namely at the position of $\overline{Bo}$= 2.87 and $\overline{Md}$=2.45. Besides these criteria for $\overline{Bo}$ and $\overline{Md}$, the e/a value of 4.24 is another criterion for the appearance of gum metals. The compositions of the gum metals are, for instance, Ti-12%Ta-9%Nb-3%V-6%Zr-1.5%O and Ti-23%Nb-0.7%Ta-2%Zr-1.2%O in mol% units. Once these gum metals undergo 90% cold working, they show super elasticity. In addition, both the elastic modulus and the linear thermal expansion coefficient remain constant in the temperature range of about 77 K to 500 K. Thus, the $\overline{Bo}-\overline{Md}$ diagram has been proved to be useful for the design of Ti alloys.

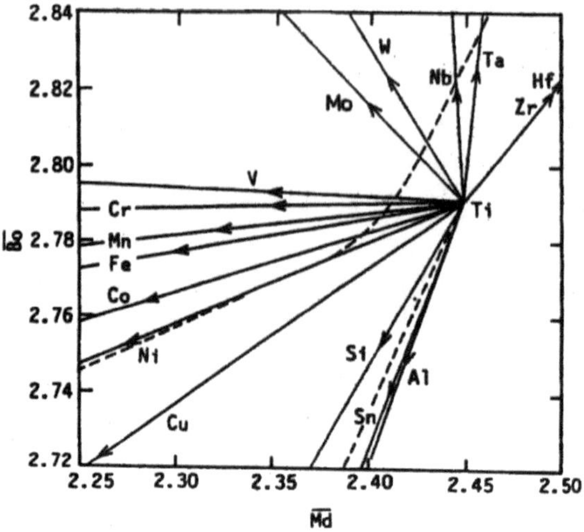

Figure 21. Alloying vector of elements, M, in Ti. Each vector starts at the position of pure Ti and ends at the position of Ti-10 mol% M.

Besides these structural alloys, corrosion resistant Ti alloys can be designed by electron theory. It is shown that the Bo is a convenient parameter for predicting the corrosion resistance.[57] For example, every Ti alloy containing higher Bo elements shows the lower critical anodic current density in the polarization curve and hence the higher corrosion resistance in both 10 mass% HCl and 10 mass% H_2SO_4 solutions at 343 K.

7. ALUMINUM ALLOYS

Aluminum alloys are used for a variety of industrial applications such as light-weight components in automobile and aeroplane systems. However, the alloy development has been carried out relying on many trial-and-error experiments, partially due to the lack of theory predicting the mechanical properties of the alloys. Despite great progress on the fundamental understanding of mechanical properties by dislocation theory, there are still large barriers to the qualitative prediction of alloy strength with the aid of theory.

A quantitative method for predicting the mechanical properties has been proposed on the basis of the molecular orbital calculation of electronic structures.[7,9] A new parameter that is the *s* orbital energy level, Mk, is introduced into this method.

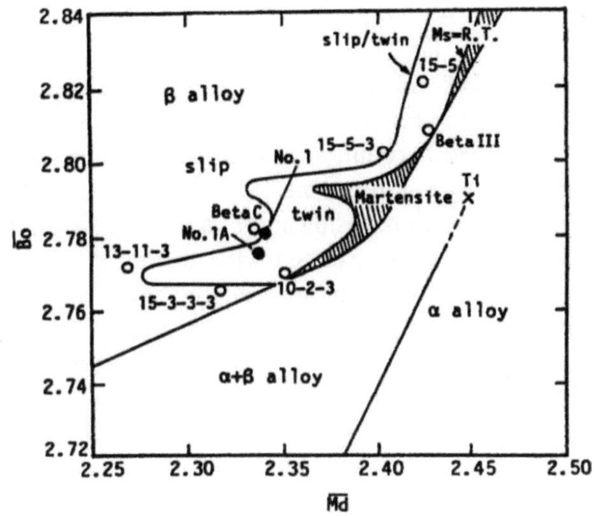

Figure 22. Twin and slip deformation regions in the $\overline{\mathrm{Bo}}-\overline{\mathrm{Md}}$ diagram. The designed alloys, No.1 and No.1A are shown with solid circles.

7.1 Correlation of mechanical properties with classical parameters

Before explaining the parameter Mk, we will first look at what kinds of parameters could be related to the mechanical properties of aluminum alloys.[7] Here, the electronegativity (Φ) and the atomic radius (R) of elements are chosen since they may represent the nature of chemical bonds between atoms in solids. For an alloy, the average difference values are defined by taking the compositional average, $\overline{\Delta\Phi}$ and $\overline{\Delta R}$, denoted as follows,

$$\overline{\Delta\Phi} = \sum_{i=1}^{n} X_i \cdot |\Phi_i - \Phi_{Al}| \tag{5}$$

$$\overline{\Delta R} = \sum_{i=1}^{n} X_i \cdot |R_i - R_{Al}| \tag{6}$$

where X_i is the mole fraction of component *i* in the alloy, Φ_i and R_i are the electronegativity and the atomic radius of component *i* respectively, and the summation extends over the components, *i*=1,2,…., n.

The atomic-radius difference, $|R_i - R_{Al}|$, is an indication showing the magnitude of the local lattice strain around an alloying element i in aluminum, and hence $\overline{\Delta R}$ means the overall average value of the lattice strain in the alloy. Following dislocation theory, such lattice strain will interact with the strain field around a dislocation and will stabilize a pinning state of the dislocation by solute atoms. This is the so-called Cottrell effect that disturbs dislocation motion, resulting in an increase in strength.

On the other hand, the electronegativity difference, $|\Phi_i - \Phi_{Al}|$, is an indication of the amount of the charge transfer between an i atom and its surrounding aluminum atoms, and hence $\overline{\Delta\Phi}$ means the overall average value of the transferred charges between alloying elements and aluminum atoms in the alloy. Such charge transfer affects the chemical bond strength between atoms, and probably causes a certain change in the shear modulus by alloying. Thus, $\overline{\Delta\Phi}$ may be associated with the shear modulus effect on the strain energy of dislocations.

The tensile strengths of commercially available wrought aluminum and aluminum alloys are plotted in Fig. 23 (a) for $\overline{\Delta\Phi}$ and (b) for $\overline{\Delta R}$. The standard strength data are taken from the Metals Handbook and the Aluminium Handbook. In the figure every alloy is identified by four digits, following the Aluminium Association Designations. It is apparent from this figure that the tensile strength increases with increasing $\overline{\Delta\Phi}$ and $\overline{\Delta R}$ even though the data are somewhat scattered, in particular, for $\overline{\Delta\Phi}$ as shown in Fig. 23.a.

7.2 Alloying parameter, Mk

The scattering observed in Fig. 23 may arise partially from the fact that both these classical parameters are assigned to pure elements and any alloying effects with aluminum are not involved in them. In order to reduce the scattering it is desirable to introduce a new alloying parameter which explicitly reflects the alloying behavior of elements in aluminum, while retaining a correlation with both the electronegativity and the atomic radius of the elements. In case of transition metals such a parameter is indeed the *d*-orbital energy level, Md, as described earlier. However, in the case of *s*, *p* metals like aluminum, the Md is no longer valid, but instead the *s*-orbital energy level, Mk, is a new parameter that meets these requirements. This Mk parameter can be obtained using an fcc cluster, MAl_{18}, which contains an alloying element, M, and its surrounding 18 aluminum atoms as shown in Fig. 24.a, and consequently alloying effects are inevitably involved in this parameter.[9] As an example, the calculated energy level structure is shown in Fig. 24.b. The $12a_{1g}$ level for M=Li, Be and $13a_{1g}$ level for M=Na, Mg, Al

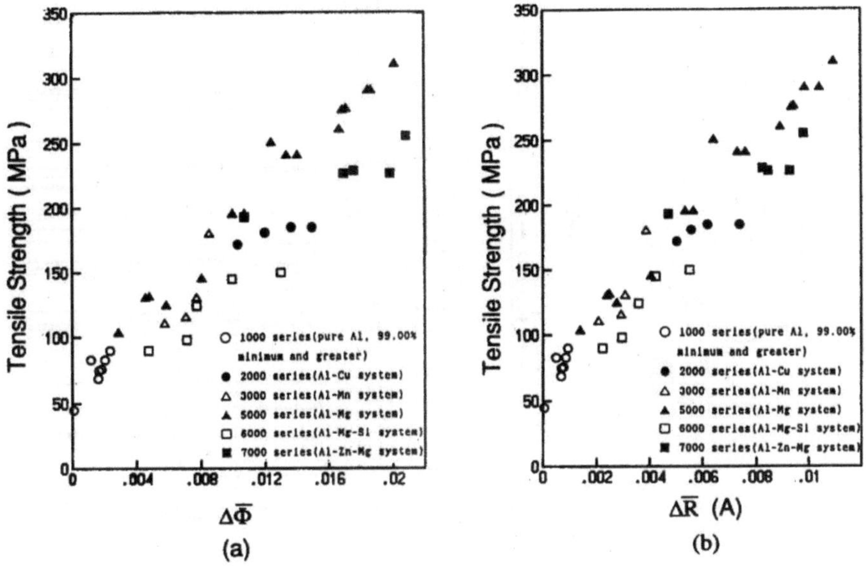

Figure 23. Correlation of the tensile strength with (a) $\Delta\overline{\Phi}$ and (b) $\Delta\overline{R}$ for the O temper alloys.

and Si, are the Mk levels where the M-s component is relatively high. This Mk parameter correlates with the electronegativity and the atomic radius of the elements, as does the Md parameter.[7]

The *p*-orbital energy level may be considered instead of the *s*-orbital energy level, but a spherical *s*-orbital is probably better than a directional *p*-orbital for the purpose of treating the mechanical properties of aluminum alloys.

7.3 A proposed method for the estimation of mechanical properties

In Table 5, the Mk parameters are listed for various alloying elements in aluminum.[7] Most of the elements have a higher Mk value than Al except for Si, Zn, Ga and Ge. For an alloy, two kinds of average values are defined as follows,

$$\overline{\mathrm{Mk}} = \sum_{\mathrm{i\text{-}1}}^{\mathrm{n}} \mathrm{X}_i \cdot \mathrm{Mk}_i \tag{7}$$

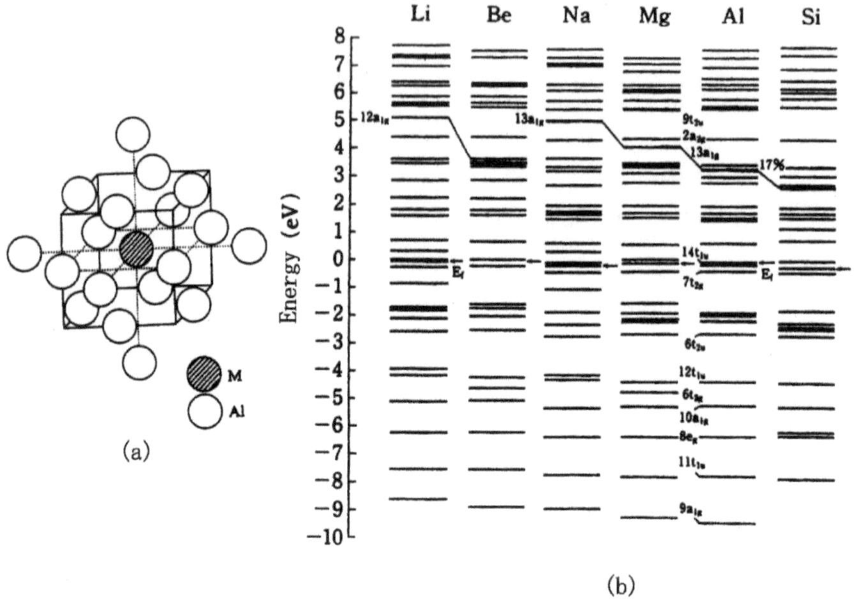

Figure 24. (a) Cluster model and (b) energy level structures for various alloying elements in alumnum. The arrows in the Figure denote the Fermi energy level, Ef.

$$\overline{\Delta Mk} = \sum_{i-1}^{n} X_i \cdot \left| Mk_i - Mk_{Al} \right| \qquad (8)$$

where Mk_i is the Mk value for component *i* listed in Table 5. In the case when all the elements in the alloy have a higher (or lower) Mk value than Al, these two averages differ only a constant bias of Mk_{Al}, 3.344 eV, and hence there is no essential difference between them. However, in the case when elements of higher and lower Mk values are mixed in the alloy, the two averages give different meanings. For example, for an Al-Zn-Mg alloy the larger Mk for Mg is somehow compensated by the smaller Mk for Zn in the calculation of $\overline{Mk}$, whereas both Mg and Zn elements act in an additive way in the calculation of $\overline{\Delta Mk}$. Thus, a certain interaction between Mg and Zn atoms is counted in $\overline{Mk}$, but not in $\overline{\Delta Mk}$. As explained below, $\overline{Mk}$ is the better average for non-heat treatable alloys, but $\overline{\Delta Mk}$ is the better average for heat treatable alloys containing precipitate phases.

Table 5. List of the Mk values for alloying elements in aluminum.

Element	*Mk* (eV)	Element	*Mk* (eV)
Li	5.096	Mn	4.443
Be	3.650	Fe	4.328
Na	5.036	Co	4.314
Mg	4.136	Ni	4.248
Al	3.344	Cu	4.037
Si	2.680	Zn	3.290
K	6.196	Ga	3.013
Ca	5.550	Ge	2.614
Sc	5.200	Zr	5.433
Ti	5.009	Nb	5.227
V	4.782	Mo	5.079
Cr	4.601		

7.4 Estimation of the mechanical properties of aluminum alloys

7.4.1 Non-heat treatable alloys

In Fig. 25 the tensile strength is plotted against (a) $\overline{\text{Mk}}$ and (b) $\overline{\Delta\text{Mk}}$ for a variety of wrought aluminum and aluminum alloys which are annealed enough to obtain the lowest strength.[7] This treatment is called the O temper, according to the Temper Designation System. In the figure, the unit (eV) of $\overline{\text{Mk}}$ and $\overline{\Delta\text{Mk}}$ is omitted for simplicity.

Compared to $\overline{\Delta\text{Mk}}$ shown in Fig. 25.b, $\overline{\text{Mk}}$ shown in Fig. 25.a exhibits a better correlation with the tensile strength. Compared to the plots for $\overline{\Delta\Phi}$ and $\overline{\Delta R}$ shown in Fig. 23, the scattering of the data is reduced significantly in the plots for $\overline{\text{Mk}}$. Also, the yield strength of these alloys varies linearly with $\overline{\text{Mk}}$.[7]

Cold working increases the strength of wrought aluminum alloys. This is called strain hardening. The effects of the Hn4 and Hn8 tempers on the tensile strength are shown in Fig. 26. Here, Hn8 temper means cold working to make about a 75% area reduction of the specimen, and Hn4 temper means cold working to guarantee a tensile strength intermediate between the O and the Hn8 temper specimens. The difference in strength between the O temper and the Hn4 (or Hn8) temper is attributable to the increase in strength due to the strain hardening.

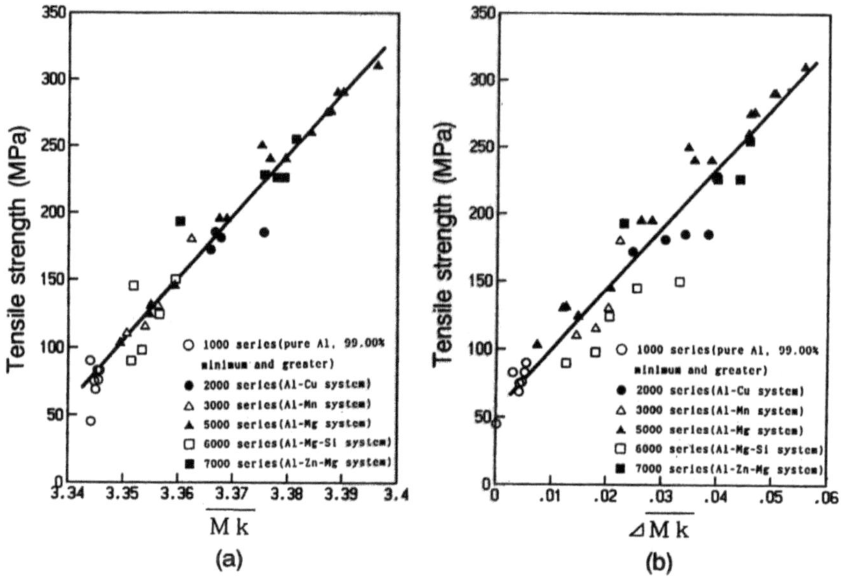

Figure 25. Correlation of the tensile strength with (a) $\overline{\mathrm{Mk}}$ and (b) $\overline{\Delta\mathrm{Mk}}$ for the O temper alloys.

7.4.2 Heat treatable alloys

Aluminum alloys are strengthened by precipitates dispersed in the matrix. This is the so-called precipitation hardening. The tensile strengths of the T6 temper alloys are plotted in Fig. 27 (a) for $\overline{\mathrm{Mk}}$ and (b) for $\overline{\Delta\mathrm{Mk}}$.[7] Here, T6 temper means that the alloy is solution-treated at high temperatures and then aged at a certain temperature (for example, 420 K – 480 K). In contrast to the O temper alloys shown in Fig. 25, the scattering of the data is smaller in the plots of $\overline{\Delta\mathrm{Mk}}$ than $\overline{\mathrm{Mk}}$. The reason is still unknown, even though the phase separation induced by the T6 temper may change solute distributions greatly, resulting in modifications of the solute-solute interactions in alloys.

The tensile strength varies almost linearly with $\overline{\Delta\mathrm{Mk}}$, and it is less dependent on the alloy types of Al-Cu (2000 series), Al-Mg-Si (6000 series) and Al-Zn-Mg (7000 series), despite the substantial difference in the type of precipitates among them. Thus, it is surprising that the tensile strength depends only on $\overline{\Delta\mathrm{Mk}}$ even for these two-phase alloys. Needless to say, the difference in strength between the T6 and the O temper specimens is attributable to the increase in strength due to the precipitation hardening of alloys. The yield strength also changes linearly with $\overline{\Delta\mathrm{Mk}}$.[7]

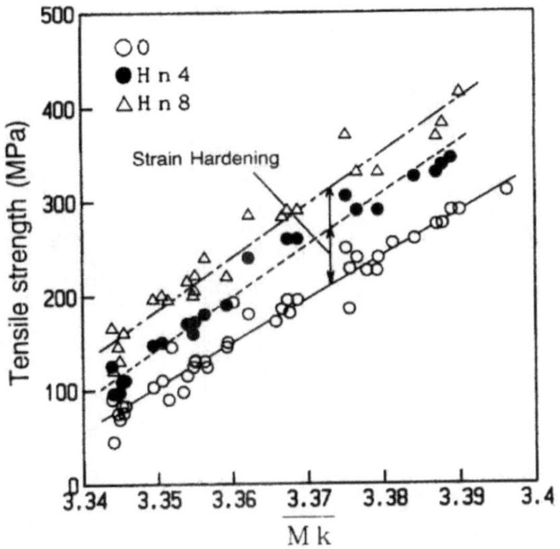

Figure 26. Correlation of the tensile strength with $\overline{\mathrm{Mk}}$ for the O, Hn4 and Hn8 temper alloys.

7.4.3 Strength map for alloy design

In order to show how to use the present Mk approach to practical alloy design, a strength map is constructed for the Al-Mg based alloys (the 5000 series alloys) as shown in Fig. 28.[7] Here, the alloying elements other than Mg are counted using the Mn equivalent. For example, the Mn equivalent (mol%) for an Al-Mg-Mn-X-Y alloy is defined as,

$$\text{Mn equivalent (mol\%)} = [\text{Mn}] + (1/\text{Mk}_{\text{Mn}})\{\text{Mk}_{\text{X}}[\text{X}] + \text{Mk}_{\text{Y}}[\text{Y}]\}, \quad (9)$$

where [Mn], [X] and [Y] are the respective compositions (mol%) of Mn, X and Y elements in the alloy. Similar equations will be readily defined even for alloys with more components. If necessary, this Mn equivalent in mol% units is easily converted to mass% units using a standard conversion method. In fact, in Fig. 28 both axes are drawn in the mass% units for the sake of convenience to represent commercially available alloys in the figure. The number 5XXX below each closed circle is the alloy name, and the number above the circle is the tensile strength (MPa). Also, the tensile elongation (%) is indicated in parenthesis. For example, the 5182 alloy located near 4.5 mass% Mg and 0.8 mass% Mn equivalent has a tensile strength of 276 Mpa and tensile elongation of 25%. Calculated iso-tensile strength lines are also drawn in the figure. It is evident from this figure that the strength of the commercially available aluminum alloys is well represented in this map.

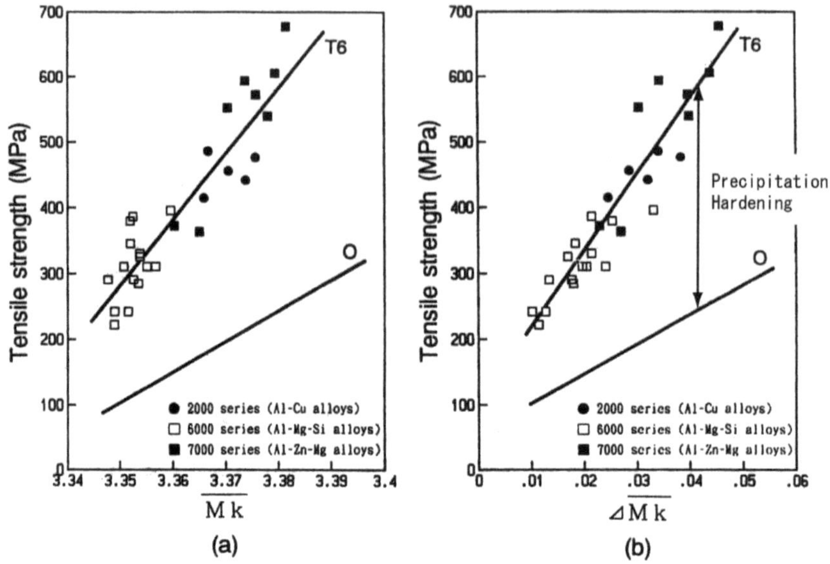

Figure 27. Correlation of the tensile strength with (a) $\overline{\mathrm{Mk}}$ and (b) $\overline{\Delta\mathrm{Mk}}$ for the T6 temper alloys.

A design of the alloy for use of body-panel in automobiles is carried out with this Mk strength map by setting a target of 294 MPA (30 kgfmm^{-2}) for the tensile strength and 30% for the tensile elongation. Here, the tensile elongation may not be treated by the present Mk approach alone, because it may vary even with the presence of minor impurities which scarcely affect the compositionally averaged $\overline{\mathrm{Mk}}$ value. In the present design the Mn equivalent is set empirically to be 0.2 mass% in order to get a longer elongation. A designed alloy is Al-5.3 mass% Mn-0.2 mass% (Mn equivalent), as indicated by a small square in Fig. 28. Experimentally, it is confirmed that this alloy indeed exhibits the target values of 294 MPa and 30%.

8. MAGNESIUM ALLOYS

Magnesium alloys are characteristic of the lowest density cases among the metallic materials for structural applications. In response to the strong demand for light-weight components in automotive systems, they have been used widely for the frame of various equipments, the wheel and the head cover of cylinders in automobile engines. Recently, their use has been further expanded into the hard disc unit of computers and the carriage of printers. In order to predict the mechanical properties, Mk parameters are determined using a hcp cluster, MMg_{18}. The Mk values are listed in Table 6 for alloying elements in magnesium.[8]

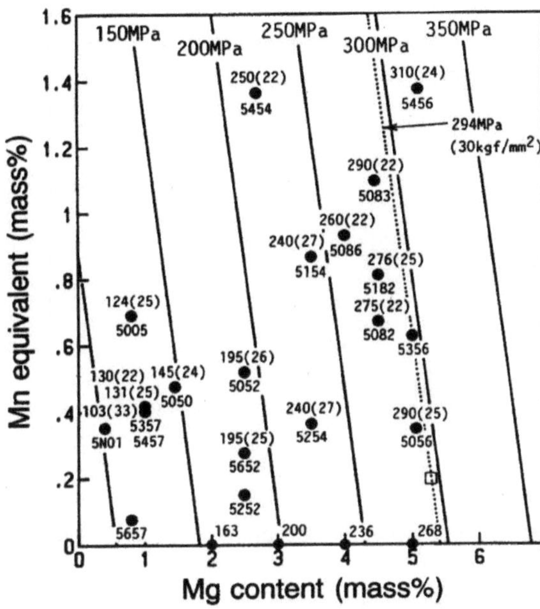

Figure 28. Strength map for Al-Mg based alloys (the 500 series alloys).

Table 6. Mk values for alloying elements in magnesium.

Element	M_k (eV)	Element	M_k (eV)
Li	4.641	Y	4.957
Be	3.891	Zr	4.846
Na	4.664	Nb	4.802
Mg	4.033	Mo	4.787
Al	3.613	Ag	4.354
Si	3.959	Cd	3.999
K	5.059	In	2.737
Ca	4.608	Sn	2.463
Ti	4.341	Sb	2.184
V	4.546	La	6.432
Cr	4.498	Ce	6.400
Mn	4.405	Pr	6.391
Fe	4.393	Nd	6.391
Co	4.352	Sm	6.398
Ni	4.304	Gd	6.404
Cu	4.141	Tb	6.406
Zn	2.727	Dy	6.413
Ga	2.403		
Ge	2.106		

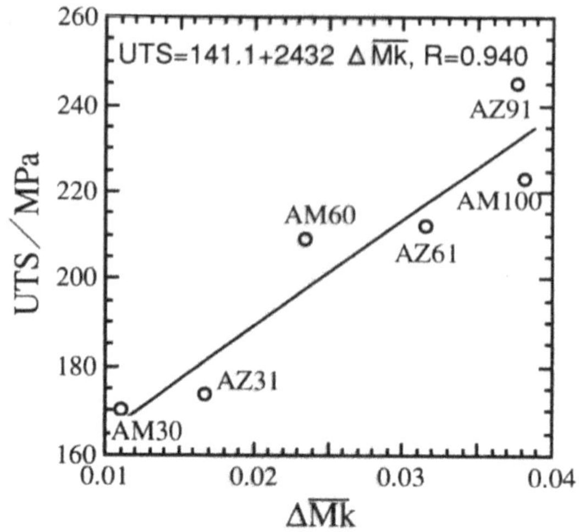

Figure 29. Correlation of the tensile strength with $\Delta\overline{Mk}$ for magnesium alloys.

8.1 Mk approach to the mechanical properties

A second phase precipitates in most magnesium alloys, so that $\Delta\overline{Mk}$ parameter is applicable to the quantitative estimation for the mechanical properties at room temperature. $\Delta\overline{Mk}$ is calculated using Eq. 8, but Mk_{Al} is replaced by Mk_{Mg} in the Mg system. In Fig. 29, the tensile strength is plotted against $\Delta\overline{Mk}$ for six commercially available alloys. There is a good correlation between the tensile strength and $\Delta\overline{Mk}$. Also, as shown in Fig. 30, the Vickers hardness measured at room temperature is plotted against $\Delta\overline{Mk}$ for various Mg-Ca-Al-Zn alloys and Mg-RE (rare-earth) alloys. The hardness changes linearly with $\Delta\overline{Mk}$, irrespective of the alloy systems.

However, $\overline{Mk}$ is a better average than $\Delta\overline{Mk}$ for the mechanical properties at high temperatures. For example, the tensile strengths of die-casting Mg-Al-RE alloys measured at 260℃ (533 K) and 316℃ (589 K), are plotted in Fig. 31 (a) against $\Delta\overline{Mk}$ and (b) against $\overline{Mk}$. There is a poor correlation in Fig. 31.a, but a strong correlation in Fig. 31.b. Also, the creep strengths measured by Waltrip[58] are plotted in Fig. 32 (a) against $\Delta\overline{Mk}$ and (b) against $\overline{Mk}$. A better correlation with the creep strength is seen in the $\overline{Mk}$ plot shown in (b) than the $\Delta\overline{Mk}$ plot shown in (a). These results imply that dislocation motion in Mg alloys at high temperatures is scarcely affected by the presence of precipitates in the matrix, but formation of dislocations and/or thermally activated motion of dislocation are responsible for the high temperature deformation in magnesium alloys.

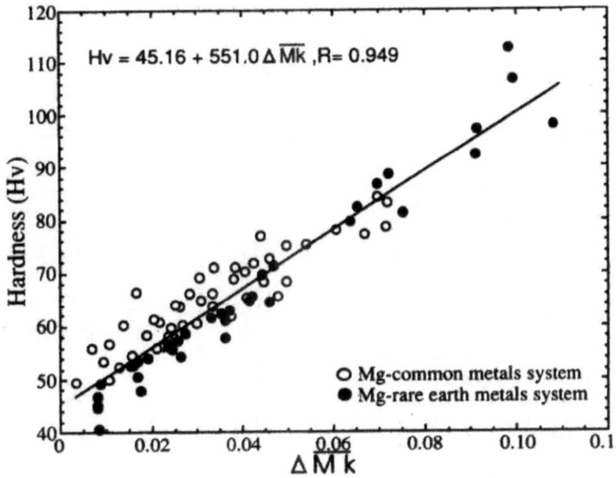

Figure 30. Correlation of the Vickers hardness with $\overline{\Delta Mk}$ for magnesium alloys.

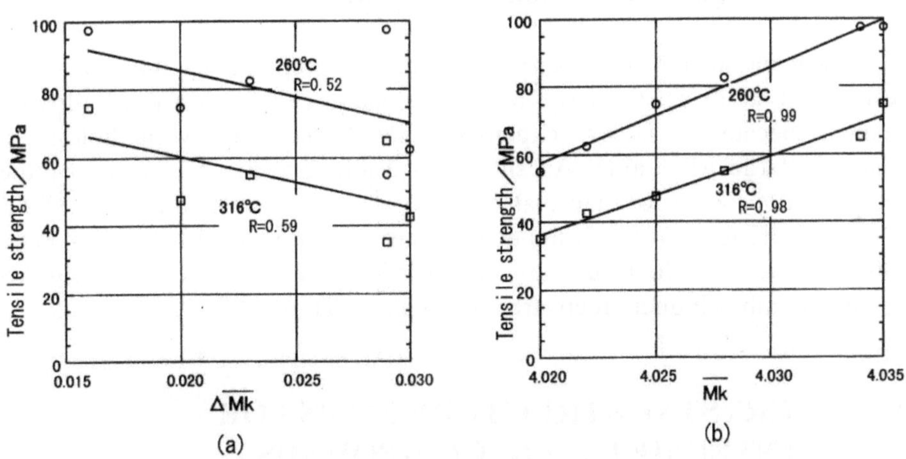

Figure 31. Correlation of the tensile strength at high temperature with (a) $\overline{\Delta Mk}$ and (b) $\overline{Mk}$ for magnesium alloys.

Thus, the alloy with large $\overline{\Delta Mk}$ value will possess a high strength at room temperature, whereas the alloy with large $\overline{Mk}$ value will possess a high strength at high temperatures. Therefore, the addition of those elements which have high Mk values (e.g., Ca, Y and RE (rare-earth)) will be suitable for increasing the alloy strengths at both room temperature and high temperatures.

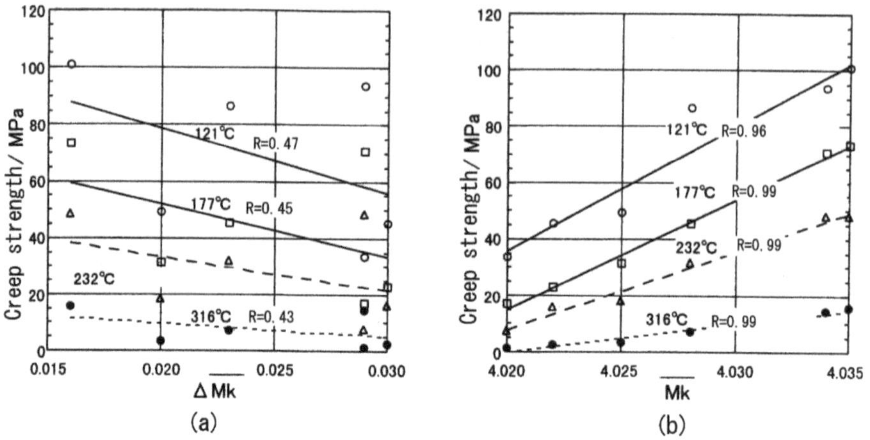

Figure 32. Correlation of the creep strength with (a) $\overline{\Delta Mk}$ and (b) $\overline{Mk}$ for magnesium alloys. R is a correlation coefficient.

8.2 Design of heat-resistant Mg alloy

A heat-resistant Mg-Al based alloy that is prepared by die-casting, is designed so as to get the large $\overline{Mk}$ value. Ca is chosen as a third alloying element, because it is less expensive than Y and RE, while holding a relatively large Mk value. As shown in Table 7.a, the $\overline{Mk}$ value of the designed alloy is larger than that of AZ91, or comparable to that of AE42. The creep strengths of the designed alloy are measured at 150℃ (423 K) and 200C (473 K), and the results are shown in Table 7.b. The designed Mg-Al-Ca alloy exhibits higher creep strengths than AZ91 and AE42.[59]

9. CRYSTAL STRUCTURE MAPS FOR INTERMETALLIC COMPOUNDS

Intermetallic compounds (e.g., TiAl,[60] $LaNi_5$) are some of the promising materials which are greatly expected to be used for various applications in the future. The Bo-Md diagram is useful for predicting the crystal structure of intermetallic compounds such as aluminides and silicides containing transition metals, M.[11] One example is shown in Fig. 33 for MAl compounds. The crystal structures are separated clearly in this diagram. As explained earlier, both the Bo and the Md parameters change following the position of elements in the periodic table. Recalling that the crystal structures of any compounds also change following the periodic table, we may say that

Table 7. (a) Chemical composition and value for designed alloys, AZ91 and AE42, and (b) their creep properties.

(a)

	Composition / mass%						
	Al	Zn	Mn	RE	Ca	Mg	$\overline{Mk}$
AZ91	8.9	0.68	0.19	–	–	bal.	3.995
AE42	4.0	–	0.22	3.4	–	bal.	4.032
Designed Alloy	3.1	–	–	–	3.2	bal.	4.033

(b)

150 ℃ –50MPa		
	Initial strain /%	Steady creep rate / %h-1
AZ91	0.1370	2.012×10^{-2}
AE42	0.1166	2.275×10^{-4}
Designed Alloy	0.1146	1.184×10^{-4}
200 ℃ –30MPa		
	Initial strain /%	Steady creep rate / %h-1
AZ91	0.0889	1.096×10^{-1}
AE42	0.0796	5.662×10^{-4}
Designed Alloy	0.0130	3.924×10^{-4}

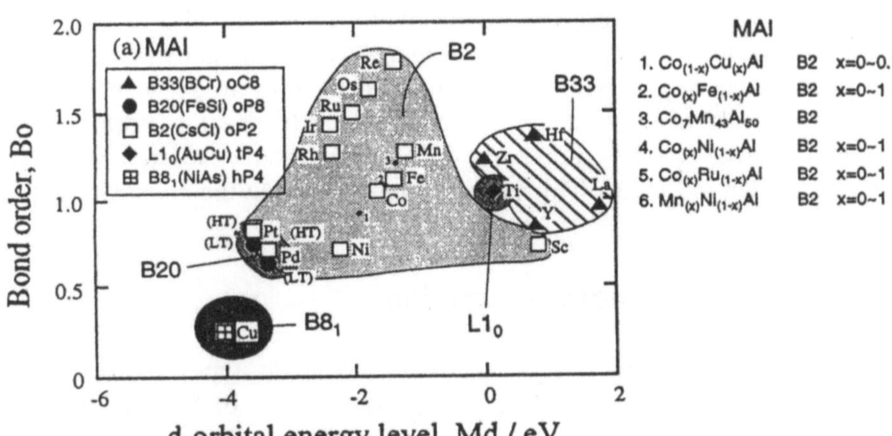

Figure 33. Bo-Md structure map for MAl, where M's are transition metals.

these Bo and Md are indeed suitable parameters for constructing structure maps. The Bo-Md structure maps have been made for a variety of compounds. The physical properties of the compounds are so sensitive to the crystal structure that these structure maps will provide us a convenient tool for the design of intermetallic compounds.[61]

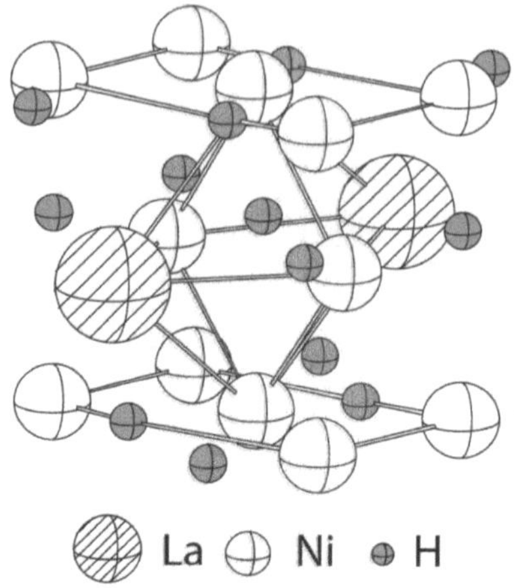

Figure 34. Cluster model used in the calculation for $LaNi_5$ system.

10. HYDROGEN STORAGE ALLOYS

Hydrogen storage alloys are key materials for the development of future clean hydrogen energy systems. A large amount of hydrogen is stored in the alloy by forming metal hydrides. Many hydrogen storage alloys have been developed, but the total amount of hydrogen stored in any alloy is still low in view of practical applications to automobiles. In order to get a clue to the development of new alloys, a series of calculations has been performed with appropriate cluster models. For example, a cluster model used for the $LaNi_5$ system is shown in Fig. 34. This is constructed based on the crystal structure of $LaNi_5H_6$.[62]

10.1 Metal-hydrogen interaction

Most hydrogen storage alloys, such as $LaNi_5$ (AB_5-type), $ZrMn_2$ (AB_2-type), TiFe (AB-type) and Mg_2Ni (A_2B-type), consist of hydride forming elements, "A", and non-forming elements, "B". Needless to say, the A element has a larger affinity for hydrogen than the B element in the binary metal-hydrogen system. Therefore, the role of each element in the hydrogen storage alloy has been understood in the following way. The hydride forming element (e.g., La) may make a strong chemical bond with hydrogen, whereas the hydride non-forming element (e.g., Ni) may work to reduce such a strong A-H bond, so that hydrogen is released readily from the A atom in the hydrogen desorption process.

However, this naive understanding is not true according to our calculations.[63-71] For example, a contour map of the electron densities is shown in Fig. 35 for the $LaNi_5H_6$ hydride on the atomic plane of containing La, Ni and H atoms. As is evident from this figure, a relatively high electron-density region near hydrogen extends to the Ni atom site, but not to the La atom site.

This result clearly indicates that hydrogen interacts more strongly with Ni atoms than La atoms in the $LaNi_5$ system,[63,64] despite the larger affinity of La atoms for hydrogen than Ni atoms in the binary metal-hydrogen system. Similar results are also obtained for the Mg_2Ni,[65] $ZrMn_2$,[66] and the TiFe system.[67,68] In all cases hydrogen interacts more strongly with B atoms (Ni, Mn or Fe) than A atoms (La, Mg, Zr or Ti).

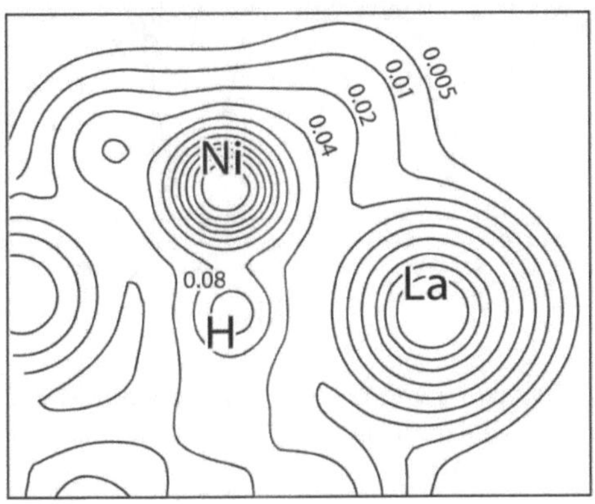

Figure 35. Contour map of the electron densities on $LaNi_5$ system. The denoted numbers indicate the numbers of electron per a.u.^{3} (1 a.u.=0.0529nm).

However, it is noticed that the B-H (e.g., Ni-H) interaction in the alloy is still weaker compared to the A-H (e.g., La-H) interaction in pure metal hydride, AH_2, (e.g., LaH_2), judging from the values of the heat of formation of the hydride. It is -209.2 kJ/mol H_2 for LaH_2 and -30.1 kJ/mol H_2 for $LaNi_5H_6$. The hydrogen desorption process can be activated readily because of such a weak B-H interaction operating in every hydrogen storage alloy.[69] On the other hand, LaH_2 is too stable to be dehydrided at moderate temperatures. In this sense, we may say that the existence of weak hydrogen-metal interactions is characteristic of hydrogen storage alloys.[71]

10.2 Roles of hydride forming and non-forming elements

However, the hydride forming element, A, does play an important role in the formation of the B-H chemical bond.[72,73] In order to make clear the roles of hydride forming and non-forming elements in hydrogen storage alloys, a series of calculations is performed using a Ni-based octahedral cluster,

M_2Ni_4H (inset in Fig. 36), where M's are various elements presented on the horizontal axis. In the calculation, the Ni-M interatomic distance is varied, depending on the atomic size of M.

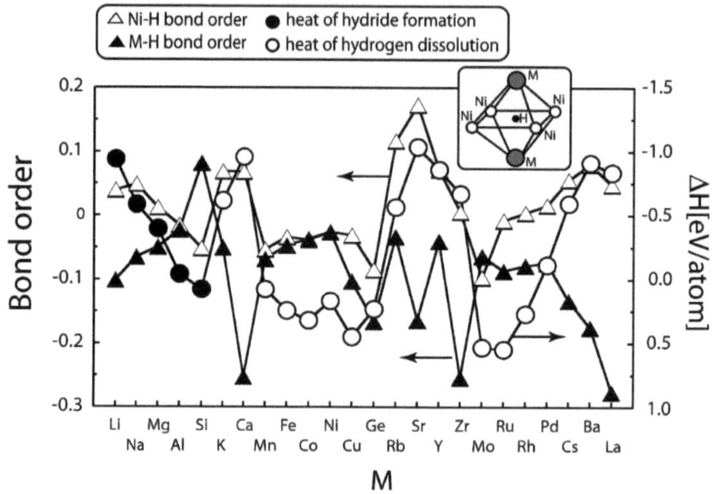

Figure 36. Bond orders between Ni and H atoms and between M and H atoms in a M_2Ni_4H cluster, and the comparison with the heat of hydride formation of M metal or the heat of hydrogen dissolution in M metal, ΔH.

From this figure, it is apparent that the Ni-H bond order is large when the M-H bond order is small. In particular, the Ni-H bond order is enhanced when M's are the hydride forming elements such as Na, Ca and La. This means that hydrogen interacts strongly with Ni atoms if the hydride forming elements exist in the neighborhood. On the other hand, when M's are the hydride non-forming elements such as Fe, Co and Ni, the Ni-H bond order is very small, so that hydrogen may not be absorbed in such an octahedral cluster. Thus, it is evident from Fig. 36 that the magnitude of the Ni-H bond order correlates well with the heat of hydride formation of M metal or the heat of hydrogen dissolution in M metal, ΔH.

Similar results are also obtained for the Ti-based and the Mg-based octahedral clusters.[71-74] Thus, there is a general trend that the B-H bond becomes strong only when the hydride forming element, A, exists adjacent to B and H atoms. Otherwise, it remains weak or even antibonding, so that it is difficult for hydrogen to be absorbed in the alloy. Thus, both A and B are indeed essential elements in hydrogen storage alloys.

The trend mentioned above is probably concerned with the fact that the B-H interatomic distance is shorter than the A-H interatomic distance, because of the smaller atomic size of element B than element A in most

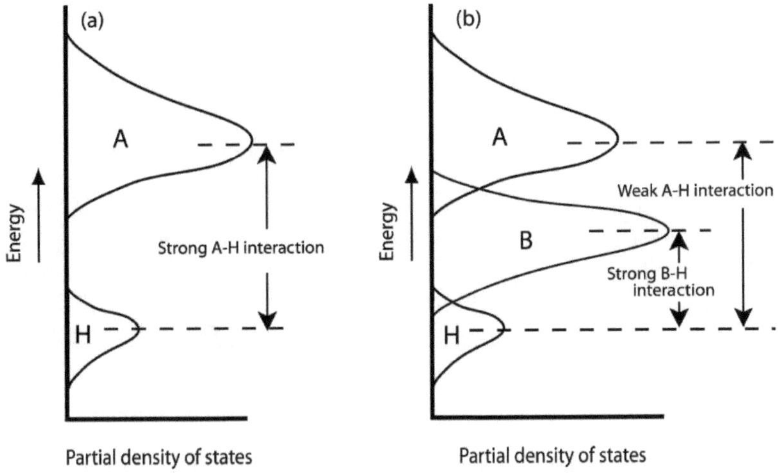

Figure 37. Schematic illustration of the energy band structures of (a) pure A metal hydride and (b) hydride formed in hydrogen storage alloy consisting of both A and B elements.

cases. Since hydrogen occupies an interstitial site at the center of a distorted octahedron or tetrahedron with unequal edge lengths, the B-H interaction is inevitably strengthened, while the A-H interaction is weakened. Thus, a polyhedron made of A and B elements provides hydrogen a unique interstitial spacing in the A-B alloy, being very different from that in pure A metal.

Furthermore, as shown in Fig. 37.b, the H-1*s* energy band appears in the energy range much closer to the band of the B element than the A element. As a result, the B-H covalent interaction becomes stronger than the A-H interaction. For example, in case of $LaNi_5H_6$, the H-1*s* band appears below the Ni-3*d* band, and the Ni-H bond is formed preferentially in the hydride. On the other hand, in the case of the AH_2 hydride (e.g., LaH_2), as shown in Fig. 37.a, the band of A element is located well above the H-1*s* band, so charge transfer occurs from A to H, which causes the ionic interaction between A and H atoms in AH_2 hydride.

10.3 Criteria for alloy design

10.3.1 Alloy cluster suitable for hydrogen storage

As explained above, the B-H bond is formed in the circumstance of A elements existing in the neighborhood. Therefore, the effective number of the B-H bonds is supposed to be large when the concentration of A element is high in the alloy. In other words, the total amount of hydrogen absorbed in

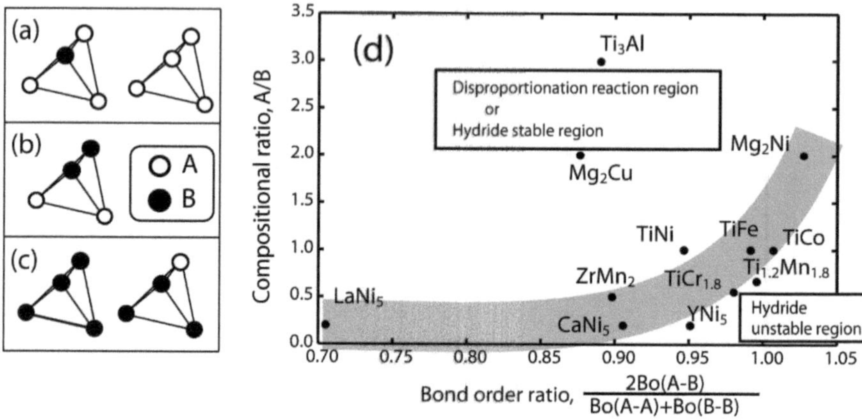

Figure 38. (a)-(c) Three types of tetrahedral clusters and (d) correlation between the A/B compositional ratio and the bond order ratio.

the alloy will increase with increasing A/B compositional ratio. However, a disproportionation reaction tends to take place if the concentration of A element is too high in the alloy.

This can be explained assuming that there are three types of tetragonal clusters in the A-B binary alloy, as shown in Fig. 38.a-c.[74,75] If the A/B compositional ratio is high, or the A-A bond is stronger than the A-B or the B-B bond, the clusters shown in Fig. 38.a are dominant in the alloy. In such a case, hydrogen will interact mainly with A elements when hydrogen is introduced into the alloy, resulting in the formation of strong A-H bonds. The disproportionation reaction, $AB + H \rightarrow AH + B$, will take place easily in it. Here, AH is so stable that the onset of this reaction is undesirable in view of the activation of the hydrogen desorption process.

On the other hand, when the A/B compositional ratio is low, or the B-B bond is stronger than the A-A or the A-B bonds, the clusters shown in Fig. 38.c are dominant in the alloy. In this case, it is very hard for hydrogen to be absorbed in such a B atom-abundant cluster.

Therefore, in order to absorb and desorb hydrogen smoothly without the onset of any disproportionation reaction, the A/B compositional ratio should be controlled in a proper manner, depending on the chemical bond strength between the A and the B elements. In other words, optimum clusters shown in Fig. 38.b may be formed in the alloy only when the bond order ratio, 2Bo(A-B)/[Bo(A-A)+Bo(B-B)], takes a suitable value for a given A/B compositional ratio.

10.3.2 Alloy compositions

A series of calculations is performed to obtain the bond order ratio, 2Bo(A–B)/[Bo(A–A)+Bo(B–B)], for various alloy systems using the tetrahedral cluster, and the results are shown in Fig. 38.d.[71,72] Here, this bond order ratio has a physical meaning similar to the ordering energy parameter, Ω_{AB} (=$V_{AB}-1/2(V_{AA}+V_{BB})$), where V_{AB}, V_{AA}, V_{BB} are the respective bond strengths between atoms given in the subscript. Needless to say, the atomic arrangement shown in Fig. 38.a or c is present in the clustering-type alloy system, but the one shown in Fig. 38.b is present in the ordering-type alloy system. The hydrogen storage alloys except for bcc V alloys[76,77] have the ordered atomic arrangement, so that their characteristics will be represented by using an ordering energy parameter, Ω_{AB}. But instead of Ω_{AB}, the bond order ratio is used here, because it is a parameter that can be estimated easily from the calculations, while still holding an approximately similar physical meaning as does Ω_{AB}.

As shown in Fig. 38.d, there exists a strong correlation between the A/B compositional ratio and the bond order ratio. For example, all the typical hydrogen storage alloys, $LaNi_5$, $ZrMn_2$, TiFe and Mg_2Ni, are located on a narrow band illustrated in the Figure. Here, for Mg_2Ni, the bond order ratio is high, because the Mg-Ni bond order is much larger than the Mg-Mg bond order. Also, for $LaNi_5$, the bond order ratio is low, because the La-Ni bond order is smaller than the La-La bond order. All the alloys located in the upper region above this narrow band tend to decompose during hydrogenation and/or tend to form a stable hydride. One example is the Mg_2Cu system, in which a disproportionation reaction, $Mg_2Cu + 2H_2 \rightarrow 2MgH_2 + Cu$, takes place and a stable hydride, MgH_2, is formed during hydrogenation. Therefore, the region lying above the narrow band is called the disproportionation reaction region or hydride stable region. On the other hand, the alloys located in the region below this narrow band, tend to form unstable hydrides. YNi_5 is an example for this case. Therefore, this region is called the hydride unstable region.

Similar results are also obtained for the octahedral clusters. Thus, the present result is independent of the clusters used for the calculation. Once A and B elements are chosen, the A/B compositional ratio can be determined from the relationship with the bond order ratio shown in Fig. 38.d.

10.3.3 Mg-based alloys

The same method is now applied to the Mg-based alloys. The results obtained from the calculation using octahedral clusters are shown in Fig. 39.

In this

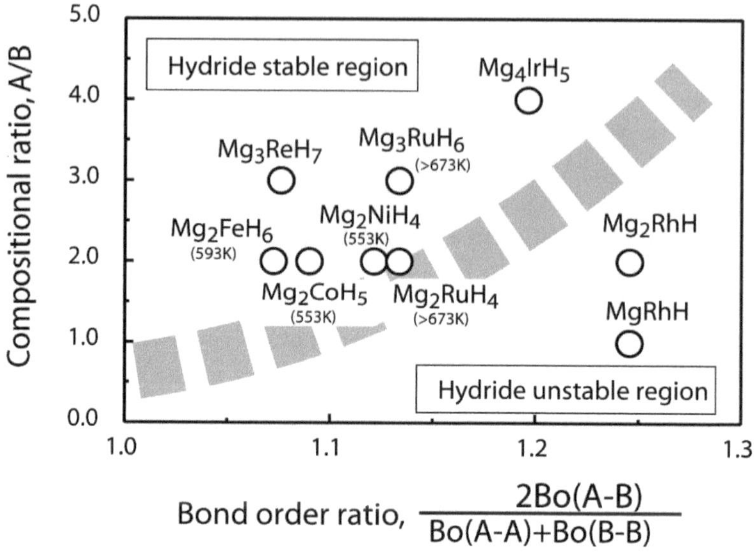

Figure 39. Correlation between A/B compositional ratio and the bond order ratio for Mg-based alloys.

figure, the decomposition temperatures at a constant hydrogen pressure of 0.1 MPa are also denoted for some hydrides. The location of the narrow band is shown here using a dotted band.

The hydriding properties of the Mg-based alloys correlate well with the bond order ratio as shown in Fig. 39. For example, for the case where the A/B compositional ratio is fixed at 2, the total amount of hydrogen stored in the hydrides changes in the order, Mg_2FeH_6 > Mg_2CoH_5 > Mg_2NiH_4 = Mg_2RuH_4 > Mg_2RhH, so that it decreases with increasing bond order ratio. This means that the hydrogenation further proceeds in such an alloy that has the lower bond order ratio. This is reasonable since the alloy with the lower bond order ratio is located in the hydride stable region, whereas the alloy with the higher bond order ratio is located in the hydride unstable region. Even for the other A/B compositional ratios, the alloys located in the region above the dotted band tend to form hydrides with high H content (e.g., Mg_3ReH_7). On the other hand, the alloys located in the region below the band tend to form hydrides with low H content (e.g., MgRhH). These results are consistent with the results shown in Fig. 38.d.

11. A UNIVERSAL RELATION IN ELECTRON DENSITY DISTRIBUTIONS IN MATERIALS

Through these investigations, electronic structures for a variety of materials have been calculated using the DV-Xα molecular orbital method. It has been discovered recently that there is a universal relation between electron density minima and atomic (or ionic) radius.[18]

For example, a calculated electron density, ρ(r), is shown in Fig. 40, where the natural logarithm of the electron density, log ρ(r), in MgO is illustrated along the line linking the first-nearest-neighbor O and Mg nuclei. Here, the logarithm is taken to detect a very small change in the ρ(r) curve. As is evident from Fig. 40, log ρ(r) decreases monotonously with the distance, r, from the oxygen nucleus, but the curve is separated into two regions in the oxygen side, each region with its own slope. The slope varies with the principal quantum number, *n*, which is a quantum number that roughly determines the energy and the size of atom. In the case of oxygen shown in Fig. 40, $n = 1$ for the inner region and $n = 2$ for the outer region. The slope was about 15.75 for the inner region composed mainly of O 1*s* electrons, and about 6.61 for the outer region composed mainly of O 2*s*, 2*p* electrons. It is known that $2Z/n$ is the slope expected from the radial distribution function of a hydrogen-like atom consisting of only one electron, where Z is the atomic number. As Z=8 for oxygen, the $2Z/n$ values are 16 for n=1 and 8 for n=2, each value being close to the respective values of oxygen described above. In particular, there is a strong resemblance in the slope for the region for n=1.

It is also found that the extent of the region (n=2) for O-2*s*, 2*p* electrons is strongly dependent on the neighboring atom in various oxides. For example, the result of log ρ(r) in Al_2O_3 is indicated by a dotted curve in Fig. 40. The region of $n = 2$ is rather limited compared to that in MgO, but the slope still remains unchanged with the neighboring metal atom. Even the valence electrons such as O-2*s*, 2*p* keep their own slope, instead of spreading over the space in a random manner. In addition, the extent of the $n = 2$ region shows the range of influence of oxygen. So, its extension means that the effective number of electrons increases at the oxygen site, resulting in a large and negative ionicity of oxygen. In this sense, the extent of the $n = 2$ region is a measure of the ionic radius as well as the ionicity.

Next, as illustrated in Fig. 40, we define the minimum electron density, ρ_{min}, and the atomic or ionic radii, r_{min}, at the position of ρ_{min}. The values of ρ_{min} and r_{min} are calculated for gases, water and solids of over 130 species, and the results are summarized in Fig. 41. In this figure the vertical axis is $\log (\rho_{min}/Z^3)$ and the horizontal axis is $2(Z/n)r_{min}$. This coordinate is chosen, because in the case of the radial distribution functions of hydrogen-like atoms, $\rho(r)/Z^3$ is a function of only one variable, $2(Z/n)r$.

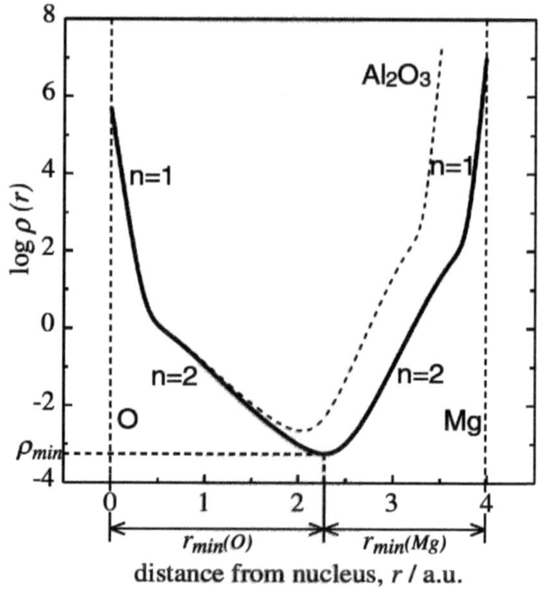

Figure 40. Representation of the natural logarithm of the electron density, log $\rho(r)$, in the line linking the first-nearest neighbor O and Mg nuclei (1 a.u.= 0.0529 nm). Minimum electron density, ρ_{min}, and ionic radii, r_{min} (O) for O and r_{min} (Mg) for Mg, are illustrated in the figure. The dotted curve is the result for Al_2O_3.

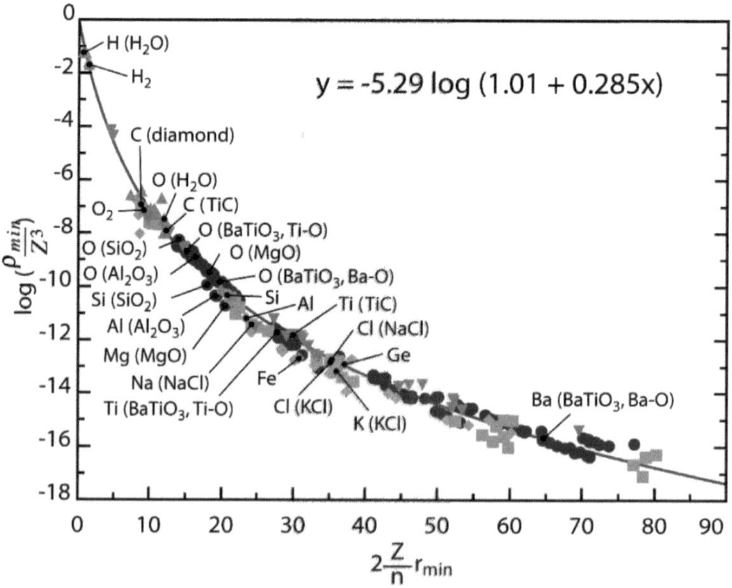

Figure 41. A universal relation between the minimum electron density, ρ_{min}, and the atomic or ionic radii, r_{min}. Notations used in the figure; for example, H_2 denotes H in H_2 gas. Ti(TiC) and C(TiC) denote Ti and C in the first-nearest neighbor Ti-C pair in TiC, respectively. Ti(or O)($BaTiO_3$, Ti-O) denotes Ti(or O) in the first-nearest neighbor Ti-O pair in $BaTiO_3$. The others are denoted in a similar way.

It is surprising that every matter falls on one curve, irrespective of gases, water and solids, despite the great difference in the nature of the chemical bond among them. For example, some of the matter named in Fig. 41 are gases (H_2, O_2), water (H_2O), diamond (C), metals (Fe, Al), semi-conductors (Si, Ge), oxides (MgO, Al_2O_3, SiO_2, $BaTiO_3$), alkali halide crystals (NaCl, KCl), and metal compounds (TiC, ZrN). Thus, this curve is indeed a universal curve common to every matter. Using least-squares analysis, it is expressed in a simple formula as,

$$\log\left[\rho_{\min} / Z^3\right] = a.\log\left[b + c.2(Z/n)r_{\min}\right] \tag{10}$$

where the fitting parameters are a = –5.29±0.066, b = 1.01±0.040 and c = 0.285±0.010. This universal relation is useful for the understanding of the nature of chemical bonds in materials.

12. CONCLUSIONS

A molecular orbital approach is useful for the design and development of not only ferrous alloys but also for non-ferrous alloys. This approach can be extended even to ceramics. In the super-computer age, such a computer-aided approach is superior to the trial-and-error approach to save time and cost necessary for materials development.

ACKNOWLEDGEMENTS

The authors acknowledge the staffs of the Computer Center, Institute for Molecular Science, Okazaki National Institute for the use of the super computer. This research was supported by the Grant-in-Aid for Scientific Research from the Ministry of Education, Culture, Sports, Science and Technology of Japan, and from the Japan Society for the Promotion of Science.

REFERENCES

1. for example, see International journal of Calphad, published by Elsevier.
2. for example, I. Yasui, *Ceramics* **25**, 507(1990).
3. for example, L.-Q. Chen, *Annu. Rev. Mater. Res.* **32**, 113 (2002).
4. M.Morinaga, N.Yukawa, H.Adachi and H.Ezaki, *Superalloys 1984*, eds. M.Gell et al., The Metall. Society of AIME (1984), p. 523.
5. M. Morinaga, N. Yukawa, H. Ezaki and H. Adachi, *Phil. Mag.* **51**, 223 and 247 (1985).

6. K. Matsugi, Y. Murata, M. Morinaga and N. Yukawa, *Superalloys 1992*, eds. S.D. Antolovich et al., The Minerals, Metals & Materials Society, (1992), p. 307.
7. M. Morinaga and S. Kamado, Modelling Simul. *Mater. Sci. Eng.* **1**, 151 (1993).
8. R. Ninomiya, H. Yukawa, M. Morinaga and K. Kubota, *J.Alloys Comp.* **215**, 315 (1994).
9. M. Morinaga, S. Nasu, H. Adachi, J. Saito and N. Yukawa, *J. Phys.:Condens. Matter* **3**, 6817 (1991).
10. M. Morinaga, N. Yukawa, H. Adachi and S. Kamado, *J. Less-Common Metals* **141**, 295 (1988).
11. Y. Harada, M. Morinaga, J. Saito and Y. Takagi, *J. Phys.: Condens. Matter* **9**, 8011 (1997).
12. K. Matsugi, Y. Murata, M. Morinaga and N. Yukawa, *Mat. Sci. Eng.* A **172**, 101 (1993).
13. R. Hashizume, A. Yoshinari, T. Kiyono, Y. Murata and M. Morinaga, *Superalloys 2004*, TMS, (2004), p. 53.
14. M. Morinaga, N. Yukawa, T. Maya, K. Sone and H. Adachi, *Sixth World Conference on Titanium*, Cannes, June6-9, 1988, Proceedings Part III, pp. 1601-1606, eds. P. Lacombe, R. Tricot and G. Beranger, Societe francaise de Metallurgie.
15. M. Morinaga, M. Kato, T. Kamimura, M. Fukumoto, I. Harada and K. Kubo, *Titanium '92, Science and Technology*, eds. F.H. Froes and I. Caplan, TMS (1993), Vol.1, p. 217.
16. M. Niinomi, D. Kuroda, M. Morinaga, Y. Kato and T. Yashiro, *Non-Aerospace Applications of Titanium*, eds. F.H. Froes, P.G. Allen and M. Niinomi, TMS (1998), p. 217.
17. M. Morinaga. R. Hashizume and Y. Murata, *Materials for Advanced Power Engineering 1994*, eds. D. Coutsouradis et al., Kluwer Academic Publishers (1994), p. 329.
18. M. Yoshino, M. Morinaga, A. Shimode, K. Okabayashi, H. Nakamatsu and R. Sekine, *Materials Transactions* **45**, 1968 (2004).
19. E.D. Ellis and G.S. Painter, *Phys. Rev.* **82**, 2887 (1970).
20. H. Adachi, M. Tsukada and C. Satoko, *J. Phys. Soc. Jpn.* **45**, 875 (1978).
21. J.C. Slater, *Quantum Theory of Molecules and Solids*, Vol. 4, McGrow-Hill, (1974).
22. M. Morinaga, N. Yukawa and H. Adachi, *J. Phys. Soc. Jpn.* **53**, 653 (1984).
23. G.C. Fletcher, *Physica* **56**, 173 (1971), and **63**, 41 (1972).
24. D. Hackenbracht and J. Kubler, *J. Phys. F* **10**, 427 (1980).
25. for example, see *Advances in Quantum Chemistry* , Vol.29, Academic Press(1997).
26. M. Morinaga, Y. Murata and H. Yukawa, *Materials Design Approaches and Experiences*, eds. J.-C. Zao, M. Fahrmann and T.M. Pollock, TMS (2001), p. 15.
27. Y. Matsumoto, M. Morinaga, T. Nambu and T. Sakaki, *J. Phys.: Condens. Matter* **8**, 3619 (1996).
28. H. Shibutani, M. Morinaga and K. Kikuchi, J. *At. Energy Soc. Japan* **40**, 70 (1998).
29. S. Inoue, J. Saito, M. Morinaga and S. Kano, *J. Phys.:Condens. Matter* **6**, 5081 (1994).
30. H. Yukawa, M. Maeda, M. Morinaga and H. Nakamatsu, *Bulletin of the Society for Discrete Variational X*α **13**, 75 (2000).
31. M. Kurihara, M. Hirata, R. Sekine, J. Onoe and H. Nakamatsu, *J. Nuclear Materials* **326**, 75 (2004).
32. W.J. Boesch and J.S. Slaney, *Met. Prog.* **86**, 109 (1964).
33. L.R. Woodyatt, C.T. Sims and H.J. Beattie, Jr., *Trans. AIME* **236**, 519 (1966).
34. W. Hume-Rothery and G.V. Raynor, *Structure of Metals and Alloys*, (London, UK: The Institute of Metals, 1954).
35. L. Darken and G.W. Gurry, *Physical Chemistry of Metals*, (New York:McGraw-Hill, Inc., 1953).
36. P. Caron, *Superalloys 2000*, eds. T.M. Pollock et al., TMS (2000), p. 737.

37. P. Caron and T. Khan, *Materials Design Approaches and Experiences*, eds. J.-C. Zhao, M. Fahrmann and T.M. Pollock, TMS (2001), p. 81.
38. M.J. Cieslak, G.A. Knorovsky, T.J. Headley and A.D. Romig, *Metallurgical Transactions* **17A**, 2107 (1986).
39. J.S. Ogborn, D.L. Olson and M.J. Cieslak, *Mater. Sci. Eng. A* **203**, 134 (1995).
40. K.S. O'Hara, W.S. Walston, E.W. Ross and R. Darolia, US Patent 5, 482, 789 (1996).
41. T. Kobayashi, Y. Koizumi, S. Nakazawa, T. Yamagata and H. Harada, *Advanced in Turbine Materials, Design and Manufacturing*, eds. A. Strang et al., Inst. of Metals (1997), p. 766.
42. P. Caron and T. Khan, *Aerosp. Sci. Technol.* **3**, 513 (1999).
43. Y. Murata, S. Miyazaki, M. Morinaga and R. Hashizume, *Superalloys 1996*, eds. R.D. Kissinger et al., TMS (1996), p. 61.
44. W.S. Walston, K.S. O'Hara, E.W. Ross,T.M. Pollock and W.H. Murphy, *Superalloys 1996*, eds. R.D. Kissinger et al., TMS (1996), p. 27.
45. Y. Murata, R. Hashizume, A. Yoshinari, N. Aoki, M. Morinaga and Y. Fukui, *Superalloys 2000*, eds. T.M. Pollock et al., TMS (2000), p. 285.
46. M. Morinaga, N. Yukawa and H. Adachi, *J. Phys. F:Met. Phys.* **15**, 1071 (1985).
47. L. Pauling, *The Nature of the Chemical Bond*, Third Edition, Cornell University Press (1960).
48. H. Ezaki, M. Morinaga, M. Kato and N. Yukawa, Acta Metall. Mater. **39**, 1755 (1991).
49. K. Jack, *Proc. R. Soc. A* **208**, 200 (1951).
50. T. Bell and W.S. Owen, *J. Iron and Steel Institute*, April (1967), p. 428.
51. M. Morinaga, N. Yukawa, H. Adachi and T. Mura, *J.Phys.F:Met. Phys.* **17**, 2147 (1987).
52. M. Morinaga, N. Yukawa, H. Adachi and T. Mura, *J.Phys.F:Met. Phys.* **18**, 923 (1988).
53. W. Schlachter and G.H. Gessinger, *High Temperature Materials for Power Engineering 1990*, eds. E. Bachelet et al., (1990), p. 1.
54. H. Ezaki, M. Morinaga, K. Kusunoki and Y. Tsuchida, Tetsu-to-Hagane, **78**, 1377 (1992).
55. F. Masuyama, *Proc. of the 78th Annual Meeting of the Kyushu Branch of the Iron and Steel Institute of Japan*, Sept. 25, 1992.
56. T. Saito, T. Furuta, J. Hwang, S. Kuramoto, K. Nishino, N. Suzuki, R. Chen, Yamada, K. Ito, Y. Seno, T. Nonaka, H. Ikehata, N. Nagasako, C. Iwamoto, Y. Ikuhara and T. Sakuma, *Science* **300**, 464 (2003).
57. M. Morishita, Y. Ashida, M. Chikuda, M. Morinaga, N. Yukawa and H. Adachi, *ISIJ International* **31**, 890 (1991).
58. J.S. Waltrip, *Proceedings of the 47th Annual World Magnesium Conference, International Magnesium Association*, (1990), pp.124-129.
59. M. Morinaga and R. Ninomiya, *Handbook of Advanced Magnesium Technolgy*, Karosu Publishing Co. (2000), pp.97-104.
60. M. Morinaga, J. Saito, N. Yukawa and H. Adachi, *Acta Metal. Mater.* **38**, 25 (1990).
61. M. Morinaga and H. Yukawa, *Advanced Engineering Materials* **3**, 381 (2001).
62. J.L. Soubeyroux, L. Pontonnier, S. Miraglia, O. Isnard and D.Z. Fruchart, *Phys. Chem. Bd.* **179**, 187 (1993).
63. H. Yukawa, Y. Takahashi and M. Morinaga, *Intermetallics* **4**, S215 (1996).
64. H. Yukawa, T. Matsumura and M. Morinaga, *J. Alloys Comp.* **293-295**, 227 (1999).
65. Y. Takahashi, H. Yukawa and M. Morinaga, *J. Alloys Comp.* **242**, 98 (1996).
66. T. Matsumura, H. Yukawa and M. Morinaga, *J. Alloys Comp.* **279**, 192 (1998).
67. H. Yukawa,T. Takahashi and M. Morinaga, *Comput. Mat. Sci.* **14**, 291 (1999).
68. T. Nambu, H.Ezaki, H. Yukawa and M. Morinaga, *J. Alloys Comp.* **293-295**, 213 (1999).
69. H. Yukawa and M. Morinaga, *Advances in Quantum Chemistry* **29**, 83 (1997).

70. T. Nambu, H. Ezaki, M. Takagi, H. Yukawa and M. Morinaga, *J. Alloys Comp.* **330**, 318 (2002).
71. M. Morinaga and H. Yukawa, *Mater. Sci. Eng. A* **329-331**, 267 (2002).
72. K. Nakatsuka, M. Yoshino, H. Yukawa and M. Morinaga, *J. Alloys Comp.* **293-295**, 222 (1999).
73. M. Morinaga, H. Yukawa, K. Nakatsuka and M. Takagi, *J. Alloys Comp.* **330**, 20 (2002).
74. K. Nakatsuka, M. Takagi, M. Nakai, H. Yukawa and M. Morinaga, *Proc. of the Inter. Conf. on Solid-Solid Phase Transformations '99(JIMIC-3)*, eds. M. Koiwa et al., The Japan Institute of Metals, (1999), pp.681-684.
75. H. Yukawa, K. Nakatsuka and M. Morinaga, *Solar Energy Materials & Solar Cells* **62**, 75 (2000).
76. T. Matsumura, H. Yukawa and M. Morinaga, *J. Alloys Comp.* **284**, 82 (1999).
77. H. Yukawa, S.Ito, D. Yamashita and M. Morinaga, *Advances in Quantum Chemistry* **42**, 263 (2003).

Chapter 9

APPLICATION OF COMPUTATIONAL AND EXPERIMENTAL TECHNIQUES IN INTELLIGENT DESIGN OF AGE-HARDENABLE ALUMINUM ALLOYS

Aiwu Zhu, Gary J. Shiflet and Edgar A. Starke Jr.
Department of Materials Science and Engineering, University of Virginia, Charlottesville, VA 22904, USA

Abstract: Age hardenable aluminum alloys are one of the traditional structural materials that are constantly under improvement to meet higher performance requirements. The practical alloys have complex microstructural features, among which the strengthening secondary phases and their dispersions are the controlling factors for major mechanical properties. Usually those phases with their specific crystal structures, elastic constants, strengths and thermal stabilities are formed in complex morphologies with particular shapes, sizes and orientations. Taking examples from Al-Cu-(Li,Mg)-based alloys, this chapter describes the quantitative characterization of the multiple secondary phases and illustrates dislocation slip simulations for evaluation of strengthening effects and for prediction of optimum precipitate structures. A combination of computational and experimental techniques will be described that are useful to study stress-aging treatments.

Keywords: Aluminum alloys, precipitation, stress-aging, quantitative microstructure characterization, dislocations, first principles calculation

1. INTRODUCTION

Aluminum alloys are classified as heat treatable and non-heat-treatable, depending on whether or not they respond to precipitation hardening.[1] The heat treatable alloys contain elements that exhibit decreasing solid solubility in aluminum as the temperature is lowered. A usual heat processing cycle includes a homogenization, followed by a solution treatment at a high temperature and rapid cooling to a low temperature to obtain a solid solution supersaturated (SSSS) with solute elements. During aging at room or at some intermediate temperature, secondary phases (abbreviated as 2^{nd}P in the following), usually consisting of the solute atoms, precipitate in the aluminum matrix strengthening the alloy.

Alloy design for age-hardenable aluminum alloys can be extremely complex. Practically some primary criteria are set for the most desired mechanical properties during the first cycle of alloy development. A balanced combination must be reached among a variety of mechanical properties including the yield strength, ductility, fracture toughness, stiffness, corrosion resistance, creep-resistance, etc. during the next test-modification cycles. On the other hand, aluminum alloys usually have a variety of microstructural features that affect or control those mechanical properties in different ways. These features include the grain structure, dislocation patterns (sub-grain structures), constituent particles, dispersoids, equilibrium 2^{nd}P and strengthening 2^{nd}P precipitates as well as their interactions. The fine dispersoids (0.01-0.2 μm), formed by transitional metal elements during ingot preheating, are at least partially coherent with the Al matrix and can act as recrystallization barriers and normally control the grain shape in wrought products. The large incoherent constituent particles (1-30 μm in diameter), formed through the liquid-solid eutectic reactions by impurities or excess solute elements during solidification or the homogenization treatment can act as void nucleation sites during deformation and hence reduce the fracture toughness. The subgrain structure has small and variable effects on the strength and strain localization, but more significantly on the precipitation of 2^{nd}Ps. Equilibrium 2^{nd}P precipitates, usually with crystal structures much different from the matrix and with partially coherent inter-phase boundaries, coarse and heterogeneously distributed, have little strengthening effect. They are primarily formed at grain boundaries (GBs), leading to formation of GB precipitate free zones (PFZ) and hence nurturing strain localization and lowering the fracture toughness. The grain structures may be controlled to a large extent by adding certain elements to form fine dispersoid phases combined with suitable thermo-mechanical processing.[2] The grain structure effects and the grain-texture induced anisotropy have been extensively studied and can be assessed based on well-developed theories and analysis tools (c.f., *e.g.* Ref. 3). As far as the strengthening effect and work hardening are concerned, the key factor is the strengthening 2^{nd}Ps that are formed during the aging treatment. They are usually coherent or semi-coherent and homogeneously distributed inside the grains. To a large extent, the solute atoms remaining in the matrix function in a similar way to those 2^{nd}P particles, and hence they can correspondingly be evaluated. Therefore, analysis of 2^{nd}P precipitation is essential for an alloy design and development; the primary alloy design goals are 1) to identify the desired

2^{nd}Ps and the optimum 2^{nd}P structures or morphology and 2) to find the right alloying elements and composition as well as the suitable thermal mechanical treatment condition to produce them. This chapter will mainly address the first issue while the second set of topics will be treated elsewhere.[4]

2. CHARACTERIZATION OF SECONDARY PHASE AND THEIR STRUCTURES

2.1 Fundamental properties

A combination of experimental techniques and computational methods can be employed to determine some fundamental properties of the 2^{nd}Ps that are usually difficult to measure accurately because of their low volume fractions and small particle sizes.

2.1.1 Crystalline structures

Crystalline structures of 2^{nd}P can be determined using X-ray diffraction (XRD), electron diffraction patterns, electron back-scattering patterns (EBSP) or quantitative HRTEM analysis, etc. The most common strengthening 2^{nd}Ps in aluminum alloys have been investigated but disparate results were obtained for some important 2^{nd}Ps that usually exist with low volume fractions and small particle sizes. For instance, the very early XRD study Perlitz and Westgren (PW)[5] suggested S (Al_2CuMg) phase in Al-Cu-Mg system has an orthorhombic structure with unit cell dimensions a=0.4 nm, b=0.923 nm, and c=0.714 nm and space group *Cmcm*, containing 16 atoms in the ratio Al:Cu:Mg=2:1:1. However since then, there has been extensive investigation that yielded results suggesting similar but different models (*c.f.*, references in Ref. 6). One of the most recent HRTEM and microanalysis studies[6] confirmed the symmetry, chemical composition and the lattice parameters of the PW crystalline structures except the atomic site-positions. Comparison between the recorded images of the S-phase laths processed by imposing crystallographic symmetries with image simulations from different models indicated that the match between experiment and simulation could be improved further from the PW model by an exchange of Mg and Cu (referred to as the RaVel model) atomic sites. This was also supported by the intensity distribution as shown in Fig. 1. To validate these proposed models, a first principles (FP) total energy calculation[7] has

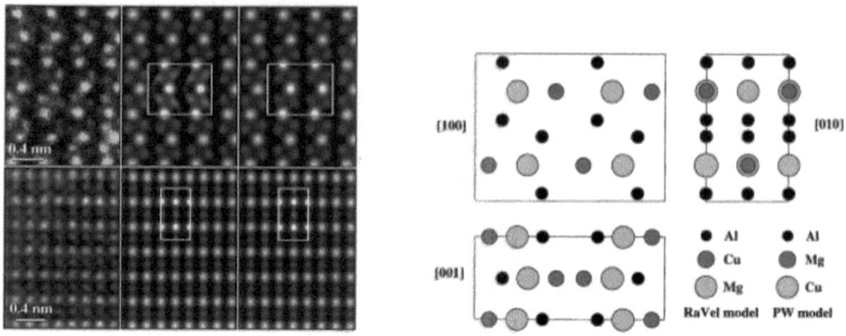

(a) (b)

Figure 1. (a) Quantitative analysis of HREM images along [100]S (first row) and [010]S (second row). First column: as-recorded images; second and third columns: processed images with insets showing image simulations based on two different models, PW and RaVel, respectively. (b) Schematic drawing of the S-phase crystal structure; the difference between the PW and RaVel model is indicated.[6]

been used to compare the energetics of the corresponding crystalline structures.[8] The calculation, verifying a similar study,[9] showed that the original PW model leads to a formation enthalpy of -0.210 eV/atom while exchange of the Cu and Mg in internal atomic coordinates, as suggested in Ref. 6, yielded a positive formation enthalpy of +0.289 eV/atom. By neglecting the entropy contribution difference (mostly from lattice vibration), this indicates that the PW model is more reasonable. Several crystal structures have also been proposed for the Ω phase. Auld suggested a monoclinic structure[10] while a widely accepted structure[11-13] is orthorhombic with a (*Fmmm*) space group. Based on analysis of convergent-beam electron diffraction (CBED) patterns, Garg and Howe[14] advocated a tetragonal structure (*I4/mcm*) that is only a metastable $\{111\}_\alpha$ variant of the equilibrium θ phase. The results of first principles calculations have ruled out the monoclinic structure because of its positive formation enthalpy.[15]

Table 1. FP (VASP) Calculated Formation Enthalpies of Proposed Crystal Structures for Ω and θ phases as compared with that of θ obtained from optimization using thermodynamic models (from Ref. 15).

Phase	Proposed structure	ΔH, eV/atom
θ	Tetragonal (I4/mcm)	-0.168
Knowles' Ω	Orthorhombic(Fmmm)	-0.167
Garg & Howe's Ω or θ_m	Tetragonal (I4/mcm)	-0.169
θ as modeled in COST507 etc.	2 sub-lattice ordered compound with composition Al2-Cu	-0.164

The energetic difference between the tetragonal and orthorhombic structures is negligible, as shown in Table 1. It shows both the tetragonal and the orthorhombic structure have the formation enthalpies that are close to the equilibrium θ phase.

2.1.2 Elastic constants

Because of their low volume fractions and small sizes, elastic constants of the 2^{nd}Ps are difficult to measure experimentally, e.g., by means of ultrasonic techniques. The 2nd order elastic constants of the crystalline solid may be defined and calculated according to the variation in the energy from its optimized lattice structure as

$$C_{ijkl} = \left(\frac{\partial E}{\partial \underline{\varepsilon}_{ij} \partial \underline{\varepsilon}_{kl}}\right)_{E=E_{\min}} \quad i,j,k,l = 1,2, \text{ or } 3 \tag{1.a}$$

where E denotes the total energy associated with the deformation and $\{\underline{\varepsilon}_{ij}\}$the strain tensor. Using "engineering" notations, Hooke's law can be written:

$$\sigma_i = \sum_{j=1}^{6} C_{ij}\varepsilon_j \tag{1.b}$$

where $\{\sigma_i\}$ is the stress tensor. By applying strains, or changes in the unit cell dimensions, in certain patterns, the variation of the energy or the resulting stress tensor can be retrieved and the elastic tensors $\{C_{ijkl}\}$ or $\{S_{ijkl}\}$ calculated.[16] Using FP calculation, one may obtain some reasonable information as verified by comparing calculated values with those experimentally measured at room temperature for known crystal structures, as shown in Table 2. The calculated elastic tensors for the θ′, Ω, θ_M, θ and S phases are listed there.

2.2 Structural Parameters

The secondary phases can be formed with various shapes, orienting on or along different crystalline planes and directions in the matrix. For the 2^{nd}Ps that are coherent or semi-coherent with the matrix lattice, their precipitate shapes can usually be approximated as spherical such as δ' phase in Al-Li based alloys, or rod-like such as S in Al-Cu-Mg, or plate-like, *e.g.*, θ′ in Al-Cu and T_1 in Al-Cu-Li alloys. A 2^{nd}P precipitate can be either shearable

Table 2. Elastic constants calculated using VASP based on the local density approximation and obtained from experiments at room temperature for some elementary crystals and 2nd phases.

Phase	Structure	Elastic Constants, 1010 N/m2	
		Calculated	Experimental
Al	fcc	C_{11}=9.30, C_{12}=6.38, C_{44}=3.13, B=7.35	C_{11}=10.8 C_{12}=6.13 C_{44}=2.85,B=7.2
Cu	fcc	C_{11}=17.9,C_{12}=12.4,C_{44}=8.03, B=14.27	C_{11}=16.84,C_{12}=12.14,C_{44}=7.54,B=13.7
Ag	fcc	C_{11}=10.93,C_{12}=8.39,C_{44}=3.78, B=9.24	C_{11}=12.4,C_{12}=9.34,C_{44}=4.61,B=10.36
Mg	hcp	C_{11}=5.39, C_{33}=4.78, C_{12}=2.39, C_{13}=3.26, C_{44}=1.75	C_{11}=5.64, C_{33}=5.81, C_{12}=2.32, C_{13}=1.81, C_{44}=1.68
θ'	Tetrahedral	C_{11}= 19.24, C_{33} = 15.62, C_{12}=2.91, C_{13} = 6.54, C_{44}=8.35, C_{66} = 4.36	n. a.
$\Omega/\theta/\theta_m$	Tetrahedral	C_{11}=18.72, C_{12}= 5.59, C_{13} = 5.36 C_{33}=18.30, C_{44} = 3.68, C_{66}= 6.27	n. a.
S	Orthorhombic	C_{11}=14.5,C_{22}=15.3,C_{33}=14.3,C_{12}= 3.0, C_{13}=6.6,C_{44}=5.3,C_{55}=7.3,C_{66}=3.3	n. a.

or unshearable for glide dislocations depending on its strength on the glide plane and the local configuration of the dislocations and other neighbor precipitates.

In general, for the most common plate-like 2[nd]Ps, each 2[nd]P can be completely or uniquely characterized by a group of structure parameters including the specific phase strength η, size distribution $\zeta(D)$ with diameter D, orientation distribution ψ(i, j, k) where {i, j, k} denotes the habit plane in the matrix lattice, aspect ratio $D{:}t$ with the thickness t, volume fraction f_v (or number density n_v). For measuring the strength of the materials using uniaxial mechanical tests, the Schmid factor for single crystals or the Taylor factor, M, for polycrystals are conveniently used to account for effect of the crystalline orientation or the texture with respect to the applied loading direction.

2.2.1 Particle Strength

If a dislocation can cut though the precipitate particles, the strength of the particle can be defined as the energy required for the glide dislocations to pass through the particle that is associated with the formation of the corresponding dislocations in the particle.

Figure 2 illustrates the atomic configuration along $(100)_\alpha$ when an edge-type dislocation with Burgers vector $\boldsymbol{b}=1/2\langle110\rangle_\alpha$ passes from an α-fcc lattice through a $L1_2$ (Cu_3Au)-type-phase interfaced with the fcc-A1.

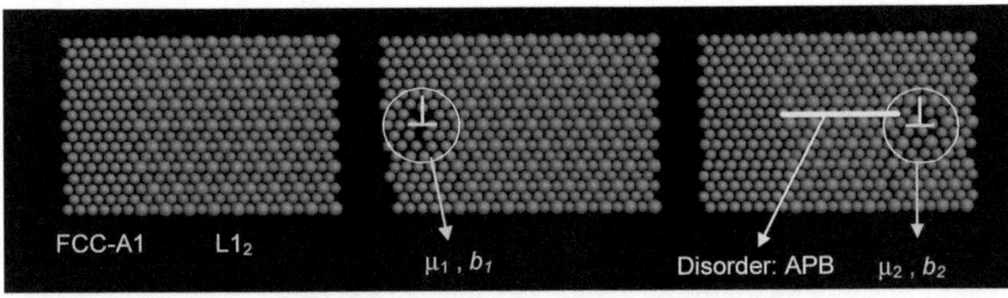

Figure 2. An edge-type dislocation generated and passing through a $L1_2$ (Cu_3Au)- type phase particle in fcc-A1 matrix.

Five possible contributions may arise from a variety of the structural changes during the process. 1) The strength associated with the creation of the antiphase boundary is $F_{m,\,order} = 2r_{cs}\,\gamma_{ap}$ where r_{cs} is the radius of the cross section of the dislocation glide plane through the 2ndP and γ_{ap} the specific antiphase boundary energy; 2) The strength associated with the creation of the new interface ledges of width of $2^{1/2}\,a_0$, $F_{m,chem} = 2b\gamma_{if}$ where γ_{if} denotes the interfacial energy; 3) The contribution due to the modulus mismatch $F_{m,\Delta G} \sim b^2|\mu_2 - \mu_1|$ where μ_1 and μ_2 are the shear moduli of the two phases; 4) The lattice misfit $\delta_{mis} = |(a_2\text{-}a_1) / a_1|$ gives rise to an strength $F_{m,\delta} \sim \mu_1 b \delta_{mis} r$ and 5) The stacking fault difference $\Delta_{sf} = |\gamma_1 - \gamma_2|$ can be accounted for by a strength $F_{m,sf} \sim \Delta_{sf} d$ where d denotes an effective length of the split dislocation in the 2ndP. The interfacial energy can be obtained by the kinetic characterization of the coarsening processes of 2ndP precipitates using conventional electron microscopy (EM). The specific antiphase boundary energy can be obtained by the experimental measurement of the critical resolved shear stresses based on the ordering strengthening model. Several theoretical approaches to the interface energies and the antiphase boundary energies have been developed. The continuum formalism based on van der Waals theory[17] employs an integral equation for the free energy of an inhomogeneous system.[18] More recent studies, taking into account lattice discreteness, are based on the Ising or lattice gas models using a variety of statistical-mechanical techniques including regular solution, CVM mean field and Monte Carlo atomistic simulations. The effective energetic interaction parameters between different atomic species are selected either by fitting to some experimentally measured properties within the CALPHAD context,[19] or by using first principles total energy calculations.[20]

Regarding more complex 2ndPs such as θ″ or θ′, S, Ω and T_1, those plate-like particles are coherent or semi-coherent with the Al matrix along one orientation (on their broad faces) and mostly incoherent on the sides. The strength of these phases is rarely investigated.

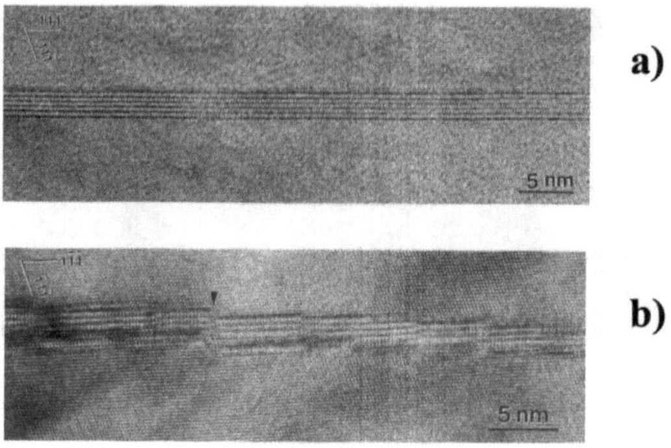

Figure 3. HRTEM image of the Ω precipitate along <011>$_{Al}$ orientation in a) the undeformed alloy. The whole image shows the uniform contrast; b) the deformed Ω phase shows APBs, steps and high strain field created by dislocation cutting.[22]

Extensive microscopy studies showed that while spherical coherent 2ndP particles like δ′ (Al_3Li-$L1_2$) can be sheared at the very beginning of plastic deformation, those plate-like phases were sheared mostly under two conditions: 1) The plate-like 2ndPs are co-existing with the shearable spherical 2ndPs *e.g.*, T1 with δ′ in Al-Cu-Li alloys;[21] 2) the deformation exhibits remarkable work hardening.[21,22] The shearing of the Ω-phase plates by glide dislocations on {111}α-planes in an Al-3.2Cu-0.45Mg-0.5Ag wt% alloy is depicted in Fig. 3. Nevertheless it is reasonable to generally treat those plate-like particles as unshearable by dislocations during the early or yielding stage of deformation because 1) the wide interaction scale (~ the plate face diameter *D*) between the dislocation and the plates greatly cut down the forces exerted by the bowed dislocation[23] and 2) those 2ndPs have higher stiffness than the aluminum fcc matrix as indicated by the FP calculated elastic constants in Table 1.

2.2.2 Morphology of 2ndPs

The 2ndP morphology can be complex. Table 3 lists some major strengthening 2ndPs in Al-Cu-Mg-X or Al-Cu-Li-X systems. Figure 4 illustrates the orientation relationship between the proposed tetragonal structure of Ω and the Al matrix. The microscopic images shown in Fig. 5 illustrates the three types of 2ndPs S, Ω and θ′ co-existing in an artificially aged Al-4Cu-0.3Mg-0.4 (wt.)% alloy.[24]

Besides the specific strength for each 2ndP, the structural parameters for morphology can be determined by a quantitative microscopy technique.[23] Usually, X-ray diffraction is used to measure the texture and the orientation distribution function (ODF) from {200}, {111} and {220} pole figures and

Table 3. The 2nd phases in Al-Cu-Mg-X (high Cu:Mg ratio) and Al-Cu-Li-X alloys.

	Nominal Composition	Morphology	Orientation	Coherency
θ′	Al_2Cu	Plates	{100} habit	Partial
θ	Al_2Cu	variable	variable	Partial
S or S′	Al_2CuMg	Rods / Laths	<100> {210} habits	Partial
δ′	Al_3Li	Spheres	Cube / Cube	Full
T_1	Al_2CuLi	Plates	{111} habits	Partial
Ω	Al_2Cu	Plates	{111} habits	Partial

the fully constrained Taylor factor *M* can thus be calculated using the popLA software package.[25] The truncation effect on the plate size distribution can be accounted for by a correction approach proposed by Crompton et al.[26] The thickness of the TEM foils can be determined by the convergent-beam electron diffraction (CBED) patterns using Kossel-Mollenstedt fringe spacing.[27] The size distribution, and hence the mean diameter, and the volume fraction of the precipitates in each sample can be obtained by TEM image analysis including size measurement and counting (using bright field or central dark field images).

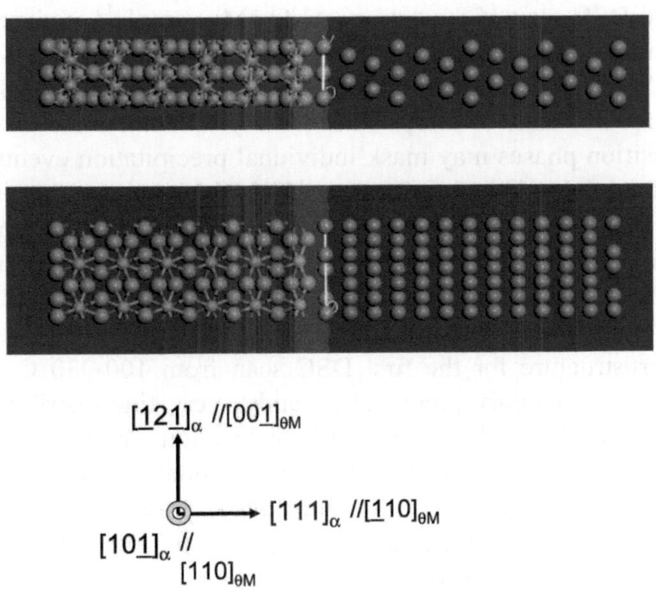

Figure 4. Supercell construction of interfacial structure between Ω<001> and Al<111>. The top is the projection on YZ-plane and the middle is on XZ-plane. The bottom shows the orientation match between the two phases.

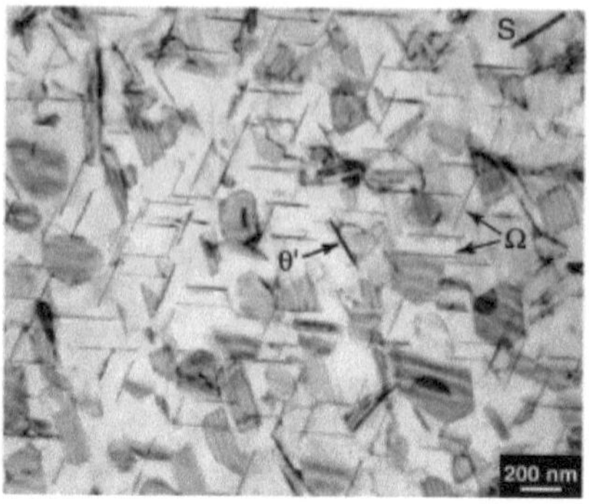

Figure 5. Bright field TEM micrograph of an Al–4Cu–0.3Mg–0.4Ag (wt. %) alloy. ST 525°C 1 h, WQ, 250°C 10 h. The electron beam is close to $<110>_{\alpha}$.[24]

2.3 Thermal Stability and Evolution of 2^{nd}Ps

The creep resistance of an age-hardenable alloy is related to the grain structure and the strengthening 2^{nd}Ps. Although fine coherent dispersoid particles, formed during the ingot preheating stage by adding certain trace elements like Zr for the β′ (Al_3Zr) phase,[28] can control recrystallization, the creep resistance primarily depends on the evolution of the strengthening 2^{nd}P(s) at elevated temperatures.

A typical DSC scan of the ternary Al-Cu-Mg sample[15] is given in Fig. 6. The orientation of the scan is such that endothermic is in the positive vertical direction. Clearly the complex nature of phase transformations needs to be considered for this class of alloys since the crystallization and dissolution of several transition phases may mask individual precipitation events. Figure 6 illustrates a subtle endothermic peak at lower temperatures that is followed by a significant exothermic peak at approximately 270°C and a moderate endothermic peak at 350°C. Subsequent DSC scans are carried out to those temperatures so that the microstructural evolution can be characterized as a function of endo- and exothermic transformations.

The microstructure for the first DSC scan from 100-230°C consists of noticeable grain boundary precipitation and a very fine distribution within the matrix. Matrix dislocations and dislocation loops appear to be preferential sites for clusters of these fine precipitates. Electron diffraction illustrates streaking along the $\{100\}_{\alpha}$ and $\{210\}_{\alpha}$ planes, which is indicative of θ''/θ' and S-phase precipitation, respectively. No other prominent streaking is observed for this condition. The DSC scan from 100-270°C, which corresponds with the deep exothermic peak, is associated with a much coarser microstructure than the previous scan.

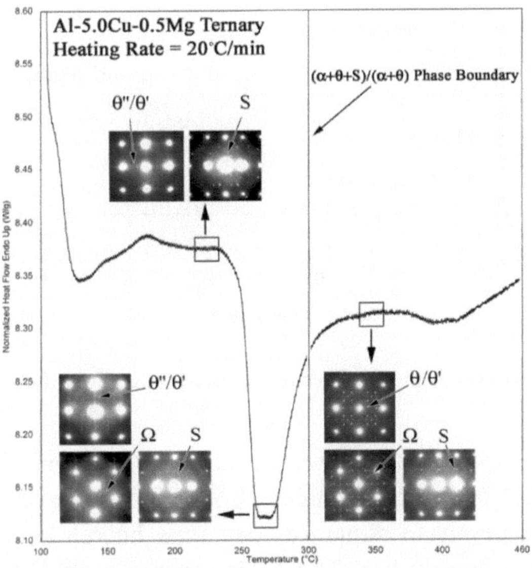

Figure 6. A DSC thermal scan with electron diffraction patterns of Al-5.0Cu-0.5Mg.[8]

Bright field images show similar grain boundary precipitation with more significant matrix precipitation. Electron diffraction again indicates the presence of precipitates with $\{100\}_\alpha$ and $\{210\}_\alpha$ habit planes, as demonstrated on Fig. 6. These patterns exhibit strong secondary diffraction spots, which is indicative of a coarser microstructure. Unlike the scan to 230°C the electron diffraction results illustrate streaking along $\{111\}_\alpha$ planes in the $[110]_\alpha$ diffraction pattern, which correlates with Ω precipitation. This observation confirms the results of Garg, et al.[29] The DSC scan over the temperature range 100-350°C again illustrates precipitation along $\{100\}_\alpha$, $\{210\}_\alpha$ and $\{111\}_\alpha$ habit planes. The streaking along the $\{111\}_\alpha$ is prominent for this thermal condition. The nature of the DSC scan over a range of temperatures may not have allowed for equilibrium conditions, and, indeed the S-phase at 350°C should be in the process of dissolution, which can be confirmed by holding at this temperature for an extended time. There is a lack of reflections or streaking indicative of S-phase precipitation for the material taken from 100-460°C. The microstructure is dominated by $\{100\}_\alpha$ and $\{111\}_\alpha$ precipitation with no sign of S-phase.

Among all those stable and metastable 2^{nd}Ps, the Ω phase is the most interesting for its high thermal stability and potential to generate high strength under certain conditions. It is observed that with small amounts of Ag addition (< 0.1 at%), Al-Cu-Mg (low Mg:Cu ratio) systems can exhibit superior yield strength.[30] As shown in Fig. 7, the Ω plates reach thickness

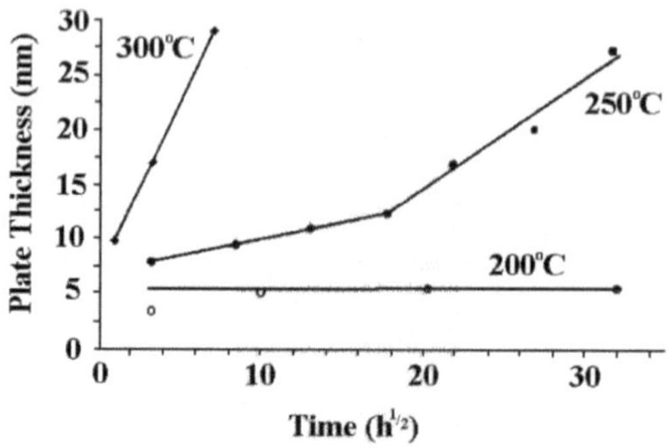

Figure 7. Coarsening of the Ω phase plates of Al-Cu-Mg-Ag alloys during aging treatment.[13]

lower than 6 nm after 100h exposure at 200°C, after which there is no detectable change in average thickness[31,32] for more than 150 days. It indicates that the Ω phase structures can thus be retained at up to 200°C without significant coarsening. The high coarsening resistance of Ω plates can be attributed to a prohibitively high barrier to migrating ledge nucleation in the strong compressive strain field normal to the broad faces of the plates. As a result, the fine and uniform dispersion of the thermally stable Ω plates can ensure the superior yield strength and hence the high creep resistance of this class of alloys.

A high resolution Z-contrast TEM image combining with EDS studies reveals more details for the enhanced coarsening resistance of the Ω plates.[24] Figure 8.a shows the micrograph of a 4 unit cell thick Ω plate in an Al-4Cu-0.3Mg-0.4Ag aged at 200°C for 100 h. Two atomic layers of enhanced intensity at the plate/matrix $(001)_\Omega \| (111)_\alpha$ interface indicates atomic segregation. This is in contrast to the monoatomic layer reported by Reich et al.[33] The layers of enhanced intensity within the plate parallel to the habit plane are separated by 0.424 nm and correspond to layers enriched in Cu. Figure 8.b shows EDS spectra obtained from the matrix, the $(001)_\Omega \| (111)_\alpha$ interphase boundary and from within the plate. The interfacial segregation is found to contain both Ag and Mg.

3. EVALUATION OF STRENGTHENING EFFECTS: DISLOCATION SLIP SIMULATION

The strengthening effects of the shearable 2[nd]Ps can be evaluated according to different models depending on the predominating strengthening mechanism(s) as mentioned previously. For shearable spherical coherent

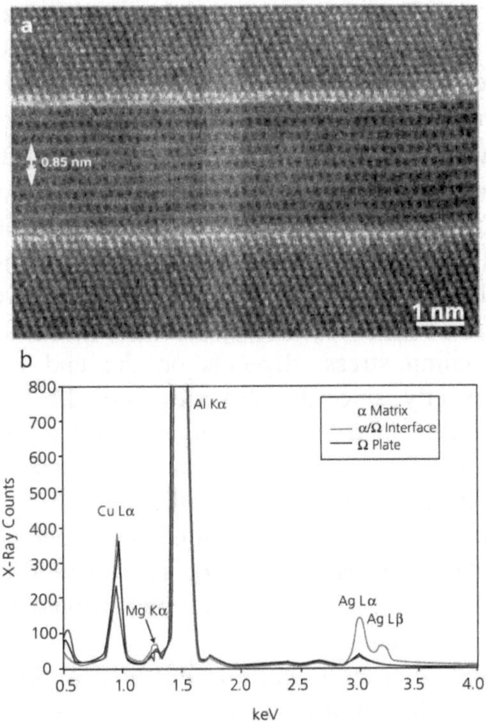

Figure 8. (a) Atomic resolution Z-contrast image of an Ω plate in Al–4Cu–0.3Mg–0.4Ag (wt. %). ST 525°C 1 h, WQ, 200°C 100 h. The electron beam is close to <110>. (b) EDS spectra obtained from the adjacent matrix, the (001)||(111) interphase boundary and within the Ω plate.[24]

2ndPs, the yield strengthening effect can be basically described as $\Delta\tau_p \sim f_v^m r_{av}^n$ with positive exponents m and n indicating higher volume fraction f_v and larger average size or radius r_{av} of the 2ndP particles lead to higher yield strength.[34] The strengthening effect of unshearable 2ndP particles has been evaluated in terms of the original or modified Orowan equations. The original Orowan-mechanism is based on an assumption that the obstacles, formed by non-shearable particles, to dislocation motion are sufficiently localized in the slip plane that they can be represented by a distribution of point forces.[35] A widely accepted equation has been derived for the increment in the critical resolved shear stress associated with the particles, i.e. "strengthening stress τ_p", as [36]

$$\tau_p = \xi \frac{Gb}{2\pi L_{cc}} \ln \frac{\chi}{r_0} \tag{2}$$

where L_{cc} denotes the planar (center-to-center) spacing between obstacles in the slip plane, G the shear modulus, and b the magnitude of the Burgers vector. r_0 and χ are the inner and the outer cut-off radii for the calculation of the dislocation line tension and the parameter ξ is related to the initial dislocation character. $\xi = 1$ or $1/(1-\nu)$ for dislocations of the edge type or of the screw type, respectively, where ν is Poisson's ratio. Practically, the particles, e.g., 2nd precipitates, have a finite size when compared with their planar spacing in the slip plane; and they may have a non-spherical or non-equiaxed shape. The critical stable configuration of a bowed-out dislocation segment that is pinned by two particles at two ends, and consequently the associated strengthening stress, depends on the end-to-end inter-particle spacing L_f. More exactly, since the obstacles formed by the particles on the slip planes are spatially distributed at random (not regularly), the critical stable configuration of a bowed-out dislocation depends on the overall geometry of the obstacles, particular on the obstacles that are relatively sparsely distributed, that the glide dislocation will meet at the instant of the slip process (as detailed below in Fig. 9). Therefore, the more reasonable evaluation is expected to go beyond the conventional treatments using a regular distribution of obstacles.

On the other hand, in order to optimize mechanical properties of practical particle-strengthened alloys, more than one type of co-existing 2ndP particles, that may have different functions, must be treated. Very often, the fine and coherent 2ndPs with uniform distribution mainly contribute to an increment of the yield strength whereas the "hard" ones may offer resistance to strain localization and premature fracture.[37] Various commercial and experimental aluminum alloys are examples.[1] The second-phase particles may differ in chemical composition, in crystallographic structure (coherent or incoherent), in particle-size, in resistance to movement of dislocations ("hard" or "soft"), in spatial distribution (e.g., homogeneous or heterogeneous), in morphology (spheres, plates or rods) and may have different habit planes. Different second-phase particles can be introduced by control of alloy chemistry, and their size, coherency, and distribution can be modulated by trace element additions[37] and by specific thermal-mechanical treatments. For instance, different precipitates such as δ', θ', and T_1 phases present in Al-Cu-Li alloys may be controlled by these methods and, therefore, the mechanical properties of the alloys may accordingly be changed. Particularly, precipitates of a bi-modal size-distribution may be produced in some binary and multi-solute alloys by double-aging treatments. Due to difficulties in quantitative characterization, only very few experiments addressing the problem of a mixture of different particles have been carried out and the findings are somewhat controversial.[38,39] Also theoretical treatments have been somewhat limited[40] and the development of a rigorous one seems to be "a formidable task".[40] There have been efforts to draw some conclusions from very limited data obtained in early computer experiments.[41,42] What seems to be widely accepted is an *ad hoc* superposition law

$$\tau_p^\alpha = \tau_1^\alpha + \tau_2^\alpha \tag{3}$$

where τ_p is the total strengthening shear stress, and τ_1 and τ_2 are the contributions to the strengthening effect due to the 1-particles (indexed as 1) and the other 2-particles (indexed as 2) respectively. The value of the exponent α is actually unknown, although it has been estimated for a few special cases.[43-45]

3.1 Simulation Methods

A dislocation simulation program was developed, based on the circle-rolling method,[41,42] to simulate a dislocation-slip process on one slip plane through obstacles formed by randomly distributed 2^{nd}P particles. Firstly, 3 dimensional structures of 2^{nd}P of one or more than one types can be reconstructed according to the structural parameters $\{\eta, \zeta(D), \psi(\text{i, j, k})\ D{:}t\ f_v$ (or n_v)} that are obtained by the quantitative characterization of each 2^{nd}P. A 2-dimensional structure of the obstacles formed by these 2^{nd}P particles that are intersected with a slip plane $(111)_\alpha$ is obtained. An initial dislocation configuration is established so that it is pinned freely at certain obstacles. For simplicity we assume, similar to the treatment in Ref. 46, that there is no stable position for the dislocation inside the obstacles. Under action of an applied resolved shear stress τ, dislocations bow between the particles, tending to overcome the particles and hence to slip on one slip plane and along one slip direction, e.g. (111) $[\underline{1}10]$ in aluminum. The line tension T_L of a bowing dislocation segment, that is pinned at two particles at the ends, depends on the end-to-end inter-particle planar spacing l_p[47]

$$T_L = \frac{\beta\, Gb^2 \ln(l_p / r_i)}{4\pi} \tag{4}$$

where G denotes the matrix shear modulus, b the magnitude of the Burgers vector, r_i the inner cut-off radius and $\beta = (1+\nu-3\nu\sin^2\theta)\ /\ (1-\nu)$. Using the value of Poisson's ratio ν for aluminum, 0.34, and averaging the dislocation orientation angle θ from 0 to 90°, the mean value of β is 0.87. The mean line tension $\overline{T}_L$, identified by inputting the mean end-to-end spacing $\overline{l}_p$, is employed for simplicity. The curvature radius ρ of the bowing dislocation segment may thus be written as

$$\rho = \frac{\overline{T}_L}{\tau\, b} \tag{5}$$

If only one dislocation slip system is active, the dislocation overcomes those particles in either of two ways. A larger applied stress causes the dislocation segments to be more bowed-out, and therefore spacing η between any two face-to-face bowed-out dislocation-segments becomes smaller, as illustrated in Fig. 9. When the two dislocation-segments are close to each other within a certain critical spacing η_c, i.e. $\eta \le \eta_c$, they annihilate and the obstacle or obstacles between them are looped away. The critical unstable local configuration of the dislocation under the applied stress will therefore depend on the curvature of the bowed-out dislocation segments and the critical spacing η_c that is related to the interaction between different parts (the two face-to-face bowed-out segments) of the dislocation. Secondly, the dislocation overcomes or cuts through an obstacle when the inclining angle ω of two *adjacent* dislocation segments becomes smaller than a critical value $\omega_c = \cos^{-1}(2\ F\ /T_L)$ where F denotes the strength of the particle when the dislocation cut through it. When hard and/or plate-like particles are predominate in the particle mixture, dislocation annihilation and looping dominates the slip process. If the self-stress effect of bowing dislocations is neglected, a simple criterion for the dislocation to overcome the particle(s) may be written,

$$\begin{cases} \eta \le \eta_c = 0 & \textit{for any two face-to-face segments} \\ \omega \le \omega_c = \cos^{-1} 2F / T_L & \textit{for two adjacent segments} \end{cases} \tag{6}$$

However since the obstacles are of finite sizes, the attractive interaction of any two face-to-face parts of the bowed-out dislocation segments will make it easier for dislocations to break way from the obstacle or the obstacles between the two segments. Therefore, the critical spacing η_c has to be greater than zero. A strict treatment of the self-interaction effect under general condition seems formidable. For an approximation, resorting to an analogous treatment as in Ref. 35, it may be assumed that the effect is accounted for by introduction of an *apparent* applied stress τ^* while the simple instability criterion is retained

$$\eta = \eta_c = 0. \tag{6a}$$

In order that Eq. 2 with $\chi = Q'$ can simply be retrieved in our simulation when point obstacles are considered, the apparent applied stress may be written as

$$\tau^* = \tau \frac{\ln(L_f / r_0)}{\ln(Q' / r_0)} \tag{7}$$

and the corresponding apparent curvature radius of the dislocation segments therefore becomes

$$\rho^* = \frac{2T_L}{\tau^* b} = \frac{Gb}{4\pi\tau} \ln \frac{Q'}{r_0} \tag{8}$$

where we have assumed for simplicity that the parameters ξ and ζ are equal. The critical spacing Q' is considered to be larger than the dimension of the obstacles,[48] and can be *4D* for circular ones.[35]

Wherever the local configuration of the dislocation becomes unstable, the dislocation moves forward, by-passing those obstacle(s) or cutting through the one obstacle until it is impeded by the next obstacle(s) resulting in a new stable configuration. A macro-slip process of the dislocation is achieved through the successive "annihilation" of the local dislocation parts; pairs of face-to-face parts of the bowed-out dislocation segments. A dislocation-slip process through the obstacles can numerically be simulated based on this analysis. The yield stress (or the particle strengthening stress τ_p) is defined as the stress at which the dislocation finds a way to slip out of a large array of obstacles. Figure 9 illustrates a dislocation slip process through a random distribution of rectangular obstacles oriented at -30, 30 and 90° with regard to the slip direction on a slip plane under an increasing applied stress.

3.2 Comparison with Experiments

3.2.1 θ′ $\{100\}_\alpha$ in Al-Cu Alloys[49]

One test of the dislocation slip simulation is to examine the strengthening effects of θ′ $\{100\}_\alpha$-phase plates in Al-0.025, 0.04 and 0.05 wt% Cu alloys. The θ′ $\{100\}_\alpha$-phase particles are assumed to be unshearable. According to the simulation, the dependence of the strengthening stress on the obstacle size is shown to be positive but less than those accounted for in the currently used "intuitive" Orowan equations. The modified Orowan equations is hence suggested for {100}-plates;

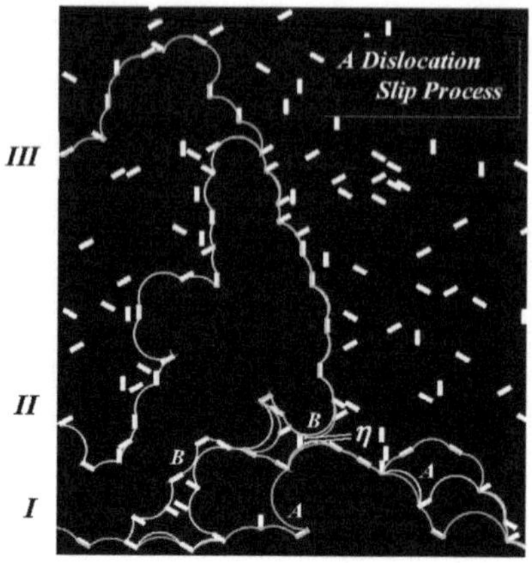

Figure 9. Equilibrium configurations of a dislocation interacting with randomly distributed rectangular obstacles under increased applied stress (*I ⇒II⇒III*) — a part of computer simulation area. "η" denotes spacing between any two face-to-face bowed-out dislocation segments.

$$\tau_p = 0.13G \frac{b}{(D_p t_p)^{1/2}} [f_v^{1/2} + 0.75(D_p / t_p)^{1/2} f_v + 0.14(D_p / t_p) f_v^{3/2}] \cdot \ln \frac{0.87(D_p t_p)^{1/2}}{r_0} \quad (9)$$

where G is the shear modulus, b the magnitude of the Burgers-vector, r_0 the interaction inner cut-off radius of the dislocation, and f_v is the volume fraction, D_p the mean diameter and t_p the mean thickness of the plates. The strengthening stress τ_p of the θ'-plates for the various specimens can be calculated from the measured yield-stress, σ, simply using the equation

$$\tau_p = (\sigma - \sigma_0)/M. \quad (10)$$

where $1/M$ is the Schmid factor for the single or macro-crystal specimens and is reciprocal to M (Taylor factor) for polycrystalline ones. The contribution σ_0 to the yield strength is partly due to Cu solid solution hardening. The value of σ_0 is set at 50 MPa for the aged Al-2.5Cu alloys, at 55 MPa for Al-4Cu and at 58 MPa for Al-5Cu[50] for the present analysis.

Using the data of the θ'-plates obtained from the quantitative characterization, the strengthening stress τ_p associated with the θ'-plates can be calculated according to Eq. 9. The constants used for the calculations are

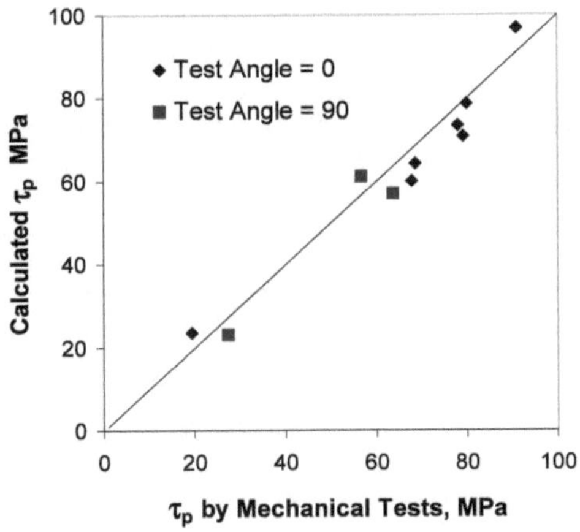

Figure 10. Comparison between the strengthening stress τ_p obtained from the compression tests of the stress-free aged AlCu single crystal specimens and those from the dislocation slip computer simulations.[49]

the shear modulus of Al, G = 28 GPa, the magnitude of the Burgers vector b = 0.286 nm and the inner cut-off radius $r_0 = b$ = 0.286 nm. As shown in Fig. 10, the evaluation of the non-preferentially oriented plate structures yields consistent results with those measured by the compression tests of the stress-free-aged specimens.

3.2.2 {δ' + T_1} phases in Al-Li-Cu Alloys

Two experimental AlLiCu alloys have been investigated in detail. One is Al-2.7Cu-2.1Li-0.6Zn-0.3Mn-0.05Zr (denoted as AF/C-489) and the other Al-2.7Cu-1.8Li-0.6Zn-0.3Mn-0.3Mg-0.08Zr (AF/C-458). Both were received in the cold-deformed and unrecrystallized state (T36 temper). They were then subjected to an aging at 150°C for 24 hours. Mechanical (tensile) tests were performed and the yield stresses σ_{YS} (0.2% offset) were measured. The contribution of the matrix to the critical resolved shear stress was estimated to be τ_{matrix} = 50 MPa. The values of the strengthening stresses due to the precipitates were determined from these mechanical tensile test data using the relationship $\tau_p = \sigma_{YS}/M - \tau_{matrix}$. Transmission electron microscopy (TEM) observations indicated that both samples contained spherical δ' (Al_3Li)- and plate-like T_1 (Al_2CuLi)-phases. The content of θ' (or θ")-phase was very minute (the volume fraction less than 0.27% in AF/C-489 and 0.023 in AF/C-458). For the δ'-particles, the order-strengthening mechanism was considered with the antiphase boundary energy γ_{apb} = 0.07 Jm^{-2},[22] and

Table 4. Precipitate structure {δ' +T_1}, obtained by the quantitative microstructure analysis and the strengthening stress τ_p measured by tensile test, by calculation via the *ad hoc* equations[37] and by computer simulation. σ_{YS} denotes the yield stresses, τ_{matrix} the contribution of the matrix to the CRSS and M the Taylor factor calculated from texture data.

Sample	Precipitate: TEM	Taylor Factor *M*	τ_p (MPa) Tensile Test	τ_p MPa by *ad hoc* Eq	τ_p (MPa) DS_S
AF/C489	δ':r=5.2nm,f_v=0.091 T_1:D=57nm,t=1.6nm f_v = 0.029	3.20	473 / 3.2 -50 = 98	126	120
AF/C458	δ':r=5.1nm,f_v=0.068 T_1:D=58nm,t=1.6nm f_v = 0.032	3.23	484 / 3.23 -50 =100	130	114

thus the particle strength $F = \pi/2\ \gamma_{apb}\ r_\delta$ where r_δ is the radius of the δ'-particles. The T_1-plates were assumed to be unshearable for individual dislocations. Inputting the data of the δ'- and T_1-phases in the two alloys for the computer simulation, the strengthening stresses τ_p can also be obtained. The results are summarized in Table 4.

The values of the strengthening stresses τ_p obtained by the computer simulation (as well as by the previous models[37]) are, in most cases, greater than those obtained from the tensile tests. This may be due to several sources. One is the possible over-evaluation of the precipitate volume fractions measured by the TEM study. Also the matrix strength (~50 MPa), consisting of contributions from remaining solute atoms, dislocations and grain-boundaries, was based on earlier estimations (c.f. citations in Ref. 37) but may vary depending on the primary processing of the alloy. The synergetic effect[51] between τ_{matrix} and τ_p was also neglected. On the other hand, errors may be brought about from some simplifications made in the simulation, e.g., only one dislocation-slip on only one active slip system. Most probably, in addition to the unpinning and the self-annihilation, dislocations may also overcome the particles in a third way, particularly by double cross slip.[51] Also, the interactions between dislocations, as well as the self-stress effects of dislocations that may ease the dislocation slip process have not been incorporated in the simulations. Nevertheless, the simulation does yield values closer to those obtained from the tests than the calculation via the previous models.[37]

3.3 Predictions of Optimum Precipitate Structures – Superposition of Strengthening Effects

3.3.1 Spherical Precipitates of Bi-modal Size Distribution

Double aging, i.e., using a low-high or high-low temperature aging sequence, can be used to produce precipitates of a bi-modal size distribution. In most cases, the resistance strength of a shearable precipitate is proportional to its radius. Suppose A-precipitates of radius r_A are so large as to be just unshearable, i.e. their resistance strength $F_A = 2.0\ \overline{T_L}$ while B-precipitates are shearable and their mean radius r_B varies from 0.1 to 1.0 r_A. Given a total precipitate volume fraction $f_v = f_{v,A} + f_{v,B,}$ where $f_{v,A}$ and $f_{v,B}$ are volume fractions of the A-precipitates and the B-precipitates, respectively, the strengthening stress τ_p due to the precipitated microstructure can be calculated according to the results above, for various specific volume fractions ($f_{v,B}\ /\ f_v$) of the B-precipitates. As depicted in Fig. 11, co-existence of the larger and unshearable A-precipitates with the smaller and shearable B-precipitates may give rise to a maximum strengthening effect in each situation. The pure "mono-dispersed" unshearable A-precipitates, produced through a conventional, single temperature "peak aging" treatment, results in a strengthening stress

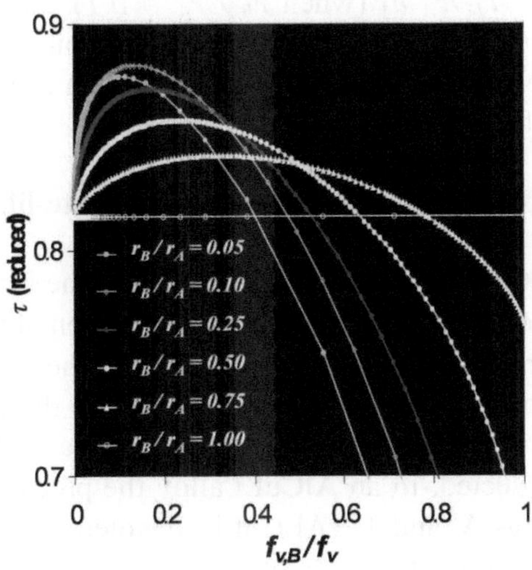

Figure 11. Dependence of the strengthening stress τ (in units of $\sqrt{3f_v/2\pi}\cdot 2\overline{T}_L/br_A$) on the specific volume fraction ($f_{v,B}\ /\ f_v$) of the B-particles.

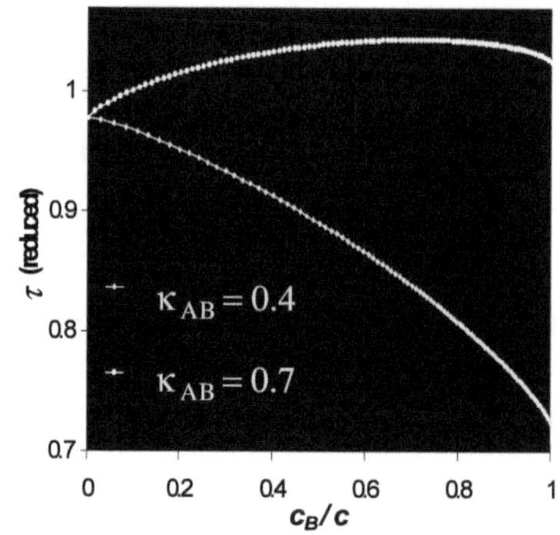

Figure 12. Dependence of the strengthening stress (in units of $\sqrt{4cV_{m,A}\gamma_A / \pi V_{m,Al}} \cdot 2T_L / bD_A$) on the Cu atomic fraction (c_B / c) which is transformed in to {111}- plates (T_1-phase).

$\tau_A = 0.820\ [(3f_v / 2\pi)^{1/2}\ T_L / r_A\ b]$. However, when a certain volume fraction of the smaller B-precipitates are mixed with the larger A-precipitates corresponding to the "peak aged" temper, a higher strengthening stress τ_p up to $0.886\ [(3f_v / 2\pi)^{1/2}\ T_L / r_A\ b]$ (when $r_B / r_A = 0.1$) can be obtained. This enhanced strengthening due to the smaller precipitate addition can be ascribed to the exponent α having a value lower than 2 in the superposition law for a "hard + soft" particle-mixture.

3.3.2 Mixture of Two types of Unshearable Plate-like Particles

An array of 10,000 plates intersecting the slip plane is used. The average lengths of the line-segments formed by the intersection are $l_A = \pi D_A / 4$ and $l_B = \pi D_B / 4$. For the {100} and {111}-plates, these line-segments are oriented at -30, 30 and 90° with reference to the slip direction [$\underline{1}$10] except one of the four {111}-variants is parallel to the slip plane. The effect of these parallel plates is neglected. In an AlCuLi alloy the precipitation of the θ' or θ'' (Al_2Cu), denoted as A, and T_1 (Al_2CuLi), denoted as B, are controlled by the total copper solute content c (in atomic fraction). The copper content c is shared by θ'- (or θ''-) and T_1-phases such that $c = c_A + c_B$ where c_A and c_B are the Cu atomic fractions transformed into A- and B-particles, respectively. The maximum c will be $c = c_0 - c_e$ where c_0 and c_e are the initial and the equilibrium Cu atomic fraction in the matrix. The volume fraction of the B-plates is $f_{v,B} = c_B\ V_{m,B} / V_{m,Al}$ where $V_{m,B}$ and $V_{m,Al}$ are the molar volume of the B-plates and of the Al matrix. The mean planar center-to-center spacing l_c, considering both A- and B-plates, may be expressed as

$$l_c^{-2} = N_{p,A} + N_{p,B} = \frac{4}{\pi} \cdot \frac{cV_{m,A}\gamma_A}{V_{m,Al}D_A^2} \cdot (1 - c_B / c + \kappa_{AB} c_B / c) \tag{11}$$

where γ_B denotes the aspect ratio ($=D_B$ / t_B), and a parameter κ_{AB} is introduced for comparing the geometrical and crystallographic characteristics of the A- and B- plates and is defined as

$$\kappa_{AB} = \zeta(\frac{D_A}{D_B})^2 / (\frac{\gamma_A}{\gamma_B} \cdot \frac{V_{m,A}}{V_{m,B}} \cdot \frac{\sin\theta_A}{\sin\theta_B}) \tag{12}$$

where θ_A and θ_B are its dihedral angles with reference to the slip plane and the coefficient ζ accounts for the effect of the orientation of the plates ($\zeta = 3$ / 4 for θ'- and T_1-precipitates).

Assuming $V_{m,A}$ / $V_{m,B}$ = 0.5, $\sin\theta_A$ / $\sin\theta_B$ = *sin 54.75°* / *sin 70.53°* = *0.866* for two situations: 1). γ_A / κ_B = *0.4* and 2). κ_A / κ_B = *0.7*. For a given total transformed copper-content c according to Eqs. 6-9, the dependence of the strengthening stress (in unit of $(4cV_{m,A}\,\gamma_A / \pi V_{m,Al})^{1/2}\, 2T_L/l_c$) on c_B / c can be obtained, as depicted in Figure 12. For condition 1) participation of T_1 of relatively large D and low D/t reduces the strengthening stress while for condition 2) it raises the strengthening stress and gives a maximum strengthening effect with a certain fraction of Cu allotment into the T_1-phase of fine dispersion. This suggests that the parameter κ_{AB} is a comprehensive index registering the superposed strengthening effects — an addition of secondary plates with higher κ_{AB} increases the probability of generating an optimum precipitate structure. If both 2[nd]Ps have relatively comparative sizes and aspect-ratios, a fine dispersion of the two co-existing 2[nd]Ps on the different habit planes is desired for the optimum strengthening effect.

Generally, the computer simulation of a dislocation passing through a random-distribution of strengthening particles of two types shows that if only one active slip-system is taken into account, the superposition law of the strengthening effects can be well described by

$$\tau_p^\alpha = n_A^{\alpha/2}\tau_A^\alpha + n_B^{\alpha/2}\tau_B^\alpha \tag{13}$$

where the exponent α varies between 1.0 (linear addition law) and 2.0 (Pythagorean addition law[52]). The simulation method proves to be an effective way to determine the exponent α and to evaluate the superposed strengthening effects of different precipitates in alloys.

4. STRESS-AGING

4.1 Background

As shown above, plate-like and rod-like coherent 2^{nd}P precipitates are formed with specific crystalline orientations, i.e., on specific habit planes, in the matrix during the aging treatment. Trace element additions can be used to enhance the production of the desired 2^{nd}P with the specific habit planes through the initial solute clustering.[4] Usually the plates are oriented homogeneously on all variants of their habit planes. For instance, the θ''/θ' phase are distributed over all three {100}-plane variants in aged Al-Cu based alloys and the Ω phase plates over all four {111}-variants in some Al-Cu-Mg based alloys. Application of a directional stress during aging may have a great impact on the orientation distribution so that the formation of the precipitates on one or two variant(s) of their habit planes is favored over others. Such a stress orienting effect of the coherent 2^{nd}P precipitation may, on the other hand, influence the yield anisotropy of the alloys. This issue is practically related to the "age forming processing" – a manufacturing method that offers a solution to the problems encountered when conventional cold forming processes are applied to integrally stiffened lightweight structures[53-55] for aerospace applications. It allows fewer parts, less assembly, lower overall weight, and lower fabrication costs, as compared with built-up skin and stringer structures that require fasteners and assembly. In principle, age forming is a stress-age hardening process. Since precipitation of hardening particles and the deformation due to stress-relaxation occur simultaneously when elastic stress is applied during an artificial aging treatment, the forming and the hardening of the structure or the components are accomplished at the same time. The final aged structure also has lower residual stresses as compared to structures formed by conventional means such as roll forming, brake forming, shot peen forming, or stretch forming. This improves long term performance of the parts since it raises the resistance to both fatigue and stress corrosion cracking.

4.2 Stress Oriented Effect on Plate Precipitates[56,57]

The effect of directional stress on the orientation distribution of precipitate structures has extensively been investigated in many age-hardenable alloy systems. Particularly for the Al-Cu system, Hosford et al.[58] found that aligned θ' plates — one or two of three (100)-variants defaulted — were generated during aging of an Al-3.98Cu alloy (all compositions in this text are in weight percent) at 210°C under either compressive or tensile stress of 48 MPa, whereas there was no noticeable orienting effect under a stress of 34.5 MPa. Eto et al.[59] could not reproduce these results under the same aging conditions using a lower copper-content—Al-3.71%Cu — alloy. Using a two-step (low-temperature + high-temperature) aging procedure they showed that the stress-orienting effect was initiated and determined during the nucleation stage and the particular variant(s) of GP zones was/were preferentially formed under the directional stress and the aligned GP zones grew to θ′-plates through GP[II] or θ″-phase that was formed below the critical temperature (453-463 K). The result, indicating that there would be no stress-orienting effect if the precipitation began directly with the θ′-phase, is contradictory to the earlier Hosford et al.'s finding.[58] Later, Skrotzki et al.[60] found that there were threshold values of the stress that must be exceeded for the formation of preferentially oriented plate-shaped precipitates in the tensile-stress aged samples of Al-5Cu and of Al-Cu-Mg-Ag alloys. Using Al-xCu alloys as model systems, a systematic experiment reveals how composition (Cu-content) and thermal mechanical conditions, i.e. aging temperature and magnitude of applied stress, affect precipitate structures in peak-aged alloys. A series of oriented single crystal or macro-crystal samples of Al-xCu alloys were prepared for use in compressive stress-aging treatments. Figure 13 depicts the TEM bright-field images and diffraction patterns of (a) conventionally-aged and (b) stress-(40 MPa) aged Al-4Cu single crystals at 201°C for 11 h. The structural parameters (diameter distribution on each of the habit plane variants {100}) of the θ″/θ′-plates produced under different conditions were determined by the above-mentioned quantitative characterization techniques. Using a <100> zone axis in the TEM, randomly oriented θ′-plates were observed with two of their three {100}-variants "edge on" in the stress-free aged specimen while only one variant of aligned θ′-plates was observed "edge on", i.e. perpendicular to the compressive load direction, in the stress-aged specimen. Quantitative examination of more than 1000 plates in 4-6 TEMs (BF or CDF images) indicated that the ratio of the number density of perpendicular plates to that of parallel ones was 50:50 for the stress-free aged sample and 99:1 for the stress-aged sample. When considering the non-edged-on {100} variant in the micrographs, the ratio of the number density of aligned plates to that of non-aligned ones was 99:2 for the stress-aged sample. This stress oriented effect was also reflected in the diffraction patterns (inserted in Figure 13) where the streaks associated with θ″ are along two orientations and are of equal intensity for the stress-free-aged sample (Figure 13.a) but only along one direction for the stress-aged sample (Figure 13.b). A parameter Γ describing the degree of alignment of the θ″/θ′-plates was introduced as

$$\Gamma = \frac{f_v^{\perp} - f_v^{//}}{f_v^{\perp} + 2f_v^{//}} \tag{14}$$

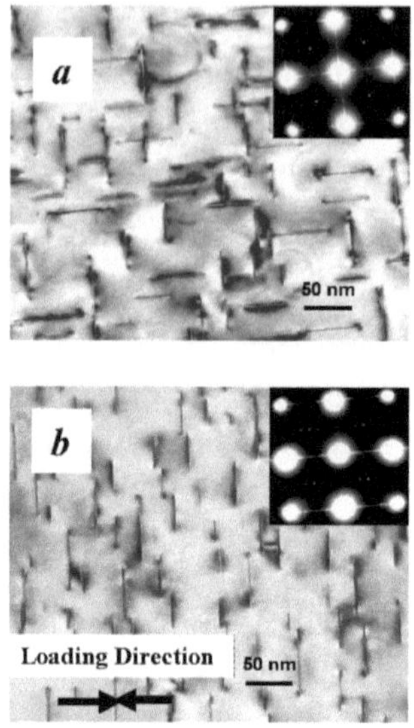

Figure 13. TEM BF images and diffraction patterns of (a) conventionally-aged and (b) stress-(40 MPa) aged Al-4Cu single crystals at 201°C for 11 h.

where $f_v^{//}$ and $f_v^{\perp}$ denote the volume fraction of the parallel plates (that were visible in TEM) and that of the perpendicular ones, and the total volume fraction f_v^T is then

$$f_v^T = 2f_v^{//} + f_v^{\perp} . \tag{15}$$

Using a difference interpolation method, at different aging temperatures, the applied stress σ_a (Γ= 0.75) required to reach a θ″/θ′-plate alignment degree Γ = 0.75 can be obtained. Figure 14 illustrates the relationship between the applied stress σ_a required for an alignment degree of Γ = 0.75 and stress aging temperature for the three alloys. From the experiments described in Ref. 56, conclusions were drawn as follows:

1. Application of an elastic stress (<< yield stress) has little or negligible impact on the overall kinetics of the aging processes;
2. Aging under an applied stress generates an aligned θ″ / θ′-precipitate structure with a higher number density and larger diameter of favored plates than those of unfavorable ones;

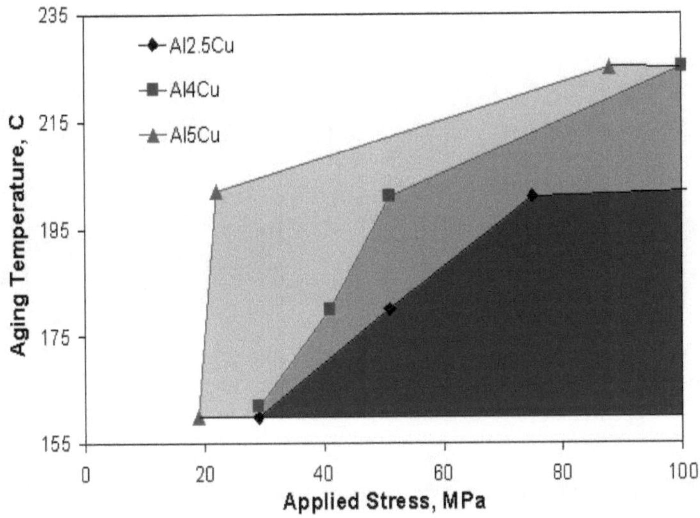

Figure 14. Stress-Temperature conditions for an alignment degree of $\Gamma = 0.75$ in compressive stress-aged Al-xCu Alloys.

3. A compressive stress leads to a higher number density of θ″/θ′-plates perpendicular to the load direction whereas a tensile stress favors the formation of θ″/θ′-plates parallel to the load direction;
4. The stress orienting effect depends on the applied stress, temperature, alloy composition and aging time.

Stress orienting phenomena were also observed for the Ω $\{111\}_\alpha$-phase in brass-textured polycrystalline Al-5Cu-0.8Mg-0.5Ag alloys under a tensile stress.[61] As shown in Figure 15, under edge-on condition for Ω phase, *i.e.* with a <1$\bar{1}$0> zone axis in the TEM, Ω plates that are closely parallel with the tensile loading direction are favored over those that are almost perpendicular.

A modified Langer-Schwartz model[62] that treats nucleation, growth and coarsening as concomitant processes was employed to simulate the evolution of the θ″/θ′-plates in Al-Cu alloys under elastic stress.[57] For the sake of simplicity, the θ″/θ′-plates are assumed as being co-existing and the transition from θ″ phase to θ′-phase to be continuous depending on their diameters. The properties such as the interfacial energy γ_{If}, lattice misfit δ_L, molar volume V_m of a second-phase plate and the equilibrium concentration c_e outside and c_p inside it, changes continuously with its size or radius R, namely,

$$\delta_L = \delta_L^{(\theta'')} + (\delta_L^{(\theta')} - \delta_L^{(\theta'')}) \cdot \frac{R}{\zeta} \ etc. \tag{16}$$

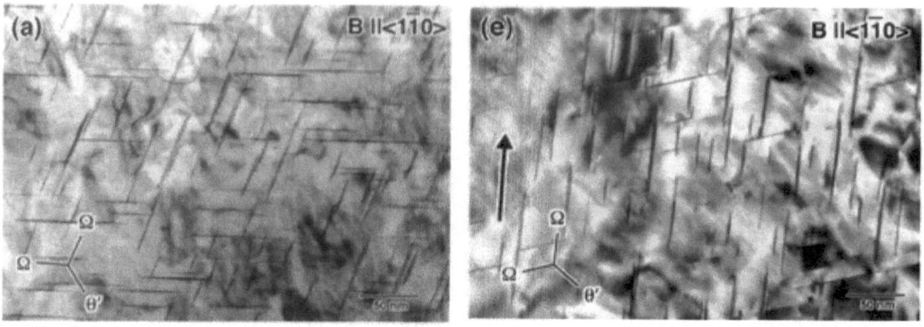

Figure 15. TEM BF images of (a) conventionally-aged and (e) tensile stress-(120 MPa) aged Al-5Cu-0.8Mg-0.5Ag polycrystalline at 200°C for 11 h. The arrow indicates the loading direction.[61]

where an adjustable parameter ς of length, may be in the order of 75 nm — the maximum radius of θ″ as indicated by X-ray analysis.[63] A similar dependence may also be applied to γ_{if}, V_m, c_e and c_p. The strain energy contribution associated with the interaction between a directional stress and the precipitate strain field due to the lattice misfit (mainly at the peripheral interface) between the θ″/θ′-plates and the Al-matrix is considered to cause differences in the nucleation energy as well as in solubility (and hence the growth rate) of perpendicular θ″/θ′-plates from those of the parallel ones and therefore to affect their precipitation processes. Even if the induced differences are small[64] for each stage, since precipitation is an accumulation process leading to coarsening, such small differences may be amplified resulting in aligned θ″/θ′-plates in the final product. Figure 16 illustrates stress-free aging and the stress-aging processes of an Al-4 wt% Cu alloy at 180°C under a compressive stress of 64 MPa. Particularly at the peak-aging stage (t_a ~ 30 h), the number density $n(^{//})$ and the volume fraction $fv(^{//})$ of the parallel plates are about 2-3 orders of magnitude less than $n(^{\perp})$ and $fv(^{\perp})$ for the perpendicular ones. Also, the average size $Da(^{//})$ of the parallel plates is remarkably lower than $Da(^{\perp})$ of the perpendicular ones. The main features of the stress orienting effects, including the effects of the direction and magnitude of the applied stress, the temperature and the composition of the alloys, observed in the experimental measurements are semi-quantitatively reproduced in the numerical simulations (c.f. details in Ref. 7).

For closer comparison with experiments, the influence of the precipitation sequence from GP zones to θ″/θ′-phases needs to be further considered in a more physically sound way. For instance, the nucleation of the θ″/θ′-phases had better to be treated as, instead of the homogenous, heterogeneous at the precursor clusters, like dislocations and solute clusters initially formed in the solid solutions.

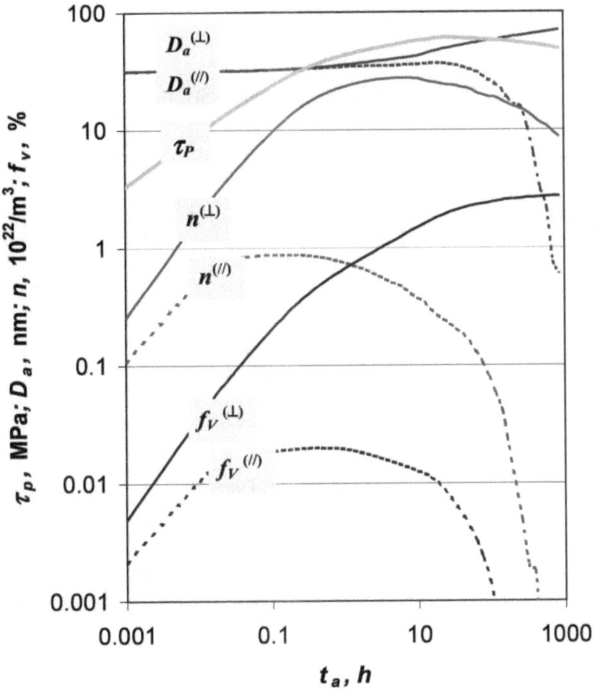

Figure 16. Calculated evolution of the θ"/θ'-plates formed during aging under a compressive stress of 64 MPa at 180°C from a supersaturated solid solution of Al-4Cu alloy. The average diameter $D_a^{(//)}$ and $D_a^{(\perp)}$, the number density $n^{(//)}$ and $n^{(\perp)}$, the volume fraction $f_V^{(//)}$ and $f_V^{(\perp)}$ of the plates on habit planes parallel and perpendicular, respectively, to the stress direction and the associated total hardening stress τ_P are depicted.

4.3 Aligned Precipitates Effects on Anisotropy[49]

The preferentially-oriented precipitate structures produced during a stress aging treatment may have a remarkable impact on the yield anisotropy of the alloys. As shown in Fig. 17, under slightly overage conditions Al-4Cu alloy samples subject to tensile stress-aging at 50 MPa have detectably lower yield stresses particularly at certain orientations than the conventionally aged ones. The tensile tests use the samples oriented with different angles between the rolling direction and the loading direction. According to the dislocation slip simulations (DSS), the equation describing the strengthening shear stress τ_p associated with aligned {100}-oriented, unshearable plates (present on only one variant of the three habit planes) was determined to be

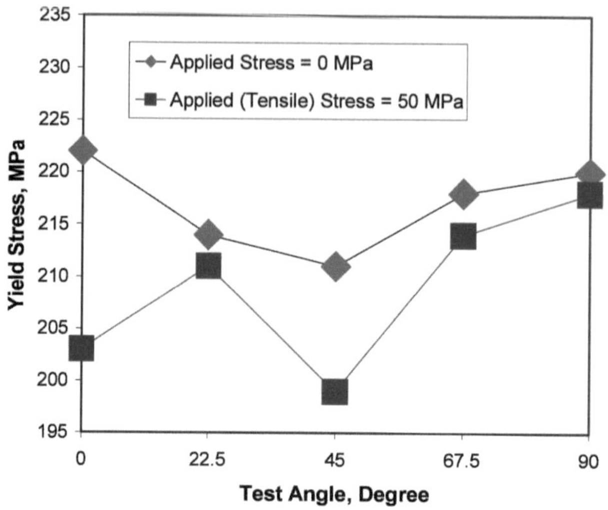

Figure 17. Tensile data of Al–4Cu in uniaxial tension as function of test direction: yield strength in slightly overaged conditions without (conventionally aged) and with an applied tensile stress of 50 MPa (stress-aged).[61]

$$\tau_p = 0.13G\frac{b}{(D_p t_p)^{1/2}}[f_v^{1/2} + 0.74(D_p/t_p)^{1/2} f_v + 0.047(D_p/t_p) f_v^{3/2}] \cdot \ln\frac{0.87(D_p t_p)^{1/2}}{r_0} \tag{17}$$

Regarding the aligned precipitate-induced yield-anisotropy, explicit solutions[65] for axis-symmetric deformation tests have been derived. For the plastic inclusion model,[66] the strengthening factor N of a plate-like precipitate, which is defined as the ratio of the effective shear strain of the precipitate to the applied strain of matrix, depends only on the angle θ between the test stress axis and the habit plane normal of the plate,

$$N = \frac{1}{2}\left(\frac{5+3\cos 4\theta}{2}\right)^{1/2} \tag{18}$$

For the elastic inclusion model,[67] the accommodation-tensor of magnitude $\|\gamma\|$, which reflects the anisotropic contribution of the plate to the critical resolved shear stress (CRSS) or the yield strength, also depends only on the angle θ,

$$\|\gamma\| = \frac{15 - 8\cos 2\theta + 9\cos 4\theta}{32} \tag{19}$$

An experiment[49] using mainly single crystal Al-xCu alloys provides a more precise analysis of the associated strengthening than possible under the conditions used previously. The specimens with clear-cut stress-oriented θ″/θ′ precipitate structures, were carefully prepared and compressive loads (rather than tensile loads) were utilized for the stress aging. Compressive tests of each pair of samples (stress-aged versus stress-free aged) indicated that the yield stresses of the stress-aged ones were always apparently lower than those of the stress-free-aged ones whether or not the test direction was along, or perpendicular to, the aging-stress direction. The differences were more pronounced than that observed in the study where a tensile stress during aging of polycrystalline specimens was used. Hardness tests indicated there was no detectable difference in precipitation kinetics between the stress-aging and conventional aging. Quantitative TEM characterization showed that after the same thermal treatments, the total volume fractions of θ″/θ′-plates produced in stress-aged samples were lower than those in conventionally aged samples. The DSSs have demonstrated that a preferentially oriented plate-structure (some variants missing) provides less strengthening than a non-preferentially plate-structure (all variants present) even for the same volume fraction, diameters and thickness. Therefore, the decrement in the yield stress of a stress-aged specimen may originate from the preferential plate-orientation effect as well as from their "detrimentally" different plate volume fraction and/or morphology when compared with those of a stress-free- aged specimen subjected to the same thermal treatment.

The quantitative analysis is summarized in Table 5 where the strengthening stress τ_p of the θ′-plates for the various specimens was calculated from the measured yield-stress, σ, simply using the equation $\tau_p = \eta \times (\sigma - \sigma_0)$ with the Schmid factor η_{schm} for the single or macro-crystal specimens. The contribution σ_0 to the yield strength is partly due to Cu solid solution hardening. The value of σ_0 was set at 50 MPa for the aged Al-2.5Cu alloys, at 55 MPa for Al-4Cu and at 58 MPa for Al-5Cu[50] for the present analysis. Using the quantitative data of the θ′-plates obtained as mentioned above, the strengthening stress τ_p associated with the θ′-plates can be calculated according to DSSs depending on whether the θ′-plates are non-preferentially oriented or aligned. The constants used for the calculations are the shear modulus of Al, G = 28 GMPa, the magnitude of the Burgers vector b = 0.286 nm and the inner cut-off radius $r_0 = b$ = 0.286 nm.

For the plastic inclusion model, according to Eq. 17, the strengthening factor N_f of every aligned {100}-plate and the averaged strengthening factor N_s of the non-preferentially oriented ones are equal at $\theta = 0$ or 90°, i.e., $N_f(\theta = 0) = N_f(\theta = 90°) = N_s(\theta = 0) = N_s(\theta = 90°) = 1$. For the elastic inclusion model, according to Eq. 18, the magnitude of the accommodation-tensor, $||\gamma||_f$, of every preferentially oriented {100}-plate in a stress-aged specimen: $||\gamma||_f(\theta = 0°) = 0.5$ and $||\gamma||_f(\theta = 90°) = 1.0$. The average magnitude of the accommodation-tensor, $||\gamma||_s$, of the non-preferentially oriented {100}-plates in a stress-free-aged specimen: $||\gamma||_f(\theta = 0°) = ||\gamma||_f(\theta = 90°) = 0.833$. In order to avoid introduction of the unknown parameters in the inclusion models, we utilized the relative difference $\delta = (\tau_{p,f} - \tau_{p,s}) / \tau_{p,f}$ between the strengthening stress $\tau_{p,f}$ of the stress-free-aged specimen and $\tau_{p,s}$ of the stress-aged specimen in each pair for the analysis. Assuming that the precipitates provide the majority of the strengthening component of the CRSS or yield stress, the relative difference δ can be evaluated according to the plastic inclusion model,

$$\delta_p = (N_f f_{v,f} - N_s f_{v,s}) / N_f f_{v,f} \tag{20}$$

or according to the elastic inclusion model,

$$\delta_e = (||\gamma||_f f_{v,f} - ||\gamma||_s f_{v,s}) / ||\gamma||_f f_{v,f} \tag{21}$$

where $f_{v,f}$ and $f_{v,s}$ are the volume fractions of the plates produced in the stress-free-aged specimen and in the stress-aged specimen, respectively. The relative difference $\delta = (\tau_{p,f} - \tau_{p,s}) / \tau_{p,f}$ can also be calculated according to the DSS. The results using these four methods are summarized in Table 5. The evaluation by DSSs yielded relative differences of δ_c very close to those of the compression tests, with a mean divergence of 7%. The calculations using the plastic inclusion model and the elastic inclusion model yielded some negative results especially when the test angle θ = 90°, with the mean divergence of 11% and 23% from those of the compression tests.

Table 5. Comparisons between Models, Simulation Equations and Experimental Results — Evaluation of the Strengthening Stress τ_p of θ′-plates produced in the stress-free aged and stress aged AlCu specimens.
Column 1: The specimen number, cf. Table 2 for the details;
Column 2: a. Strengthening stress calculated from the compression-test-measured yield-stress using $\tau_p = \sigma_p \times \eta$ where η is the Schmid factor for the single or macro- crystal specimens or 1 / Taylor factor for polycrystalline specimens; b. The relative difference δ_M between the stresses τ_p of free aged and stress aged specimens, i. e., $100\times(\tau_{p,f}-\tau_{p,s})/\tau_{p,f}$;
Column 3: The same as in Column 2 but calculated according to the computer-simulation respectively by inputting the data of the quantitative characterization of θ′-plates in the stress-free aged and stress aged specimens;
Column 4: Relative difference δ_p of the strengthening stress by Horsford's model;
Column 5: Relative difference δ_e of the strengthening stress by Bate's inclusion model.

Materials	Specimen No.	Mechanical Tests		Dislocation-Slip Sml.		Horsford's δ_p, %	Bate's δ_e, %
		τ_p, MPa	δ_m, %	τ_p, MPa	δ_c, %		
Al4Cu, Single Crystal	1f	80	-15.3	78.7	-6.9	-6.1	-33.6
	1s	67.8		73.3			
	2f	78	-13.1	73.5	-6.1	-17.1	-26.5
	2s	67.8		69			
	3f	79.2	-12.4	70.9	-13.0	-7.9	-31.9
	3s	69.4		61.7			
	4f	68.6	-12.0	64.3	-4.5	2.1	-41.8
	4s	60.4		62.4			
	5f	67.8	-22.3	60.1	-9.7	-3.4	-36.4
	5s	52.7		54.3			
Al2.5Cu SC	6f	19.4	-23.2	23.5	-10.1	7.5	-47.3
	6s	14.9		20.6			
A5Cu PC	7f	91.0	-7.5	96.8	-10.7	1.9	-41.6
	7s	84.2		86.4			
Al4Cu SC Test Angle = 90°	8f	27.4	-3.3	23.1	-1.3	6.6	28.4
	8s	26.5		22.8			
	9f	63.7	-3.9	57	-5.3	1.2	21.9
	9s	61.2		54			
	10f	56.7	-13.4	61.1	-9.0	-5.8	13.4
	10s	49.1		55.6			

5. CLOSURE

Recent advances in both computational and laboratory techniques provide great potentials to eliminate the "trial and error" method of alloy development and replacing it with a more intelligent and streamlined approach. Application of first-principles calculations and dislocation-slip simulations, combined with the high-resolution and analytical microstructural characterizations can assist in identifying the preferred 2nd phases, their morphology and distribution, resulting in alloys with high strength, fracture toughness and creep resistance.

ACKNOWLEDGEMENTS

The authors would like to thank the Office of the Naval Research for support under contract ONR N00014-91-J-1285, Dr. George Yoder, contract monitor, and the Air Force Office of Scientific Research under grant F49620-97-1-1034, Dr. Craig Hartley, Program monitor. Drs Al Csontos, Brian M Gable, Unlu Necip, Henrich Hargarter and Jin Chen were involved in different parts of the experimental work.

REFERENCES

1. E.A. Starke Jr. and J T. Staley, *Prog. Aerospace Sci.* **32**, 132 (1996).
2. J.C. Williams, E.A. Starke Jr The role of thermomechanical processing in tailoring aluminum and titanium alloys, ASM Materials Science Seminar, Metals Park, OH:ASM, p1983.
3. W. Gambin, Plasticity and textures, Kluwer Academic Publishers, Boston 2001.
4. A. Zhu, G. Shiflet, E.A. Starke Jr, in preparation for *Prog. Mater. Sci.*
5. H. Perlitz and A. Wetgren, *Ark Chem Miner. Geo.* **16B**, 1 (1943).
6. V. Radmilovic, R. Kilaas, U. Dahmen, G. J. Shiflet, *Acta mater.* **47**, 3987 (1999).
7. G. Kresse and J. Furthmueller, *J. Comput. Mater. Sci.* **6**,15 (1996).
8. A. Zhu et al, AFOSR Report 2002 (unpublished).
9. C. Wolverton, *Acta mater.* **49**, 3129 (2001).
10. J.H. Auld, *Acta Cryst.* **A28**, S98, Suppl. (1972).
11. S. Kerry and V.D. Scott, *Metals Sci.* **18**, 289 (1984).
12. K.M. Knowles and W.M. Stobbs, *Acta Cryst.* **44**, 207 (1988).
13. C. Muddle and I.J. Polmear, *Acta Metall.*, **37**, 777 (1989).
14. A. Zhu, B.M. Gable, G.J. Shiflet & E.A. Starke, *Adv. Mater. Eng.*, **4**, 839 (2003).
16. A. Zhu, B.M. Gable, G.J. Shiflet and EA Starke Jr, in "Metallurgical Modelling for Al Alloys", Ed: M Tiryakioglu and L.A. Lalli, Pittsburgh, PA, 2003, p 127.
17. van der Waals, *Z. Physik. Chem.* **13**, 657 (1894).
18. J.W. Cahn and J.E. Hilliard, *J Chem. Phys.* **28**, 258 (1958).
19. M. Asta, *Acta mater.* **44**,10,4131(1996).
20. M. Sluiter and Y Kawazoe, *Phys. Rev. B* **54**,15:10381 (1996).

21. A. Csontos and Starke Jr EA, Inter J. Plasticity, in press.
22. B.Q. Li and F.E. Wawner, *Acta mater.* **46**, 15, 5483 (1998).
23. A. Zhu and E.A.Starke, Jr, *Acta Mater.* **49**, 2285 (2001).
24. C.R. Hutchinson, X. Fan, S.J. Pennycook and G.J. Shiflet, *Acta Metall.* **49**, 2827 (2001).
25. U.F. Kocks, J.S Kallend, H.R.,Wenk, A.D. Rollett and S.I Wright, popLA-Preferred Orientation Package, Los Alamos National Laboratory, Los Alamos, NM, July, 1994.
26. J. M. Crompton, R. M. Wayhorne and G. B. Brooki, *J. Appl. Phys.* **17**, 1301 (1966).
27. D.B. Williams and C.B. Carter, Transmission Electron Microscopy, Plenum Press, New York, 1996, p174.
28. I.J. Polmear, Light Alloys: Metallurgy of the Light Metals, New York: Edward Arnold, 1989.
29. A. Garg, Y.C. Chang and J.M. Howe, *Scripta Met.* **24,** 677 (1990).
30. I.J.Polmear and M.J.Couper, *Metall Trans.* **19** A, 1027 (1988).
31. S.P. Ringer, W. Yeung, B.C. Muddle, I.J. Polmear, *Acta metall. mater.* **42**, 1715 (1994).
32. C.R. Hutchinson, X. Fan, S.J. Pennycook and G.J. Shiflet, *Acta Metall.* **49**, 2827 (2001).
33. L. Reich, M. Murayama and K. Hono, *Acta. Mater.* **46**, 17, 6053 (1998).
34. E. Hornbogen and E.A. Starke Jr, *Acta metal. mater.* **41**, 1 (1993).
35. U.F Kocks., *Mater. Sci. Eng.* **27**, 191 (1977).
36. Ashby M.F., *Acta Metall.* **14**, 679 (1966).
37. D.L Gilmore. and E.A Starke. Jr., *Metall. Mater. Trans.* **28**A, 1399 (1997).
38. B Reppich, Materials Science & Technology Vol.6, VCH, Weinheim, 1993, p311.
39. A.J. Ardell, *Metall. Trans.* A **16**, 2131 (1985).
40. H. Mughrabi., Materials Science & Technology Vol.6, VCH, Weinheim, 1993, p1.
41. A.J.E. Forman and M.J. Makin, *Philo. Mag.* **14**, 911 (1966).
42. J.W. Morris Jr and D.H Klahn, *J. Appl. Phys.* **45**, 2027 (1974).
43. B.Reppich, W Kuhlein., G. Meyer, D. Puppel, M.Schulz and G.K Schumann, *Mater. Sci. Eng.* **83**, 45 (1986).
44. E. Nembach and M. Martin, *Acta Metall.* **28**, 1069 (1980).
45. E Nembach and M Neite, *Prog. Mater. Sci.* **29**, 177 (1985).
46. A. Melander, *phys. stat. sol.* (a) **43**, 223 (1977).
47. L.M.Brown and R.K.Ham, in Strengthenin Methods in Crystals, ed. A.Kelly and R.B.Nicholson, John Wiley&Sons, Inc. New York, 1971, p. 15.
48. M.F. Ashby, Physics of Strength and Plasticity, MIT press, Cambridge, Mass., 1969, p113.
49. A. Zhu, J.Chen & E.A.Starke, Jr, *Acta Materialia* **48**, 2239 (2000).
50. J.R. Davis, Aluminum and Aluminum Alloys, The Mater. Information Soc., 1993, p649.
51. E.Nembach, *Acta metall. mater.* **40**, 3325 (1992).
52. T.J.Koppenaal and D. Kuhlmann-Wilsdorf, *Appl. Phys. Lett.* **4**, 59 (1964).
53. M.C.Holman, J. Mechanical Working Technology **20**, 477 (1989).
54. H.M. Brewer Jr. and M.C.Holman, *World Aerospace Structure Technology* '90, 1990, p.4.
55. J.M.Newman, , M.D. Goodyear, J.J. Witters., J.Veciana and G.K. Platts, *Proc 6th Conf. on Aluminum-Lithium Alloys*, DGM, 1992, p.1371.
56. A. Zhu and E.A.Starke, Jr, *Acta Materialia* **49**, 2285 (2001).
57. A. Zhu and E.A.Starke, Jr, *Acta Materialia* **49**, 3063 (2001).
58. W.F. Hosford and S.P Agrawal, *Metall. Trans. A* **6**, 487 (1975).
59. T. Eto T., A. Sato A. and T.Mori, *Acta Metall.* **26**, 499 (1978).
60. B. Skrotzki B., G.J. Shiflet and E.A. Starke Jr., *Metall. Mater. Trans. A* **27**, 3431 (1996).
61. H. Hargarter H., M. T. Lyttle and E.A Starke Jr., *Mater. Sci. Eng. A* **257**, 87 (1998).
62. J.S. Lange and A. J Schwartz, *Phys. Rev. A* **21**, 928 (1980).

63. J.M. Silcock, T.J. Heal, H.K. Hardy, *J. Institute of Metals*, 1953-1954, **82**, 239.
64. G. Sauthoff, *Zeit. Metallkdte* **66**, 106 (1975).
65. M.T. Lyttle and J.A. Wert, *Metall. Mater. Trans.* A **30**, 1283 (1999).
66. W.F. Hosford and R.H. Zeisloft., *Met. Trans.* **3**, 113 (1972).
67. P.Bate, W.T. Roberts and D.V. Wilson, *Acta Metall.* **29**, 1797 (1981).

Chapter 10

MULTISCALE MODELING OF INTERGRANULAR FRACTURE IN METALS

Vesselin Yamakov[1], Dawn R. Phillips[2], Erik Saether[3] and Edward H. Glaessgen[3]
[1] National Institute of Aerospace, Hampton, VA 23666; [2] Lockheed Martin Space Operations, Hampton, VA 23681; [3] NASA Langley Research Center, Hampton, VA 23681

Abstract: Multiscale modeling methods for the analysis of fracture in metallic microstructures are discussed. Molecular dynamics models are used to analyze grain-boundary sliding and fracture in an aluminum bicrystal model. A bilinear traction-displacement relationship that may be embedded into cohesive zone finite elements for microscale problems is extracted from the nanoscale molecular dynamics results.

Keywords: multiscale modeling; molecular dynamics; cohesive zone model; grain-boundary sliding; aluminum; fracture.

1. INTRODUCTION

Classical fracture mechanics is based on the comparison of computed fracture parameters to their empirically determined critical values. Although this paradigm has been extremely successful for modeling crack growth at structural scales, it does not describe the fundamental processes that govern fracture. Ultimately, an understanding of events and processes that occur at length scales on the order of 10^{-9} to 10^{-3} m is needed to fully understand fracture.

1.1 Coupling Methods

Multiscale modeling provides an efficient means of interrogating deformation and fracture of metallic materials from the micro- to the nano-scales. Many multiscale modeling strategies have been explored by mechanists in recent years; a few of them are discussed here. Coupling methods[1-3] involve *a priori* dividing a problem into spatial regions based on the simulation technique to be used and tying those regions together with interface boundary conditions. For crack problems, a recently advocated approach combines *ab initio* quantum mechanics with molecular dynamics (MD) and finite element (FE) continuum mechanics.[1,2] Quantum mechanic *ab initio* calculations are used to describe accurately the specific interactions between atoms at the crack tip, while atoms in the vicinity of the crack tip are simulated by molecular dynamics (MD). In MD, empirical potentials are fitted to reproduce the interatomic forces in the bulk material, but are not very accurate for atoms near surfaces or in highly defected regions (i.e., where the atomic surrounding is very different from that in the bulk material). Further away from the crack, outside the damage zone, finite element continuum mechanics is used to simulate the elastic response of the surrounding medium. While the theory behind coupling methods is quite elegant, some challenges arise including the artifacts introduced by the appearance of interfaces separating regions of different length and time scales, complications in mesh generation in the finite element portion of the model, and the method's inadaptability to changing deformation and loading states.

1.2 Quasicontinuum Methods

Quasicontinuum methods[4,5] employ a continuum mechanics framework with a finite element discretization wherein each element is formed by nodes that coincide with "representative" individual atoms of the material. In regions where the atomistic behavior is essential to understanding the problem, the finite element mesh is refined to the level where all atoms are considered as representative and coincide with nodes. In regions where a coarse mesh is appropriate, an individual element may encompass many atoms within its interior in addition to the atoms that coincide with the nodes. Also, it is assumed that the element constitutive relations are obtained directly from atomistic calculations. The Cauchy-Born hypothesis is used to relate homogeneous deformations of the continuum to the superposed lattice to obtain the element strain energy by summing up the contributions of the interatomic potentials of the contained atoms.[4,5] In general, quasicontinuum methods aim to overcome the problem with the interfaces in the coupling methods by using a single simulation technique providing gradual transition between length scales. The drawback is that in the atomic region, quasicontinuum methods cannot represent correctly the kinematics of atomic interactions and are incapable of nucleating and propagating atomic-scale defects such as vacancies, dislocations, deformation twins, etc., which play an essential role in the plastic behavior of the material.

1.3 Equivalent Continuum Mechanics Methods

The equivalent continuum mechanics method[6] is similar to the quasicontinuum methods, but uses the meshless local Petrov-Galerkin (MLPG) method rather than finite elements. In general, meshless methods are developed to overcome some of the disadvantages of the finite element method, such as the need to interpolate discontinuous secondary variables across interelement boundaries and the need for remeshing in large deformation problems.

1.4 Constitutive-Relation Based Scaling

The coupling, quasicontinuum and equivalent continuum mechanics methods each have applications for which they are well suited; however, they all have one common and pragmatic drawback – they are very complex in implementation and are computationally intensive. That is, they are best suited for modeling relatively small domains containing a single crack-like defect or a dislocation. This chapter presents another concept of a multiscale strategy – a strategy that may allow larger volumes of material with more complex local architecture and responses to be modeled. This concept is to use a lower level simulation technique, such as atomistic MD, to extract the type and form of the constitutive relation used in a higher level simulation, such as continuum FE. Unlike in the coupling methods[1-3] discussed above, the two methods, MD and FE, do not connect directly in a simulation. The information from the lower to higher level is transferred through a constitutive equation. The constitutive equation has to reflect the most dominant characteristics of the individual deformation mechanisms revealed at the lower level and reproduce their effects on the system correctly at the continuum level. This concept requires that the two simulation techniques be used first separately on different models. The model for the lower level method has to be designed to extract the necessary parameters for the constitutive relation as efficiently and accurately as possible. For this reason, usually an idealized analytical model is built, which may not represent a realistic experimental situation, but does reflect correctly the surrounding boundary and environmental conditions under which the constitutive equation is to be applied in the continuum model.

Because of the rapid advances in the computer technology, recently it became possible that a lower level method can simulate a large enough system, which starts to attain the features of the next level-up simulation. For example, a multi-million atom MD system of a solid body already behaves as a continuum elasto-plastic material subsiding its discrete atomic structure. Consequently, once the parameters of the constitutive equation for the continuum model are determined by the atomistic simulation, a test model can be designed, which can be simulated separately by both the atomistic MD and the parameterized continuum FE method. A comparison between these two simulations on the same type of microstructure would show the adequacy of the chosen constitutive relation and the effectiveness of its parameterization.

The proposed strategy for multiscale modeling thus consists of several stages: (i) choosing the most suitable continuum type of simulation method (FEM, for example) capable of simulating the desired microstructure; (ii) embedding the most relevant physical mechanisms through constitutive relations in the continuum model; (iii) designing a lower level simulation (MD or *ab initio* simulation) on an analytical idealized system to define the parameters of the constitutive equation; and (iv) verifying the parameterization and the implementation of the constitutive equation on a test microstructure that can be simulated separately by the two methods. If at stage (iv) a good agreement is found between the two simulations, then the constitutive equation has passed the lower level features of the system to its continuum equivalent successfully.

To study the particular case of *intergranular* fracture in aluminum, the multiscale modeling strategy proposed here is based on MD and the finite element method (FEM) as the building blocks. The MD simulations are used to determine the mechanisms of plastic deformation and fracture of aluminum around a grain boundary (GB) at atomic scales (Fig. 1.a). This information is then recast to obtain an averaged continuum traction-displacement relationship (Fig. 1.b) to represent the cohesive interaction along a characteristic length of the GB. The traction-displacement relationship is incorporated through the implementation of cohesive zone models (CZM) (Fig. 1.c)[7-10] into an FE model to study the behavior at the grain scale (Fig. 1.d). In this way, the CZM elements represent the mechanical response of the GBs, while the constitutive properties of the crystalline grains are represented by finite elements with anisotropic elastic properties effectively matched to the elastic constants, defined by the interatomic potential used in the MD simulation (Fig. 1.e).

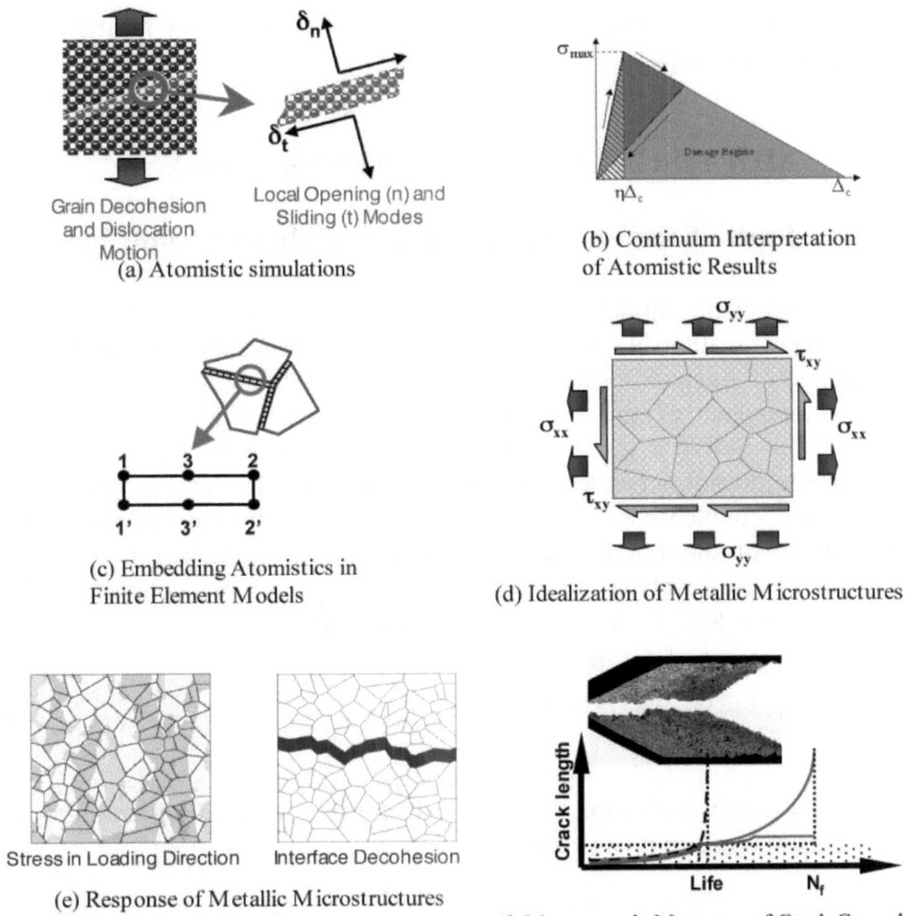

(a) Atomistic simulations

(b) Continuum Interpretation of Atomistic Results

(c) Embedding Atomistics in Finite Element Models

(d) Idealization of Metallic Microstructures

(e) Response of Metallic Microstructures Under Mechanical Loads

(f) Macroscopic Measures of Crack Growth

Figure 1. Multiscale modeling with cohesive zone models.

Thus, an integrated multiscale modeling strategy for understanding the connection between macroscopic fracture and the details of the underlying atomistic failure mechanisms emerges (Fig. 1.f). The range of relevant length scales achieved by this strategy can span from the atomistic (10^{-9} m) as shown in Fig. 1.a to the macroscopic ($>10^{-3}$ m) as shown in Fig. 1.f.

To address the ideas depicted in Figs. 1.a-b, in this chapter, two MD models are discussed to extract the traction-displacement relationship for a GB in two distinctive processes of deformation: GB sliding and decohesion. A way to implement the MD atomistic information into a CZM model in the FE method is presented and discussed. Previous attempts to this implementation are also mentioned. When successful, this procedure allows the FEM to be used as a part of a multiscale modeling strategy to extrapolate the MD-derived information to larger scales to study the failure properties of polycrystalline materials.[9]

2. MULTISCALE MODELING STRATEGY

2.1 Finite Element Modeling with Cohesive Zone Models

Cohesive zone models approximate the interfacial atomic traction-displacement relationship using bulk material properties and can model the appearance of fracture surfaces in a continuum.[9] Although the normal and shear components of the traction and displacement may be considered separately, as in Ref. 8, a more realistic approach may be to consider that the components do not act independently of each other. In the Tvergaard and Hutchinson[7] coupled cohesive zone model (CCZM), the normal and shear components of the traction and displacement are combined into single measures, τ and λ, respectively, so that the responses are coupled.[10] The CCZM given in Ref. 10 defines a traction potential,[11]

$$\Phi(\delta_n,\delta_t) = \delta_n^c \int \tau(\lambda)d\lambda \tag{1}$$

where λ is a non-dimensional measure of the relative opening and sliding displacements (δ_n and δ_t) as defined by

$$\lambda = \left[\left(\frac{\delta_n}{\delta_n^c} \right)^2 + \left(\frac{\delta_t}{\delta_t^c} \right)^2 \right]^{1/2} \tag{2}$$

δ_n^c and δ_t^c are the critical values for the opening and sliding modes, respectively, and a fracture surface is assumed to have formed once λ reaches a value of unity. The use of cohesive zone models in finite elements permits highly discontinuous displacement behavior to be exhibited in a single discretized model without remeshing using an incremental piecewise linear solution algorithm. The displacement discontinuity may be defined in three separate regimes that depend on the relative CZM element nodal displacements. First, the CZM element functions as a multipoint constraint to enforce node coincidence at the GB between connected continuum elements. Secondly, during strain softening of the opening of the CZM element, the effective stiffness is reduced as a function of the relative displacements of initially coincident nodes. Finally, after the boundary has opened and the effective stiffness becomes zero, further increments may cause local load redistribution resulting in separated surfaces coming back into contact. In this regime, the CZM element functions as a contact element to prevent interpenetration.

As a consequence of the unique numerical features of the CZM element, a number of issues related to convergence exist and must be addressed prior to performing a successful analysis. One issue involves the magnitude of the initial "penalty stiffness" that is used to enforce the multipoint constraint. This quantity is problem-specific and can lead to convergence difficulties if it causes the global stiffness matrix to become ill-conditioned.[12] Also, if interpenetration is detected after a boundary has opened, the abrupt replacement of zero stiffness terms between surface elements with penalty stiffness values can cause divergence of the solution. Another issue pertains to the integration scheme used to compute the stiffness coefficients for the CZM element. In some instances, a Gaussian quadrature scheme can contribute to poor convergence compared to a Newton-Cotes scheme.[13] Solution convergence is also dependent on the size of the CZM elements compared to the extent of the process zone surrounding the fracture front.[14] An additional issue pertains to the possible extensive use of CZM elements within a model. If no *a priori* knowledge of crack paths is available, CZM elements may be placed between all continuum elements such that the crack path will be automatically determined by the predicted stress fields. However, if the initial or "penalty" stiffness is set to a small value, this can introduce an artificial flexibility into the model that can alter load transfer and displacements.

2.2 Molecular Dynamics Modeling

To effectively model polycrystalline material at the atomic scale, large numbers of atoms are needed. Approximations describing the potential energy of these interactions have been developed in the form of empirical and semi-empirical potentials.[15] These potentials are defined for atom-pair and many-body force interactions. A MD simulation is made tractable for large systems of atoms by treating each atom as a point mass, summing forces due to interactions with surrounding atoms, and using Newton's Second Law over a succession of time steps to obtain trajectories specifying atom positions and velocities.

In the case of metals, metallic bonding is formed through delocalized electrons that are effectively shared between all atoms in the crystal lattice. Thus, simple atom-pair potentials cannot be used to account for all of the mutual interactions between the electron clouds surrounding the atoms.

Modifications, such as cluster potentials, pair functionals, and cluster functionals, have been made to pair potentials to develop general classes of interatomic potentials. The embedded-atom method (EAM), which uses a pair functional and treats each atom as being embedded in a field of electrons created by the surrounding atoms, has been successful in simulating the atomic bonding in metals.[16] Used in a MD simulation, this potential is capable of reproducing very closely various physical properties of the metal, such as lattice constant, cohesive energy, stacking fault and surface energy, vacancy formation energy, melting temperature, elastic constants, etc. As a consequence, a MD simulation is capable of giving a close prediction of various mechanical properties like yield stress,[17] material strength and toughness,[2,18] etc.

For the purpose of quantifying the mechanical properties of a GB for implementation in a CZM element, an idealized atomistic model of a bicrystal with periodic boundary conditions to mimic the bulk material is used. The model uses the EAM potential fitted by Mishin et al.[19] to reproduce very closely the physical properties of Al, such as lattice constant a_0 = 0.405 nm; cohesive energy, elastic modulus, etc.[19] This makes the potential well suited for modeling fracture. In this model, two crystals are joined together to form a flat GB interface. The relative misorientation between the crystals defines the crystallographic type of the GB, on which its mechanical properties are strongly dependent.[20] For the specific purpose of this study, the misorientation angle is chosen to be 89.42° (so that a microstructure of four-fold 90 degree symmetry can be constructed) around the [1 1 0] crystallographic axis, common for both crystals. The GB thus formed is a <110> ∑99 symmetric tilt GB. The atomic structure of this high-angle GB in Al is closely investigated in Ref. 21. The grain boundary energy γ_{GB} is estimated from the Al potential to be 0.65±0.05 J/m^2 at 100 K. This large excess GB energy (i.e., above the perfect cohesive crystal) facilitates the GB decohesion.[20]

Two variations of this bicrystal model are used. In the first, the microstructure has nearly cubic dimensions (Fig. 2) with Crystal I oriented at [7 7 10] and Crystal II oriented at [7 7 10] occupying the upper and lower half of the cube, respectively, and forming a flat ∑99 GB at their interface. Because of its cubic shape the model is highly suitable to study the elasto-plastic properties of a GB in a fully three-dimensional perspective. Unfortunately, because the system volume increases as the cube of the system size, large systems (> 100 nm) soon become intractable.

To reach large in-plane system size, another model is used (Fig. 3). The model presents the same microstructure as the bicrystal cubic model discussed previously (Fig. 2).

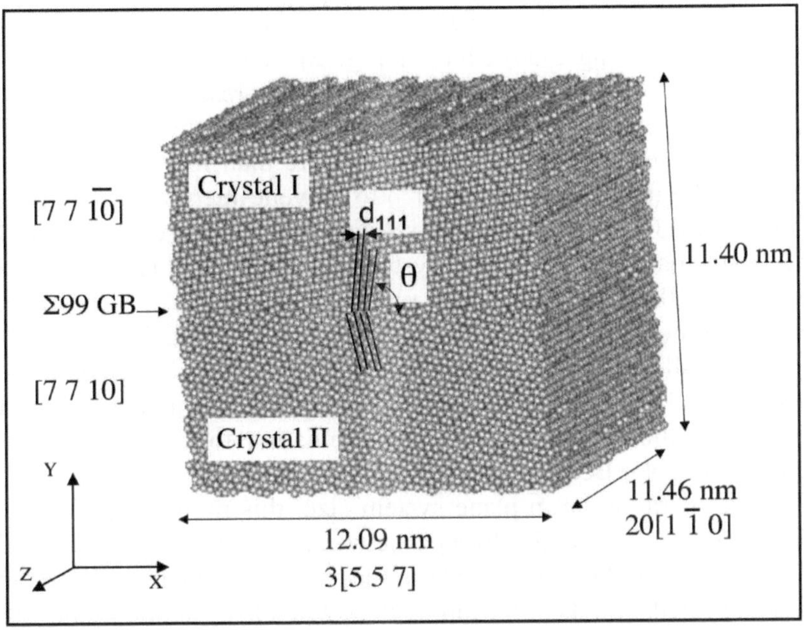

Figure 2. Bicrystal cubic microstructure used to obtain the sliding resistance of Σ99 symmetric tilt GB.

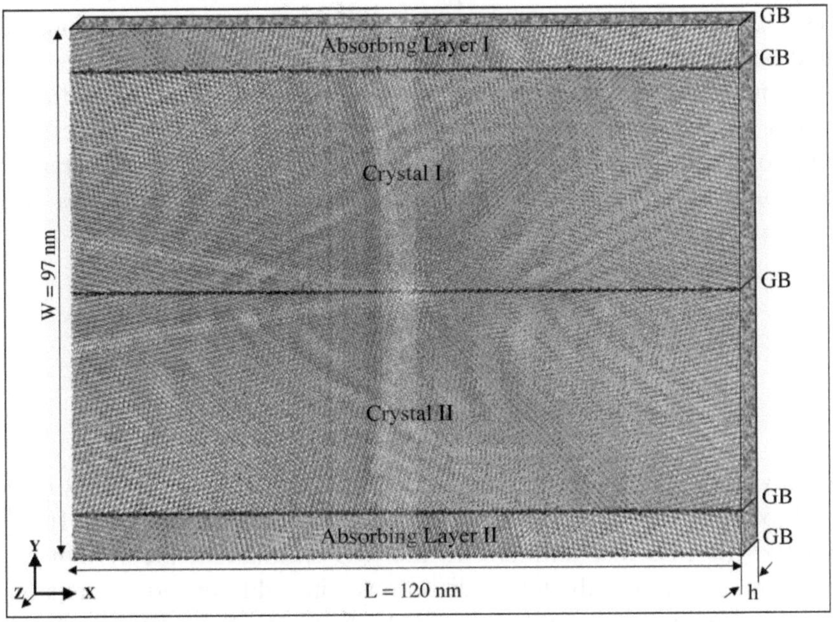

Figure 3. Bicrystal columnar microstructure used to study fracture along Σ99 symmetric tilt GB.

The major difference from the cubic model is that the thickness *h* in the third dimension (along *z*-direction in Fig. 2) is limited to only 10{220} atomic planes or $h = 10/\sqrt{2}a_0 = 2.86$ nm. This results in a model of a columnar, quasi two-dimensional microstructure. The thickness of 2.86 nm is chosen because it is still large enough (more than four times larger than the range of the interatomic potential[19]) to prevent interference of the atoms with their periodic images along the *z*-direction, thus preserving the local three-dimensional physics in the system. In addition, the <110> texture makes two of the four {111} slip planes in the fcc lattice available for dislocation glide and cross-slip events, as in a fully 3D space.[22,23] The configuration allows for the system size in the *x*- and *y*-directions to extend up to 120 and 97 nm, respectively, while limiting the number of simulated atoms to 1,994,000, thus allowing the simulation to be carried out on a modest 16 processor Beowulf cluster.

Because of the large in-plane system size, this model is used to study fracture processes at the GB formed between "Crystal I" and "Crystal II" (Fig. 3), which is of the same type as in the bicrystal model. An intergranular crack is nucleated, and its propagation along the GB is monitored throughout the simulation, as will be discussed in the next section.

The two narrower layers on each side of the bicrystalline system (Absorbing Layer I and II in Fig. 3) serve as shock-wave absorbers,[24] where a damping friction coefficient is applied to the atoms to absorb the phonon waves[24] expected to be generated from the crack tips. The additional GBs formed between these layers and Crystals I and II act as absorbers for the dislocations that may be emitted from the crack tips during propagation. The inclusion of these shock-wave and dislocation absorbers suppresses the creation of periodic images of the disturbances emitting from the crack tips, thus avoiding the unrealistic influence that these disturbances would otherwise have on the crack propagation.

3. ANALYSIS

3.1 Grain-Boundary Sliding

In this study, the bicrystal cubic system discussed in Section 2.2 (Fig. 2) is used to study the sliding properties and the shear strength of <110> ∑99 symmetric tilt GBs in aluminum. In the MD simulation, the system box is allowed to shear in all three directions, in addition to expansion or contraction, making it possible to estimate the shear strain directly. To study the sliding resistance along the two perpendicular sliding directions, [557] and $[1\bar{1}0]$ (Fig. 2) in the GB plane, a constant shear strain rate of

$\dot{\varepsilon}_{xy(xz)} = 1.8 \times 10^8\ s^{-1}$ is applied, which produces a sliding velocity at the GB equal to 1 m/s. The stress response of the GB to sliding in both directions is given in Fig. 4. The stress, after an initial elastic increase, starts to fluctuate with the sliding distance, forming peaks and valleys very coherent in the case of sliding along the [557] direction, and more distorted for the sliding in the [110] direction. The distance between the peaks can be easily correlated to the interplane distances d_{111} and d_{110} of the dominant atomic planes {111} and {110} in the crystal lattice. For example, the (111) planes are crossing the GB interface at an angle $\theta \approx 80^0$ in lines perpendicular to the [557] direction. Sliding along [557] makes the edges of the (111) planes from the two crystals to overlap at the GB every $d_{111}/\sin\theta = 0.24$ nm, where θ is the angle between the (111) plane and the GB plane (Fig. 2), producing peaks in the stress. Similar is the situation with the (110) planes when the sliding is along the $[1\bar{1}0]$ direction, though these peaks are not as deep and as regular as in the first case (see Fig. 4).

The stress-displacement curves, shown in Fig. 4 for the two perpendicular sliding directions of the ∑99 GB in aluminum, can be used to fit a traction-displacement relationship for a CZM element in the form shown in Fig. 5. For sufficiently large GBs in the continuum limit, the atomic planes overlapping will become incoherent due to random imperfections in the GB structure, and the stress peaks, visible in the idealized bicrystal MD model (Fig. 4), will smooth out to produce an averaged stress response to sliding. This is expressed as a constant, displacement independent traction τ_p in the CZM curve, which takes the mean value of the fluctuations seen in the MD curve.

3.2 Grain-boundary Decohesion: Molecular Dynamics and Finite Element Relationship

The goal of the approach described in this section is to build an atomistic MD model that is compatible with the underlying framework of a CZM element.[25] A properly posed MD simulation will allow the extraction of the CZM decohesion law under local conditions similar to those experienced by the CZM element in an FE polycrystalline model. The CZM decohesion curve is constructed to reflect the response of the CZM element to an approaching and propagating crack.[25] Thus, the MD model is a model of crack propagation and not a model of interfacial separation (material strength).

To ensure that the FE simulation would incorporate the MD results successfully, first, the FE model has to be tuned to reproduced correctly the far-field elastic stress around a crack in the MD system before starting the procedure of extracting the CZM constitutive relation.

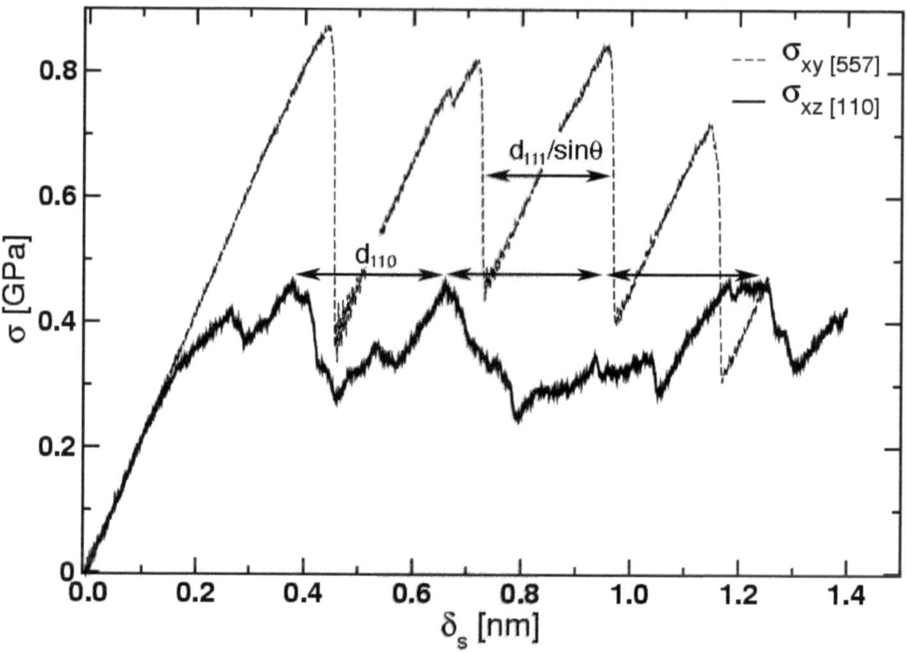

Figure 4. Shear stress vs. sliding distance for Σ99 symmetric tilt GB.

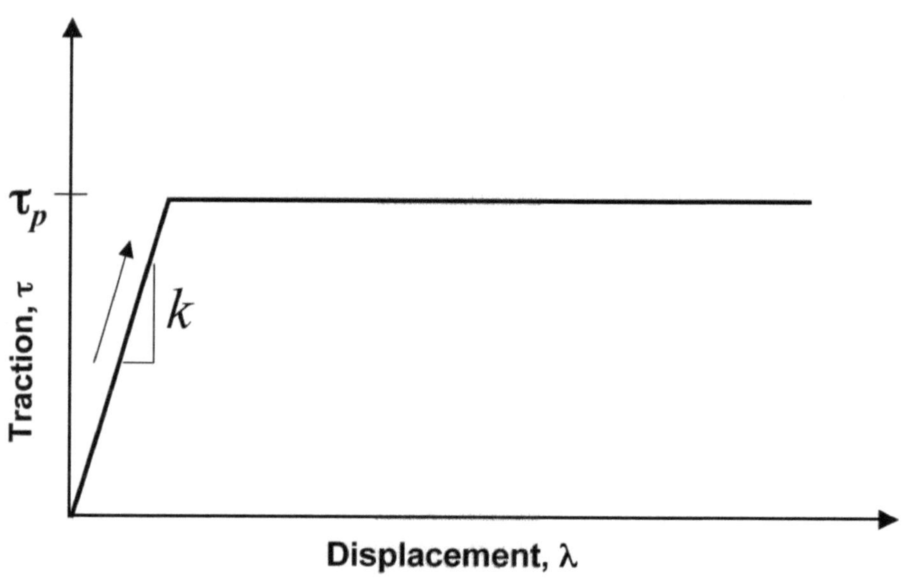

Figure 5. Traction-displacement relationship for CZM representing sliding resistance of a GB.

For this purpose, an elastically equivalent FE reproduction of the MD model, with anisotropic elastic constants extracted from the MD interatomic potential, is developed. A close match between the MD and FE results would ensure that the CZM curve will be extracted under equivalent loading conditions in the MD simulation.

3.2.1 Molecular Dynamics Results

The MD simulation set-up for analyzing the GB decohesion process implements the columnar quasi two-dimensional microstructure model discussed in Section 2.2 and shown in Fig. 3. The simulated system is preloaded by applying constant external hydrostatic loads ($\sigma_{xx} = \sigma_{yy} = \sigma_{zz} = \sigma$) at a constant temperature of 100K. The temperature 100K is chosen to suppress grain boundary and surface diffusion. After reaching mechanical equilibrium, the system size is fixed and the follow up study is performed under constant strain conditions. The hydrostatic preloading helps to eliminate plasticity effects not related to the crack, such as spontaneous dislocation nucleation from the GB,[22] which could otherwise dominate the deformation in this (essentially) two-dimensional configuration.

The crack is initiated by screening the interatomic potential between atoms on both sides of the Crystal I-Crystal II GB plane (Fig. 3) along a region of 5.7 nm along the middle of the GB. From this initial crack, the GB opens, and the crack starts to propagate in both directions along the GB interface. Figure 6 shows MD snapshots of internal cracks that have grown in a plate preloaded at initial hydrostatic stress of σ = 3.5, 3.75, 4.0, and 4.25 GPa. Common neighbor analysis (CNA)[26] is used to identify atoms in different crystallographic states: fcc, hcp, and non-crystalline atoms (gray, dark grey, and black in Fig. 6, respectively). Atoms with more than 1/3 of their nearest neighbors missing are identified as surface atoms (green in Fig. 6).

In each of the cases presented, the crack propagation is not symmetric in the $+x$ and $-x$ directions along the GB (as defined in Fig. 3). The propagation in the $-x$ direction produces two twin patterns ① also shown in Fig. 6 that grow almost symmetrically in the two joined crystals and retard the $-x$ direction propagation of the crack. Also observed in the snapshots are stacking faults ② and partial dislocations ③ that originate near the $-x$ direction tip of the crack. By contrast, the propagation in the $+x$ direction is more brittle, accompanied by only a few perfect dislocation emissions ④.

As a result, the crack propagates much further along the GB in the $+x$ direction. This asymmetric fracture of the GB crack is a result of the specific orientation of the {111} slip systems in the two crystals relative to the GB plane.

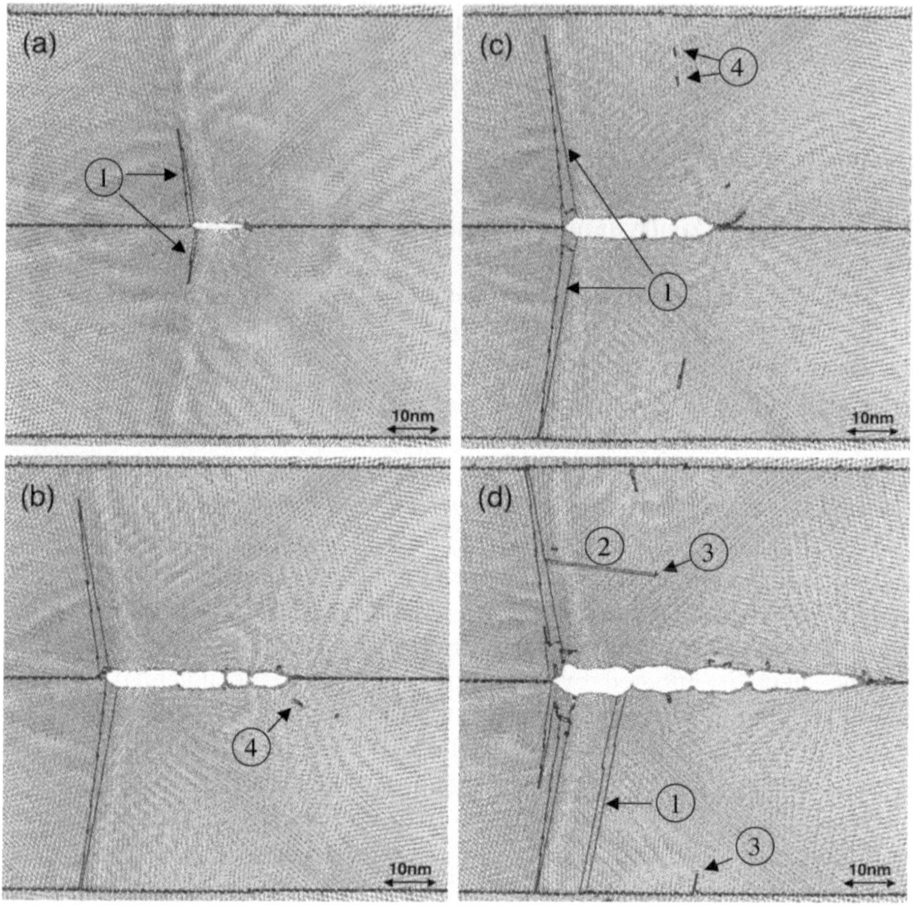

Figure 6. MD snapshots of cracks, which have propagated in the MD system, shown in Fig. 3, prestrained at four different initial hydrostatic loads: σ = 3.5 (a), 3.75 (b), 4.0 (c) and 4.25 GPa (d). As in Fig. 2, CNA is used to identify atoms in different crystallographic states: fcc (gray background), hcp (dark gray), and non-crystalline atoms (in black). Atoms with more than 1/3 of their nearest neighbors missing are identified as surface atoms (edges of the white voids). Thus, the straight lines of dark gray atoms are either twin boundaries (1) or stacking-faults (2) in a partial (3) or perfect dislocation (4).

This orientation favors twinning in the -*x* direction and decohesion in the +*x* direction. This asymmetry in the crack propagation can be validated by the theoretical model of Tadmor and Hai.[27]

In all four snapshots in Fig. 6 the crack, after growing to a certain size, becomes in equilibrium with the elasto-plastic deformation of the surrounding crystal lattices. In other words, the deformation twinning, developed at one of the crack tips, together with the dislocation emission from the other crack tip, plastically accommodates enough elastic strain to

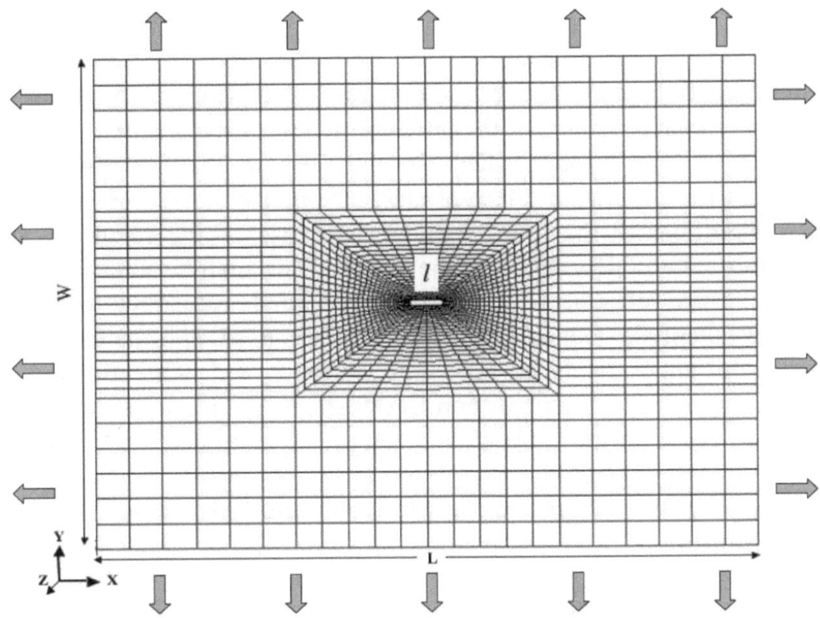

Figure 7. Finite element mesh of model with built-in lenticular crack.

stop the crack growth. This elasto-plastic equilibrium is achieved at different crack lengths, depending on the preload and the crack dynamics during growth. The complex interrelation between the preload and the crack dynamics makes the prediction of the equilibrium crack length nontrivial. For example, contrary to the expectations, the crack at 4.0 GPa preload has grown smaller than the crack at 3.75 GPa and at 4.25 GPa. The reason for this is that the higher growth rate at 4.0 GPa, compared to 3.75 GPa, initiates more dislocation emissions resulting in more active plastic accommodation at the crack tips, which stops the growth sooner than at 3.75 GPa preload. In the case of 4.25 GPa, the initially stored strain energy is high enough for the crack to overcome the plastic dissipation for a longer time and propagate further.

3.2.2 Finite Element Model

The corresponding FE configuration is presented in Fig. 7. The dimensions of the system are scaled to recover the proportions of the MD system. The model contains a built-in lenticular slit of varying relative length, $0.05 \leq l/L \leq 0.91$, to study the evolution of the system at different stages of crack growth. Although the lenticular slit is not a crack in the true continuum mechanics sense, it is an analog to the crack in the MD configuration and hence is used to represent a crack-like discontinuity (hereafter referred to as a crack for simplicity). To avoid stress singularities in the FE system, the crack tips have a finite, but very small initial radius that corresponds to an initial crack opening of approximately 1 nm, i.e., slightly larger than the range of the interatomic forces (0.67 nm).

There is no elastic equivalent for the GBs in the FE system. It is assumed that the somewhat different elastic response of the GB layers does not alter significantly the crack behavior because of the relatively small volume ratio of the GB volume in the system. The matrix is assumed to be anisotropic linear elastic, with constants assigned according to the MD interatomic potentials. Generalized plane strain is used, and the calculations are carried out under displacement control in the *x*- and *y*-directions. At this stage of the study no dynamic or non-linear effects are incorporated, and the results are relevant for infinitely slow crack propagation under small strain.

3.2.3 Comparison between Finite Element and Molecular Dynamics Models

Plastic processes (twinning and dislocation emission) at the crack tips have a pronounced effect on the stress distribution around the growing crack. This effect is revealed by comparing the stress distributions obtained from the MD and the FE simulations of the two elastically equivalent models described previously. The MD simulation necessarily incorporates all the plastic processes together with the elastic response of the bicrystal, while the FE model is used to predict the elastic response separate from the plastic components of the solution. The MD stress distribution is determined after the atomistic system reaches elasto-plastic equilibrium and the crack stops growing, while the FE stress distribution is determined when the system is in elastic equilibrium. The comparison between the MD and FE results will identify the role of the plastic and dynamic effects in the studied process of fracture. Two-dimensional (*x*-*y*) stress-maps for the MD results are created by averaging the local virial stress[28] over a chosen volume of $6a_0$ x $6a_0$ x h in the *x*-, *y*-, and *z*-directions, respectively (2.43 x 2.43 x 2.86 nm^3 volume including 864 atoms) and over 16 ps of simulation time after reaching elasto-plastic equilibrium. The virial theorem, which is the basis for calculating the virial stress, provides the oldest and most frequently used expression for relating forces and motion within an atomic system to a continuum stress. In the limit of time and ensemble averages at equilibrium, the virial stress coincides with the Cauchy stress used in the continuum mechanics.[29]

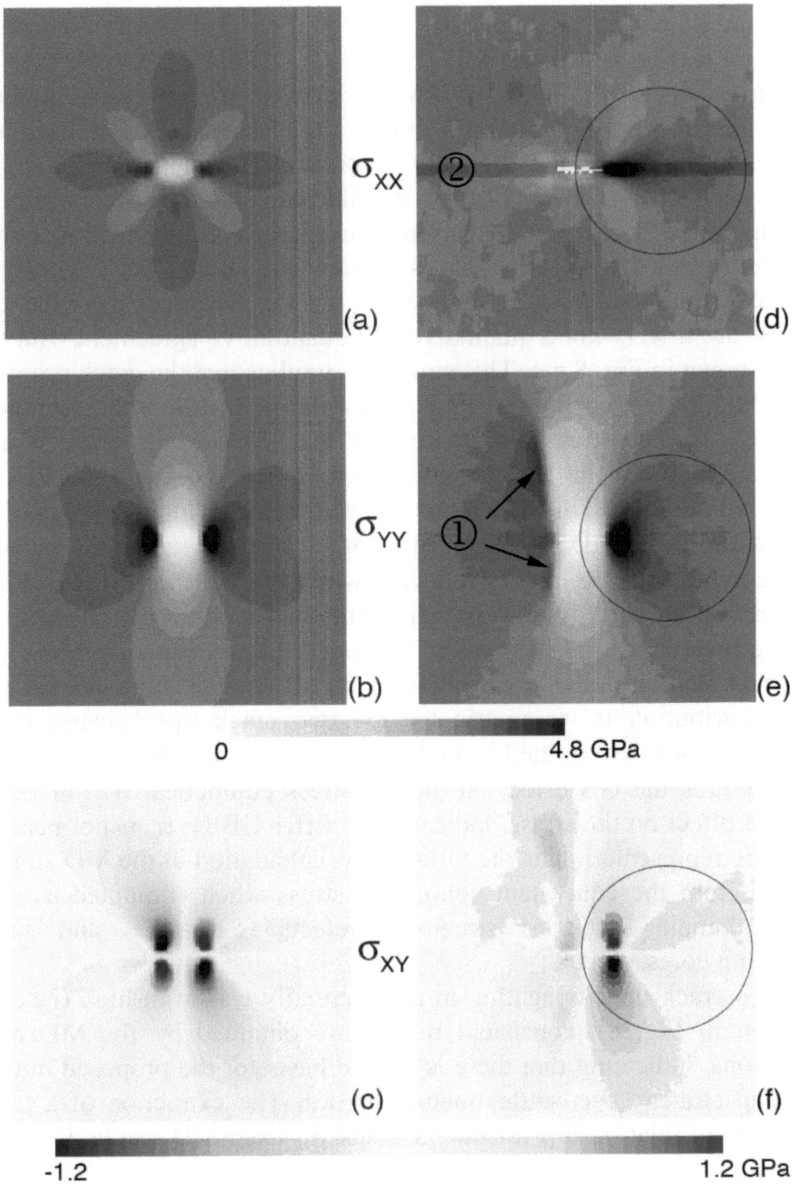

Figure 8. Stress contours from the FEM model (a-c) (corresponding to Fig. 7) and the corresponding stress maps (d-f) from the MD model (corresponding to Fig. 6.a at 3.5 GPa prestrain load) for σ_{xx}, σ_{yy}, and σ_{xy} stress components. Positive and negative stress are indicated by red and blue, respectively. In (a), (b), (d), and (e), the negative stress in blue is defined as the stress in tension. In (c) and (f), positive and negative shear corresponds to shear directions relative to the GB interface as shown at the two sides of the σ_{xy} stress indicator.

Figure 8 presents a comparison between the stress distributions obtained from the FE (Fig. 8.a-c) and MD (Fig. 8.d-f) simulations for the case of the small prestress of 3.5 GPa, which produces a 10 nm crack in the MD simulation. In the MD model, the crack is in elasto-plastic equilibrium with its neighboring material. The size of the crack is sufficiently small compared to the system size, and the elastic stress field calculated by the corresponding FE simulation does not experience edge effects from the boundary conditions, thus this system behaves as an infinite plate. The MD results for stress distribution shown in Fig. 8.d-f for the $+x$ crack tip (the circled regions) are in very good qualitative and quantitative agreement with the FE results shown in Fig. 8.a-c. This similarity implies that the crack propagation in the MD system in the $+x$ direction is almost brittle with essentially no plasticity. For the stress distribution for the $-x$ crack tip, there is a large difference between the MD results shown in Fig. 8.d-f and the FE results shown in Fig. 8.a-c. This difference shows that the twinning observed at the $-x$ crack tip in the MD system significantly alters the stress; the σ_{xx} and σ_{xy} stress components are largely relieved, while the σ_{yy} stress ((1) in Fig. 8.e, compare with (1) in Fig. 6.a) is redistributed away from the crack tip along the twin boundaries (see Fig. 6.a). This redistribution explains the very slow crack propagation in the $-x$ direction of the MD model. As expected, the FE stress distribution is symmetric for the two crack tips because they are elastically equivalent in the FE model (Fig. 8.a-c). In addition, the MD stress field captures the GB effect on the σ_{xx} stress component ((2) in Fig. 8.d). This GB effect on the stress, indicating a stiffer GB layer, is not necessarily a realistic representation as the virial stress calculation in the MD simulation deviates from the equivalent continuum stress when computed over small atomic domains such as structural defects (e.g., GBs and surfaces, dislocation cores, etc.).[29]

For a crack tip propagating in an essentially elastic fashion (i.e., the $+x$ crack tip in Fig. 8), consistent results are obtained by the MD and FE simulations, indicating that there is a solid basis for the proposed multiscale modeling strategy for brittle fracture modes. The extraction of a traction-displacement relationship for the $+x$ crack tip from MD results is discussed next.

3.2.4 Defining a Traction-Displacement Relationship from MD

Grain-scale simulations that use CZM to study fracture (such as those presented in reference 10) typically use heuristically derived relationships and input values to define the CZM. In such an approach, these values tend to be only gross estimates of parameters such as GB yield stress; however, a cohesive formulation can be part of an effective physics-based approach if constitutive parameters are used that optimize the similitude between atomistic simulation and continuum finite element results. The model can be developed from results of the MD analyses allowing the physical insight of MD to be embedded in the more computationally efficient FE models.

The traction-displacement function of a CZM describes how the traction τ developed at the crack surface depends on the crack opening λ (see Fig. 1.b). In general, traction-displacement functions must be defined for both normal and tangential components of traction. Here, the special case of normal opening (mode I) under hydrostatic load is considered. The tensile component of the normal stress near the debonding GB interface from the MD simulation, σ^s_{yy}, can be recast as the normal traction τ_n in the CZM. To obtain the mode I traction-displacement relationship, the MD simulation box is divided into thin subdomains of length Δ, as shown in Fig. 9. A set of test volumes or cohesive zone volume elements (the dotted areas in Fig. 9) of dimensions Δ x 2Δ x h centered at the debonding interface are defined. For each subdomain along the GB interface, the normal opening of the crack δ_n as a function of x is estimated, and the local normal stress σ_{yy} is averaged to obtain σ^s_{yy} as a function of x. The opening of the crack δ_n as a function of x can be estimated by calculating the radius of gyration R_g (or the angular momentum) of each subdomain. For example, during crack propagation, the opening δ_n at the middle of the dashed subdomain in Fig. 9 will grow, thereby changing R_g of this subdomain. If a uniform mass distribution is assumed over the rest of the subdomain, R_g is given as the second moment of the mass distribution on the subdomain

$$R_g^2 = \frac{\int_{\delta_n/2}^{W/2} y^2 dy}{\int_{\delta_n/2}^{W/2} dy} = \frac{1}{12}\left(W^2 + W\delta_n + \delta_n^2\right). \tag{3}$$

Another way to calculate R_g is by counting the atoms that belong to the subdomain:

$$R_g^2 = \frac{1}{N}\sum_{i=1}^{N}(y_i - y_C)^2 = \frac{1}{N}\sum_{i=1}^{N} y_i^2 - \left(\frac{1}{N}\sum_{i=1}^{N} y_i\right)^2 \tag{4}$$

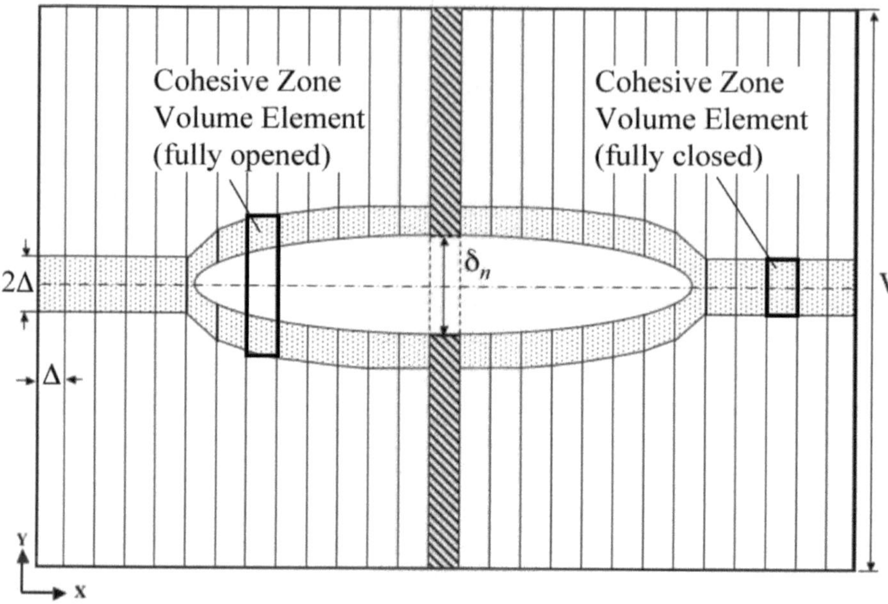

Figure 9. Schematic diagram of the slicing of the system volume in the MD simulation and defining the representative regions for extracting the parameters for the cohesive-zone interface elements in a continuum simulation as explained in the text.

where y_i is the y-coordinate of atom i, N is the number of atoms in the subdomain, and $y_C = \frac{1}{N}\sum_{i=1}^{N} y_i$ is the y-coordinate of the mass center of the subdomain. Knowing R_g from Eq. 4, the crack opening at the subdomain is the positive root for δ_n in Eq. 3, which becomes

$$\frac{1}{12}\left(W^2 + W\delta_n + \delta_n^2\right) = \frac{1}{N}\sum_{i=1}^{N} y_i^2 - \left(\frac{1}{N}\sum_{i=1}^{N} y_i\right)^2. \tag{5}$$

Thus, calculating $\sigma^s_{yy}(x)$ and $\delta_n(x)$ for each cohesize zone volume element placed at position x along the debonding interface, one can develop the functional relationship $\sigma^s_{yy}(\delta_n)$, which represents the traction-displacement curve of a CZM volume element (CZVE) of size Δ x 2Δ x h. By definition, Δ must be small enough so that the stress variation over the CZVE is much smaller than the stress itself. In the present simulation, $\Delta \approx 1.9$ nm, which, with $h = 2.86$ nm, gives a volume of 2 x 1.9^2 x 2.86 = 20.7 nm^3 and contains about 1250 atoms that contribute to the stress-displacement response of one CZVE. There are 63 CZVE along the 120 nm long GB interface in the system (Fig. 3). During the simulation, the $\sigma^s_{yy}(\delta_n)$ state is scanned every 4 ps for a period of 200 ps, resulting in 3000 values from which to determine the $\sigma^s_{yy}(\delta_n)$ dependence for each simulation run.

Fifteen hundred $\sigma^s_{yy}(\delta_n)$ values are taken from the CZVE on the right half of the model for the case of 4.25 GPa preload and are presented as points in Fig. 10. These points represent the behavior of the crack as it propagates in the +*x* direction (see Fig. 8). The +*x* direction tip propagates more than 50 nm (Fig. 3), causing 25 CZVE to experience a complete transition from a fully closed to a fully opened state. The $\sigma^s_{yy}(\delta_n)$ traction-displacement curve for the 4.25 GPa preload is extracted by taking a moving average of the $\sigma^s_{yy}(\delta_n)$ values over $0.0 \le \delta_n \le 4.0$ nm. The spread of the data points shows that at this small length scale of 2 nm there is a large dispersion around the mean traction-displacement curve. This process is repeated for the cases of 3.75 GPa and 4.0 GPa preload, and the resulting $\sigma^s_{yy}(\delta_n)$ curves are also shown in Fig. 10 (note that the corresponding $\sigma^s_{yy}(\delta_n)$ values are omitted from the figure). The curves in Fig. 10 show a bilinear type of constitutive relation for the CZVE similar to the CZM suggested by Camacho and Oritz[8] and indicate that for the case of almost brittle crack propagation in the +*x* direction for a Σ99 symmetric-tilt GB in FCC metals, this bilinear model is a good assumption. In addition, Fig. 10 shows that $\sigma^s_{yy}(\delta_n)$ changes little with the prestress of the system over the range 3.75 GPa $\le \sigma \le$ 4.25 GPa and is approximately $\sigma_p \approx 5.0$ GPa at $\delta_n^p \approx 0.4$ nm. Full opening (debonding) of the interface occurs at $\delta_n^c \approx 2.5$ nm, when σ^s_{yy} becomes zero.

4. DISCUSSION

This chapter discusses a multiscale modeling strategy that may be used to study metallic fracture. It is based on embedding atomic-scale information into finite element simulations by means of cohesive zone models. The ability of the cohesive zone models to naturally incorporate grain-boundary sliding and decohesion, as shown in this study, makes them a powerful tool to represent the grain boundary behavior in an elasto-plastic FE model that can successfully predict deformation and fracture in a polycrystalline microstructure. Through the cohesive zone model's constitutive law, the actual parameters of the different plastic deformation processes can be embedded, which in turn, can be quantified by MD simulation, rather than being deduced empirically from mechanical testing.

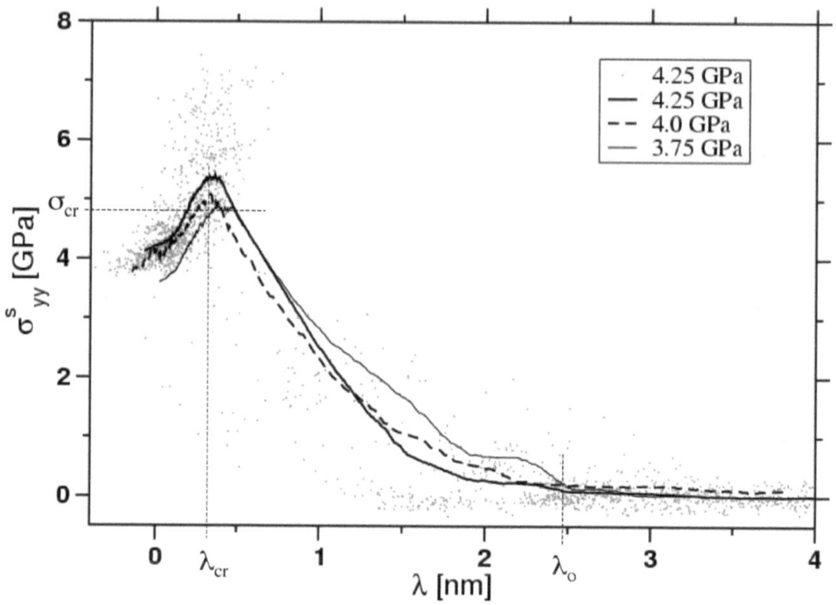

Figure 10. Surface stress vs. crack opening curves, $\sigma^s_{yy}(\delta_n)$, characterizing the propagation of the cleavage tip in the $+x$ direction for three prestrain loads.

The success of the presented multiscale modeling strategy depends crucially on the ability to extract reliable parameters for the constitutive relation, which in this case is the traction-displacement relationship for the CZM elements. The traction-displacement relationship has to incorporate the mechanical properties of the GB in order to reproduce the GB response during the deformation of the polycrystal. This work illustrates the way to extract these properties for two specific modes of deformation, GB sliding and normal opening, for one chosen high-angle GB in aluminum. Being very specific, this data, shown in Figs. 4 and 10, is still not enough to construct a complete FE model for intergranular fracture in a general Al polycrystal. Additional work is needed to study the dependence of the traction-displacement relationship on GB crystallography and structure, as well as on different modes of loading. The outlined simulation conditions of hydrostatic preloading at cryogenic (100 K) temperature differ substantially from the more common condition of uniaxial loading at room temperature. While the chosen conditions facilitate the MD study by stabilizing the system at nanometer-nanosecond length and time scales, an important issue is how the obtained results would relate to more general situations. The estimated peak stress of debonding $\sigma_p \approx 5.0$ GPa has to be considered in view of the nanometer dimensions of the simulated crack. The scaling of this value for the case of a micron or millimeter size crack that may span many grains and exhibit more complex deformation mechanisms is also a relevant issue to be addressed.

Modeling of materials failure by CZM elements has been advanced to the level to be able to perform large scale simulations of fracture in polycrystals. Recently, Zavattieri et al.[30] have studied fracture and failure in alumina ceramic microstructures subjected to multi-axial dynamic loading. The effective size of the studied polycrystalline specimen was 0.54 mm, reaching macroscopic scales. A bilinear traction-displacement relationship was used and parameterized to the empirical data for alumina. Zavattieri and Espinosa[31] have used a modification of the same model to study interface effects of the alumina specimen in contact with steel plates. Wei and Anand[32] have used a modified CZM model to study intergranular fracture in nanocrystalline Ni. In their FEM simulation the CZM element accounted for both reversible and irreversible inelastic sliding-separation deformations at the grain boundaries prior to failure. The parameterization of the model was, again, performed by using available experimental data on nanocrystalline Ni with grain size of 15-40 nm. Iesulauro et al. have applied the CZM technique to simulate fatigue crack initiation in Al polycrystals.[10] The use of MD simulations to parameterize the traction-displacement curve was suggested, but the actual parameterization was performed by using the known macroscopic yield properties of aluminum.

In view of the above noted large-scale models,[10,30-32] of material fracture implementing empirically parameterized CZM elements, it is possible that a methodology to connect atomistic MD models with the CZM technique, as presented in this chapter, would allow more realistic simulations that may lead to actual predictions of the failure properties of a large class of materials and microstructures, even when experimental data is not available or is difficult to obtain.

ACKNOWLEDGEMENTS

V. Yamakov and D. R. Phillips were sponsored through cooperative agreement NCC-1-02043 with the National Institute of Aerospace and contract NAS1-00135 with Lockheed Martin Space Operations, respectively.

REFERENCES

1. J. Q. Broughton, F. F. Abraham, N. Bernstein and E. Kaxiras, Concurrent coupling of length scales: methodology and application, *Phys. Rev. B* **60**, 2391-2403 (1999).
2. F. F. Abraham, N. Bernstein, J. Q. Broughton and D. Hess, Dynamic fracture of silicon: concurrent simulation of quantum electrons, classical atoms, and the continuum solid, *Mater. Res. Soc. Bull.* **25**, 27-32 (2000).

3. N. M. Ghoniem and K. Cho, The emerging role of multiscale modeling in nano- and micro-mechanics of materials, *CMES: Comp. Model. Eng. Sci* **3,** 147-173 (2002).
4. E. B. Tadmor, M. Ortiz and R. Phillips, Quasicontinuum analysis of defects in solids, *Phil. Mag. A* **73**, 1529-1563 (1996).
5. R. E. Miller and E. B. Tadmor, The quasicontinuum method: overview, applications, and current direction, *J. Comp. Aided Mat. Design* **9**, 203-239 (2002).
6. S. Shen and S. N. Atluri, Multiscale simulation based on the meshless local petrov-galerkin (MLPG) method, *CMES: Comp. Model. Eng. Sci.* **5**, 235-255 (2004).
7. V. Tvergaard and J. W. Hutchinson, The relation between crack growth resistance and fracture process parameters in elastic-plastic solids, *J. Mech. Phys. Solids* **40**, 1377-1397 (1992).
8. G. T. Camacho and M. Ortiz, Computational modeling of impact damage in brittle materials, *Int. J. Solids Struct.* **33**, 2899-2938 (1996).
9. P. Klein and H. Gao, Crack nucleation and growth as strain localization in a virtual-bond continuum, *Eng. Fract. Mech.* **61**, 21-48 (1998).
10. E. Iesulauro, A. R. Ingraffea, S. Arwade, and P. A. Wawrzynek, Simulation of grain boundary decohesion and crack initiation in aluminum microstructure models, in *Fatigue and Fracture Mechanics: 33rd Volume, ASTM STP 1417,* (W. G. Reuter and R. S. Piascik, Eds., American Society for Testing and Materials, West Conshohocken, PA, 2002).
11. V. Tvergaard and J W. Hutchinson, The influence of plasticity on mixed-mode interface toughness, *J. Mech. Phys. Solids* **41**, 1119-1135 (1993).
12. J.C.J. Schellekens and R. de Borst, On the numerical integration of interface elements, *Int. J. Numer. Meth. Enng.* **36**, 43-66 (1992).
13. P.P. Camanho and C.G. Davila, Mixed-mode decohesion finite elements for the simulation of delamination in composite materials, NASA/TM-2002-211737.
14. A. Turon and C.G. Davila, An engineering solution for using coarse meshes in the simulation of delamination using cohesive zone models, NASA/TM-2005 (to be published).
15. D. Raabe, *Computational Materials Science: The Simulation of Materials Microstructures and Properties*, (Wiley-VCH, Weinheim, 1998).
16. M.S. Daw, S.M. Foiles and M.I. Baskes, The embedded-atom method: a review of theory and applications, *Mat. Sci. Reports* **9**, 251-310 (1992).
17. J. Schiotz and K.W. Jacobsen, A maximum in the strength of nanocrystalline copper, *Science* **300**, 1357-1359 (2003).
18. F. F. Abraham, The atomic dynamics of fracture, *J. Mech. Phys. Solids* **49**, 2095-2111 (2001).
19. Y. Mishin, D. Farkas, M. J. Mehl and D. A. Papaconstantopoulos, Interatomic potentials for monoatomic metals from experimental data and ab initio calculations, *Phys. Rev. B.* **59**, 3393-3407 (1999).
20. D. Wolf, Structure-energy correlation for grain boundaries in fcc metals – III. Symmetrical tilt boundaries, *Acta Metal.* **38**, 781-790 (1990).
21. U. Dahmen, J. D. Hetherington, M. A. O'Keefe, K. H. Westmacott, M. J. Mills, M. S. Daw and V. Vitek, Atomic structure of a Σ99 grain boundary in Al: a comparison between atomic-resolution observation and pair-potential and embedded-atom simulations, *Phil. Mag. Lettrs*. **62**, 327-335 (1990).
22. V. Yamakov, D. Wolf, S. R. Phillpot, A. K. Mukherjee and H. Gleiter, Dislocation processes in the deformation of nanocrystalline Al by molecular-dynamics simulation, *Nature Materials* **1**, 45-48 (2002).

23. V. Yamakov, D. Wolf, S. R. Phillpot and H. Gleiter, Dislocation-dislocation and dislocation-twin reactions in nanocrystalline al by molecular-dynamics simulation, *Acta Mater*. **51**, 4135-4147 (2003).
24. P. Gumbsch, S.J. Zhou and B.L. Holian, Molecular dynamics investigation of dynamic crack stability, *Phys. Rev. B* **55**, 3445-3455 (1997).
25. C.G. Dávila, Mixed-mode decohesion elements for analysis of progressive delamination. 42nd AIAA/ASME/ASCE/AHS/ASC Structures, Structural Dynamics, and Materials Conference and Exhibit, Seattle, WA April, 16-19 2001, article: AIAA-01-1486.
26. J.D. Honeycutt and H.C. Andersen, Molecular dynamics study of melting and freezing of small Lennard-Jones clusters, *J. Phys. Chem.* **91**, 4950-4963 (1987).
27. E. B. Tadmor and S. Hai, A Peierls criterion for the onset of deformation twinning at a crack tip, *J. Mech. Phys. Solids* **51**, 765-793 (2003).
28. M. Parrinello and A. Rahman, Polymorphic transitions in single crystals; a new molecular dynamics method, J. *Appl. Phys.* **52**, 7182-7190 (1981).
29. J.A. Zimmerman, R. E. Jones, P.A. Klein, D.J. Bammann, E.B. Webb III and J.J. Hoyt, Continuum definitions for stress in atomistic simulation, SAND Report, SAND2002-8608.
30. P.D. Zavattieri, P.V. Raghuram and H.D. Espinosa, A computational model of ceramic microstructures subjected to multi-axial dynamic loading, *J. Mech. Phys. Solids* **49**, 27-68 (2001).
31. P.D. Zavattieri and H.D. Espinosa, An examination of the competition between bulk behavior and interfacial behavior of ceramics subjected to dynamic pressure-shear loading, *J. Mech. Phys. Solids* **51**, 607-635 (2003).
32. Y.J. Wei and L. Anand, Grain-boundary sliding and separation in polycrystalline metals: application to nanocrystalline fcc metals, *J. Mech. Phys. Solids* **52**, 2587-2616 (2004).

Chapter 11

MULTISCALE MODELING OF DEFORMATION AND FRACTURE IN METALLIC MATERIALS

Diana Farkas[1] and Jeffrey M. Rickman[2]
[1]*Department of Materials Science and Engineering, Virginia Tech, Blacksburg, VA 24061, USA;* [2]*Department of Materials Science and Engineering and Department of Physics, Lehigh University, Bethlehem, PA 18015, USA*

Abstract: This chapter discusses multiscale modeling methods for the study of deformation and fracture in metallic materials. Both atomistic and dislocation dynamics modeling are outlined in the context of problems in materials failure. In particular, molecular statics and dynamics models are described as applied to polycrystalline samples of random grain orientations and grain sizes in the nanometer regime. The application of dislocation dynamical models to mesoscale dislocation dynamics is then presented. Finally, dislocation pattern formation at various length scales is discussed as an illustration of multiscale modeling.

Keywords: multiscale modeling; molecular dynamics; nanocrystalline structures; dislocation dynamics; fracture.

1. INTRODUCTION

The processes involved in the deformation and fracture of metals span multiple length and time scales and involve interactions among various solid-state defects. For example, the plastic deformation behavior of a metal is tied to the generation and subsequent motion of mutually interacting dislocations, leading to the formation of dislocation substructures having disparate length scales. Other material failures, such as those due to fracture, occur by the nucleation of cracks whose growth kinetics are governed by atomic-level processes near the crack tip, as well as by crystallography and macroscopic elastic energy considerations.

Multiscale computer simulation is playing an increasingly important role in elucidating the failure mechanisms underlying deformation and fracture. In particular, atomistic simulation permits one to model processes at the nanometer length scale and picosecond time scale that are relevant at a crack tip (e.g., dislocation emission, blunting) under different loading conditions and to describe the dislocation patterns formed during plastic fracture. At larger length scales and longer time scales dislocation dynamics (DD) simulations, wherein line defects rather than atoms are the fundamental objects, are employed to describe the overdamped motion and entanglement of dislocation lines associated with highly-worked metals. As the presence of point and extended defects can inhibit dislocation motion and, hence, influence yield strength, DD simulation is also used to describe, for example, line-defect interactions with solutes, precipitates and grain boundaries.

In this article we summarize the application of these techniques to the modeling of deformation and fracture in metals, and discuss practical strategies for linking atomic-level simulation and DD with each other and with the continuum. We begin with a discussion of both atomistic and mesoscale DD simulation methodologies in the context of metallic deformation. Examples are given to highlight insights obtained from these simulations. We then discuss dislocation pattern formation at different scales using atomistic and DD simulations. Finally, we conclude with a brief summary of the contents of this chapter.

2. ATOMISTIC SIMULATION

2.1 Overview

Atomistic simulation studies of deformation and fracture are aimed both at addressing practical problems in materials engineering and at enhancing our understanding of the fundamental mechanisms underlying these phenomena. For example, one goal of such studies is the development of computational tools to predict the fracture toughness of materials as a function of relevant experimental parameters, including: composition, microstructure, temperature, and loading conditions. Computer simulation of fracture and deformation in single crystals has also provided new insight into the stability of crack propagation, the phenomenon of lattice trapping, and the origins of both brittle and ductile behavior. Most recently, simulation studies of nanocrystalline solids have proven to be important research tools for investigating fracture and deformation mechanisms in these materials. Indeed, large-scale simulations, enabled by increasing computational power,[1,2,3] can shed new light on phenomena peculiar to the nanoscale and permit comparison with experimental observations.[4,5]

The emergence of new phenomena at the nano-scale[6,7] creates a great need for theory, modeling and large–scale computer simulation to guide materials development and to understand better material behavior in this regime. For example, computer simulation has played an important role in the development of new nano-scale materials and devices, using the "materials by design" concept.[8,9] However, although considerable progress has already been made in understanding the mechanical response of relatively simple systems, the development of a theory describing structure/property connections at the nano-scale remains as a central challenge in this field.

Before describing the atomistic simulation methodology, we wish to emphasize two points. First, in addition to first principles (FP) *ab initio* and large–scale computational approaches, other phenomenological models are needed to connect atomic-level descriptions with the mesoscale and the continuum. The problem of metallic deformation often involves microstructural features (e.g., interfaces, triple junctions), as well as an interplay among other lattice defects. Various scales of theory, simulation and modeling are thus essential in understanding behavior at the nanoscale.[10,11,12] In short, what is required is a multiscale modeling approach to the problem. In this section we focus on the application of atomistic modeling to materials deformation problems.

Second, it is worth noting that the reliability of any atomic-level description depends critically on the accuracy of the potential chosen to represent interatomic interactions.[13] Most classical potentials have been developed to reproduce a material's equilibrium bulk properties, which depend mostly on the shape and curvature of the potential near its minimum. By contrast, the behavior of a material under loading depends critically on its mechanical response in regions where chemical bonds are stretched, sheared or compressed, and bonding geometry thereby distorted, so that interactions are governed by the shape of the potential far away from its minimum.

Despite these concerns, we note that the behavior of many metallic materials, particularly those with FCC structure, are well described with reasonable accuracy by computationally efficient many-body semi-empirical potentials such as the Embedded Atom Method (EAM)[14] and Effective Medium Theory (EMT).[15] These potentials have been developed by fitting to FP calculations, as described by Mishin *et al.*,[13] Ercolessi *et al.*[16] and more recently by Mendelev *et al.*[17]

2.2 Atomistic Simulation Methodology

After selecting an interatomic potential, a simulation study of deformation and fracture in a polycrystal proceeds by first identifying the initial configuration of a digital sample to be deformed or fractured and then prescribing the loading geometry and the appropriate boundary conditions. If the initial configuration corresponds to a polycrystal, a model grain structure can be constructed using a variety of algorithms, including the Voronoi construction.[18] To cause the sample to fracture or deform, we apply a strain or stress and then maintain, increase or decrease it as the system evolves in time. This goal is accomplished through the use of constraints, which are typically applied via the system's boundary conditions.

Two atomistic simulation techniques can be employed in simulations of deformation and fracture, namely molecular statics and molecular dynamics. Molecular statics (MS) is designed to determine the lowest energy configuration of a system subject to applied constraints. In this approach, every atom within the simulated system interacts with its neighbors according to an interaction potential, and the presence of a defect typically induces forces in an unrelaxed system. The non-constrained atoms are moved using an iterative relaxation process to bring the system to a minimum energy configuration. For example, the conjugate gradient minimization technique involves the movement of atoms along the direction of the steepest gradient of the energy function, i.e., in the direction of greatest energy decrease. Thus, in each single iteration step, the atom is displaced in the direction of the resultant force applied by its neighbors as well as in a direction perpendicular to its previous displacement. The energy is computed after each iteration, and equilibrium is achieved when the forces on each atom are below a minimum desired value.

Once the system has equilibrated, the applied strain, stress or stress intensity is varied and the system is again relaxed to elastic equilibrium. This procedure is repeated until the process being studied, e.g. deformation or the motion of a crack across the sample, reaches the desired point. The MS method represents the quasistatic evolution of the system under a slowly varying strain, but has the disadvantage that it does not treat atomic vibrations and the effects of finite temperature. Since no atomic vibrations due to thermal activation are taken into account, the results obtained only represent the material behavior at 0 K.

The molecular dynamics (MD) methods allows the introduction of strain rate and temperature as simulation variables. In a MD simulation, the initial configuration includes the position, mass, and velocity of each atom. The initial velocities are typically selected from a random distribution (e.g. the Maxwell-Boltzmann distribution) associated with a specific initial temperature; if a different random velocity distribution is used, the velocities will evolve toward the Maxwell-Boltzmann distribution in the first steps of the simulation.

A simulation proceeds by calculating the forces acting on each atom from its neighbors and integrating the equations of motion to show the actual time evolution of the system. The associated integration time step, Δt, must be small, typically in the femtosecond range. To control the temperature in the simulation, one can use a thermostat algorithm such as the Nose-Hoover device. Alternatively, one can rescale the velocities of all atoms to consistently correspond to the desired temperature.

If periodic boundary conditions are used, stress or hydrostatic pressure can be similarly controlled through the use of a barostat which allows the simulation cell size, and thus the attendant strain, to fluctuate. Care should be taken when using both thermostats and barostats since they may introduce fluctuations in pressure and temperature. The application of a barostat can be implemented separately in the three different directions of space. This is extremely important in deformation studies with controlled strain. For example, the strain may be applied in one direction and the barostat in the other two directions to maintain zero pressure and therefore obtain the correct system response in a plastic regime.

The main drawback of the molecular dynamics technique in this context is the short time scale accessible. For studies of deformation this shortcoming implies that the corresponding strain rates are extremely high and, since the strain rate may affect the deformation mechanism, it follows that MD, although of substantial utility, can give only part of the deformation picture. On the other hand, the molecular statics alternative yields quasi-equilibrium configurations at various stress intensities, and is therefore a better model of stable crack growth.

More generally, due to their size and time-scale limitations, atomistic simulations can provide only limited information about the fracture and deformation behavior of a bulk solid. Accurate atomistic simulations need to be linked with simulations at other length scales to form a more complete picture. This may be accomplished, at least in part, by employing some relatively simple strategies. These are as follows: (a) incorporating interatomic potentials based on calculations performed at the quantum theory level; (b) linking the output of atomic-level simulation with the input to larger scale simulations (i.e., dislocation dynamics), and imposing continuum-level boundary conditions on an atomistic simulation cell.

2.2.1 Connection with *ab initio* calculations

Classical interatomic potentials describe the energy associated with chemical interactions among atoms of the same or differing species as a function of interatomic distance in a simplified form that is computationally efficient for implementation. One way to bridge length scales is, in a sense, to use interatomic potentials derived from FP simulations of impurity effects, mostly using *ab initio* density functional theory in the linear augmented plane wave (LAPW) approach.[13] These calculations can be performed for cluster sizes of 10 to 50 atoms, and they must be bridged in some way to techniques at a larger scale.

Accurate modeling of a crack tip region, or any other location where strains are large, requires good fitting of the potential to high-energy configurations. It is therefore important to use a description of interatomic interactions that, though empirical, can be reliable for these off-equilibrium configurations. Experimental information is usually linked to situations that only deviate slightly from equilibrium, mostly in the elastic regime. With the exception of the activation energy for diffusion, there are few bulk experimental properties that can provide information on energetic interactions far from equilibrium, thereby making FP calculations central to the development of accurate interatomic potentials. In particular, accurate representations of surface energy as a function of crystallographic orientation and unstable stacking fault energies are critical. For example, the former property determines, to a large degree, the types of cleavage planes observed. Consequently, a viable interatomic potential should enable one to capture these extended defect energetics.

2.2.2 Interface with dislocation dynamics and the continuum

There are various strategies for linking atomic-level simulations of deformation and fracture with the mesoscale and the continuum. From a mesoscale point of view, plastic deformation in a crystalline material proceeds by simultaneous sliding on available slip planes. More specifically, slip systems on which the resolved shear stress exceeds a certain threshold value are assumed to be active, and dislocation sliding on these systems determines the overall plastic deformation of the body. Thus, atomistic simulation can provide critical input data to mesoscale, dislocation dynamics (DD) simulations including, for example, critical resolved shear stresses and dislocation mobilities. Given the degree of spatial and temporal coarse graining inherent in DD simulations, this data must be supplied to the simulation from an external source. A more complete discussion of the DD methodology is given in Sec. 3 below.

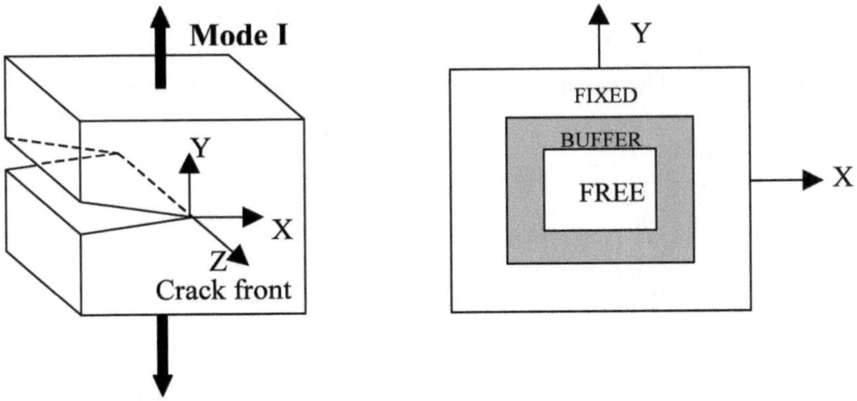

Figure 1. Example illustration of block geometry for a fracture simulation.

One important link between atomistic simulation of deformation and fracture and the corresponding continuum description is via boundary conditions. In particular, continuum fracture mechanics provides the boundary conditions necessary for fracture simulations in many cases, and appropriate parameter values can be found by an iterative process. In the case of plastic deformation, the goal is to find a self-consistent solution for which the criteria for the onset of plasticity used in the macroscopic calculations of the boundary conditions are consistent with the corresponding atomistic results. We note that this is a very efficient way to interface atomistic and continuum calculations that uses mostly existing code, without the need for the development of so-called handshaking procedures.

As an illustration of this procedure, Fig. 1 shows a schematic of the application of boundary conditions to an atomic-level fracture simulation cell. In this figure we indicate the possibility of introducing a buffer region of atoms that is at an intermediate distance from the crack tip. The buffer atoms do not move independently, but their positions can be adjusted according to the forces they experience from the free atoms. While the buffer region is not necessary, it permits the use of smaller simulation cells while mitigating spurious effects caused by the fixed region. If a buffer region is not used, larger simulation cells will be necessary to avoid effects from the fixed region.

The introduction of the crack is performed using the continuum solutions for all the regions indicated in Fig. 1. The role of the continuum solution in this context is twofold. First, it serves as an initial guess for the relaxed atomic configuration in all the regions of the simulation cell. Second, it provides boundary conditions that are maintained in the fixed region far from the crack tip. In the simplest case, the boundary conditions are deduced from the displacement field, **u**(**r**), of a semi-infinite crack in an isotropic medium. The components of the displacement field are:[19]

$$u_x = \frac{K}{2\mu}\sqrt{\frac{r}{2\pi}}\cos\frac{\theta}{2}\left(2-4\nu+2\sin^2\frac{\theta}{2}\right) \tag{1}$$

$$u_y = \frac{K}{2\mu}\sqrt{\frac{r}{2\pi}}\sin\frac{\theta}{2}\left(4-4\nu-2\cos^2\frac{\theta}{2}\right) \tag{2}$$

where r and θ are referred to a coordinate system located at the crack tip. Using this isotropic approximation, simulations of more complicated polycrystalline and multiphase systems can also be performed since the continuum solutions, to first approximation, are independent of the detailed crystal configuration in the atomistic simulation block. Finally, we note that since the continuum solution should be valid far from the crack tip, the atomic positions in the fixed region are invariant during the energy minimization procedure in molecular statics or during a certain number of molecular dynamics time steps in the MD technique. As the simulation proceeds, the loading can be increased, and this is accomplished by an appropriate modification of the boundary conditions.

Since the MD technique follows the actual forces on the atoms as they migrate, fracture mechanisms can be determined by direct observation, without the need for any *a priori* assumptions. In a typical MD simulation, an increase in stress intensity (up to about three times the Griffith value) leads to crack advance that is monitored over many time steps. As the simulated strain rate is artificially high, MS simulations of the same samples are conducted to assess the impact of strain rate on observed mechanisms. In many cases, energy minimization studies confirm fracture mechanisms observed in MD.

Dynamic fracture simulations show both crack advance and blunting behavior. The latter behavior is exhibited in Fig. 2 which shows typical examples of atomistic simulations of fracture in the initial condition (left), where the displacements around the crack tip follow those of the continuum theory, and after the crack has blunted (right). Such simulations have been used in the study of fracture in Fe, Cu and Al single crystals.[20-26]

2.3 Case Study: Fracture of Nanocrystalline Ni

As a case study, we present here some results on the fracture behavior of nanocrystalline Ni. The techniques for fracture simulations in fully three dimensional polycrystalline Ni are described by Farkas et al.[18,27] Figures 3 and 4 show some results obtained using similar techniques. Figure 3 (left) shows the emission of a dislocation loop (seen in dark) from the crack tip. Blunting in the crack tip is also observed as well as the emission of straight partial dislocations that travel through the grain towards a nearby grain boundary. This latter mechanism is also seen in Figure 3 (right).

Furthermore, twinning is observed in the vicinity of the crack tip as a deformation mechanism. This is illustrated in Figure 4 (left). Finally, the observations of the crack propagation mechanisms show the formation of nano-voids ahead of the main crack, as shown in Figure 4 (right). In addition, we have performed fracture simulations in columnar crystals with a common <110> orientation. These results are shown in Figure 5, where the formation of a nano-void/nano-crack ahead of the main crack can be observed. These results also enabled us to observe that grain boundaries can arrest the crack, and cause blunting, as it propagates. This occurs when the boundaries are oriented nearly perpendicular to the direction of crack propagation.

The results point to a mechanism that is mostly intergranular, governed by the formation of nanocracks ahead of the main crack. This seems to be a general feature of nanocrystals, since it was observed also in bcc materials Fe and Mo.[28,29] The appearance of twinning[30] also seems to be a general phenomenon.

Figure 2. Illustration of a fracture simulation in the initial condition (left), and after the crack has blunted (right). Shading is correlated with the stress field.

Quantitative evaluation of simulation data allows comparison between simulation and experiment. One important quantity is fracture resistance. By plotting the applied stress intensity as a function of crack tip position one can obtain crack resistance curves, such as are used in continuum fracture mechanics.

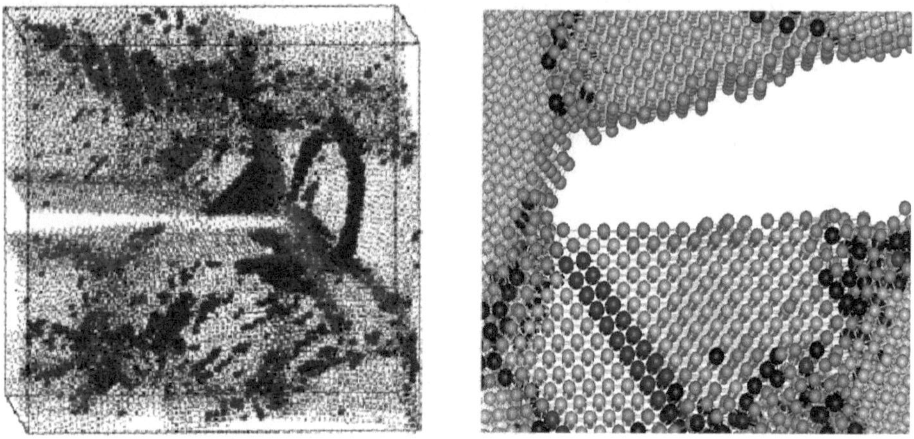

Figure 3. An example simulation of the fracture of nanocrystalline Ni with a 10 nm grain size. Grain boundary and surface atoms are shown in green. The presence of stacking faultsis indicated by the dark symbols. A dislocation loop can be seen on the left.

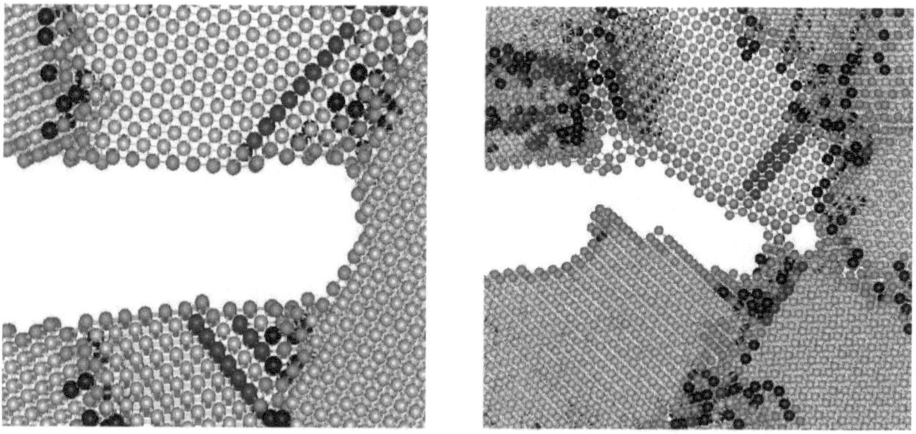

Figure 4. Detail of the twinning process and the emission of Shockley partials in 5 nm grain size Ni (left). Formation of a nano-void ahead of the main crack (right).

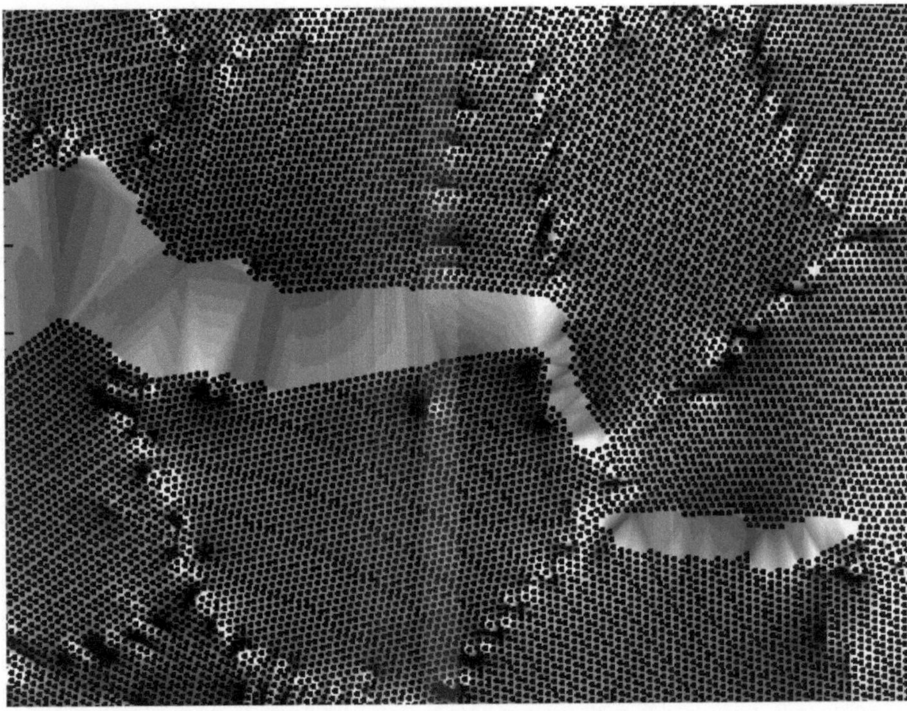

Figure 5. Fracture of nano-crystalline Ni in a columnar grain structure with a <110> common columnar axis. Note the formation of a nano-crack ahead of the main crack.

These curves give information on how the crack advances as increasingly higher loading is applied, including the effects of the plastic deformation processes that occur simultaneously with crack advance. These curves are particularly useful in studying effects of various parameters of the crack configuration and loading conditions. In the example of nanocrystalline Ni, this technique can be used to study the effect of grain size on fracture resistance. The results are shown in Figure 6, where increased fracture resistance is shown with crystallographic orientation, temperature, or grain size in polycrystalline decreasing grain size.

3. DISLOCATION DYNAMICS SIMULATIONS OF DEFORMATION

In cases where substantial plastic deformation has occurred, the dislocation density can be rather large, and it becomes increasingly difficult

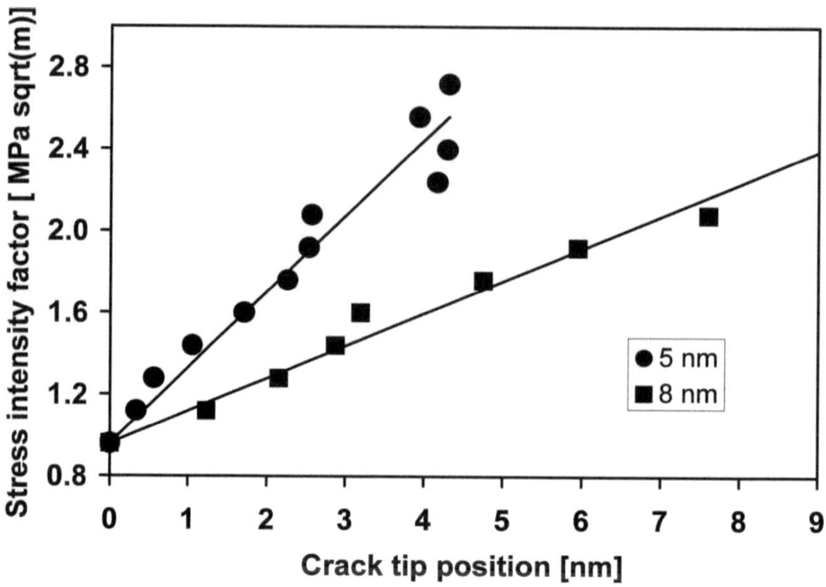

Figure 6. Fracture resistance curves from simulation studies of nanocrystalline Ni with two different average grain sizes (5 nm and 8 nm).

to model dislocation interactions and motion on the atomic level. An alternative approach to characterizing deformation in this limit is to construct a mesoscale model wherein the defects themselves are the fundamental objects. In this context, one would therefore like to develop a DD model capable of describing the spatio-temporal evolution of a collection of dislocations. Because of the homogenized nature of such a model, a connection with the corresponding atomic-level description requires that information regarding possible slip systems, dislocation mobilities, etc., obtained at the atomic-level serve as input at the mesoscale. In this Section we briefly outline the DD methodology and related issues.

A history of the theory of dislocation dynamics goes back to the early work of Frank[31-33] and others who deduced the inertial response for isolated edge and screw dislocations in an elastically isotropic medium. In the last few years DD simulations have emerged as useful tools for examining the formation and subsequent evolution of dislocation microstructures.[34-38] Indeed, recent computational advances have permitted large-scale dislocation dynamics DD simulations.[39-42] Such simulations enable a description of the net plastic response of a material in terms of an average over the motion of all constituent dislocations.

Several dislocation dynamics simulation methods, in both two and three dimensions, have been developed that account for complex dislocation geometries and/or the motion of multiple, interacting dislocations. Most notable, perhaps, are DD simulation methods based upon front-tracking techniques. Three-dimensional simulations based upon these methods were first performed by Kubin and co-workers[43-44] and later by other researchers.[45-47] In these approaches, dislocation lines are discretized into individual segments, and the calculated Peach-Koehler force[48] on each segment due to all other segments leads to tracked segment motion. Such methods enable realistic simulations of plastic deformation, although the attendant computational cost is high given the necessity of force calculation and tracking for each segment.

One recent alternative to these approaches, namely a DD methodology within a level-set framework,[49,50] mitigates many of the shortcomings inherent in other techniques in that dislocation lines need not be discretized or remeshed as they evolve. In this approach dislocation lines in three dimensions are represented as the intersection of zero contours of scalar functions. The resulting level-set functions are then evolved using a smoothed velocity field, with line motion implicitly determined by the evolution of the functions. The inherent simplicity of the level-set approach makes it easily generalizable to include complex microstructural features. We note that several of its advantages (see below) are shared with so-called phase-field methods for dislocation dynamics.[51-53]

3.1 DD methodologies

3.1.1 Front–tracking approaches

Various methodologies exist for simulating the motion of mutually interacting dislocations in two spatial dimensions. As these methodologies have been discussed in detail elsewhere, we provide only an overview below. For example, a typical, force-based simulation of long, straight dislocation lines proceeds by placing N such defects in a cell, and then moving them in response to the attendant Peach-Koehler forces to reach a lower energy state of the system. In this approach energy is not conserved, and one eventually reaches a constrained equilibrium state. Typically, one assumes that the motion of dislocations is overdamped, and thus there is a connection between dislocation velocity $\mathbf{v}$ and force $\mathbf{F}$ on a dislocation via the Einstein relation $\mathbf{v}= M\mathbf{F}$. (The mobility M is taken to be a scalar here, though a mobility tensor can be employed to distinguish between glide and climb motion.) Since the long-ranged stresses associated with a collection of dislocations give rise to a Peach-Koehler force, a proper treatment of the boundary conditions is essential to avoid spurious dislocation substructures.[38]

One alternative to this Langevin dynamical approach is a kinetic, lattice Monte Carlo simulation[54] wherein one constructs an energetics in terms of the dislocation coordinates. The defect interaction energy determines, in turn, the transition rates in a master equation that governs the evolution of the system. This approach has the advantage that other defects (e.g., misfitting solute atoms) can be easily incorporated into the model once their associated eigenstrains are prescribed. Thus, one can, for example, investigate strain-aging and the creation of Cottrell atmospheres as well as dynamical pinning phenomena associated with the Portevin-LeChatelier effect. Furthermore, fully periodic boundary conditions are readily implemented as one can sum analytically the relevant interaction energies generated by all of the images of the objects in the central simulation cell. In brief, such studies have elucidated different regimes of dislocation motion in the presence of impurities and suggested coarse-graining strategies that permit the elimination of some degrees of freedom from the description.

In three dimensions, topological constraints complicate a description of the motion of a collection of mutually interacting dislocation loops. More specifically, dislocation lines cannot end in the bulk, and so loop closure must be enforced as the system evolves. One especially fruitful approach to this problem has been to adopt a parametric representation in which a given dislocation loop is described in terms of a collection of nodes connected by spline functions.[55] The equations of motion for the loops are obtained by a variational method, and a fast summation technique is employed to calculate the needed forces acting on loop segments. The utility of this methodology has been demonstrated in several applications including those involving loop interactions, dislocation junction formation and dislocation generation mechanisms (e.g., Frank –Read source).

3.1.2 Level-set approach

The level-set formalism, due originally to Osher and Sethian[56] and applied to a wide range of physical problems,[57] enables the simulation of complex interactions among dislocations and other defects resulting in non-trivial dislocation conformations.[49,50] In this approach, the aforementioned level-set functions are evolved using a smoothed velocity field so that dislocation line motion is implicitly determined by the evolution of two level-set functions. The calculation of the required stress fields follows from linear elasticity theory[58] upon employing a FFT method, assuming periodic boundary conditions. As the trajectories of individual dislocation segments are not tracked, the level-set approach is well-suited to modeling complex line shapes associated with dislocation multiplication and annihilation. Thus, this method is capable of representing the salient features of three-dimensional dislocation motion, including dislocation glide, cross-slip, climb, and interactions with second-phase particles.

The level-set method in this context has several advantages over the more traditional front tracking methods. In particular, in this approach, dislocation conformations need not be discretized into a series of straight line or polynomial segments, and the dislocation lines need not be remeshed as the conformation evolves. Topological events, such as those associated with the formation of dislocation loops and the evolution of dislocation reactions, occur within the level-set framework without the necessity of imposing arbitrary collision rules. Consequently, the method has been successfully applied to the modeling of Frank-Read sources, and therefore loop generation, as well as strengthening mechanisms involving dislocation/precipitate interactions (e.g., the classical Orowan bypass mechanism and those involving cross slip).

3.2 Coarse-grained DD: link with the continuum

3.2.1 Overview

While large-scale DD simulations, containing on the order of 10^5-10^6 dislocations are beginning to provide a wealth of information about dislocation energetics and dynamics, it is worth noting that the macroscopic deformation response of well-worked materials with dislocation densities ranging between 10^{10}-$10^{14}/m^2$ can be accurately described by a smaller number of macrovariables. This reduction in the number of degrees of freedom required to characterize plastic deformation implies that a homogenization, or coarse graining, of variables is appropriate over some range of length and time scales. Indeed, there is experimental evidence that, at least in some cases, the mechanical response of materials depends most strongly on the macroscopic density of dislocations[59] while, in others, the gross substructural details may also be of importance. A coarse-grained counterpart to DD is desirable to formulate constitutive relations, based on relatively few macrovariables, for use in continuum-level descriptions of mechanical behavior. A self-consistent homogenization scheme, starting with simulations at the mesoscale, would provide a direct link between DD and continuum plasticity. With this in mind, we highlight below one such self-consistent coarse-graining strategy.

3.2.2 Methodologies

The formulation of coarse-grained description of an ensemble of mutually interacting dislocations is an ongoing enterprise presenting numerous conceptual and computational challenges. For example, with regard to the time evolution of this system, various workers have formulated kinetic models of the dislocation density based upon a separation of time scales.[60-62] Complementary investigations of spatial coarse graining have sought to identify a macrovariable set that captures complex dislocation-dislocation interactions. We summarize below one such promising investigation.

A recently developed, real-space approach to a coarse-grained dislocation energetics involves a subdivision of the system into volume elements and multipole expansions of the dislocation density.[63] More specifically, it was found that the interaction energy between well-separated volume elements can be written as a sum of multipolar interactions. The interaction energy between lowest-order moments can be transformed into that of Kosevich[64] for infinitesimal volume elements and, more generally, it is found that a multipolar energy expansion converges quickly (i.e., usually at dipole or quadrupole order) for well-separated elements. With this expansion it was found that the energetics of a collection of randomly distributed dislocation loops can be expressed in terms of the dislocation density and several low-order spatial derivatives. Building upon earlier work by Rickman and Viñals,[61] who developed a noise-free conserved dynamical model for the evolution of the dislocation density tensor in an elastically isotropic medium, the current aim is to use this energy expression to obtain coarse-grained equations of motion.

4. DISLOCATION PATTERN FORMATION

In both atomistic and DD simulations dislocation interactions are anisotropic and long-ranged and line defect motion is constrained by crystallography. On the atomic level, such constraints are determined in part, in the absence of climb, by the different (Peierls) barriers to motion associated with different crystal planes while, at the mesoscale, allowed slip systems can be prescribed based on atomic-level barrier height calculations. The results of simulations at both scales indicate that dislocation interactions and kinetic constraints lead to the formation of patterns at multiple length scales as dislocations organize into relatively low-energy substructures, consistent with experimental findings.[65,66] In this Section we briefly review pattern formation in atomistic and DD simulations.

4.1 Atomistic simulation

In a polycrystalline metal undergoing plastic deformation, various dislocation nucleation mechanisms can be activated including, for example, dislocation emission from grain boundaries with concomitant changes in boundary structure and location. While the details of such defect nucleation mechanisms are largely unclear, it is clear that the location for dislocation emission is boundary dependent and determined by the magnitude of local stresses. Once emitted from grain boundaries, dislocations interact within individual grains and move so as to lower the overall system energy. The resulting, spatially confined patterns are therefore dictated by a complex interplay among several factors, including the number of active slip systems and the grain size. The grain size, in particular, determines the grain perimeter and, hence, the defect nucleation rate and the distribution of obstacles to dislocation motion.

Dislocation emission and subsequent pattern formation can be seen in the results of MD simulations of the deformation. Simulations of tensile deformation of nanocrystalline Ni were performed using columnar grain structures with a common <110> type axis. The technique for generating polycrystalline samples and introducing the deformation is described by Farkas and Curtin 2005.[67] A variety of grain sizes have been used and dislocation patterning was observed. The dislocations were emitted from the grain boundaries and travel through the grains. Figure 7 shows this process for a sample with an average grain size of 20 nm. The figure shows the process in three different magnifications. The details of the structure of a single emitted dislocation can be seen in the picture to the left, where the individual atoms are shown. The picture in the right shows several grains and dislocation patterning can be seen, for example in the area marked by the circle. This area is an example of a 90 degree dislocation wall. The formation of these walls can be studied using dislocation dynamics techniques, as described in the following section.

4.2 DD simulation

Dislocation polygonization leading to substructure has also been observed and characterized in DD simulations. More specifically, standard correlation function and cluster analyses have been employed to contrast low-energy dislocation microstructures from a random distribution of objects. Such analyses reflect the underlying anisotropic nature of dislocation-dislocation interactions.

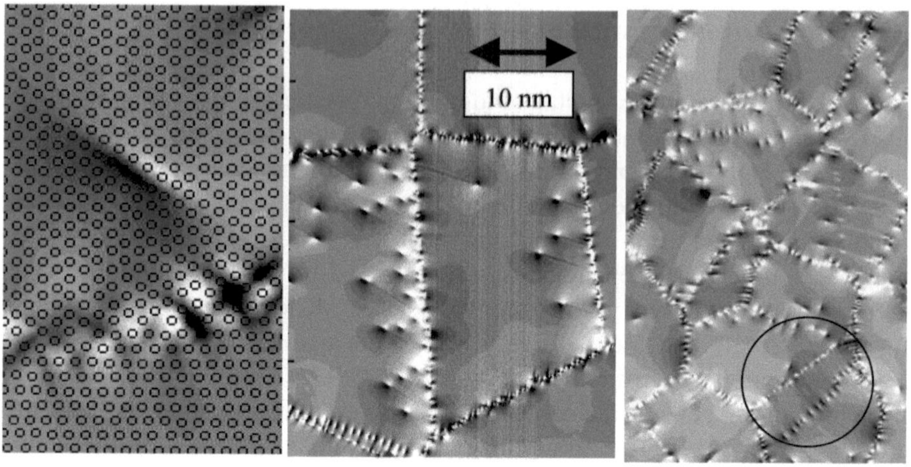

Figure 7. Deformation of nanocrystalline Ni in a columnar grain structure with a <110> common columnar axis. Dislocations emitted from the grain boundaries after 5% tensile deformation (center), and details of the emission at 2% deformation (left), and dislocation patterning observed after 8% deformation (right).

Consider, for example, a low-energy configuration of straight edge dislocations having no net Burgers vector resulting from a force-based DD simulation as described in Sec. 3.1. A snapshot of dislocation positions is shown in Figure 8. The tendency to form dislocation walls can be quantified via the calculation of an orientationally weighted pair correlation function[35,38] from a large-scale, two-dimensional mesoscale simulation of edge dislocations, as shown for example in Fig. 8. As is evident from the figure, both 45° and 90° walls are dominant (with other orientations also represented), consistent with the propensity to form dislocation dipoles with these relative orientations. The length of these walls can also be estimated from the correlation length inherent in the latter figure or, in a complementary fashion, using the aforementioned cluster analysis as described by Wang et al.[38]

5. SUMMARY

Atomistic and dislocation dynamics simulations provide complementary avenues for modeling deformation and fracture behavior in metals. In this chapter we have outlined the essential ingredients for each methodology as they relate to the description of materials failure. We have also indicated how these disparate approaches can be coupled to obtain a consistent,

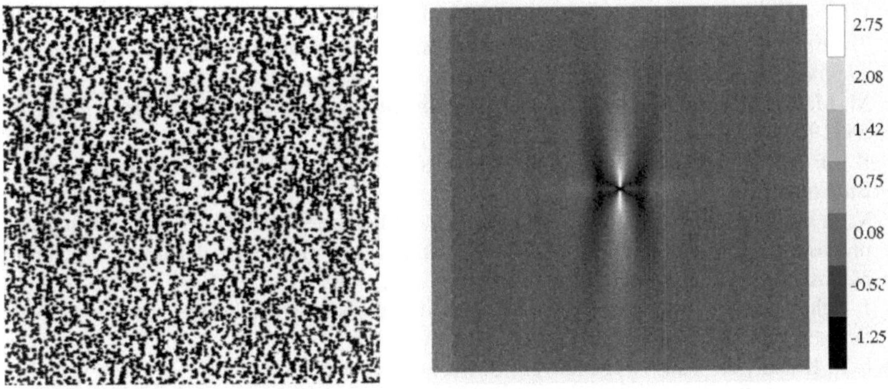

Figure 8. (a) A snapshot of a low-energy dislocation configuration in two dimensions. Some dislocation clustering is evident here, as confirmed by correlation function and cluster analyses. (b) An angular pair-correlation function. In the white (black) region there is a relatively high probability of finding a dislocation with positive (negative) Burgers vector, given that a dislocation with positive Burgers vector is located at the origin.

multiscale picture of mechanical behavior. This linkage also involves a connection with continuum mechanics, via imposed boundary conditions in atomistic simulation and via the development of coarse-grained constitutive relations from the results of DD. Finally, we have highlighted polygonization in simulation results using both approaches to show that dislocation pattern formation occurs at different scales.

ACKNOWLEDGEMENTS

The authors would like to thank Dr. R. LeSar and Dr. H. Van Swygenhoven for many helpful discussions. JMR would like to thank the Air Force Office of Scientific Research for its support under grant number FA9550-05-1-0082. DF acknowledges support from NSF, Materials Theory program.

REFERENCES

1. M. J. Buehler, A. Hartmaier, H. J. Gao, M. Duchaineau, F. F. Abraham, Atomic plasticity: description and analysis of a one-billion atom simulation of ductile materials failure *Computer Methods in Applied Mechanics and Engineering* **193** (48-51) 5257-5282 (2004).
2. F. F. Abraham, N. Bernstein, J. Q. Broughton, and D. Hess, Dynamic fracture of silicon: concurrent simulation of quantum electrons, classical atoms, and the continuum solid. *MRS Bulletin* **25**, No. 5, 27-32. (2000)

3. F.F. Abraham, Very large scale simulations of materials failure, *Philosophical Transactions of the Royal Society of London Series A-Mathematical Physical and Engineering Sciences* **360**, 367-382 (2002).
4. M. Marder, Molecular dynamics of cracks. *Computing in Science and Engineering* **1**, 48-55 (1999).
5. R.L.B. Selinger and D. Farkas (eds.), Atomistic theory and simulation of fracture. *MRS Bulletin*, **25**, No. 5.
6. V. Yamakov, D. Wolf, S.R. Phillpot, A.K. Mukherjee and H. Gleiter, Dislocation processes in the deformation of nanocrystalline Al by molecular-dynamics simulation, *Nature Materials* **1**, 45-48 (2002).
7. J. Schiotz and K.W. Jacobsen, A maximum in the strength of nanocrystalline copper, *Science* **300**, 1357-1359 (2003).
8. K.S. Kumar, H. Van Swygenhoven, S. Suresh, Mechanical behavior of nanocrystalline metals and alloys, *Acta Materialia* **51**, 5743-5774 (2003).
9. H. Van Swygenhoven, M. Spaczer, A. Caro, D. Farkas, Competing plastic deformation mechanisms in nanophase metals, *Phys. Rev. B* **60**, 22-25 (1999).
10. R.E. Miller, and E.B. Tadmor, The quasicontinuum method: overview, applications, and current direction, *J. Comp. Aided Mat. Design* **9**, 203-239 (2002).
11. E.B. Tadmor, M. Ortiz and R. Phillips, Quasicontinuum analysis of defects in solids, *Phil. Mag. A* **73**, 1529-1563 (1996).
12. N.M. Ghoniem, K. Cho, The emerging role of multiscale modeling in nano- and micro-mechanics of materials, *CMES: Comp. Model. Eng. Sci* **3,** 147-173 (2002).
13. Y. Mishin, D. Farkas, M.J. Mehl and D. A. Papaconstantopoulos, Interatomic potentials for monoatomic metals from experimental data and ab initio calculations, *Phys. Rev. B.* **59**, pp. 3393-3407 (1999).
14. M.S. Daw and M.I. Baskes, Semiempirical, quantum mechanical calculation of hydrogen embrittlement in metals. *Phys. Rev. Lett.* **50**, 1285-8 (1983).
15. K.W. Jacobsen, J.K. Nørskov, and M.J. Puska, Interatomic interactions in the effective-medium theory. *Phys. Rev.* B 35, 7423-42 (1986).
16. X.Y. Liu, F. Ercolessi and J.B. Adams, Aluminium interatomic potential from density functional theory calculations with improved stacking fault energy. *Modelling and Simulation in Materials Science and Engineering* **12**, 665-670 (2004).
17. M.I. Mendelev, S. Han and D.J. Srolovitz, Development of new interatomic potentials appropriate for crystalline and liquid iron. *Phil. Mag.* **83**, 3977-3994 (2003).
18. D. Farkas, H. Van Swygenhoven and P.M. Derlet, Intergranular fracture in nanocrystalline metals. *Phys. Rev. B* **66**, 060101-4(R) (2002).
19. G.C. Sih and H. Liebowitz, *Fracture: An Advanced Treatise*, Vol. II, edited by H. Liebowitz, Academic Press, New York, **69**, 189 (1968).
20. K.S. Cheung and S. Yip, Brittle-ductile transition in intrinsic fracture behavior of crystals. *Phys. Rev. Lett.* **65**, 2804–2807 (1990).
21. B. DeCelis, A.S. Argon, and S. Yip, Molecular dynamics simulation of crack tip processes in alpha-iron and copper. *J. Appl. Phys*. **54**, 4864-78 (1983).
22. C. Shastry and D. Farkas, Molecular statics simulation of fracture in α-iron. *Modelling Simulation Mater. Sci. Engng.* **4**, 473-92 (1996).
23. S.J. Zhou, P.S. Lomdahl, R. Thomson, and B.L. Holian, Dynamic crack processes via molecular dynamics. *Phys. Rev. Lett.* **76**, 2318–2321 (1996).
24. S.J. Zhou, P.S. Lomdahl, A. F. Voter, B.L. Holian, Three-dimensional fracture via large-scale molecular dynamics. *Engineering Fracture Mechanics* **61**, 173-187 (1998).
25. D. Farkas, Twinning and recrystallization as crack tip deformation mechanisms during fracture, *Phil. Mag.* **85** (2-3): 387-397 Sp. Iss. (2005).

26. D. Farkas, M. Duranduru and W.A. Curtin, Multiple-dislocation emission from the crack tip in the ductile fracture of Al, *Phil. Mag.* **81** (5): 1241-1255 (2001).
27. D. Farkas, S. Van Petegem and P.M. Derlet, Dislocation activity and nano-void formation near crack tips in nanocrystalline Ni, *Acta Mater.* **53**, 3115-3123 (2005).
28. A. Latapie and D. Farkas, Molecular dynamics investigation of the fracture behavior of nanocrystalline α-Fe, *Phys. Rev. B* **69**, 134110 (2004).
29. S.L. Frederiksen, K.W. Jacobsen and J. Schiotz, Simulations of intergranular fracture in nanocrystalline molybdenum, *Acta Mater.* **52**, 5019-5029 (2004).
30. E.B. Tadmor and S. Hai, A Peierls criterion for the onset of deformation twinning at a crack tip, *J. Mech. Phys. Solids* **51**, 765-793 (2003).
31. F.C. Frank, On the equations of motion of crystal dislocations, *Proc. Phys. Soc.* **62A**, 131-134 (1949).
32. J.D. Eshelby, Supersonic dislocations and dislocations in dispersive media, *Proc. Phys. Soc.* **B69**, 1013-1019 (1956).
33. T. Mura, Continuous distribution of dislocations, *Phil. Mag.* **8**, 843-857 (1963).
34. J. Lepinoux and L.P. Kubin, The dynamic organization of dislocation structures: a simulation, *Scripta Metall.* **21**, 833 (1987).
35. A.N. Gulluoglu, D.J. Srolovitz, R. LeSar and P.S. Lomdahl, Dislocation distributions in two dimensions, *Scripta Metall.* **23**, 1347 (1990).
36. G. Canova and L. P. Kubin, *Continuum Models and Discrete Systems 6* Vol. 2, ed. By G. A. Maugin (Harlow, UK: Longman)
37. E. van der Giessen and A. Needleman, Discrete dislocation plasticity: a simple planar model, *Modelling Simul. Mater. Sci. Engng.* **3**, 689 (1995).
38. H.Y. Wang, R. LeSar and J.M. Rickman, Analysis of dislocation microstructures: impact of force truncation and slip systems, *Phil. Mag. A* **78**, 1195-1213 (1998).
39. E. van der Giessen and A. Needleman, Micromechanics simulations of fracture, *Ann. Rev. Mater. Res.* **32**, 141 (2002).
40. R. Madec, B. Devincre and L. Kubin, Simulation of dislocation patterns in multislip, *Scripta Mater.* **47**, 689-695 (2002).
41. M. Rhee, D. Lassila, V.V. Bulatov, L. Hsiung, T.D. de la Rubia, Dislocation multiplication in bcc molybdenum: a dislocation dynamics simulation, *Phil. Mag. Lett.* **81**, 595 (2001).
42. M. Koslowski, A. Cutiño and M. Ortiz, A phase-field theory of dislocation dynamics, strain hardening and hysteresis in ductile single crystals, *J. Mech. Phys. Solids* **50**, 2597 (2002).
43. L.P. Kubin and G.R. Canova, *Electron Microscopy in Plasticity and Fracture Research of Materials*, eds. U. Messerschmidt, *et al.* (Akademie Verlag, Berlin, 1990), p. 23.
44. L. P. Kubin, G. Canova, M. Condat, B. Devincre, V. Pontikis and Y. Brechet, *Solid State Phenomena* **23/24**, 455 (1992).
45. H.M. Zbib, M. Rhee and J.P. Hirth, On plastic deformation and dynamics of 3D dislocation, *Int. J. Mech. Sci.* **40**, 113 (1998).
46. M. Rhee, H.M. Zbib, J.P. Hirth, H. Huang, T. de la Rubia, Models for long-/short-range interactions and cross slip in 3D dislocation simulation of bcc single crystals, *Modelling Simul. Mater. Sci. Eng.* **6**, 467 (1998).
47. K.W. Schwarz, Simulation of dislocations on the mesoscopic scale. I. methods and examples, *J. Appl. Phys.* **85**, 108-119 (1999).
48. M. Peach and J.S. Koehler, The forces exerted on dislocations and the stress fields produced by them, *Phys. Rev.* **80**, 436-439 (1950).

49. Y. Xiang, L.T. Cheng, D.J. Srolovitz and W.E., A level set method for dislocation dynamics, *Acta mater.* **51**, 5499-518 (2003).
50. Y. Xiang, D.J. Srolovitz, L.T. Cheng, and E. Weinan, Level set simulation of dislocation-particle bypass mechanism, *Acta Mater.* **52**, 1745-1760 (2004).
51. Y.U. Wang, Y.M. Jin, A. Cuitiño and A.G. Khachaturyan, Nanoscale phase field microelasticity theory of dislocations: model and 3D simulations, *Acta Mater.* **49**, 1847-1857 (2001).
52. Y.M. Jin and A.G. Khachaturyan, Phase field microelasticity theory of dislocation dynamics in a polycrystal: model and three-dimensional simulations, *Phil. Mag. Letters* **81,** 607-616 (2001).
53. S.Y. Hu and L.-Q. Chen, Solute segregation and coherent nucleation and growth near a dislocation - a phase-field model for integrating defect and phase microstructures, *Acta Mater.* **49**, 463-472 (2001).
54. Y. Wang, D.J. Srolovitz, J.M. Rickman and R. LeSar, Dislocation motion in the presence of diffusing solutes: a computer simulation study *Acta Mater.* **48**, 2163 (2000).
55. N.M. Ghoniem, S.-H. Tong, and I.Z. Sun, Parametric dislocation dynamics: a thermodynamics-based approach to investigations of mesoscopic plastic deformation, *Phys. Rev. B* **61**, 913- 927 (2000).
56. S. Osher and J.A. Sethian, Fronts propagating with curvature- dependent speed: algorithms based on Hamilton-Jacobi formulations, *J. Comput. Phys.* **79**, 12 (1988).
57. P. Burchard, L.T. Cheng, B. Merriman, and S. Osher, Motion of curves in three spatial dimensions using a level set approach, *J. Comput. Phys.* **170**, 720 (2001).
58. L.D. Landau and E.M. Lifshitz, *Theory of Elasticity*, 3rd ed., (Pergamon Press, New York, 1986).
59. A. Turner and B. Hasegawa, Mechanical testing for deformation model development, *ASTM* 761 (1982).
60. D.L. Holt, Dislocation cell formation in metals, *J. Appl. Phys.* **41**, 3197 (1970).
61. J.M. Rickman and Jorge Viñals, Modeling of dislocation structures in materials, *Phil. Mag. A* **75**, 1251 (1997).
62. M.C. Marchetti and K. Saunders, Viscoelasticity from a microscopic model of dislocation dynamics, *Phys. Rev. B* **66**, 224113 (2002).
63. R. LeSar and J.M. Rickman, Incorporation of local structure in continuous dislocation theory, *Phys. Rev. B* **69**, 172105 (2004).
64. A.M. Kosevich, *Dislocations in Solids*, ed. by F. R. N. Nabarro (North-Holland, New York, 1979), p. 37.
65. J.P. Hirth and J. Lothe, *Theory of Dislocations* (Krieger, Malabar, Florida, 1982).
66. D.A. Hughes, D.C. Chrzan, Q. Liu, and N. Hansen, Scaling of misorientation angle distributions, *Phys. Rev. Lett.* **81**, 4664-4667 (1998).
67. D. Farkas and W.A. Curtin, Plastic deformation mechanisms in nanocrystalline columnar grain structures, *Materials Science and Engineering A* **412**, 316-322 (2005).

Chapter 12

FRONTIERS IN SURFACE ANALYSIS: EXPERIMENTS AND MODELING

Daniel Farías[1], Guillermo Bozzolo[2,3], Jorge Garcés[4] and Rodolfo Miranda[1]

[1]*Departamento de Física de la Materia Condensada and Instituto Nicolás Cabrera, Universidad Autónoma de Madrid, Madrid, Spain;* [2]*Ohio Aerospace Institute, 22800 Cedar Point Rd., Cleveland, OH, 44142 USA;* [3]*NASA Glenn Research Center, Cleveland, OH 44135, USA;* [4]*Centro Atómico Bariloche, 8400 Bariloche, Argentina*

Abstract: Due to the extensive research performed over the past three decades, we have gained considerable insight into the properties of metal and semiconductor surfaces. Much less effort has been devoted to the study of more complex surfaces, like alloy and oxide surfaces, even though they are technologically relevant. In this chapter we will discuss how the combination of experimental techniques (STM and H_2 diffraction) and computational modeling can be applied to characterize the composition, structure and electronic properties of some complex surfaces, like O/Ru(0001), NiAl(110), Fe/Cu(100), Fe/Cu(111) and $Fe_4N(100)$.

Keywords: Scanning Tunneling Microscopy; Hydrogen scattering; Computer Simulations; Semi-empirical methods.

1. INTRODUCTION

The astonishing advances in both experimental and theoretical techniques that happened during the past twenty five years, have taken surface science from adolescence to maturity.[1] Now we are able to determine accurately the composition, the crystalline and electronic structure and the dynamics of simple, single-crystal surfaces, adsorbed layers or ultrathin films prepared in UHV or in electrochemical cells.[2] The emphasis in surface science is shifting to the study of more complex systems, including organic, biological or astrophysical systems or technologically-relevant surfaces and interfaces.[3] Theoretical modeling has increased in power to the point that we are approaching a true convergence of experimentation and simulation able to open a broad range of scientific opportunities in nanoscience and nanotechnology at the intersection of physics, chemistry, biology and engineering.

This chapter illustrates how the problems related to the surfaces of some complex solids can be studied with an unprecedented level of accuracy by the appropriate combination of experimental techniques and computer simulations. We will concentrate on a combination of real space (STM) and reciprocal space (H_2 diffraction) experimental techniques blended with appropriate theoretical tools, as applied to the characterization of composition, structure and electronic properties of some oxide and alloy surfaces.[4] Particular attention will be paid to the surface reconstruction of magnetic materials, as well as dynamic problems such as adsorption or intermixing during heteroepitaxial growth. The future is full of exciting challenges for surface scientists and this work tries to show that working out a clever combination of experimental and modeling techniques is a key ingredient for success.

2. EXPERIMENTAL RESULTS

The He and H_2 diffraction experiments reported here were performed with the apparatus described in detail in Ref. 5, which has recently been transferred to the Surface Science Lab at the Universidad Autónoma de Madrid. Briefly, it consists of a three-stage differentially pumped beam system and an 18 inch diameter scattering chamber. The free jet expansion is produced through a nozzle of d=10 μm diameter. The nozzle temperature T_0 can be varied between 300 K and 700 K, allowing a variation of the H_2 incident energy between 75 and 160 meV. After expansion, the beam is collimated by a 0.5 mm diameter skimmer and traverses two differential pumping stages before entering the sample chamber. The beam was mechanically chopped with a magnetically coupled rotary motion feed through in the third stage to allow phase sensitive detection. The measurements reported here were performed with a source pressure P_0 = 50 bar behind the nozzle. The velocity spread of the H_2 beam was estimated in 6% under these conditions, as judged from the angular resolution observed in the H_2-diffraction spectra as compared to He-diffraction data. Only low energy rotational states of H_2 are significantly populated in the beam:[6] the population fractions in states with angular momentum quantum number J=0, 1, 2, and 3 are 0.16, 0.67, 0.10, and 0.07 at E_i = 100 meV, and 0.10, 0.55, 0.15, and 0.18 at E_i = 150 meV. Because of the large spacing between the vibrational levels of the H_2 molecule, it can be safely stated that more than 99% of the molecules are in the υ=0 state for incident energies E_i< 200 meV.

The base pressure in the chamber was typically ~7 x 10^{-11} mbar, reaching ~5 x 10^{-10} mbar, with the H_2 beam on. The crystal was mounted on a manipulator that allows azimuthal rotation of the sample as well as heating to 1200 K and cooling to 90 K. The angular distribution of the scattered H_2 molecules was analyzed with a quadrupole mass spectrometer mounted on a two-axis goniometer. This arrangement allows rotations of 200°C in the scattering plane - defined by the beam direction and the normal to the surface - as well as +/-15° normal to the scattering plane for a fixed angle of incidence. In particular, this allows the recording of the incident beam intensity with high accuracy, by simply removing the sample from the He-beam and putting the detector behind. The full width at half maximum (FWHM) of the specular H_2 beam is 1.4°. All data presented here refer to a scattering geometry in which the H_2 beam impinges perpendicular to the [001] symmetry direction of the substrate.

The Scanning Tunnelling Microscopy (STM) results were obtained with a variable temperature STM microscope placed in a UHV chamber with base pressure in the low 10^{-11} mbar regime. The sample can be cooled down to 40 K. The chamber contains a rear view LEED optics, ion sputtering gun, mass spectrometer, a number of different evaporators and a radio-frequency plasma source. The polycrystalline W tips of the STM were routinely cleaned by ion bombardment and field emission. All images reported here were recorded in the constant current mode.

The NiAl(110) and Cu(100) samples were prepared in UHV by cycles of sputtering with 1 keV Ne-ions at 500 K and annealing to 1150 K.[7] The Ru(0001) crystal was cleaned by cycles of Ar+ sputtering and annealing followed by oxygen exposure and heating to high temperature.[8] As a criterion of the cleanliness of the samples used in the diffraction machine, we take an optimum (~ 40%) He reflectivity and sharp diffraction peaks, as well as the ability to reproduce the previously reported hydrogen ordered adlayers, as these are very sensitive to the presence of impurities on the surface.

3. THEORY: THE BFS METHOD

The BFS method for alloys[9] is based on the simple concept that the energy of formation of a given atomic configuration is the sum of the individual atomic contributions, $\Delta H = \Sigma\varepsilon_i$. Each contribution by atom i, ε_i, can be calculated as the sum of two terms: a strain energy, ε_i^S, computed in the actual lattice as if every neighbour of the atom i were of the same atomic species i, and a chemical energy, ε_i^C, computed as if every neighbour of the atom i were in an equilibrium lattice site of a crystal of species i, but retaining its actual chemical identity. The computation of ε_i^S, using Equivalent Crystal Theory (ECT),[10] involves three pure element properties for atoms of species i: cohesive energy, lattice parameter and bulk modulus.

These three parameters for each of the constituent elements are needed in the general derivative structure as the final alloy. Consequently, when studying Fe-Cu fcc-based alloys, both elements would need to be parameterized as if they were fcc. The chemical energy, ε_i^C, accounts for the corresponding change in composition, considered as a defect in an otherwise pure crystal. The chemical 'defect' deals with pure and mixed bonds, therefore, two additional perturbative parameters (Δ_{AB} and Δ_{BA} where A, B = Fe, Cu) are needed to describe these interactions. A reference chemical energy, ε_i^{Co}, is also included to insure a complete decoupling of structural and chemical features. All the needed BFS parameters are determined with the Linearized-Augmented Plane Wave method (LAPW).[11] From the theoretical standpoint, the immiscibility of Fe and Cu in the bulk presents a challenge, due to the lack of accurate experimental evidence that would help validate the parameterization used. In this work, LAPW calculation of the equilibrium properties of Fe and Cu in the fcc phase, as well as the equilibrium properties of Fe_3Cu ($L1_2$), FeCu ($L1_0$) and $FeCu_3$ ($L1_2$) ordered alloys was performed in order to define the BFS perturbative functions. Although these binary phases do not exist, their properties can be readily computed via first-principles methods. Moreover, this calculational scheme enables an accurate determination of the concentration dependence of the BFS parameters Δ. The functions used in this work are given by $\Delta_{FeCu}(x_{Fe}) = -0.119136\ x_{Fe} + 0.210957$ and $\Delta_{CuFe}(x_{Cu}) = -0.0690423\ x_{Cu} + 0.125062$, where x_{Fe} (x_{Cu}) denotes the fraction of nearest-neighbors of a given Cu (Fe) atom that are Fe (Cu) atoms. Finally, the strain and chemical energies are linked with a coupling function g_i, which ensures the correct volume dependence of the BFS chemical energy contribution. Therefore, the contribution of atom i to the energy of formation of the system is given by

$$\varepsilon_i = \varepsilon_i^S + g_i\left(\varepsilon_i^C - \varepsilon_i^{C_0}\right) \tag{1}$$

Table 1 lists the necessary parameters for applying the BFS method to the Fe-Cu system. We refer the reader to Ref. 9 for a detailed discussion of the BFS method, its definitions, operational equations and their implementation.

Table 1. LAPW[11] results for the lattice parameter, cohesive energy, and bulk modulus for Fe and Cu and the resulting equivalent crystal theory (ECT)[10] parameters p, α, l and λ.

	Lattice parameter (Å)	Cohesive energy (eV)	Bulk modulus (GPa)	ECT parameters			
				p	α (Å^{-1})	l (Å)	λ (Å)
Fe	3.4452	4.270	298.40	6	2.91489	0.21238	0.59679
Cu	3.6325	3.558	141.79	6	2.87172	0.27390	0.76966

4. RESULTS

4.1 Oxygen on Ru(0001): What do we "see" with the STM on oxide surfaces?

Oxide surfaces are extremely important in a wide range of applications, from supported metal catalysts to lubrication, adhesion or gas sensors.[4] Recently, they have attracted additional interest due to the increasing use of ultrathin insulating films in magnetic nanodevices. In different areas of research, e.g. catalysis, high T_c superconductors, colossal magnetoresistence materials, it is essential to distinguish metal from oxygen sites at surfaces of transition metal oxides in order to understand processes such as dissociation of molecules, spatial inhomogeneities in the superconducting gap or the role of impurities in the formation of striped phases.[12]

Although STM topographs are routinely used to characterize surfaces at the atomic level, they do not always simply reflect the real position of surface atoms.[13-16] If we restrict ourselves to adsorbed oxygen layers or oxide surfaces, several studies[17-23] have reported atomically resolved STM images. The correct identification of mechanisms[23] or active sites[20] for reactions, often related to surface defects (e.g. oxygen vacancies), depends, however, on the precise assignment of the atomic features detected in the STM image, which is only possible by comparison with calculated images. The calculation of STM images requires, in turn, the knowledge of the geometry and electronic structure of the surface, on the one hand, and a *realistic modeling* of the STM tip, including its geometry and chemical composition, on the other.[24]

In many cases the assignment of the atomic-scale features in STM images of oxides to metal or oxygen ions is not unique, since it is only based on calculations of the surface Local Density Of States (LDOS)[18,20] and do not consider a realistic tip structure. For instance, bright rows of atomic features are seen in STM images of $TiO_2(110)$, which have alternatively been interpreted as reflecting the positions of oxygen[18] or metal[19] atoms. Although in this case it is accepted now that rows of *metal* atoms are imaged *bright* in empty-state imaging mode, for $RuO_2(110)$, where similar bright rows are seen, they are interpreted as due to bridging *oxygen* atoms.[20,21] In both cases the geometry of the mentioned oxide surfaces is dominated by prominent rows of oxygen atoms protruding from the surface. It would be important to find a suitable model system where a detailed comparison of calculated STM images with a large experimental data basis could help us to advance in understanding.

Figure 1. The upper panel shows a 10 nm x 5 nm STM image of a clean Ru(0001) surface recorded at a sample voltage of -23 mV with a gap resistance of 2 MΩ. The lower panel shows a 10 nm x 5 nm STM image of the O(2x2) superstructure on Ru(0001) recorded with a sample voltage of -0.1 V and a gap resistance of 200 MΩ. The oxygen atoms appear as holes with a circular shape in these conditions.

The adsorption of oxygen on Ru(0001) is such a model system. There are several ordered phases with increasing oxygen content which have been determined structurally with high accuracy mostly by LEED and other diffraction techniques. At a coverage of 0.25 ML, an ordered (2x2) pattern is formed by oxygen atoms sitting on the threefold hcp sites of the Ru(0001) surface, i.e. continuing the lattice of Ru, 1.3 Å above the Ru nuclei.[25]

At low bias voltage and low tunnelling resistance the clean Ru(0001) surface is imaged as an hexagonal array of round protrusions separated by 2.7 Å, as shown in the upper panel of Figure 1. Scanning Tunnelling Spectroscopy (STS) experiments and *ab initio* calculations indicate that the states that dominate the current at standard distances are due to a surface resonance of p_z character, located close to the Fermi energy and spatially localized on top of the Ru atoms.[24] The lower panel of Figure 1 shows an

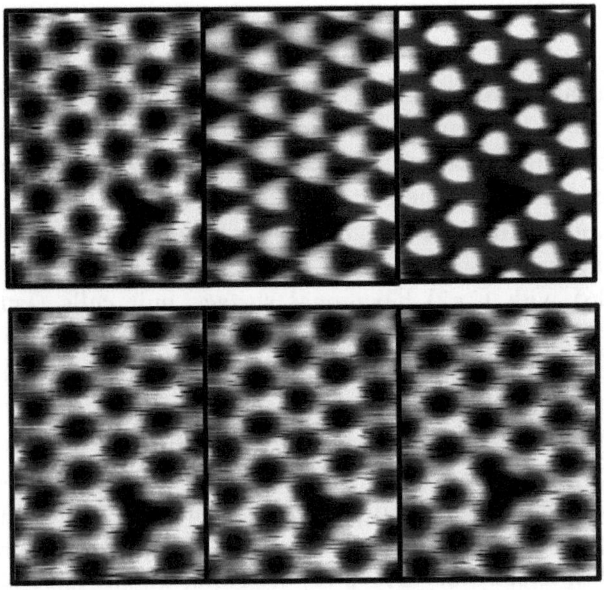

Figure 2. Series of 2.2 nm x 3.2 nm dual-mode STM images of O(2x2)/Ru(0001). The upper panel shows images recorded at constant sample voltage of -30 mV and decreasing gap resistances (from left to right, 100, 10 and 2.5 MΩ). The reference images in the lower panel were all taken at a sample voltage of -30 mV and a constant gap resistance of 30 MΩ.

STM image of a compact O adlayer with (2x2) periodicity with respect to the substrate recorded at 300 K and standard tunnel resistance. The O atoms are visualized as *circular depressions*. The bright regions in the image correspond to (mobile) oxygen vacancies in the (2x2) superstructure, i.e. clean Ru patches. Images taken at lower tunnelling resistances, in which both the bare Ru(0001) surface and islands of the O(2x2) superstructure appear simultaneously with atomic resolution, allow us to determine the oxygen registry with respect to the Ru lattice. In these conditions of low gap resistance, however, the oxygen atoms are seen as *triangular* depressions.

In order to eliminate the possibility of unwanted changes in the STM tip during scanning, dual-mode STM images were recorded by measuring the forward and backward linescans with a different sample voltage. A representative selection of dual mode STM images recorded for increasing tunnel currents (and fixed voltage, -30 mV) is shown in Figure 2. The lower panel displays the reference (forward) channel. All reference images are identical, except for a minor drift. The upper panel displays three images illustrating the reversible transition observed in the appearance of the "atomic" features. As mentioned before, an hexagonal array of circular depressions assigned to oxygen atoms is seen at large gap resistances. Upon decreasing the gap resistance, the depressions acquire progressively a triangular appearance.

The electronic structure of the tip and the sample surface were calculated[24] using the Density Functional Theory (DFT) package VASP,[26] which describes ion-electron interactions by ultrasoft pseudopotentials.[27] The exchange correlation potentials were calculated using a generalized gradient method.[28] The clean Ru(0001) surface was modeled by a three dimensional supercell consisting of 7 atomic layers (separation distance c/2, with c/a=1.586) with (2x2) hexagonal surface unit cells (lattice constant a=2.723 Å) and a seven layer vacuum range.[24] The Ru atomic positions of the surface layer were relaxed in three directions. The preferred adsorption site of O was found to be the hcp site directly above second layer Ru atoms, at a vertical distance of 1.16 Å, in agreement with LEED.[25] The tip structure was modeled[24] with the same DFT method by a pyramid of seven W(110) layers; the apex atom being either W or O. In these simulations all ionic positions of the two top layers were also fully relaxed. From the calculation of surface and tip electronic structures, the Kohn-Sham states of surface and tip electrons were obtained, and then used as the input for the STM simulation. The tunnelling current was calculated in the Bardeen approach.[29] This perturbative method limits the validity of the calculation to tip-sample distances larger than 4-5 Å, which is also the limit for the majority of experimental scans described here. The surface irreducible Brillouin zone was sampled with 20 *k*-points. The energy cutoff in the plane wave expansion is typically 30 Ry (400 eV).

Figure 3 shows measured (left) and simulated (right) STM images for two representative gap resistances.[24] When imaged with a W tip, the O atoms appear as depressions, while Ru is seen bright. As gap resistance decreases by one order of magnitude, the changing shape of the features associated with O *and* Ru is nicely reproduced by the simulations. Their shape is *circular* when the tip is relatively far away from the surface, and *triangular* when the tip is closer to the surface. This change is mostly due to the different geometry of Ru p_z orbitals, with rotationally symmetric lobes pointing outwards, and hybridized s/p_{xy} orbitals, with threefold rotational symmetry, with respect to the adsorption site. At large distances, the circularly shaped Ru p_z orbitals contribute most of the current, while closer to the surface the contribution from s/p_{xy} orbitals increases and originates the triangular shape. It has to be highlighted that cuts of the LDOS at different constant values reproduce accurately the measured STM images *only* at large gap resistances. The inclusion of a realistic tip structure and the use of the Bardeen approach in the calculations is essential to obtain quantitative agreement between simulated and measured images in the studied range.[24]

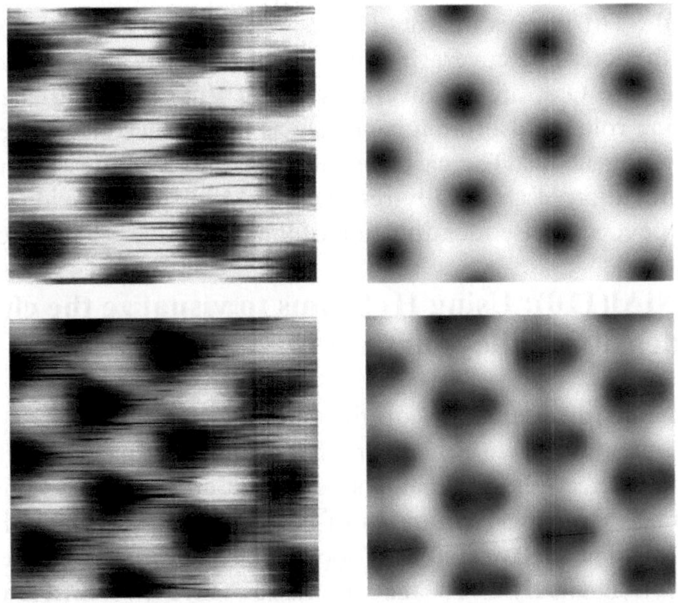

Figure 3. Comparison of experimental (left) and simulated (right) STM images of O(2x2)/Ru(0001). In both cases the sample voltage was -30 mV. The simulations have been performed with tunnelling currents of 0.03 nA (above) and 0.3 nA (below) and agree with the experimental ones for 300 MΩ (above) and 30 MΩ (below) gap resistances. The maximum corrugation is 0.5 Å in all cases.

Thus, oxygen atoms on O(2x2)/Ru(0001) appear as distinct *depressions*, about 0.5 Å deep in STM images taken with a clean W tip in a wide range of tunnelling parameters,[24] while the bright features seen in the images correspond to the Ru sites. This *negative* corrugation for the adsorbates is due to a significant lowering of the LDOS at the Fermi level in the neighborhood of the O sites[30] with respect to the Ru sites. Quantitatively, the reduction in the LDOS above the site of the O atom is such that, in order to keep the tunnelling current constant, the tip-sample distance has to decrease in spite of the protruding geometry of oxygen.

The *electronic* effects that are able to hide the real surface geometry in STM images of the O(2x2)/Ru(0001) phase, are still dominant for the O(2x1)/Ru(0001) phase in a wide range of gap resistances.[31] The *bright* rows seen in STM images of the O(2x1) structure taken with a clean W tip corresponds to rows of Ru atoms,[31] although the apparent corrugation in the O(2x1) phase is smaller by a factor of two than the one for the O(2x2) phase. At even higher O coverages, corresponding to the 3O(2x2) (0.75 ML) and O(1x1) structures (1 ML), the calculated STM images are determined by the surface geometry, with the *bright* atomic-size features in these cases being associated to the protruding *oxygen* atoms. This would also be the case for the reported STM images of $RuO_2(110)$.[20,21] In this system the inversion in contrast with increasing coverage of oxygen, from Ru being imaged bright at low coverage, to O at higher coverages and at $RuO_2(110)$ surfaces, reflects the increasing role of adsorption geometry as the surface electronic structure becomes homogeneous. The case described here exemplifies the need of an accurate simulation of atomically-resolved STM images of oxide surfaces in order to achieve a precise identification of the respective atomic species.

4.2 NiAl(110): Using H_2 beams to visualize the charge density of alloy surfaces

An accurate determination of the surface composition for alloys is extremely important since a number of their applications, from bimetallic catalysts to defect-resistant alloys depend crucially on the precise composition. Although STM images of bimetallic alloy surfaces have been shown to distinguish two atomic species, the results presented in the previous section clearly show that, in cases where two different kind of atoms are present on a surface, the interpretation of STM images is never straightforward. Which one are the "brighter" or "darker" features observed in STM images strongly depend on the chemical composition of the tip, *which is very often unknown in the experiments.* In this section we will discuss a very different approach, which may be useful to visualize the surface charge density of complex surfaces, namely the diffraction of monoenergetic H_2 beams. We will see that, in contrast to the tip-surface problem in STM, the potential involved in the H_2-surface interaction may be rather complex, but *it is always well defined,* and can be calculated with high accuracy using *ab initio* techniques.

Helium atom scattering is a well established tool to investigate the structure and dynamis of solid surfaces.[32] Because of the low energies used (10–250 meV), the incident atoms probe the topmost layer of the substrate surface in an absolutely nondestructive manner, and are equally applicable to insulators, semiconductors and metals. In the case of He atoms at low energies, the scattering is predominantly elastic and, as the de Broglie wavelengths are of the order of ~ 1 Å, diffraction effects dominate. Diffraction experiments give information on both the dimensions of the unit cell, via the angular location of the diffraction peaks and, by analyzing diffraction intensities, on the *corrugation of the surface charge*, which often provides direct pictures of the geometrical arrangement of the surface atoms. To understand the way in which this information is recovered from experimental data, a brief description of the He–surface interaction potential is needed.

At distances not too far away from the surface, the impinging He atoms feel an attraction due to van der Waals forces. Closer to the surface, they will be repelled due to the overlap of their electronic wavefunctions with those of the surface atoms (Pauli exclusion); this causes the repulsive part of the interaction potential to rise steeply. It can be shown that the classical turning points are farther away for He atoms impinging on top of the surface atoms than for He atoms impinging between them; this gives rise to a periodic modulation of the repulsive part of the potential. The locus of the classical turning points follows a surface of constant total electron density ρ(r), very much like the one imaged by STM, where every point constitutes a scattering center; the resulting scattering surface is called the *corrugation function* ζ (R) (R denotes a two-dimensional vector in the surface plane). The classical turning points are located at 3-4 Å away from the surface atom cores; this implies that with thermal He atoms densities between 1×10^{-4} and 2×10^{-3} au can be mapped. Therefore, the arrangement of the surface atoms can be obtained by relating the experimentally determined corrugation function ζ (R) to self-consistent calculations of surface charge-density profiles, ρ(r).

This analysis is also valid for diffraction of Ne atoms. However, owing to its larger polarizability, a larger corrugation amplitude is seen with Ne as compared to He atoms. As an example, we show in Figure 4 the best-fit corrugation functions obtained for the clean NiAl(110) surface with Ne and He diffraction.[33] On NiAl(110) the Al atoms are at the external surface with Ni being located in the layer below, shifted with respect to the Al atoms (see the scheme at the bottom left of Figure 4). Note that a significantly larger corrugation amplitude is seen with Ne (0.11 Å), than with He (0.035 Å) diffraction, a result also reported for several fcc(110) surfaces.[32] A very remarkable result is that both Ni and Al atoms are visible in the Ne-derived corrugation, whereas only the topmost Al atoms appear in the corrugation determined from He-diffraction data.

The situation is already much more complex if we consider diffraction of H_2 molecules, whereby its stronger interaction with the surface makes diffraction intensities larger than for He. This is because the six-dimensional character of the H_2-surface problem comes into play and, in particular, the coupling with the dissociative adsorption channels. In spite of this, recent work[34-36] has shown that a detailed analysis of the angular distribution of reflected molecules in scattering experiments along well defined incidence directions provides detailed information on the molecule/surface dynamics and, therefore, on the corresponding six-dimensional potential energy surface (PES). A few experimental attempts along this line have been already published for H_2/Pd(111) and H_2/Pt(111).[37-40]

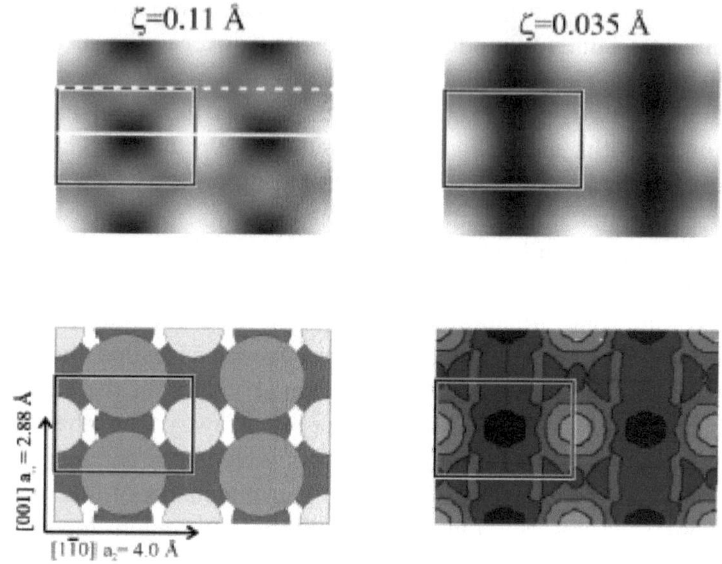

Figure 4. Top: Gray scale representation of the best-fit corrugation functions obtained for the clean NiAl(110) surface using Ne (left) and He (right) diffraction. Note the larger corrugation amplitude seen with Ne diffraction. Maxima (bright features) indicate the positions of the topmost Al atoms, as concluded from a comparison with a contour of constant surface electronic charge density (bottom right) determined from *ab initio* calculations. The geometry of the surface is also shown (left bottom) with Al atoms as large darker circles and Ni atoms as smaller, lighter circles.

Classical calculations performed for the H_2/Pd(111) system using an *ab initio* determined PES, showed that the classical turning points at E_i ~100 meV, are located only ~ 1.5 Å above the topmost surface atoms.[37-39] This means that the H_2 molecules are sampling a larger corrugation than He, and even Ne, atoms. As a consequence, H_2 diffraction might be a powerful tool to unravel structural details of complex surfaces. Note that according to the picture outlined above, the corrugation is expected to increase with increasing kinetic energy, because high-energy molecules penetrate deeper into the electron-density profile.

To check whether H_2 diffraction may be suitable to determine the structure of alloy surfaces, we performed experiments on the NiAl(110) surface at several incident energies. Results corresponding to E_i= 45 meV and 150 meV are presented in Figure 5.

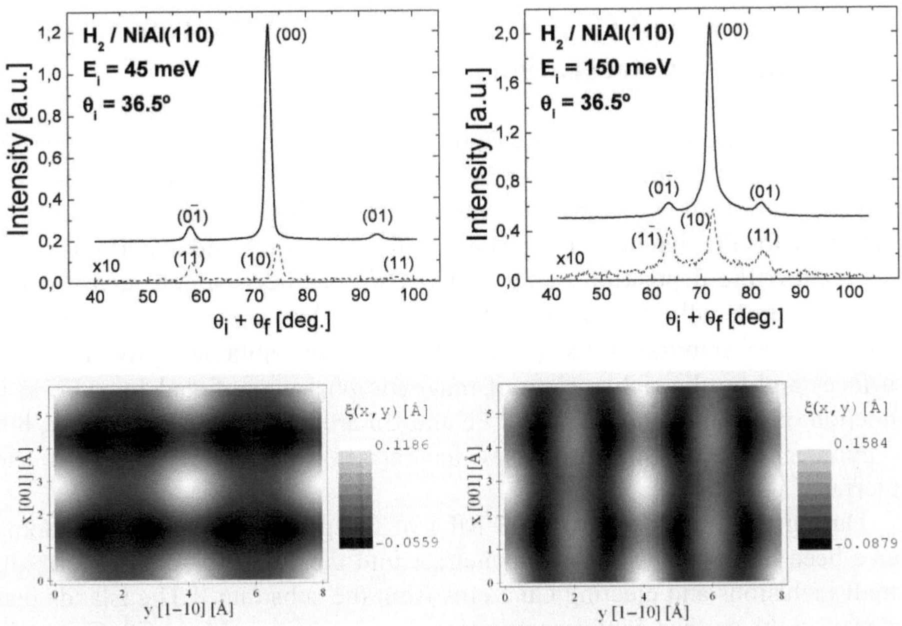

Figure 5. Top: In-plane (solid lines) and out-of-plane (dash lines) H_2 diffraction spectra from NiAl(110) and two different incident energies. Bottom: Gray scale representation of the best-fit corrugation functions derived assuming a three-dimensional scattering potential. Note that both first and second layer atoms are seen with H_2 beams.

The appearance of pronounced in-plane as well as out-of-plane diffraction peaks at the lowest incident energy investigated, demonstrates that the incoming H_2 molecules are sampling a highly-corrugated two-dimensional surface. To recover the corrugation function from the data, we considered the scattering problem as being three-dimensional, and carried out close-coupling calculations using a realistic, soft potential. Assuming a three-dimensional PES is equivalent to neglect the two degrees of freedom which determine the orientation of the molecule with respect to the surface, plus the one corresponding to the H-H separation. This assumption is well justified in the case of the NiAl(110) surface, since the barrier for dissociation of H_2 is very large (~ 1 eV).[41] The best results were obtained using a Morse potential, with parameters D = 35 meV (well depth) and α = 0.925 $Å^{-1}$. As we can see in Fig. 5, the corrugation functions determined in this way strongly depend on the energy of the incoming molecules. Their behaviour is closer to Ne than He probes, i.e. both Ni and Al atoms from the first two atom layers are visible. These results are very promising, and suggest that diffraction of H_2 beams might be used to determine the charge-density profiles of alloy and oxide surfaces.

4.3 Growth of Fe on Cu(100) and Cu(111): Intermixing and step decoration

Alloying also occurs directly upon heteroepitaxial growth of metals on metal substrates even if both metals are *immiscible* in the bulk. An accurate characterization of the degree of intermixing during growth is particularly important for the magnetic properties of ultrathin films and superlattices. In this context the deposition of Fe on Cu(100) has been widely investigated both experimentally and theoretically.[42-48] This has been motivated by the possibility of stabilizing fcc (γ) Fe at 300 K by epitaxial growth on Cu surfaces and by the rich variety of magnetic phases predicted for γ-Fe as a function of its lattice constant.[46-48] Fe and Cu are immiscible in the bulk, but a number of atomistic processes leading to intermixing occur at the interface.[49-56]

During the initial stages (0.03 ML) of the growth at 300 K, Fe atoms have been observed by STM to penetrate into the Cu(100) surface, forming small inclusions and ejecting Cu atoms from the substrate.[49] The islands that appear on the surface contain a mixture of ejected Cu and Fe, with Cu partly decorating the surface and edges of the islands and are frequently several layers high.[51] The steps of the substrate are partially etched and decorated by Fe. All these facts were previously reported for the initial stages of the growth of Co[57-60] and Fe[52-54] on Cu(111), plus the observation of the appearance of vacancy islands on the Cu terraces, as a result of the etching processes.[52,58] These surface etching effects were interpreted as caused by the stress induced by the presence of Fe atoms *within* the surface plane, which is released by kink and vacancy island formation.[50,57,60] Furthermore, upon mild annealing, or simply as a function of time at 300 K, Cu has been observed to segregate to the external surface covering the Fe (or Co) islands.[43,44] At even higher temperatures Cu segregates massively to the surface, with the onset for segregation depending on the orientation of the Cu surface.

A comprehensive theoretical study of the growth mode of Fe on Cu(100) and Cu(111) at low coverages and over a wide range of temperatures has been performed using the previously described BFS method for alloys. We will show that the complex behavior observed in experiment for these systems can be rationalized by three basic mechanisms already present at very low coverages.

The results for a single Fe atom on a Cu (100) surface, as seen in Table 2 predict surface interdiffusion, in apparent contradiction with the bulk immiscibility of Fe and Cu. To explain the observed behavior, Figure 6 shows the chemical energy terms for both the Fe atom and the ejected Cu atom, as the Fe atom goes to deeper layers. Both are positive, consistent with the expected Fe-Cu bulk immiscibility. However, as the Fe atom enters the surface layer, there is a substantial decrease in BFS strain energy. This indicates that the behavior of both Fe/Cu(100) and Fe/Cu(111) can be understood as a competition between two effects: chemical immiscibility and strain reduction. In spite of the chemical immiscibility, Fe will accept interdiffusion into Cu, as it greatly reduces its strain energy. Chemical energy gains of the order of 0.28 eV are also obtained if the ejected Cu atom migrates somewhere else on the surface. This is in agreement with *ab initio* calculations[50] predicting that Fe adsorption *in or below* the surface layer is preferred over Fe placed on the surface. The most stable configuration consists of a Fe atom in the first layer below the surface, with the ejected Cu atom somewhere else on the surface. Furthermore, Fe-Fe bonds are favored over Fe-Cu bonds, inducing, upon increasing the amount of Fe deposited, the formation of a (buried) compact Fe film over interdiffusion or surface alloying.

Table 2. Difference in energy (in eV/atom) between the lowest energy configuration (Fe atom in Cu bulk) and the other possible locations of a single Fe atom on Cu(100). The second and third columns list the BFS energies for unrelaxed and relaxed configurations, respectively. The last column displays *ab initio* results[50] for (periodic) unrelaxed configurations.

Configuration/Cu(100)	Unrelaxed	Relaxed	ab initio (unrelaxed)
Fe(O)	0.519	0.549	0.650
Fe(S)Cu(O)$_f$	0.586	0.609	
Fe(S)Cu(O)$_1$	0.562	0.578	0.42
Fe(1b)Cu(O)$_f$	0.092	0.098	
Fe(1b)Cu(O)$_2$	0.143	0.140	
Fe(2b)Cu(O)	0.002	0.016	
Fe(bulk)	0	0	0

With respect to step decoration, a simple analysis indicates that migration of deposited Fe atoms leads to interesting features in the structure and composition of nearby Cu steps. Figure 7 shows several possible adsorption sites for Fe along the edges of a rough Cu island on a (100) substrate. The BFS chemical contribution both for Fe and the surrounding Cu atoms is positive, but the increased coordination introduces significant strain energy gains, offsetting the unfavourable chemical energy which, alone, would inhibit step decoration. Each site is characterized by the number of nearest- and next-nearest-neighbours of the Fe atom, (n,m), and the difference in energy (in eV) with respect to the case where the Fe atom is located elsewhere in the overlayer, away from the Cu island, clearly showing that

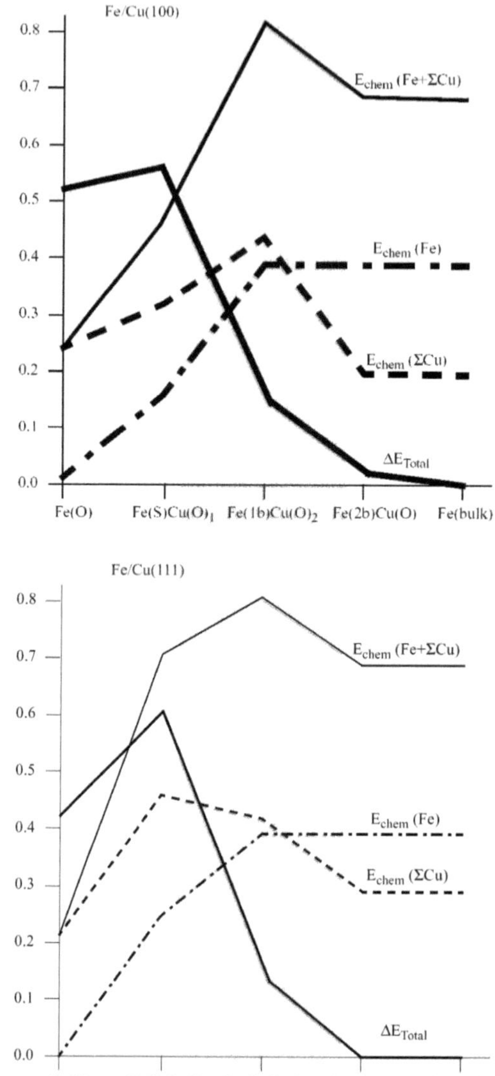

Figure 6. Solid lines: Difference in total energy for one Fe atom with respect to the final state (Fe atom in the bulk) for different locations of the Fe atom on Cu(100) (top) and Cu(111) (bottom) surfaces. The evolution of the chemical energy of the Fe atom (dashed-dot line), of the whole cell (thin solid line) and of the Cu atoms around the Fe atom (dashed line) are also shown.

the energy gain is maximized for larger values of n and m. Comparing these results with those discussed in Figure 6, it is clear that the processes of Fe/Cu mixing in the surface are clearly favored over the simple interdiffusion of Fe into the bulk.

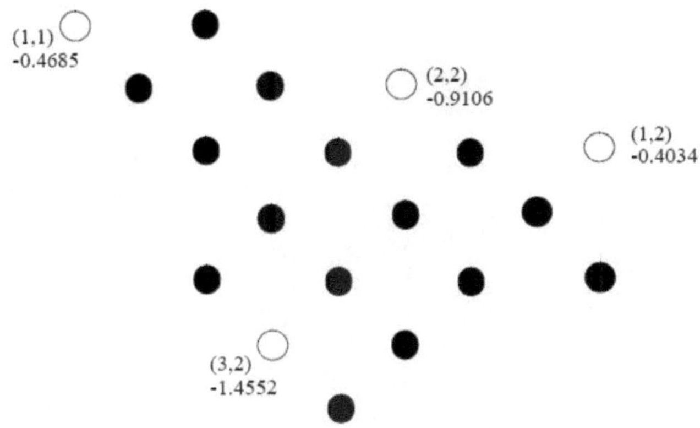

Figure 7. Energetic of different adsorption sites along the edges of a Cu island on Cu(100). For each site, the number of Cu nearest- and next-nearest-neighbors (n,m) are indicated, as well as the difference in energy (in eV) with respect to an isolated Fe atom somewhere else in the overlayer.

In order to address the observed step-edge erosion of the Cu substrate upon Fe deposition,[53] configurations including steps in the [100] and [110] directions on Cu (100) have been prepared. The step along the compact direction [110] is stable against vacancy formation, while the step along the open [100] direction is not, and vacancies appear along the step edge. This process requires approximately an activation barrier of 0.5 eV/atom. Adding Fe to the steps does not alter the trend mentioned before, but it further favors erosion in steps aligned in less compact orientations. In the [100] direction, the system lowers its energy if Cu atoms diffuse to neighboring sites, thus decreasing the strain energy of the Fe atom (clearly, much higher than when located on the [110] direction). Therefore, steps oriented along non compact directions will experience increased disorder as the Fe coverage increases, as observed experimentally.[52, 53]

It is interesting to note that BFS calculations indicate large energy gains when ejected Cu atoms nucleate to form an island (0.75 eV/atom gain for 12 Cu atoms randomly placed in the overlayer). This trend towards Cu island formation, also found in *ab initio* calculations[50] is a third ingredient relevant to the understanding of the existence and composition of the observed Cu-Fe islands. Experiments show that for 1 ML of Fe, the most favorable configuration upon rising slightly the temperature consists of the Fe ML being completely *covered* by a Cu monolayer. Figure 8 shows results of Monte Carlo simulations performed using the BFS method to simulate this

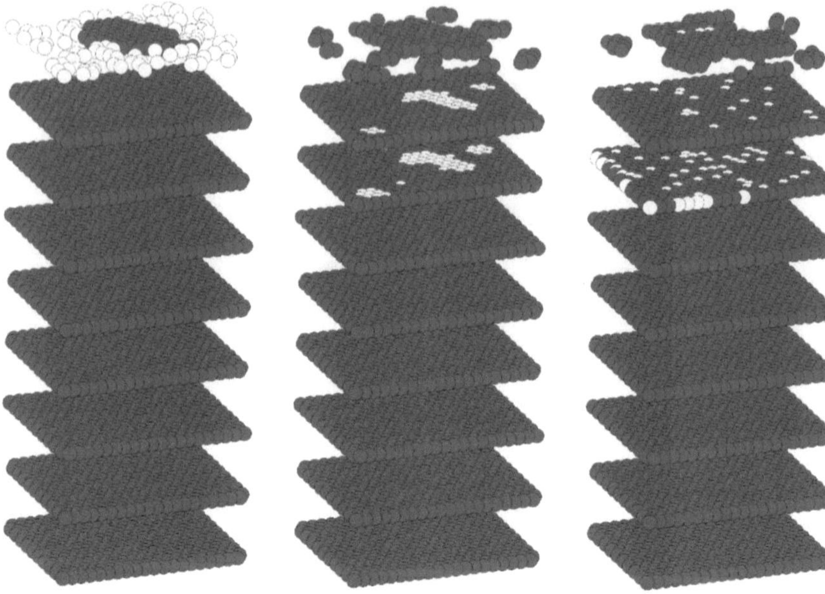

Figure 8. Initial and final states of Monte Carlo simulations performed at different temperatures, corresponding to a coverage of Fe of 0.21ML on Cu(111).

observation for Fe on Cu(111). The starting point is the cell shown on the left, with a compact Cu island and Fe atoms randomly distributed around it. The simulation uses the BANN algorithm,[9] where only exchanges between nearest-neighbors are allowed and the acceptance and rejection of the exchange depends on the available thermal energy. Only the adlayer, surface layer, and first layer below the surface are activated for exchanges. While the previous results indicate that step decoration would be preferred, the nature of the interdiffusion behavior discussed in Table 1 would favor further Fe interdiffusion into the bulk, once exchanges between adlayer and surface sites are allowed, thus leading to the behavior shown in the intermediate simulation cells of Figure 8. However, Fe nucleation under the Cu island, consistent with the preference for Fe-Fe bonds over Fe-Cu bonds, is clearly dominant over isolated Fe atoms in solution in the bulk. The lower surface energy of Cu also contributes to the burying of the Fe islands.

In summary, theoretical simulations performed with the BFS method for the deposition of a single Fe atom identify the three basic mechanisms driving the formation of the complex Fe/Cu interface: 1) Fe adsorption below the Cu surface and ejection of Cu atoms to the surface; 2) diffusion of Cu atoms away and nucleation of Cu islands to reduce the (repulsive) BFS chemical energy and 3) decoration of steps and island edges by Fe to decrease its BFS strain energy by surrounding itself with Cu. The preferential segregation of Cu to cover the Fe islands formed at higher coverages that occur at higher temperatures can also be well reproduced in Monte Carlo simulations using the BFS method.

4.4 Fe_4N(100) on Cu(100): a magnetism-driven surface reconstruction

Surface reconstructions have been the focus of intense research for more than thirty years.[61,62] They influence surface reactivity, determine kinetic pathways and control electronic charge and spin transport across interfaces. It is generally accepted that they are energetically driven by reducing the density of dangling bonds in semiconductors or by maximizing the surface density in many clean metal surfaces.[63] However, in spite of the efforts, the driving forces of many surface reconstructions have not been identified yet. As we move towards the study of more complex solids, new types of reconstructions appear. Here we describe one recent case in which the working together of experiments and theoretical simulation was essential to reach an initial understanding of the problem.

There is an increasing interest in magnetic nitrides driven by their chemically inert surfaces and their robust magnetic properties. Films of Fe_4N up to a thickness of 200 ML can be grown on Cu(100) by exposing the substrate at 700 K simultaneously to a flux of atomic N from a plasma (radio frequency) source *and* a flux of Fe from an electron gun evaporator. This results in the growth of single-crystal, epitaxial films of γ'-Fe_4N(100),[64] as deduced from X-Ray Diffraction, Conversion Emission Mössbauer Spectroscopy, Magneto-Optic Kerr Effect, and Rutherford Backscattering analysis[65] The films are ferromagnetic below 760 K.

The (100) surface of magnetic γ'-Fe_4N films shows a (2x2)*p*4*g* reconstruction detected by Low Energy Electron Diffraction (LEED) and STM[66] as shown in Fig. 9. The *p*4*g* superstructure is characterized by the absence of the four first half-order spots of the simple (2x2) LEED pattern. This indicates the existence of two perpendicular glide planes which produce the systematic extinction of the half order spots along <110> directions.

Low Energy Ion Scattering (LEIS) was employed to find out the atom positions at the surface by measuring the azimuthal dependence of the intensity of the Ar, Fe and N peaks in LEIS measurements.[66] The atomic positions for Fe and N were determined by comparing the measured LEIS spectra with simulations carried out with the code MATCH.[66] The results indicate a lateral shift in the Fe positions of the order of 0.4 Å along the <011> directions, while the N atoms are relaxed outwards 0.27 Å above the Fe atoms plane. The resulting surface arrangement is shown in Figure 10 and

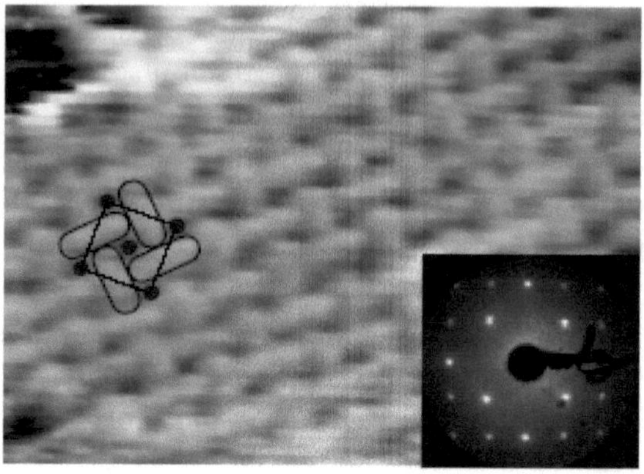

Figure 9. Atomically resolved STM image of the (2x2) p4g reconstruction of γ′-Fe_4N(100) recorded with sample bias voltage of +0.3 V and tunnelling current of 1.2 nA. The unit cell is indicated by the dotted line. The inset reproduces the LEED pattern at an electron energy of 110 eV of this surface.

it can be visualized as if the square units of four Fe atoms would rotate alternatively clockwise and counterclockwise around the surface normal, or else as if the Fe atoms would dimerize in two perpendicular directions.

The experimental results were compared with first principles, theoretical calculations performed in the context of Density Functional Theory[67] using the SIESTA[68] method. For the exchange correlation potential a Generalized Gradients Approximation[69] was adopted. The norm conserving pseudopotentials used follow the Troullier-Martins scheme[70] in the non local form proposed by Kleinman and Bylander[71] and with partial core corrections. The experimentally observed atomic positions in this complex reconstruction are properly accounted for by these first-principles total energy calculations,[66] which further reveal that the (2x2)*p*4*g* reconstruction originates from the perpendicular crossing of one-dimensionally dimerized Fe chains at a magnetically-ordered topmost Fe_4N_2 plane decoupled from the bulk.

In order to understand the nature of the reconstruction and its driving force, an isolated 2D layer of Fe_4N_2(100) was chosen as an starting model system.[66] Its calculated equilibrium lattice parameter was 3.65 Å, i.e the layer would be under substantial tensile strain when placed at the lattice parameter imposed by the bulk γ′-Fe_4N crystal. Figure 10 shows that the total energy displays a well defined minimum as a function of the *lateral* displacements of the Fe atoms. The layer, thus, would spontaneously lower its symmetry giving rise to a (2x2)*p*4*g* reconstruction. This is also true if the calculation is done at the 2D layer equilibrium lattice constant (3.65 Å). This

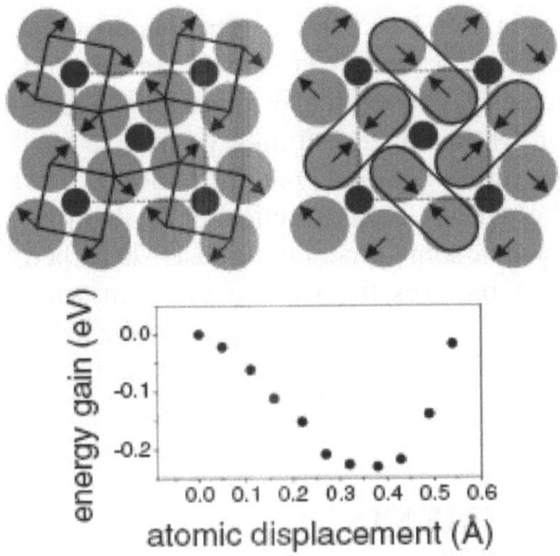

Figure 10. Energy gain per unit cell of a 2D $Fe_4N_2(100)$ layer as a function of the lateral displacement of the Fe atoms, which leads to a (2x2)*p*4*g* reconstruction. The starting configuration corresponds to the calculated equilibrium lattice parameter of bulk Fe_4N. The resulting atomic model for the (2x2)*p*4*g* reconstructed surface is also shown. The N atoms are shown as small spheres and the Fe atoms as large spheres. The arrows indicate the direction of displacement of the Fe atoms.

shows that strain in the layer is *not* responsible for the reconstruction. The predicted lateral displacement of the Fe atoms is 0.36 ± 0.05 Å, in excellent agreement with the experiment.[66] The corresponding energy gain is 220 meV per unit cell.

One possibility of getting the reconstruction on a bulk-terminated slab is by saturating the *subsurface* layer with N, i.e. by placing a Fe_4N_4 layer underneath the Fe_4N_2 surface. In this case the surface displays a clear energetic preference for the (2x2)*p*4*g* reconstruction, with a lateral displacements of the Fe atoms of 0.30-0.45 Å in agreement with the 2D Fe_4N layer calculation of Figure 10 and a vertical N-Fe distance around 0.2 Å. The subsurface layer shows an important buckling of about 0.6 Å. The physical nature of this *p*4*g* reconstruction is *not* similar to others observed previously. In the present case the surface needs to be magnetic and somehow decoupled from the bulk, and the driving force is a *dimerization* of the Fe atoms in two mutually perpendicular directions, as revealed by the atomic positions resulting from a calculation for a 2D $Fe_4N_2(100)$ layer with all the Fe and N atoms *fixed,* except those along one particular <011> row.[66] The calculated minimum of energy occurs when the Fe atoms form pairs with a displacement of 0.4 Å. The calculations indicate that this dimerization is the result of a 1D, Peierls-like, distortion of the lattice that reduces the minority spin DOS at the Fermi level. The majority spin band lies well below the Fermi energy. When all the atoms are allowed to move, pairing occurs in two mutually perpendicular directions, resulting in dimers of alternate orientation whose superposition gives rise to the "weaving" pattern seen in STM images of this (2x2)*p*4*g* reconstruction.

In summary, the surface reconstruction of $Fe_4N_2(100)$, although of the same *p*4*g* symmetry that some others previously described,[72,73] is of a rather different origin: i) it requires the surface to be magnetic and effectively decoupled from the rest of the crystal, which can be achieved by saturating the subsurface layer with additional N and ii) it involves two 1D, Peierls-like, pairing of Fe atoms perpendicular to each other.

ACKNOWLEDGEMENTS

This research was supported by the DGI (Spain), Projects No. FIS-2004-01026, NAN2004-08881-C02-01, Comunidad de Madrid (contracts No. GR/MAT/0155/2004 and S-0505/MAT-0194) and the “Programa Ramon y Cajal”. We thank our coworkers at LASUAM for fruitful discussions and their help in preparing some of the figures presented in this chapter.

REFERENCES

1. See for instance the volume *Surf. Sci.* **500**, (2002).
2. See the series of articles in *Nature* **437**, (2005) pp. 638-671.
3. J.J. de Miguel, J. Camarero and R. Miranda, *J. Phys.: Condens. Matter* **14**, 6155 (2002).
4. S.A. Chambers, *Surf. Sci. Rep.* **39**, 105 (2000).
5. T. Engel and K. H. Rieder, *Surf. Sci.* **109**, 140 (1981).
6. M. Faubel et al., J. Chem. Phys. **101**, 8800 (1994); D. R. Miller, in *Atomic and Molecular Beam Methods,* edited by G. Scoles (Oxford University Press, New York, 1988), Vol. I, pp. 14–53.
7. D. Farías, R. Miranda and K. H. Rieder, *J. Chem. Phys.* **117**, 2255 (2002).
8. A.L. Vázquez de Parga, O. S. Hernán, R. Miranda, A. Levy Yeyati, N. Mingo, A. Martín-Rodero, and F. Flores, *Phys. Rev. Lett.* **80**, 357 (1998).
9. G. Bozzolo and J. Garcés, in *"The Chemical Physics of Solid Surfaces",* D. P. Woodruff ed., Vol. 10, Elsevier (2002).
10. J. R. Smith, T. Perry, A. Banerjea, J. Ferrante and G. Bozzolo, *Phys. Rev. B* **44**, 6444 (1991).
11. P. Blaha, K. Schwartz, J. Luitz, WIEN97, Vienna University of Technology. Updated Unix version of the copyrighted WIEN code, P. Blaha, K. Schwartz, P. Sorantin, S. B. Trickey, *Comput. Phys. Commun.* **59**, 399 (1990).
12. E.W. Plummer, R. Ismail, A. Matzdorf, A.V. Melechko, J.P. Pierce and J. Zhang, *Surf. Sci.* **500**, 1 (2002).
13. N. D. Lang, *Phys. Rev. Lett.* **56**, 1164 (1986).

14. J. Jacobsen, B. Hammer, K. W. Jacobsen and J. K. Norskov, *Phys. Rev. B* **52**, 14954 (1995).
15. P. Sautet, J. Dunphy, D.F. Ogletree and M. Salmeron, *Surf. Sci.* **295**, 347 (1993).
16. W. Hofer, *Prog. Surf. Sci.* **71**, 147 (2003).
17. U. Diebold, J.F. Anderson, K-O. Ng and D. Vanderbilt, *Phys. Rev. Lett.* **77**, 1322 (1996).
18. D. Novak, E. Garfunkel and T. Gustafsson, *Phys. Rev. B* **50**, 5000 (1994).
19. U. Diebold. J. Lehman, T. Mahmoud, M. Kuhn, W. Hebenstreit, G. Leonardelli, M. Schmidt and P. Varga, *Surf. Sci.* **411**, 137 (1998).
20. H. Over, Y. D. Kim, A.P. Seitsonen, S. Wendt, E. Lundgren, M. Schmidt, P.Varga, A. Morgante and G. Ertl, *Science* **287**, 1474 (2000).
21. H. Over, A.P. Seitsonen, E. Lundgren, M. Schmidt and P. Varga, *Surf. Sci.* **515**, 143 (2002).
22. R. Schaub, P. Thostrup, N. Lopez, E. Laesgaard, I. Stensgaard, J. Norskov and F. Besenbacher, *Phys. Rev. Lett.* **87**, 266104 (2001).
23. R. Schaub, E. Wahlstrom, A. Ronau, E. Laesgaard, I. Stensgaard and F. Besenbacher, *Science* **299**, 377 (2003).
24. F. Calleja, A. Arnau, J.J. Hinarejos, A.L. Vázquez de Parga, W.A. Hofer, P.M. Echenique and R. Miranda, *Phys. Rev. Lett.* **92**, 206101 (2004).
25. M. Lindroos, H. Pfnür, G. Held and D. Menzel, *Surf. Sci.* **222**, 451 (1989).
26. G. Kresse and J. Hafner, *Phys. Rev. B* **47**, 558 (1993*); ibid* **49**, 14251 (1994).
27. G. Kresse and J. Furthmüller, *Comput. Matter Sci. B* **6**,15 (1996); and *Phys. Rev. B* **54**, 11169 (1994).
28. J. P. Perdew et al., *Phys. Rev. B* **46**, 6671 (1992).
29. J. Bardeen, *Phys. Rev. Lett.* **6**, 57 (1961).
30. N. Lang, *Comments Cond. Mat. Phys.* **5**, 253 (1989).
31. C. Corriol, F. Calleja, A. Arnau, J.J. Hinarejos, A.L. Vázquez de Parga, W.A. Hofer, and R. Miranda, *Chem. Phys. Lett.* **405**, 131 (2005).
32. D. Farías and K. H. Rieder, *Rep. Prog. Phys.* **61**, 1575 (1998).
33. D. Farías, M. Patting and K. H. Rieder, *J. Chem. Phys.* **117**, 1797 (2002).
34. M. Bertino and D. Farías, *J. Phys. C* **14**, 6037 (2002).
35. W. A. Diño, H. Kasai, A. Okiji, N.B. Arboleda, K. Fukutani, T. Okano, T. Okano, D. Farías and K. H. Rieder, *J. Vac. Soc. Japan* **46**, 391 (2003).
36. D. Farías, R. Miranda, K. H. Rieder, W. A. Diño, K. Fukutani, T. Okano, H. Kasai and A. Okiji, *Chem. Phys. Lett.* **359**, 127 (2002).
37. C. Díaz, H. F. Busnengo, P. Riviere, A. Salin, F. Martín, P. Nieto and D. Farías, *Physica Scripta* T**110**, 394 (2004).
38. D. Farías, C. Díaz, P. Nieto, A. Salin and F. Martín, *Chem. Phys. Lett.* **390**, 250 (2004).
39. D. Farías, C. Díaz, P. Riviere, H.F. Busnengo, P. Nieto, M.F. Somers, G.J. Kroes, A. Salin and F. Martín, *Phys. Rev. Lett.* **93**, 246104 (2004).
40. P. Nieto, E. Pijper, D. Barredo, G. Laurent, R.A. Olsen, E.J. Baerends, G.J. Kroes, and D. Farías, *Science* **312**, 86 (2006).
41. P. Riviere, H.F. Busnengo, and F. Martín, *J. Chem. Phys.* **121**, 751 (2004).
42. J. Thomassen, F. May, M. Wuttig and H. Ibach, Phys. Rev. Lett. **69**, 3831 (1992).
43. S. Müller et al., *Phys. Rev. Lett.* **74**, 765 (1995).
44. A. Biedermann, M. Schmid and P. Varga, *Phys. Rev. Lett.* **86**, 464 (2001).
45. A. Biedermann, R. Tscheling, M. Schmid and P. Varga, *Phys. Rev. Lett.* **87**, 086103 (2001).
46. V.L. Moruzzi, P.M. Marcus and J. Kübler, *Phys. Rev. B* **39**, 6957 (1989).
47. D. Spisák and J. Hafner, *Phys. Rev. B* **61**, 16129 (2000).
48. D. Spisák and J. Hafner, *Phys. Rev. Lett.* **88**, 056101 (2001).

49. D. Chambliss and K. Johnson, *Phys. Rev. B* **50**, 5012 (1994).
50. D. Spisák and J. Hafner, *Phys. Rev. B* **64**, 205422 (2001).
51. F. Dulot, B. Kierren and D. Malterre, *Surf. Sci.* **494**, 229 (2001).
52. M. Klaua, H. Hoeche, H. Jenniches, J. Barthel and J. Kirschner, *Surf. Sci.* **381** (1997) 106.
53. M. Passeggi Jr., J.E. Prieto, R. Miranda and J.M. Gallego, *Surf. Sci.* **462**, 45 (2000).
54. M. Passeggi Jr., J.E. Prieto, R. Miranda and J.M. Gallego, *Phys. Rev. B* **65**, 035409 (2001).
55. M. T. Kief and W. F. Egelhoff Jr, *Phys. Rev. B* **47**, 10785 (1993).
56. Th. Detzel and N. Memmel, *Phys. Rev. B* **49**, 5599 (1994).
57. J. de la Figuera, J.E. Prieto, C. Ocal and R. Miranda, *Phys. Rev. B* **47**, 13043 (1993).
58. J. de la Figuera, J.E. Prieto, C. Ocal and R. Miranda, *Surf. Sci.* **307-309**, 538 (1994).
59. J. de la Figuera, J.E. Prieto, G. Kostka, S. Müller, C. Ocal, R. Miranda, and K. Heinz, *Surf. Sci.* **349**, L139 (1996).
60. L. Gómez, C. Slutzky, J. Ferrón, J. de la Figuera, J. Camarero, A.L. Vázquez de Parga, J.J. de Miguel, and R. Miranda, *Phys. Rev. Lett.* **84**, 4397 (2000).
61. S. Titmuss, A. Wander and D. A. King, *Chem. Rev.* **96**, 1291 (1996).
62. R. Matzdorf, Z. Fang, J. Zhang, T. Kimura,Y. Tokura, K. Terakura and E. W. Plummer, *Science* **289**, 746 (2000).
63. A. Zangwill, "Physics at Surfaces" (Cambridge University Press, 1988).
64. J.M. Gallego, S. Y. Grachev, D. M. Borsa, and D.O. Boerma, D. Ecija, and R. Miranda *Phys. Rev. B* **70**, 115417 (2004).
65. J.M. Gallego, S. Y. Grachev, M.C.S. Passeggi Jr., F. Sacharowitz, D. Ecija, R. Miranda, and D. O. Boerma, *Phys. Rev. B* **69**, 121404(R) (2004).
66. J.M. Gallego, D. O. Boerma, R. Miranda and F. Ynduráin, *Phys. Rev. Lett.* **95**, 136102 (2005).
67. W. Kohn and L. J. Sham, *Phys. Rev.* **140**, A1133 (1965).
68. J. M. Soler, E. Artacho, J. D. Gale, A. Garcia, J. Junquera, P. Orderjon and D. Sanchez-Portal, *J. Phys. Condens. Matt.* **14**, 2745 (2002).
69. J. P. Perdew, K. Burke, M. Ernzerhof, *Phys. Rev. Lett.* **77**, 3865 (1996).
70. N. Troullier and J. L. Martins, *Phys. Rev. B* **43**, 1993 (1991).
71. L. Kleinman and D. M. Bylander, *Phys. Rev. Lett.* **48**, 1425 (1982).
72. J. H. Onuferko, D. P. Woodruff, and B. W. Holland, *Surf. Sci.* **87**, 357 (1979).
73. H. Onishi, T. Aruga, and Y. Iwasawa, *Surf. Sci.* **283**, 213 (1993).

Chapter 13

THE EVOLUTION OF COMPOSITION AND STRUCTURE AT METAL-METAL INTERFACES: MEASUREMENTS AND SIMULATIONS

Richard J. Smith
Physics Department, Montana State University, Bozeman MT 59717, USA

Abstract: In this chapter the growth of transition metal (TM) films, Fe, Co, Pd, Ni, and Ti, on surfaces of Al(110), Al(001), and Al(111) is discussed. Attention is drawn to similarities and differences arising from the substrate orientation. The tendency for Ti to grow as an overlayer while the other metals form interface alloys is discussed. Monte Carlo computer simulations using EAM potentials are presented for Ni on the three orientations of Al. Results of model calculations based on the BFS method are shown to nicely categorize the growth behavior of the five metals on Al in terms of the contribution of each atom to the overall formation energy. The technique of MeV ion backscattering and channeling has been used to study the interface structure and stoichiometry for the films on the Al surfaces. In some cases x-ray photoemission spectroscopy, low-energy (keV) He^+ ion scattering, and low-energy electron diffraction are used to obtain additional information about the structure of the TM-Al interfaces.

Keywords: Interface alloys, metal-metal interfaces, transition metals, Al, Fe, Co, Pd, Ni, Ti, EAM, BFS, ion channeling, MC simulations

1. INTRODUCTION

This is an exciting time in materials physics as we develop the ability to grow layered structures, only nanometers thick, in a controlled way, leading ultimately to new devices and new physics.[1,2] We have developed appropriate deposition techniques, and many of the tools necessary to look at and manipulate individual atoms on a surface, but the fundamental physics controlling thin film growth is not yet entirely understood. In spite of our skills we are still lacking a comprehensive model that can be used to reliably predict the atomic structure at a metal-metal interface. The problem becomes even more acute for the characterization of buried interfaces, not generally accessible to the arsenal of surface science techniques.[3] Ideally, we would like to grow structures with flat, chemically abrupt interfaces for electronic and magnetic devices, but we are hindered by the reality of rough surfaces, and the tendency of some materials toward island growth. The interfaces should at least be thinner than the films making up the structure if the physical properties of the device are to be determined by the characteristics of the films rather than by those of the interface. We have a limited selection of materials with good lattice matching, frequently resulting in layers with defects or strained layers of limited thickness. Furthermore, we must design interfaces that are stable at device growth temperatures where interdiffusion at the interface may compromise the device performance. When natural processes restrict the growth of sharp interfaces, a "work around" must be found. We have recently demonstrated one such approach, the use of an atomically thin interlayer of Ti at the interface between Fe and Al(001).[4] We believe that the Ti interlayer stiffens the Al surface by forming strong, intermetallic Ti-Al bonds. This structure inhibits the interdiffusion of Fe and Al that occurs at room temperature for the bimetallic interface with no interlayer, and remains effective up to about 400 °C. While this qualitative description is credible, we would like to have a more fundamental understanding of how the interlayer works, and a capability to predict possible interlayer materials for any film/substrate combination.

To make further progress in achieving at least chemically abrupt, if not flat, epitaxial interfaces, experiments are needed to better characterize the structure and composition at the buried interface. These results should in turn provide a database for testing our understanding of the physical phenomena competing during growth, and our ability to model and predict interface formation.

In this chapter we report on the characterization of the interface structure and composition for several transition-metal/aluminum interfaces grown at room temperature. The examples are chosen to demonstrate a range of behaviors, from epitaxial overlayers to disordered alloy interfaces. In most cases, we do not discuss the atomistic details of interface structure at low adatom coverage, something better left to microscopic probes such as the STM. Rather, we attempt to show how the evolution of composition at the buried interface depends on substrate orientation, and how the apparent stoichiometry of the interface alloy evolves with adatom coverage, something generally beyond the reach of tunneling microscopy. The emphasis here is to show what has been determined by probing the buried interface with ion channeling techniques, and in this way to identify problems in need of computer modeling and theoretical calculations to improve our understanding of interface evolution for metal-metal interfaces.

2. STRUCTURE OF METAL-METAL INTERFACES

The challenge of developing stable interfaces for thin-film metal multilayer structures can be seen if we take a brief preview of some of our results. Films of Pd, Ni, Fe and Co all form alloys at the interface with an Al single crystal substrate.[5-8] At room temperature this alloy can be several nm thick. On the other hand, Ti and Ag adatoms form epitaxial overlayers on Al single crystal surfaces.[9,10] For all of these cases (except Ag), the surface energy of the deposited metal is more than twice that of the Al surface,[11] as shown in Table 1. This implies that the surface energy can be reduced if Al atoms alloy with the adatoms and move to the surface. On the other hand, a thermodynamically stabile interface may result if there is a negative interface energy, e.g. in the form of a strong interface bond.[12] In addition, for epitaxial growth the lattice mismatch should be small so as to avoid significant strain in the overlayer. Table 1 also shows the lattice constant and crystal structure for the elements discussed here, and the formation energy for the various binary aluminide compounds (Al-X). Our results show that a model that considers only the formation energy of the intermetallic bond is inadequate since Ti, with an aluminide formation energy that is intermediate between that for Fe and Ni, forms an epitaxial overlayer on Al while Fe and Ni react with the substrate to form alloys. Also, among these three adatoms, the bcc Fe lattice has the best lattice matching with fcc Al if the unit cells are rotated by 45° with respect to each other, but the FeAl compound still forms. We have suggested that the stability of the Ti overlayer can be attributed to a combination of substrate strain associated with indiffusion of a relatively large adatom, and the kinetic barrier present for this indiffusion at room temperature.[4,13] We have also shown that at temperatures above 400 °C, the Ti atoms diffuse readily into the Al substrate, and ultimately occupy substitutional sites in the Al lattice.[14] On the other hand, Fe, Ni, Pd and Co atoms are relatively small and diffuse into the Al surface via interstitial trajectories already at room temperature, disordering the Al lattice, and forming interfacial compounds that have photoelectron binding energies similar to the TM-aluminides with the B2 CsCl structure.[15]

To form a stable interface we seek *chemical* and *structural* stability over a range of temperatures. In referring to *chemical stability* at the interface we are thinking about the tendency toward compound formation typical for many transition metal aluminides. For example, a thin layer of Ti may form a stable compound at the Al surface, but a third material deposited on the Ti

Table 1. Lattice constant and surface energy for selected elements, and the formation energy for Al-X compounds.

Elem (X)	a_{nn} (nm)[a]	E^s (J/m^2)[b]	ΔH_{AlX} (eV/at)[b]
Al	0.404 fcc	1.160	–
Ag	0.408 fcc	1.250	-0.04[c]
Ti	0.295 hcp	2.100	-0.39
Fe	0.286 bcc	2.475	-0.26
Ni	0.352 fcc	2.450	-0.61
Co	0.251 hcp	2.550	-0.56
Pd	0.388 fcc	2.050	-0.95

[a] Ref. 16.
[b] Experimental, Ref. 11.
[c] R. Hultgren, R. L. Orr, P. D. Anderson and K. K. Kelley, *Selected Values of the Thermodynamic Properties of Binary Alloys* (Wiley, New York, 1963).

layer could have larger formation energy with Ti. As a result, the competition for the Ti atoms at the interface could compromise the desired interface stability. *Structural stability* of the interface is more likely, but not guaranteed, if the lattice mismatch at the interface is small. For example, the interatomic spacing of Ti is close to that of Al (about 3% lattice mismatch), while the interatomic distance at the Fe(001) surface is nearly identical to that for Al(001) (< 1% mismatch).[16] Yet, we observe that Fe forms a disordered interface alloy approximately 1.5 nm thick at the Al(100) surface while Ti forms an fcc epitaxial overlayer on Al(100). Finally, the interface structure must have *thermal stability* and remain intact while the structure is heated, either during growth or during subsequent device operation. We would hope to have some activation barriers, associated with strain or chemical bonding, which would stabilize the interface. As mentioned above, we have found that the Ti adlayer on Al(110) is stable at the surface up to approximately 350 °C. On the more close-packed surfaces of Al(001) and Al(111) the Ti remains at the surface until the substrate temperature reaches about 400 °C.[14] The interlayer structure of Fe/Ti/Al shows evidence of Al diffusion to the surface at about 200 °C, but the bcc Fe film on the interlayer retains its structure until about 400 °C.[4]

Finally, it is important to recognize that a solution found for a stable interface during one growth process may not work in other growth situations. An especially relevant case for our work is seen in the recent report of Buchanan et al.,[17] who used grazing incidence x-ray reflectometry to measure interface mixing at room temperature in a series of aluminum-transition metal (TM) bilayers deposited with dc magnetron sputtering. A wide range of intermixing is seen, and it is not symmetrical in that the amount of mixing for films of the form TM-on-Al is often quite different from that for Al-on-TM, and usually considerably larger. The authors also note that the width of the intermixed region in the sputtered films is considerably larger than the widths reported by us for TM films deposited using physical vapor deposition on single crystal substrates. The authors suggest that the differences might be due in part to grain boundary diffusion in the sputtered films. Consequently, if one wants to fully understand the role played by the materials themselves in interface alloying, separate from the contributions of defects in the evaporated films, it becomes even more important to carry out studies of the type described in this chapter. In contrast to our studies on single crystal surfaces, the results of Buchanan et al. may be more directly related to thin film applications in industries using sputter deposition. On the other hand, our results should more easily lend themselves to direct comparison with the results of computer modeling for understanding the contribution of various alloying mechanisms during interface formation. Clearly, both types of studies are needed.

3. EXPERIMENTAL TECHNIQUES

Our primary tools for determining the geometric structure and composition of the metallic interfaces are high-energy ion backscattering and channeling (RBS/c or HEIS), x-ray photoemission spectroscopy (XPS), low-energy electron diffraction (LEED), and low-energy ion scattering spectroscopy (LEIS). We characterize the evolution of the interface structure by measuring with ion backscattering the number of substrate atoms displaced from bulk lattice sites as a function of various deposition parameters, such as film thickness and substrate temperature. We measure *for the same surface* the core-level photoemission spectra to determine any chemical shifts in the atomic binding energies that might be used to identify the compounds formed at the interface. Since the depth resolution of HEIS is not very good for disordered surfaces, we complement the measurements with LEIS to determine surface composition with monolayer sensitivity. This combination of HEIS, XPS and LEIS is especially advantageous for obtaining both structural and chemical information about the buried interface. To relate our ion scattering and photoemission measurements to structural models for the interface, we compare Monte Carlo (MC) computer simulations of the interface formation with our experimental results. Snapshots of the evolving interface provide more than a visual clue to what is happening at the interface in that we use them to calculate ion scattering and photoelectron yields. Thus we are testing *experimentally* the reliability of using MC simulations to predict interface evolution.

High-energy ion scattering, when used in the channeling mode, provides a powerful tool to probe the geometric structure of solid surfaces and thin films.[18] In addition, ion backscattering provides a direct means for accurately

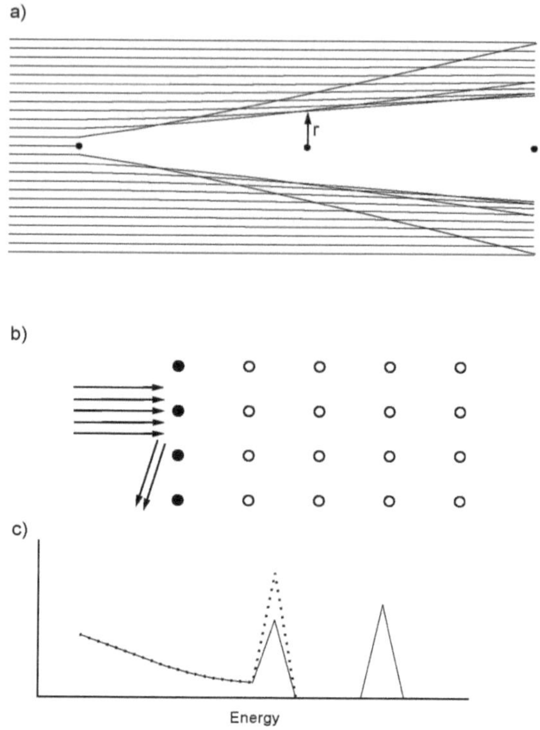

Figure 1. Schematic drawing showing (a) the shadow cone concept for ion scattering, (b) the incident ion beam aligned with a single-crystal surface having a single layer of adatoms (solid circles), and (c) the surface peak of backscattered ions for the clean substrate (dotted curve), the reduction of the surface peak associated with shadowing of the substrate atoms by a layer of adatoms, and the formation of an additional peak at larger energy due to the heavier adatoms.

measuring the coverage of atoms deposited on the surface. In the channeling geometry the ion beam is incident along a low-index direction of the single-crystal target, and most of the ions channel into the substrate along the relatively open areas between the parallel rows of atoms extending from the surface into the solid, as shown schematically in Fig. 1.b. Ions that backscatter from the first few atoms in each row give rise to the surface peak (SP) observed in the backscattered ion energy distribution curve (Fig. 1.c), while the small-angle forward scattered ions form a shadow cone that extends along the row into the solid (Fig. 1.a). For a static model of an ideally terminated bulk lattice, the first atom of a row (surface atoms) shadows subsequent atoms along that row. If deposited metal atoms disorder the surface, substrate atoms are uncovered and the backscattered ion yield from substrate atoms increases. If on the other hand, deposited metal atoms sit on lattice sites at the surface, the substrate atoms are shadowed even more, and the backscattered ion yield decreases for substrate atoms. The measured surface peak areas are converted to area densities of visible target atoms (atoms/cm^2) using the Rutherford scattering cross section, the solid angle subtended by the detector, and the time-integrated incident ion current. Because the Rutherford scattering cross sections are known, quantitative structural information about the surface may be extracted from the data by comparing measured channeling yields with calculated scattering yields obtained from computer simulations of the channeling experiment for various atomic configurations.[19]

Low-energy (keV) ion scattering allows us to study the surface composition for the growing film and to investigate possible interdiffusion or alloy formation at the surface with monolayer depth resolution.[20] While LEIS is not a quantitative technique like RBS, the extreme surface sensitivity made possible by using ion detection, combined with the capability of RBS for providing absolute metal coverage, makes LEIS a valuable component of these studies. Electron diffraction (LEED) allows us to characterize the long-range order of the surface while channeling and shadowing of the high-energy He^+ ions provides quantitative information about local order for adatoms above substrate lattice sites.

The experiments were carried out in an ultra-high vacuum (UHV) surface analysis chamber that includes facilities for HEIS, LEIS, XPS and LEED.[21] A chamber base pressure of 1.6×10^{-8} Pa was obtained after baking the system. The Al single crystals were cut and polished to within 1° of the respective crystallographic planes, as measured by x-ray diffraction. The crystals were then chemically etched for 15 seconds in an aqueous solution containing HCl(1.5%), HF(1.5%) and HNO_3(2.5%), and mounted in the UHV chamber on a thick Mo block attached to a three-axis goniometer for channeling measurements. The temperature of the Mo block was monitored using a calibrated Pt resistor mounted inside of the block. After every experiment, and before metal deposition, the samples were sputter-cleaned with 1-1.5 keV Ar^+ ion bombardment for several hours with the sample at room temperature, followed by annealing the sample at 450 °C for 15 minutes. The cleaning procedure was repeated until the photopeak associated with aluminum oxide was completely removed from the XPS spectrum. The O 1*s* photopeak could not be used to reliably monitor the Al surface oxide because the XPS analysis area included a small portion of the Mo sample holder surrounding the Al crystal. However, in some experiments, LEIS measurements made with a sub-millimeter size He^+ ion beam, confirmed that after cleaning the sample negligible oxygen remained on the surface when the Al oxide peak was absent from the XPS spectrum. After cleaning the sample, a collimated beam of MeV He^+ ions, passing through an aperture of 1.2 mm^2 area, was used to carry out the ion channeling measurements. The sample was aligned with the ion beam incident along the low-index surface normal direction by minimizing the backscattered ion yield in a small energy window behind the surface peak of the backscattered ion spectrum. During metal deposition channeling measurements were performed with the ion beam incident perpendicular to the surface. The standard dose of incident He^+ ions for one spectrum was 1.6×10^{15} ions/cm^2, a sufficiently low dose of ions to avoid surface damage. A solid-state detector was used to collect the backscattered He^+ ions at a scattering angle of 105°. Metal films were vapor-deposited on the Al crystal surfaces from resistively heated wire filaments (0.25mm diameter, 3 strands). Deposition rates in these experiments were between 0.2 and 1.0 ML/min, as specified for each experiment below. One monolayer (ML) equivalent coverage as used here is equal to an atomic density of 0.862 x 10^{15} atoms/cm^2 on the Al(110) surface, 1.22 x 10^{15} atoms/cm^2 on the Al(001) surface, and 1.41 x 10^{15} atoms/cm^2 on the Al(111) surface, respectively. The adatom coverage after evaporation was determined with the sample rotated out of the channeling alignment. Measurements with the random-alignment geometry eliminate possible errors associated with the shadowing of adatoms in determining the film coverage. The uncertainty in the ion scattering yields reported here is estimated to be < 6% with the largest contribution to the uncertainty coming from the determination of the detector solid angle, and smaller contributions coming from uncertainties in the integrated charge, the scattering angle, and the determination of the surface peak area.

LEIS measurements were made using a VSW-100 hemispherical analyzer (100 mm radius) and a 1 keV He^+ ion beam. The LEIS scattering angle was 128° and the angle of incidence was typically 45° from the surface normal, accept when doing angular yield scans. The XPS spectra were recorded using the VSW-100 analyzer in a fixed-analyzer-transmission mode with pass energy set to 50eV and a scan rate of 0.1 eV/sec. A Mg K_α (1253.6 eV) x-ray source with 300 W power was used to generate x-rays. To reduce uncertainty in the ion channeling yields measured as a function of adatom coverage, the channeling and XPS measurements and sequential evaporations were all made without moving the sample. This restriction meant that the photoelectrons entered the analyzer with a polar emission angle of 30° from the sample normal, and the adatom flux was incident on the surface at an angle of approximately 45° from the surface normal.

4. RESULTS AND DISCUSSION

4.1 Fe films on Al(001) and Al(110)

4.1.1 Al(001) surface

Figure 2 shows two ion scattering spectra collected after sputter-cleaning and annealing the Al(001) sample. The upper spectrum (open circles) was taken with the 0.56 MeV ion beam incident on the sample in a random direction near normal incidence, while the lower spectrum (solid circles) was taken with the beam incident along the [001] direction, i.e normal to the surface. The measured value for the normalized minimum yield, χ_{min}, was 3.7%, measured at an energy just to the left of the surface peak. The calculated value for χ_{min} is 3.6%, obtained using a one-dimensional rms atomic vibration amplitude of 0.105 Å at room temperature, and a crystal lattice constant of 4.05 Å.[22] This good agreement is important since it serves as a measure of the bulk crystal quality. Larger measured values of χ_{min} result from increased ion dechanneling which can be associated with defects resulting from sputtering and sample preparation. The measured surface peak area (SPA) in the aligned spectrum corresponds to 9.14 x 10^{15} atoms/cm^2, or 7.5 ML, visible to the incident ion beam. The calculated yield is 8.4 ML, obtained using the VEGAS simulation code with uncorrelated thermal vibrations for the atoms and no enhancement of vibration amplitudes at the surface.[19]

In Fig. 3 we show the Al and Fe surface peaks before and after a deposition of 0.87 ML of Fe. From Fig. 3 we see that the Al surface peak area increases after the Fe deposition. This indicates that more Al atoms are visible to the incident ion beam in the presence of Fe atoms at the surface. That is, surface Al atoms have moved from their equilibrium positions and reduced the shadowing of subsurface Al atoms. If Fe atoms formed an ordered overlayer directly above Al atoms, as might be expected due to the excellent lattice match with Al, then we would expect to see a reduction in the SPA of Al due to the Fe atoms shadowing the Al atoms from the incident He ions. Instead we see an increase in the SPA of Al suggesting that Fe and Al atoms are intermixing at the interface, and Fe atoms are displacing the Al atoms from their equilibrium positions. A triangular background subtraction method was used to remove the background signal under the surface peak.[23] After this background removal the surface peak areas are used to calculate the number of Al and Fe atoms visible to the ion beam. The Fe SPA gives a direct measure of the surface Fe coverage throughout the experiment.

In Fig. 4 we plot the number of Al atoms/cm^2 (Al-SPA) visible to the incident He^+ ion beam as a function of Fe coverage (Fe-SPA). The deposition rate in these experiments was typically 0.83 ML/min. Such plots are very useful for characterizing the evolution of the metal-metal interface, and have been used by others to develop models for alloy formation at interfaces.[24] Throughout this chapter we will display the ion channeling results in this form so as to allow quick comparisons between different binary systems. In Fig. 4 we see that more Al atoms become visible to the incident ion beam as the Fe coverage increases. There are two main regions to note in this figure. In the first region the number of visible Al atoms increases sharply with the Fe coverage up to about 5 ML.

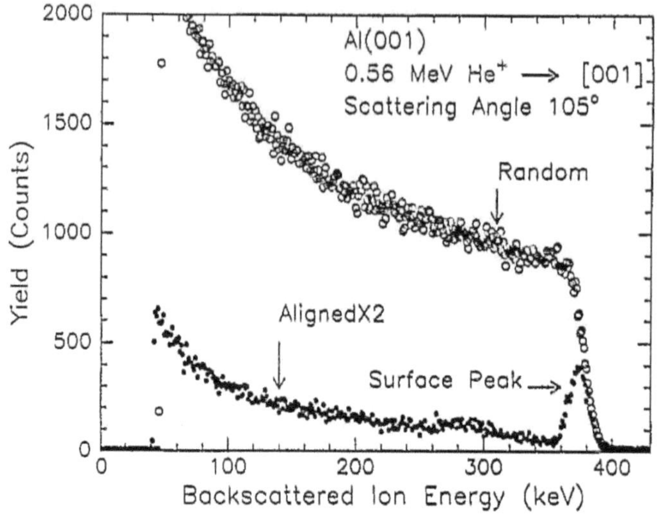

Figure 2. He^+ backscattered ion spectra for 0.56 MeV ions incident on the clean Al(001) surface. Open circles represent the spectrum taken with a random angle of incidence near the normal direction, and solid circles show the results for channeling along the [001] direction, multiplied by 2.

The slope of the curve in this region is 1.15 Al atom per Fe atom, suggesting that approximately one Al atom is displaced for every Fe atom deposited on the surface. Therefore the average stoichiometry of the mixed interface would be 1:1, i.e. FeAl. This is reasonable since the heat of formation of FeAl is -28 kJ/mol, and FeAl is a stable phase in the Fe-Al phase diagram with the relatively simple B2 (CsCl) structure, and cubic lattice constant of 2.91Å.[15]

At larger Fe coverage the curve in Fig. 4 seems to saturate with a zero slope. These observations suggest that the deposited Fe atoms interact with Al atoms on the Al(001) surface and form a mixture of FeAl at the interface up to about 5 ML Fe coverage, and that an Fe film ultimately covers the mixed surface for larger coverage. This is in good agreement with the results of Anderson and Norton[25] for which alloy formation was reported up to about 4 ML of Fe coverage, and growth of an Fe(001) overlayer was seen for higher coverage. A similar analysis was carried out for the area of the Fe surface peak. A comparison between the HEIS spectra collected with the ion beam incident in a channeling and a random geometry shows negligible Fe shadowing for Fe coverage less than 4.5 ML. A small, but measurable amount of Fe shadowing is observed for coverage greater than 5 ML.

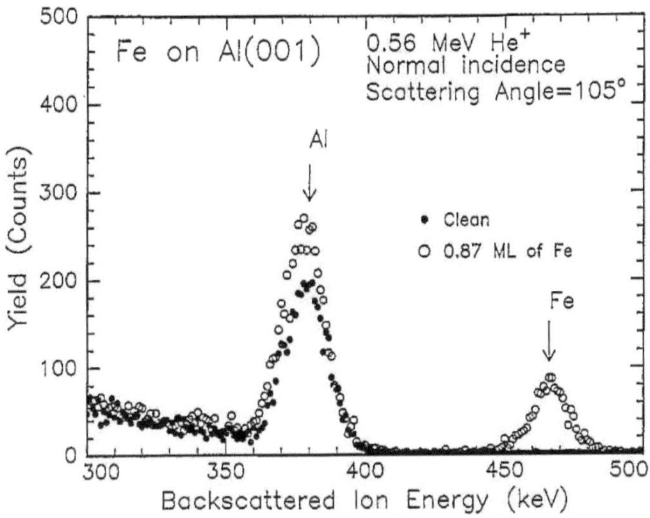

Figure 3. Ion scattering channeling spectra for the clean Al(001) surface (solid circles) and for 0.87 ML Fe deposited on the Al(001) surface (open circles). The energies for backscattering from Al and Fe surface atoms are indicated by the arrows.

This small difference in the Fe hitting probability suggests that oriented Fe islands may be growing on the interface alloy for larger Fe coverage. A more detailed discussion of these results can be found in Ref. 7.

4.1.2 Al(110) surface

We performed a similar analysis for the deposition of Fe on Al(110). For this system it is again clear that strong intermixing occurs at the Fe-Al interface rather than forming a simple Fe overlayer as expected based on the small lattice mismatch. In Fig. 5 we show the characteristic growth curve for Fe on Al(110), determined by measuring the areas of the Al and Fe surface peaks in the channeling spectra. From this figure we see that more Al atoms become visible to the incident ion beam as the Fe coverage increases up to about 9 ML (7.76×10^{15} atoms/cm^2). There are three main regions to recognize in this figure. In the first region the number of visible Al atoms increases sharply with the Fe coverage up to 2 ML. The slope of the curve in this region is 3.2 meaning that each Fe atom effectively displaces about 3 Al atoms. This is suggestive of forming an $FeAl_3$-like compound in the first region. $FeAl_3$ is a stable phase in the Fe-Al phase diagram with a formation energy of -28 kJ/mol.

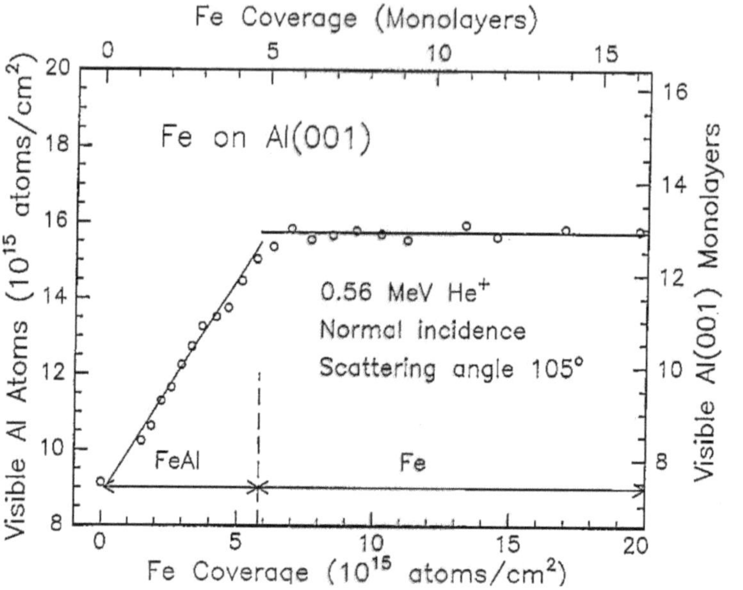

Figure 4. Visible Al atoms, at 0.56 MeV incident ion energy, as a function of Fe coverage deposited at room temperature on Al(001). Two regions of film growth are indicated. The solid lines are linear fits to the data in the two regions.

In the second region, up to 9 ML Fe coverage, the number of visible Al atoms increases with a slope of only 0.96 Al atom per Fe atom, suggesting that an FeAl-like compound is being formed. After 9 ML of Fe coverage the curve seems to saturate along a line with zero slope. It appears that the displacement of Al atoms has stopped after 9 ML Fe, and an Fe overlayer begins to form. These observations suggest that Fe atoms are mixing with Al atoms on the Al(110) surface at two different rates, forming a surface alloy at the interface, until an Fe metal film ultimately covers the mixed interface. The mixing continues up to about 9 ML of Fe coverage at room temperature. However, it is important to note that it is difficult to confirm with ion scattering alone the actual stoichiometry of a compound when the layer is thin. There may be some displacements of Al atoms just below the alloy interface. Also, near-surface dechanneling below the interface can occur as a result of ion beam defocussing by near-surface Fe atoms. We have reported this phenomenon previously for Ni films on Al(110),[26] and review these results later in this chapter. A more detailed discussion of these results, and a summary of results for Fe-Al interfaces from the literature can be found in Ref. 7.

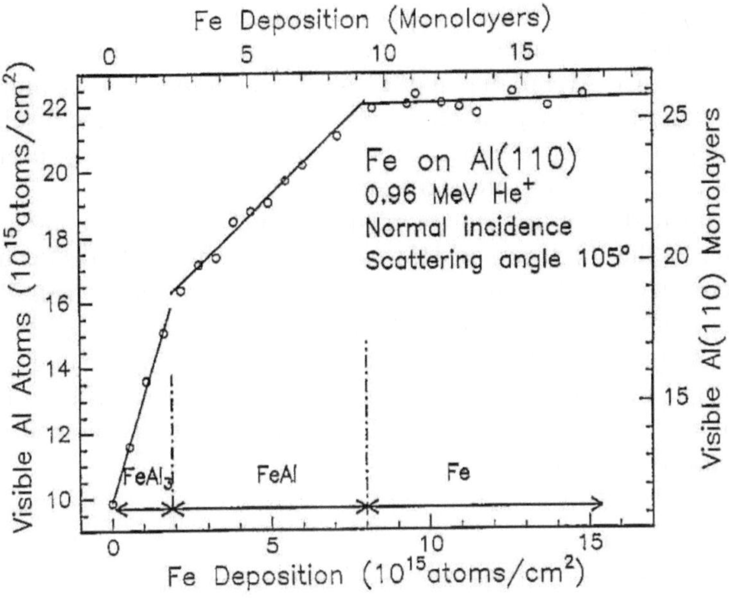

Figure 5. Visible Al atoms, at 0.96 MeV incident ion energy, as a function of Fe coverage on the Al(110) surface. The solid lines are linear fits to the data within the three coverage regimes indicated by the vertical lines.

4.2 Co films on Al(100) and Al(110)

4.2.1 Al(100) surface

In Fig. 6 we plot the characteristic growth curve for Co films deposited on the Al(100) surface, that is, the number of visible Al atoms as a function of Co coverage, as determined from the Al and Co surface peak areas in channeling spectra similar to those shown in Fig. 3. Initially, the number of visible Al atoms increases rapidly with Co coverage up to about 3 ML. The slope of the curve in this region is equal to 1.98. Subsequently, the slope is almost zero, indicating that Co metal has started to grow on top of the mixed interface. Furthermore the low energy side of the Co surface peak does not have any broadening up to 9 ML of Co coverage, ruling out the possibility of any diffusion of Co atoms deep into the bulk, that is to depths greater than the depth resolution of our detector (135 Å). The presence of the Co atoms at the surface causes the minimum yield behind the Al surface peak to increase gradually. This may be due to small deflections of some He$^+$ ions as they traverse the disordered interface layers.[6,26] We infer that the deposited Co atoms are staying near to the surface region, at least within the depth resolution of our solid-state detector.

The large increase in the number of Al atoms visible to the incident ion beam, measured as a function of Co coverage and shown in Fig. 6, indicates that there is Co-Al mixing at the interface, although it is limited to a relatively thin layer. In a conventional interpretation[24,27] of the results in Fig. 6, the slope of 1.98:1 would suggest the formation of $CoAl_2$, a phase that does not exist in the bulk phase diagram.[15]

A compound with this stoichiometry might form in the non-equilibrium conditions at the solid surface. However, based on the XPS results discussed elsewhere,[8] we conclude that a CoAl-like compound is formed. The CoAl compound has the highest heat of formation for compounds in the phase diagram, and it has a relatively simple crystal structure so we might expect it to form first. A very similar situation was observed and has been discussed in detail for the case of Ni on Al(110).[6] We present a summary of those arguments in section 4.4.

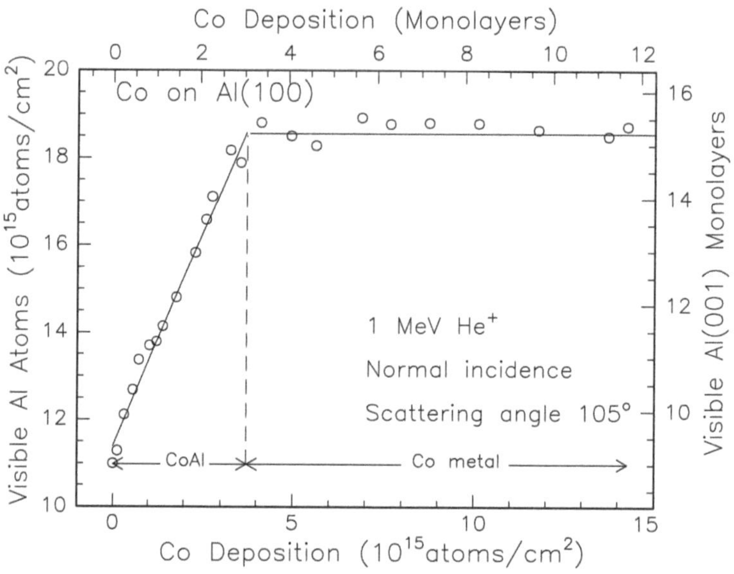

Figure 6. Visible Al atoms at 1 MeV incident ion energy, as a function of Co coverage deposited at room temperature on Al(001). Two regions of film growth are indicated. The solid lines are linear fits to the data in the two regions.

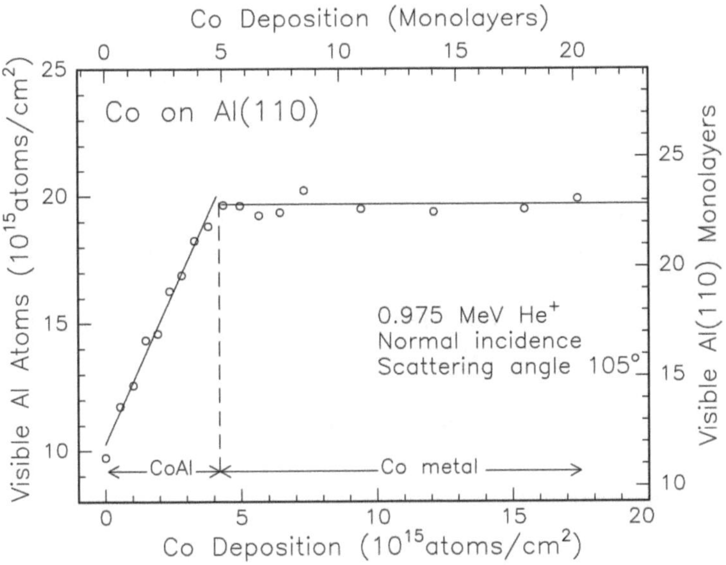

Figure 7. Visible Al atoms, at 0.975 MeV incident ion energy, as a function of Co coverage deposited at room temperature on the Al(110) surface. The solid lines are linear fits to the data within the two coverage regimes indicated in the figure.

4.2.2 Al(110) surface

A similar analysis was performed as a function of Co coverage on the Al(110) surface. Figure 7 shows the characteristic growth curve for this surface, plotting the number of Al atoms visible to the incident ion beam as a function of Co coverage.

From the figure it is clear that there are two distinct regions of growth. Initially, the number of visible Al atoms increases at a rate of 2.3 Al atoms per deposited Co atom up to about 5 equivalent (110) monolayers. This is suggestive of forming a Co_2Al_5-like compound in the first region. However, based on our results for Co on Al(001) and for Ni on Al(110),[6] we believe that it is difficult to confirm the stoichiometry of the alloy with the ion scattering results alone. Near-surface dechanneling below the interface can occur as a result of ion beam defocusing by the relatively dense Co-Al mixture at the surface. After 5 ML of Co coverage the curve seems to saturate with a zero slope. In other words, the displacement of Al atoms has stopped after a coverage of 5 ML, and a Co metal overlayer grows on top of the mixed interface. Again, no evidence was seen for diffusion of Co deep into the Al(110) substrate for deposition at room temperature. Also, as soon as a small amount of Co was deposited on the Al(110) surface the LEED spots completely disappeared. No LEED pattern was observed for any Co coverage up to 15 ML for Co deposited on the Al(110) surface.

Perhaps the most remarkable result for Co deposition on Al is the similarity of the characteristic curves for the two surfaces shown in Figs. 6 and 7. The interface evolution appears to stop at a coverage of about $4x10^{15}$ at/cm^2 on both surfaces. Such a similarity occurs also for Pd on Al (Section 4.3). However, the situation for Ni on Al is quite different (Section 4.4). The metals Ni and Co, when combined with Al, have a combination of properties that are very similar, as seen in their respective phase diagrams. Thus we might expect similar ion channeling results when depositing these metals on Al surfaces. However, as discussed below, the two systems exhibit some remarkably different behavior. Such results beg for more studies and model calculations to fully understand these interfaces.

4.3 Pd films on Al(100) and Al(110)

4.3.1 Al(001) surface

In Fig. 8 we plot the characteristic growth curve for Pd films on Al(001), showing the number of Al atoms visible to the incident ion beam as a function of Pd coverage, determined from the areas of the surface peaks in channeling spectra similar to those shown in Fig. 3. There are two main regions to note in this figure. First, the Al surface peak area exhibits a liner increase with a slope of approximately 1 up to 5 ML of Pd coverage. After this coverage, there is an apparent saturation stage during which the Al surface peak area remains unchanged. The increase in the number of visible Al atoms with increasing Pd coverage indicates again that more Al atoms are visible to the ion beam in the presence of Pd atoms at the surface. That is, surface Al atoms have moved from their initial equilibrium positions and have reduced the shadowing of substrate Al atoms. This observation rules out the formation of an ordered Pd overlayer at lattice sites directly above Al surface atoms.

A similar plot was made for the number of Pd atoms visible to the incident ion beam in the normal-incidence channeling direction as a function of Pd coverage. The coverage was measured with the ion beam incident in a random direction so as to avoid any Pd-Pd shadowing. Very little shadowing of Pd atoms is seen in the channeling alignment, suggesting that there is no alignment of Pd atoms along the [001] direction, normal to the surface. X-ray photoemission measurements were also carried out for Pd deposition on Al(001) and Al(110), discussed below. In both cases it was found[5] that for deposition of Pd, up to 5 ML on Al(001), the Pd $3d_{5/2}$ photopeak exhibited a chemical shift of approximately 1.7 eV to higher binding energy. This shift has been shown to be characteristic of the formation of AlPd. A much larger shift of 2.5 eV occurs in the formation of

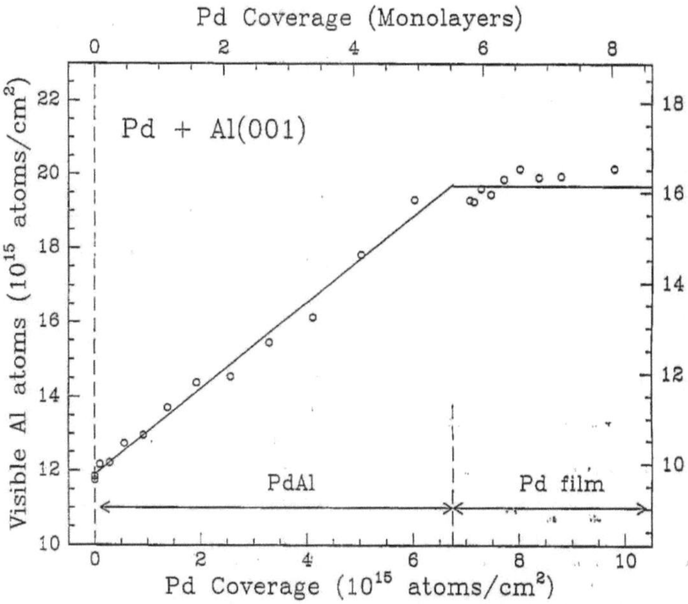

Figure 8. Number of Al atoms visible to the incident ion beam as a function of Pd coverage on Al(001). The solid lines are the least-squares fit to the data points in the two regions indicated in the lower part of the figure.

Al_3Pd.[28] Thus, for the case of Pd deposition on Al, both the XPS and ion channeling results are in agreement in indicating the formation of AlPd at the interface between Pd films and Al(001) single crystal substrates.

4.3.2 Al(110) surface

The results of ion channeling measurements for Pd deposition on Al(110) are shown in Fig. 9, again using the areas of the surface peaks to determine the number of Al atoms visible to the incident ion beam. Only two distinct regions of growth are evident in the figure, an initial region with a slope of approximately 1.2 Al atoms per deposited Pd atom, and a region of apparent saturation where the area of the Al surface peak no longer increases. As mentioned in the preceding section, XPS measurements for Pd film growth on both Al(001) and Al(110) clearly support the formation of an interface compound with the AlPd stoichiometry and a chemical shift of 1.7 eV. The main difference between the two substrates is that the initial growth region for Pd on Al(110) continues to about 9 ML of Pd deposition, as contrasted with saturation after 5 ML deposition on the Al(001) surface. Another

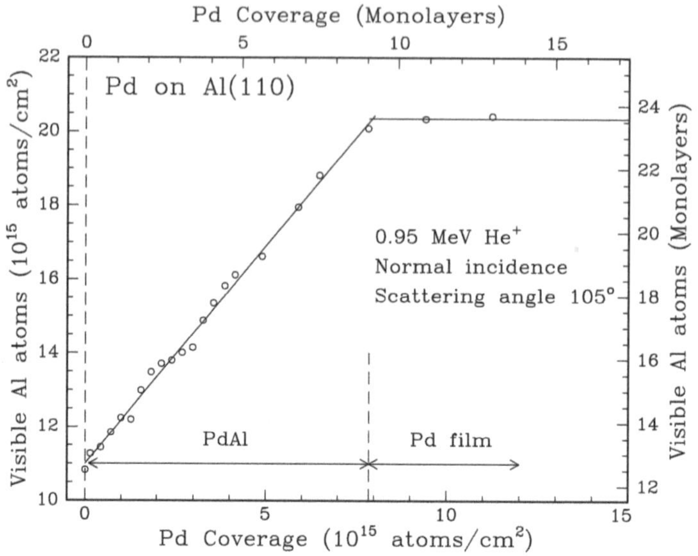

Figure 9. Number of Al atoms visible to the incident ion beam as a function of Pd coverage on the Al(110) surface. The solid lines are the least-squares fit to the data points in two regions. Two stages of film growth are indicated.

significant difference is that a small amount of Pd appears to be diffusing deep into the Al(110) substrate, even for deposition at room temperature. This may be the result of the more open Al(110) surface in combination with the quite large chemical formation energy for AlPd (see Table 1). A similar measurement of Pd shadowing vs Pd coverage did not reveal any significant Pd shadowing. We concluded that there is no Pd-Pd or Al-Pd alignment normal to the Al(110) surface.

4.4 Ni Films on Al(001), Al(110), and Al(111)

4.4.1 Al(111) surface

In Fig. 10 we show the characteristic growth curve as measured with ion channeling for Ni deposition on Al(111), using an incident ion beam of 0.95 MeV He^+. The Al surface peak area increases with two different slopes up to a coverage of 5.5 ML Ni. The increase means that the deposited Ni atoms react with the Al atoms of the substrate and displace Al atoms from their initial positions, which suggests alloy formation. The plateau with a

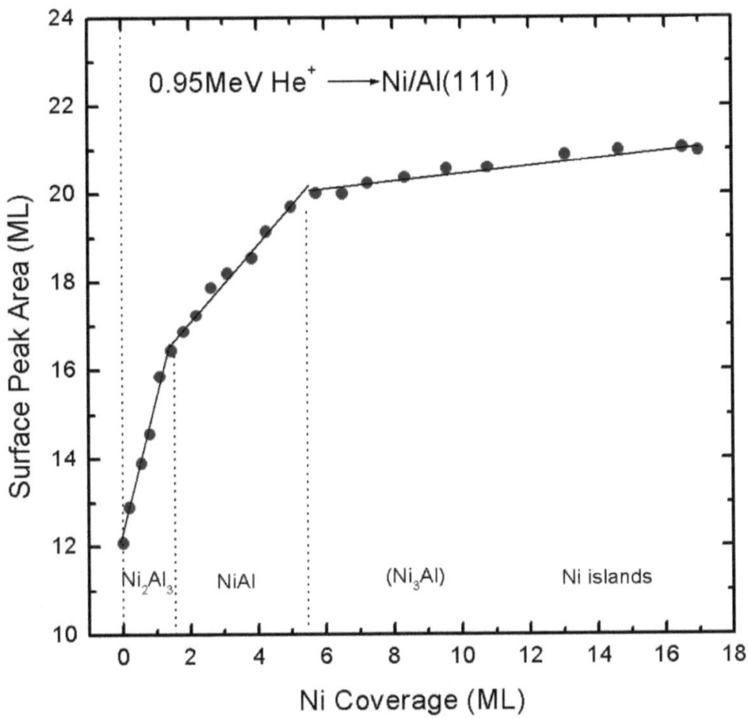

Figure 10. Number of Al atoms visible to the incident ion beam as a function of Ni coverage on the Al(111) surface. The solid lines are the least-squares fit to the data points in two regions. Two stages of film growth are indicated.

slight increase after the deposition of 5.5 ML Ni means that the deposited Ni atoms no longer displace Al atoms from their initial positions. In the initial stage of deposition up to 5.5 ML of Ni, the slopes are approximately 3 and 0.9. This suggests that one Ni atom effectively displaces 3 and 0.9 Al atoms of the substrate, respectively. From these slopes one might infer the formation of $NiAl_3$ and NiAl compounds. However, analysis of compound formation as measured with XPS suggests a different interpretation.[6] The measured shifts in binding energies for the Ni $2p_{3/2}$ photopeak, as well as the position of the satellite feature in the XPS spectrum, suggest that during the initial deposition up to 1.5 ML, a Ni_2Al_3-like compound is forming on the surface. During the next growth phase up to 5.5 ML, the XPS results agree with the formation of a NiAl-like compound. Another interesting observation for Ni growth on Al(111) comes from LEIS measurements[6] that show very little Ni on the Al surface up to a coverage greater than 5 ML (measured with RBS), further supporting the idea of Ni-Al alloy formation at the surface with Al migrating to the surface because of its lower surface energy.

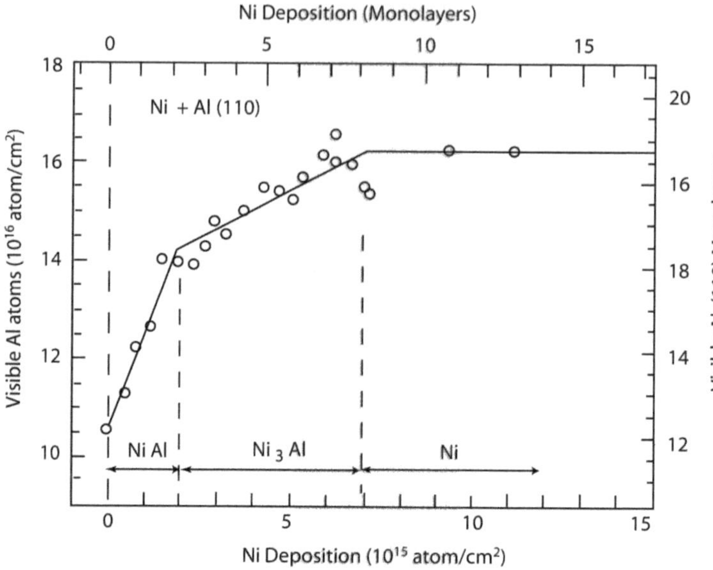

Figure 11. Number of Al atoms visible to the ion beam as a function of Ni coverage on the Al(110) surface. The solid lines are the least-squares fit to the data points in three different regions. Three different alloy compositions are also indicated.

4.4.2 Al(110) surface

In Fig. 11 we plot the characteristic growth curve for Ni deposition on the Al(110) surface. We see that more Al atoms become visible to the incident ion beam as the Ni coverage increases up to about 8 ML. Furthermore, the interface appears to have three stages of growth. Initially, the slope of the curve is about 1.97, consistent with each Ni atom effectively displacing about two Al atoms. This is suggestive of forming a $NiAl_2$-like compound during this first stage. However, we have pointed out that the slope alone is not a reliable indicator of the stoichiometry of the interface alloy in this channeling configuration.[6] From photoemission data for the shift in binding energy of the Ni $2p_{3/2}$ core level, as well as Monte Carlo simulations like those described in Section 5, it appears that the stoichiometry of the interface may be closer to that for NiAl up to a coverage of 2 ML. The CsCl crystal structure of NiAl is the least complex structure in the phase diagram and has a relatively high formation energy (see Table 1). The initial growth stage is followed by a second stage where the scattering yield increases at the slower rate of 0.35 Al atom per deposited Ni atom. Such behavior might result if diffusion of additional Al atoms with Ni deposition is limited by the relatively dense NiAl-like phase of stage one. A more Ni-rich compound would then form at the interface. Photoemission results for the chemical shift of the Ni 2*p* core level are consistent with the formation of Ni_3Al during this stage of growth.[6] Finally, after 8 ML of Ni coverage, additional Ni deposition does not result in more visible Al atoms, indicating that the interdiffusion at the interface has stopped and Ni metal is presumably growing on the alloy at the interface. Photoemission measurements at this point in the experiment show a Ni core-level binding energy similar to that for bulk Ni metal.

4.4.3 Al(001) surface

Figure 12 shows the characteristic growth curve for Ni deposition on Al(001) as measured with ion channeling. It is perhaps the most surprising of the three Ni surfaces, and is unlike any of the results shown above. The number of visible Al atoms first increases, suggesting the formation of a disordered interface alloy, but then decreases back to the value for the clean surface. The peak area is then unchanged until about 6 ML of Ni have been deposited on the surface. At that coverage, the Al surface peak area begins to increase again as if an interface alloy is being formed, and more like the results for Al(110) and Al(111). These results were obtained for two different incident ion energies (0.98 and 0.49 MeV) and so are not thought to be associated with any unusual channeling behavior for this surface. No shadowing of Ni atoms is observed up to about 3 ML, but the number of visible Ni atoms is reduced by about 15% in the channeling direction normal to the Al(001) surface for Ni coverage between 5 and 11 ML. In addition, the Al surface peak begins shifting to lower energy for Ni coverage above 4 ML, a sign of Ni film growth burying the Al surface atoms.

Together with these unusual ion channeling results for Ni on Al(001), the XPS and LEED measurements tend to support a growth model in which B2 NiAl(001) grows on the Al(001) surface during the first several monolayers of Ni deposition.[29] At higher coverage (11 ML) there is still LEED evidence of ordered Ni film growth. Bulk Ni is fcc with a lattice constant of 3.52 Å (see Table 1). The NiAl structure is bcc (CsCl) with a lattice constant of 2.89 Å, very close to the interatomic distance for Al atoms on the Al(001) surface, 2.86 Å. The LEED pattern for the clean Al(001) surface is completely destroyed with no sign of diffraction spots after only 0.5 ML Ni coverage. A 1x1 LEED pattern becomes visible again for Ni coverage greater than 2 ML, and persists but with less well-defined spots up to 11 ML Ni. XPS measurements for one ML Ni indicate a chemical shift for the Ni $2p_{3/2}$ core level similar to that measured in NiAl.[28] In addition, photoelectron diffraction measurements show a forward scattering peak for Ni at 45° polar angle along the Al[110] azimuth, and another relatively weak peak at 53° along the Al[100] azimuth. These are consistent with the Ni atoms being in

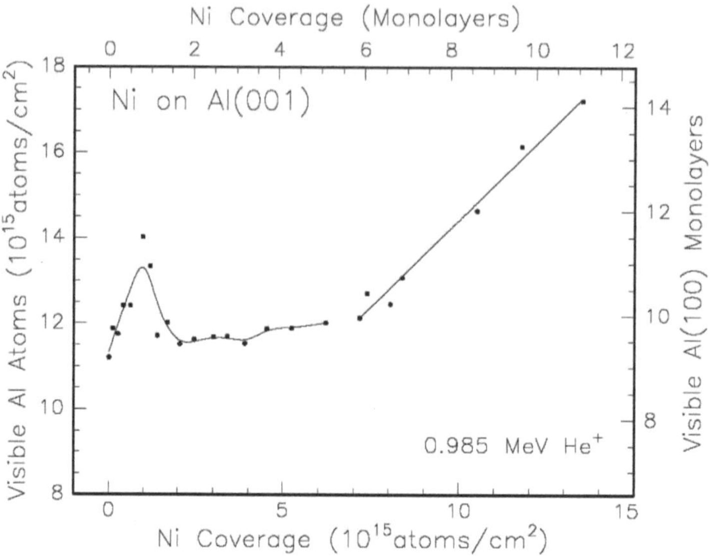

Figure 12. Number of Al atoms visible to the incident ion beam as a function of Ni coverage on the Al(001) surface. The solid lines are the least-squares fit to the data points in three different regions.

a bcc structure, such as NiAl. Finally, the attenuation of the Al XPS signal is gradual with Ni coverage, in agreement with a model where Ni and Al interdiffuse as the NiAl layer increases in thickness. Since the diffusion of Al at room temperature is limited, a Ni film eventually grows on the surface of the NiAl layer. After 11 ML Ni deposition, the photoelectron diffraction peaks that characterize the bcc structure are absent, and the LEED pattern is not well defined. The above interpretation for Ni growth on Al(001) is further supported by the MC simulations discussed in Section 5.

4.5 Ti films on Al(001), Al(110), and Al(111)

4.5.1 Al(001) surface

Contrary to the cases shown above for Fe, Co, Pd, and Ni deposition on Al surfaces, Ti deposition leads to shadowing of the substrate Al atoms, and growth of an fcc Ti film. The open circles in Fig. 13 show the characteristic growth curve for Ti deposition on Al(001). For the first half ML of Ti coverage the trend in the SP area is not very clear, and may be assumed to be constant to within experimental uncertainty.

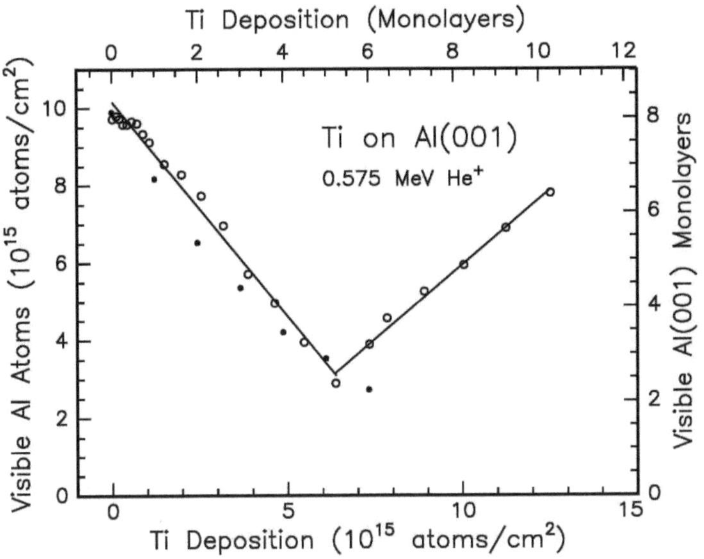

Figure 13. Number of Al atoms visible to the incident ion beam as a function of Ti coverage on the Al(001) surface. The solid lines are the least-squares fit to the data points in the two different regions.

However, after this coverage and up to about 5 ML of deposited Ti, a decrease in the Al SP area was observed, and attributed to shadowing of Al by Ti adatoms. Because the radius of the shadow cone is quite narrow (~ 0.1 Å), the Ti atoms must be within a few tenths of an Angstrom of the fcc Al lattice sites to shadow the Al atoms to this degree. The solid circles show the results of computer simulations of the ion scattering experiment using the VEGAS code.[19] The VEGAS program calculates the probability for an incident ion to hit various atoms in a snapshot of the target crystal. The hitting probability for a particular atom is based on the Gaussian distribution of vibration amplitudes for that atom evaluated at the location of the ion trajectory. The total hitting probability is a summation of these individual amplitudes summed over the ion trajectory. The calculation is performed in a Monte Carlo approach, typically using 10^4 incident ions per square Angstrom, with randomly generated atomic vibration amplitudes for each binary encounter along the ion trajectory. In these simulations the Ti atoms were arranged in a flat overlayer and placed on Al fcc lattice sites above the Al surface. Lattice parameters and atomic vibration amplitudes for bulk Al were used in the simulations.[22] In view of the good agreement between experiment and calculations, this shadowing behavior has been attributed to the growth of an fcc Ti layer up to the critical thickness of 5 ML.[9]

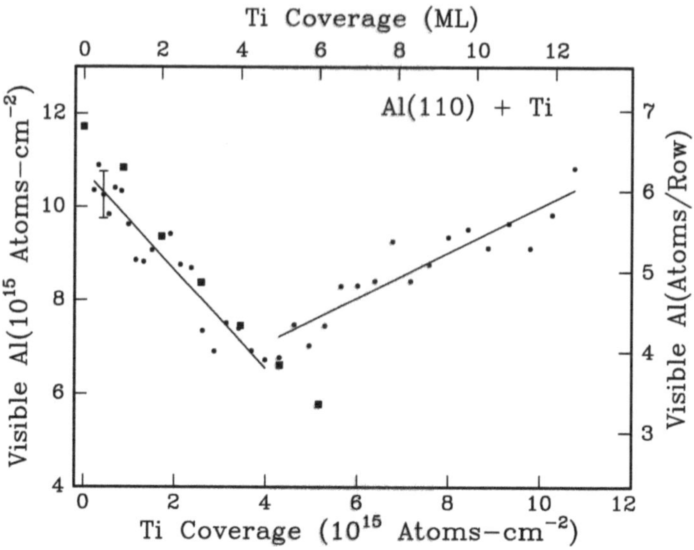

Figure 14. Visible Al atoms, at 0.96 MeV incident ion energy, as a function of Ti coverage deposited at room temperature on Al(110). The error bar represents an uncertainty of ±5% in the measurement of SP areas. Two regions are indicated. The solid lines are linear fits to the data, intended to guide the eye. The filled squares indicate the expected yield for a flat, pseudomorphic Ti film as calculated using the VEGAS simulation code.

A quantitative LEED analysis also concluded that Ti grows on Al(001) in an fcc structure.[30] At higher Ti coverage an increase in the Al peak area is observed. However, it is important to note that the Al SP area does not increase in these experiments to a value larger than that for the clean surface. This suggests that there is very little disruption of the substrate even above the critical thickness for the Ti film. The increase in Al scattering yield after 5 ML is presumably the result of accumulated strain energy in the film that gradually results in dislocations or other structural defects that cause Ti atoms to uncover the substrate atoms.

4.5.2 Al(110) surface

The characteristic growth curve for Ti deposition on the Al(110) surface is similar to that presented above for the Al(001) surface.[9] As shown in Fig. 14, the Al surface peak areas initially decreases as a function of Ti coverage, indicative of shadowing of the Al substrate atoms by the accumulating Ti atoms on the surface, and the growth of a pseudomorphic fcc overlayer. This interpretation is supported by results from computer simulations (VEGAS) of the ion scattering yield for a model of layer-by-layer growth of fcc Ti shown by the solid squares in Fig. 14.

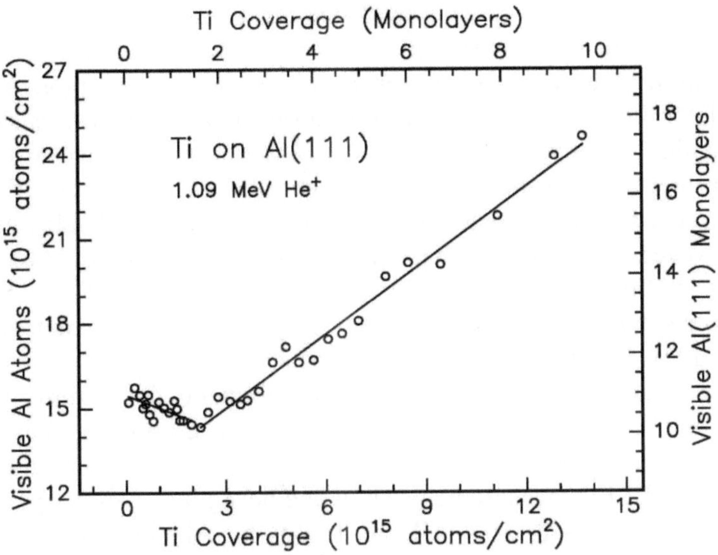

Figure 15. Visible Al atoms, at 1.09 MeV incident ion energy, as a function of Ti coverage deposited at room temperature on the Al(111) surface (open circles). The solid lines are linear fits to the two regions indicated, and are provided to guide the eye.

Note that at this higher ion beam energy, the amount of Al shadowing by Ti is less than that seen in Fig. 12 for Al(001), primarily because the Ti shadow cone radius is smaller at the higher energy. As more Ti is deposited on the substrate, exceeding a critical thickness of about 5 ML, the initial growth regime is followed by one in which the number of visible substrate atoms begins to increase. The increase continues up to a total coverage of 12.5 ML of Ti, at which coverage the experiment was terminated. Again we note that the scattering yield from Al atoms never exceeds that for the clean surface. A quantitative LEED study for this system was attempted, but LEED patterns useful for quantitative analysis were not obtained for films of more than a few monolayers thickness.[30] That work did confirm the pseudomorphism of a single Ti monolayer on Al(110).

4.5.3 Al(111) surface

The characteristic growth curve for Ti film growth on the Al(111) surface is shown in Fig. 15. The incident He^+ ion energy is 1.09 MeV. A small amount of Al shadowing occurs for the first 1.5 ML of Ti coverage where the number of visible Al atoms decreases slightly. However, at larger Ti coverage the Al yield begins to *increase*, and continues to do so in a linear fashion up to about 10 ML of Ti coverage where the experiment ended. This behavior is remarkably different from that seen for the Al(001) and Al(110) surfaces discussed above. The evidence of shadowing up to a critical thickness of 5 ML is not observed, and the ion yield from Al atoms rapidly increases *above* that for the clean surface. Increases in the Al ion yield were observed for Pd, Fe, Co, and Ni deposition on Al surfaces, as described earlier in this chapter, and attributed to alloy formation at the transition metal-Al interface. An increased scattering yield might also occur if the Ti adatoms caused Al atoms below the interface to move off of substrate lattice sites. However, we do not believe that such displacements would continue to occur in a linear fashion over such a large range of Ti coverage. Furthermore, the photoemission results for this system do not support a model of alloy formation at the interface.

We have carried out LEIS and XPS measurements for Ti films grown on Al(111) and conclude that an hexagonal film grows on the surface.[9] Photoelectron diffraction measurements show clearly the formation of an hcp Ti layer. LEED measurements show the hexagonal symmetry at the end of our experiment with 10.4 ML of Ti on the Al(111) surface, and ion channeling rocking curves demonstrate shadowing of Ti atoms for ions incident normal to the Al(111) surface. We are at present unable to explain the ion channeling results shown in Fig. 15, and will look to computer modeling and additional experiments for an answer.

The thermal stability of the Ti films grown on the three low-index Al surfaces has also been measured recently.[31] Surface sensitive LEIS measurements were used to monitor surface composition as a function of temperature. Ion channeling and backscattering were used to monitor total Ti composition and to look for diffusion of Ti into the Al substrate as the temperature was increased. It was found that Ti diffuses into the relatively open Al(110) substrate above 350 °C, but is stable to 400 °C on the more close-packed Al(001) and Al(111) surfaces. When the Ti atoms move into the Al substrate, they occupy substitutional sites in the Al lattice.

5. COMPUTER MODELING OF INTERFACE EVOLUTION: NI ON AL SURFACES

5.1 Monte Carlo Snapshots with VEGAS simulations

Unlike the case of Ti film growth on Al substrates, where adatoms shadow substrate atoms and reduce backscattered ion yield, our results for the growth of Fe, Ni, Co, and Pd films show increased numbers of visible Al atoms with adatom coverage. In such circumstances, when there is no

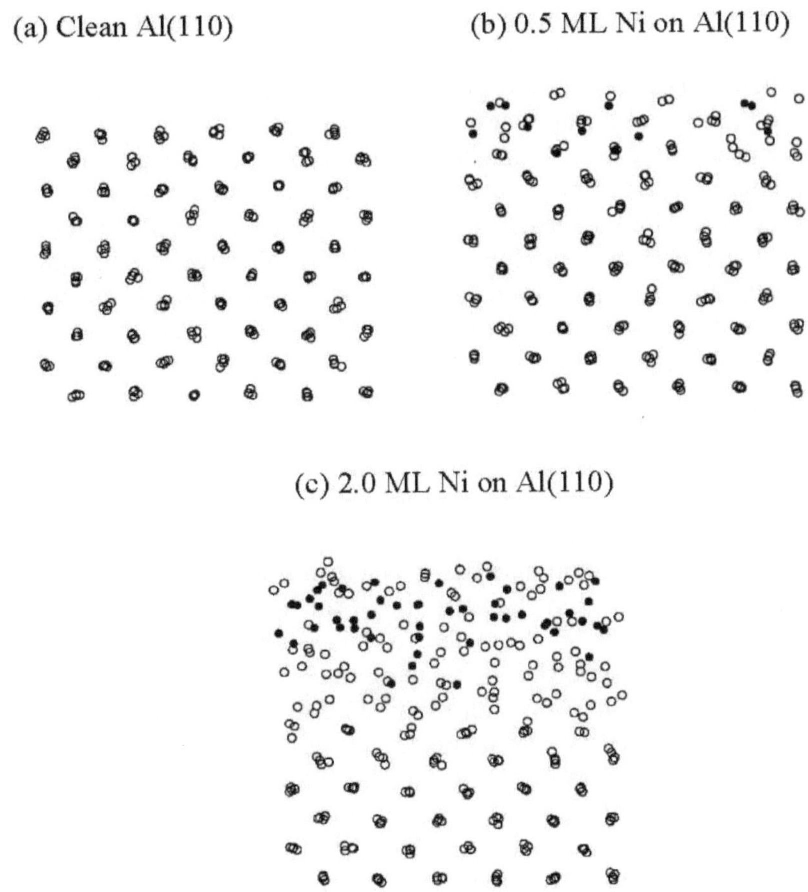

Figure 16. Side view of the snapshots of the simulated Al(110) surfaces for different Ni coverages as indicated. The positions of the atoms are projected onto the (001) plane. Open (solid) circles indicate the positions of Al (Ni) atoms. Only the top 10 layers of the simulated crystal are shown.

shadowing of substrate atoms, the high-energy ion scattering measurement provides very little structural information apart from the number of displaced substrate atoms. To gain further understanding of the buried interface structure we have performed Monte Carlo simulations of the interface evolution using embedded atom potentials.[32] Results of simulations for Ni on Al(110) have shown a behavior that agrees remarkably well with the measured ion scattering yields for this interface.[6]

In our approach, snapshots of the evolving interface are generated, similar to those shown in Fig. 16. In these side views of the Al(110) crystal the positions of the Al atoms (open circles) are projected onto the (001) plane. The displacements of the atoms from their equilibrium lattice sites associated with thermal vibrations are apparent in the snapshots, as well as the increasing disorder at the surface as Ni atoms (dark circles) are added. Several hundred snapshots are generated for each Ni coverage. We then simulated the ion scattering measurement by directing He ions at the atoms in the snapshots and calculating the hitting probability using the VEGAS code. We have reported previously[6,26] that the calculated results are in very good agreement with the measured results up to a coverage of about 4.5 ML. In particular, the change in slope around 2 ML Ni coverage in Fig. 11 is reproduced. We were thus motivated to use the snapshots to gain further insight regarding the structure of the Ni-Al interface. We found that the evolution of the interface composition in the simulations is more consistent with the growth of NiAl from zero coverage up to 2 ML, having a stoichiometry of 1:1, followed by the growth of a more Ni-rich phase. In these simulations, the slope observed below 2 ML (see Fig. 11) for the number of visible Al atoms appears to be an artifact of the channeling process, since the hitting probability in the simulations for the top 15 layers of the snapshots, where the disorder is occurring, increases with a slope of approximately unity. The increased slope in the simulations, i.e. 2:1 vs 1:1, is apparently the result of channeled ions passing through the disordered region and being deflected slightly, leading to dechanneling a few nm below the surface (layers 16 to 30). With the limited depth resolution of the solid state detector, the ions dechanneled in this subsurface region are indistinguishable from those backscattered in the disordered region. A smaller value of the slope (0.31) is seen to originate solely from the Ni-induced disordered region in the top 15 layers of the simulated crystal, in good agreement with the measured slope (0.38).

The success of these simulations for Ni on Al(110) has encouraged us to pursue the comparison of measurements and computer modeling of other metal-metal interfaces, a process that only requires having appropriate potentials for the metallic constituents in the system. The comparisons for Ni on Al(110) indicated that the ion channeling measurements with limited depth resolution, typical of channeling data obtained with solid-state detectors, do not by themselves allow for a decisive conclusion about the stoichiometry at a growing alloy interface, although the information obtained is still quite useful.

5.2 Orientation dependence of the interface evolution

In view of the successful simulations of interface evolution and ion channeling measurements for Ni deposition on Al(110), as described above, we proceeded with simulating the deposition of Ni on Al(001) and Al(111). The goal was to contrast the three surfaces and to look for substrate orientation dependence in the channeling and shadowing as presented in section 4.4. For the simulations, Ni atoms were introduced above the Al surface in increments of 4 or 5 atoms so as to avoid any Ni-Ni interactions above the surface. We had previously seen that placing a completed monolayer of Ni on Al(110) led to a metastable equilibrium in which the Ni layer remained in tact on the surface, in contrast to the experimental observations. Subsequent MC simulations, adding the Ni atoms a few at a time, led to interface disorder as seen in the snapshots of Fig. 16. Embedded atom potentials[32] were used for the Ni and Al atoms.

Figure 17 shows three side views (left) of snapshots for the three low-index Al orientations studied. The Ni coverage is 1 ML for each configuration. Top views for the same three configurations are shown on the right side of the figure. The individual configurations were allowed to equilibrate prior to adding any Ni atoms, and were further equilibrated following the addition of individual Ni atoms, typically a few atoms at a time at random coordinates above the surface. Looking at the snapshots with the eye is analogous to the ion channeling experiment. If an atom in the configuration is closer to the eye than other atoms, e.g. a surface or top atom, it will conceal the atoms directly behind it from your vision. Similarly, that atom will shadow the atoms behind it in the channeling experiment, and reduce the hitting probability for the shadowed atoms. From the side views in Fig. 17, all three of the configurations appear to have considerable disorder near the surface and we might expect increased hitting probability for Al atoms, as seen in the experiments for Al(110) and Al(111), Figs. 11 and 10, respectively. However, the top views in Fig. 17 suggest another story. The Al(111) and Al(110) configurations appear to have considerable disorder with Al atoms off of lattice sites, some apparently in interstitial positions. For Al(001), however, several of the Ni atoms have occupied positions directly above the rows of Al atoms, shadowing them from the eye, and presumably from the incident ions in the experiment. The shadowing of Al by Ni atoms for Al(001) in these snapshots is confirmed by numerical simulations of the channeling experiment, similar to the procedure described earlier for Ni on Al(110) and shown in Fig. 11. In section 4.4.3 we suggested that the Ni atoms are forming the B2 CsCl bcc structure. This can be accomplished on Al(001) if Ni atoms occupy the 4-fold hollow site above the surface plane, or if the Ni atom displaces an Al atom in the first layer below the surface, i.e. below the 4-fold hollow site. An interlayer relaxation of about 30% would then bring the top three layers into the B2 structure. Recall that a tetragonal distortion along the [001] direction is sufficient to transform the fcc lattice into a bcc lattice.

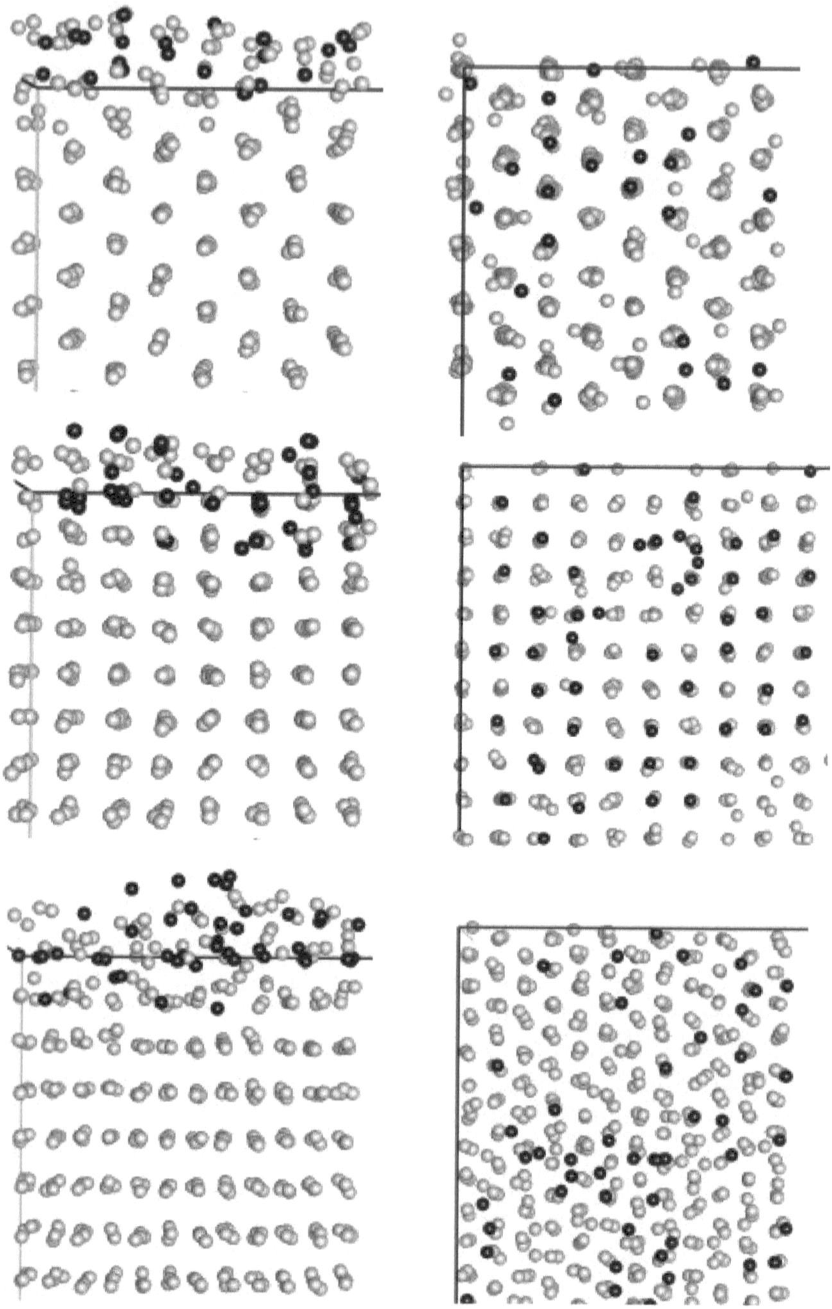

Figure 17. Side view (left) and top view (right) of MC snapshots for 1 ML Ni on: Al(110) top; Al(001) center; Al(111) bottom.

This type of distortion is also supported by the configurations of Fig. 17 where it appears that adding a monolayer of Ni results in the least vertical expansion of the crystal. That is, for Al(110) and Al(111), adding the Ni monolayer results in Al atoms with larger coordinates above the original surface (horizontal solid line) than appears to be the case for Al(001). In summary, the embedded atom method computer simulations seem to support a model of ordered growth of B2 NiAl on Al(100).

6. ATOMISTIC MODELING OF METAL-METAL INTERFACES USING THE BFS METHOD

The purpose of atomistic modeling for alloys, is not to reproduce every detail of the real process, as done in computer simulations, but to identify the main features and driving mechanisms of specific behavior. In working with metal-metal interfaces, we want the model to be simple in its application, universal with respect to the parameter set used for each element, and versatile in terms of the limited number of restrictions imposed. The BFS method for alloys satisfies most of the requirements imposed on such quantum approximate methods, and has been used to model numerous metal-metal systems.[33] A very brief description of the method is given here. More details regarding the method, its definitions, operational equations, and implementation can be found in the references. Most recently, we have used the BFS method to study the Ag-Al system, and find several points of agreement between model calculations and experiments for Ag deposition on Al surfaces, and Al on Ag surfaces.[34]

Within the BFS framework, the energy of formation of a given atomic configuration is the sum of individual atomic contributions. Individual atom energies are the sum of two terms: a strain energy, computed in the actual lattice of the configuration as if every neighbor of the atom were of the same type, and a chemical energy, computed as if every neighbor of the contributing atom were in an equilibrium lattice site for the selected atom, but retaining its actual chemical identity. The strain energy can be thought of as arising from a change in electron density at the atom of interest associated with changing the distance to neighboring atoms. In a similar way, the chemical energy arises from a change in electron density at the site of interest associated with a change in the chemical identity of neighboring atoms. The two contributions are linked by means of a coupling function that guarantees agreement with first-principles calculations, used in the parameterization of BFS, and BFS predictions at or around equilibrium. The connection between electron density and energy is achieved using equivalent crystal theory and the universal binding energy relation.

Table 2. Contributions to the formation energy within BFS[a].

Element	E(eV/at)	Element	E(eV/at)	ΔH(eV/at)
Ti	-1.48	Al	+0.64	-0.42
Fe	+1.12	Al	-1.64	-0.26
Ni	+0.20	Al	-1.42	-0.61
Co	+0.38	Al	-1.50	-0.56
Pd	+0.33	Al	-2.22	-0.95

[a] G. Bozzolo, private communication

In BFS calculations, atoms of different species in an alloy are characterized by how they raise or lower the energy of formation, ΔH, of the alloy. For example, consider the formation energies shown in Table 2. Each row in the table lists the elements in a binary compound and their respective contributions to ΔH, assuming the B2 structure. For FeAl, all Fe atoms have a positive contribution to ΔH (+1.12 eV/at), meaning that Fe does not favor having Al neighbors nearby. Aluminum atoms have a negative contribution to ΔH (-1.64 eV/at), and thus favor forming the FeAl alloy. The total ΔH is negative (-0.26 eV/at) because the Al contribution dominates over the Fe contribution, and we refer to Al as the alloying element of the compound. The remaining rows in the table show the calculated contributions to ΔH for other metal-metal interfaces we have studied experimentally. Note that in every case where we concluded alloy formation at the interface (Ni, Co, Pd, Fe), the film material repels the Al atoms, but Al atoms favor alloy formation and attract the deposited atoms. However, the case of Ti growth on Al is characterized by Al atoms repelling the Ti atoms, while Ti favors alloying (negative contribution to ΔH) in this compound. We have shown that Ti grows as an epitaxial fcc film on Al, with minimal alloying at the interface. The table also shows how ΔH alone does not predict the observed results of our experiments, because all five TM-Al pairs have $\Delta H < 0$, yet ΔH for Ti-Al, the only case of an epitaxial overlayer, falls in the middle of the range of values for the alloy-forming pairs. Up to this point in our work, only the larger atomic radius for Ti was recognized as distinguishing Ti-Al for different behavior. The BFS framework described here gives us a new perspective for these bimetallic systems, i.e. the asymmetry in the contributions to ΔH, a property we describe as the preference of an atom to attract or repel neighboring atoms. In addition to this chemical preference to alloy, or not alloy, atomic size may still be a factor in that diffusion of a large atom like Ti into the Al lattice may require an exchange process. Smaller atoms may be able to enter the lattice as interstitials. Thus, there will be both kinetic and chemical factors in determining alloy formation at an interface. We have shown, for example, that at elevated temperatures the Ti atoms also diffuse into the Al lattice.[31]

7. CONCLUSION

We have shown how high energy ion scattering can be used to study metal-metal interfaces, including cases of epitaxial growth of an overlayer, and formation of a disordered alloy at the interface. For Ti growth on Al, the adatoms are located on lattice sites above the surface, resulting in geometric shadowing of the substrate Al atoms and a decrease in backscattered ion yield. On the other hand, Fe, Co, Ni, and Pd deposition cause surface substrate atoms to be displaced, uncovering substrate atoms below the surface and leading to increased backscattered ion yield. We have also shown an example of how metallic interfaces can be modeled with atomistic simulations and the results subjected to experimental testing. Many of these studies would benefit considerably from additional model calculations to provide further insight regarding the structure and composition of the buried interfaces.

ACKNOWLEDGEMENTS

The technical support of Norm Williams is greatly appreciated. This work was supported by the National Science Foundation. I am particularly grateful to the several graduate students who have contributed to this work over the past 15 years, including: Xu Mingde, Adli Saleh, V. Shutthanandan, N. Shivaparan, V. Kraseman, G. White, M. A. Teter, B.-S. Park, R. Hutchinson, and R. Reibel. Several postdoctoral scholars and sabbatical visitors have also contributed, including: C.N. Whang, M. Kim, Y.W. Kim, C. V. Ramana, D.-Y. Jeon, and B.-S. Choi. Many undergraduate students have assisted in the ion beams lab at Montana State University, and in the completion of this work. Of particular note are P. Masse and N. Winward who assisted with the computer simulations.

REFERENCES

1. E.G. Bauer, B.W. Dodson, D.J. Ehrlich, L.C. Feldman, C.P. Flynn, M.W. Geis, J.P. Harbison, R.J. Matyi, P.S. Peercy, P.M. Petroff, J.M. Phillips, G.B. Stringfellow, and A. Zangwill, *J. Mater. Res.* **5**, 852 (1990).
2. P. Grunberg, *Phys. Today* **54**, 31 (2001).
3. A.G. Naumovets, and S.G. Davison, *Prog. Surf. Sci.* **67**, 9-16 (2001).
4. C.V. Ramana, P. Masse, Bum-Sik Choi, and R.J. Smith, *Phys. Rev. Lett.* **90**, 066101 (2003); C.V. Ramana, N. Winward, Bum-Sik Choi, and R.J. Smith, *Surf. Sci.* **551**, 189-203 (2004).
5. R.J. Smith, A.W. Denier van der Gon ,and J.F. van der Veen, *Surf. Sci.* **223**, 103 (1990); V. Shutthanandan, Adli A. Saleh, and N.R. Shivaparan, R.J. Smith, *Surf. Sci.* **350**, 11-20

(1996); N.R. Shivaparan, V. Shutthanandan, and V. Krasemann, R.J. Smith, *Surf. Sci.* **373**, 221 (1997).
6. V. Shutthanandan, Adli A. Saleh, and R.J. Smith, *J. Vac. Sci. Tech. A* **11**, 1780 (1993); V. Shutthanandan, Adli A. Saleh, and R.J. Smith, *Surf. Sci.* **450**, 204 (2000); Y.W. Kim, G.A. White, R. Reibel, and R.J. Smith, *Surf. Rev. Lett.* **6**, 781 (1999).
7. N.R. Shivaparan, V. Krasemann, V. Shutthanandan, and R.J. Smith, *Surf. Sci.* **365**, 78 (1996).
8. N.R. Shivaparan, M.A. Teter, and R.J. Smith, *Surf. Sci.* **476**, 152-160 (2001).
9. Adli A. Saleh, V. Shutthanandan and R.J. Smith, *J. Vac. Sci. Technol. A* **11**, 1982 (1993); Adli A. Saleh, V. Shutthanandan, and R.J. Smith, *Phys. Rev. B* **49**, 15 (1994); Adli A. Saleh, V. Shutthanandan, N. R. Shivaparan, R.J. Smith, T.T. Tran, and S.A. Chambers, *Phys. Rev. B* **56**, 9841 (1997); Y.W. Kim, G.A. White, N.R. Shivaparan, M.A. Teter, and R.J. Smith, *Surf. Rev. Lett.* **6**, 775 (1999).
10. R.J. Smith, C.V. Ramana, Bum-sik Choi, Adli A. Saleh, N.R. Shivaparan, and V. Shutthanandan, *Appl. Surf. Sci.* **219**, 28-38 (2003).
11. F.R. de Boer, R. Boom, W.C.M. Mattens, A. R. Miedema, and A.K. Niessen, *Cohesion in Metals: Transition Metal Alloys* (North Holland, Amsterdam, 1988).
12. J.H. van der Merwe and E. Bauer, *Phys. Rev. B* **39**, 3632 (1989).
13. R.J. Smith, Adli A. Saleh, V. Shutthanandan, N.R. Shivaparan, and V. Krasemann, Mat. Res. Soc. Symp. Proc. **399**, 135 (1996); R.J. Smith, N.R. Shivaparan, V. Krasemann, V. Shutthanandan, and Adli A. Saleh, *J. Korean Phys. Soc.* **31**, 448 (1997).
14. Adli a. Saleh, Ph.D. Thesis (Montana State University, 1994); C.V. Ramana, Bum Sik Choi, R.J. Smith, Byoung Suk Park, Adli A. Saleh, Dong Ryul Jeon, R. Hutchison, and S.P. Stuk, *J. Vac. Sci.Tech. A* **21**(4), 1326-1331 (2003).
15. M. Hansen, *Constitution of Binary Alloys,* ed. K. Anderko (McGraw-Hill, New York, 1958).
16. C. Kittell, *Introduction to Solid State Physics* (Wiley, New York, 1968).
17. J.D.R. Buchanan, T.P.A. Hase, B.K. Tanner, P.J. Chen, L. Gan, C.J. Powell, and W.F. Egelhoff, Jr., *Phys. Rev. B* **66**, 104427 (2002); J.D.R. Buchanan, T.P.A. Hase, B.K. Tanner, P.J. Chen, L. Gan, C.J. Powell, and W.F. Egelhoff, *J. Appl. Phys.* **93**, 8044 (2003).
18. L.C. Feldman, J.W. Mayer, and S.T. Picraux, *Materials Analysis by Ion Channeling* (Academic Press, New York, 1982).
19. J. W. Frenken, R.M. Tromp, and J. F. van der Veen, *Nucl. Instrum. Meth. B* **17**, 334 (1986).
20. H. Niehus, W. Heiland, and E. Taglauer, *Surf. Sci. Rep.* **17**, 213-303 (1993).
21. R.J. Smith, C.N. Whang, Xu Mingde, M. Worthington, C. Hennessy, M. Kim, and N. Holland, *Rev. Sci. Instr.* **58**, 2284 (1987).
22. D.S. Gemmel, *Rev. Mod. Phys.* **46**, 129 (1974).
23. I. Stensgaard, L.C. Feldman and P.J. Silverman, *Surf. Sci.* **77**, 513 (1978).
24. J.F.Van Loenen, M.Iwami, R.M.Tromp, and J.F.van der Veen, *Surf. Sci.* **137**, 1 (1984).
25. G.W.Anderson and P.R.Norton, *Surf. Sci.* **336**, 262 (1995).
26. V. Shutthanandan, Adli A. Saleh, A.W. Denier van der Gon, and R.J. Smith, *Phys. Rev. B* **48**, 18292 (1993).
27. L.C. Feldman and J.W. Mayer, *Fundamentals of Surface and Thin Films Analysis*, (North Holland, Amsterdam, 1986).
28. F. U. Hillerbrecht, J. C. Fuggle, P. A. Bennett, Zygmunt Zolnirek, and Ch. Freiburg, *Phys. Rev. B* **27**, 2179 (1983).
29. V. Shutthanandan, Ph. D. Thesis, Montana State University, 1994.
30. S.K. Kim, F. Jona, and P.M. Marcus, *J. Phys: Condens. Matter* **8**, 25 (1996).

31. C.V. Ramana, Bum Sik Choi, R.J. Smith, R. Hutchison, S. P. Stuk, Byoung Suk Park, Adli A. Saleh, and Dong Ryul Jeon, *J. Vac. Sci. Technol. A* **21**, 1326 (2003).
32. A.F. Voter, S.P. Chen, in *Characterization of Defects in Materials*, R.W. Siegel, J.R. Weertman, R. Sinclair (Eds.), MRS Symp. Proc. Vol 82 (Materials Research Society, Pittsburgh, PA, 1986) p. 175.
33. G. Bozzolo, J. E. Garcés, R. D. Noebe, P. Abel, and H. Mosca, *Prog. Surf. Sci.* **73**, 79 (2003); G. Bozzolo and J. E. Garcés in *Surface Alloy and Alloy Surfaces*, The Chemical Physics of Solid Surfaces, Vol. 10, D. P. Woodruff, ed. (Elsevier, Amsterdam, 2001).
34. G. Bozzolo, J. E. Garcés, and R.J. Smith, *Surf. Sci.* **583**, 229-252 (2005).

Chapter 14

MODELING OF LOW ENRICHMENT URANIUM FUELS FOR RESEARCH AND TEST REACTORS

Jorge Garcés[1], Guillermo Bozzolo[2,3], Jeffrey Rest[4] and Gerard Hofman[4]
[1] Centro Atómico Bariloche, 8400 Bariloche, Argentina; [2] Ohio Aerospace Institute, Cleveland OH 44142, USA; [3] NASA Glenn Research Center, Cleveland, OH 44135, USA; [4] Argonne National Laboratory, 9700 S. Cass Ave., Argonne, IL 60439, USA

Abstract: During the last decade, a considerable international effort to develop a low enriched uranium fuel for research and test reactors with high uranium density has been underway. UMo-based alloys are the best candidates for achieving this conversion although several failures in U-Mo dispersion fuel plates like pillowing and large porosities have been reported during irradiation experiments, introducing obstacles to further developments. In this chapter we apply the BFS method to model the behavior of the interface Al/UMo_x and the interdiffusion of additives, as Si and Ge, added to the Al matrix, in order to identify the driving forces responsible for the observed effects. The basic features characterizing the real system are identified in this modeling effort as are: the trend to interfacial compound formation, the Al "stopping power" of increasing Mo concentration, the depletion of Si in the Al matrix and the reduced diffusion of Al into UMo with high Si concentration. These and other basic questions must be answered in order to have a better understanding of the basic behavior of this fuel previous to its qualification. While the approach presented in this chapter is relevant to other applications as well, it is important to highlight the influence that modeling techniques can have in problems of high technological importance, and the benefits arising from virtual experiments and detailed understanding from simple atomistic approaches.

Keywords: LEU fuels, UMo alloys, uranium, aluminum, interfaces, atomistic modeling, Quantum approximate methods, interfaces, Monte Carlo simulations

1. INTRODUCTION

Research and development programs based on the synergy between theory, modeling, and experiment, are growing both in number and strength.

Their growth, however, relies heavily on a number of factors which favor certain fields of research more than others, as the theoretical complexities, the availability of computational modeling tools, and the ease to acquire experimental evidence rarely converge to provide a uniform platform for any given research program, thus limiting the added value that arises from an optimum balance between them. One area of relevance where this is apparent is in the development of nuclear fuels for research and test reactors that meet current needs and concerns at a global scale. Since its implementation in 1978, the Reduced Enrichment for Research and Test Teactors (RERTR) program[1] gradually became an international effort directed towards the use of low enriched uranium (LEU, $^{235}U < 20$ at.%) in fuels for all new research reactor designs worldwide and for conversion of existing reactors from higher enrichments. The US Global Threat Reduction Initiative, established in 2004, provides new impetus for RERTR fuel development activities and resulting core conversions.

Due to the decrease in ^{235}U enrichment on conversion to LEU, the total density of uranium atoms in the fuel must be increased in order to maintain an appropriate neutron flux. The development of high density U-alloys with an increased concentration of U is one of the key ingredients for high neutron flux research reactors with LEU fuel. Suitable fuels have been qualified for most research and test reactors but those operating at very high power densities cannot be converted to LEU with commercially available uranium silicide-based (U_3Si_2) fuel.[2]

In order to meet fissile atom density requirements at a fuel particle volume fraction of 55% or less with existing fuel configurations and fabrication technology, fuel for high power test reactors requires fuel particles with uranium densities greater than 15000 kg U/m^3. There are two types of materials that approach or meet this density criterion: metallic uranium alloys and the U_6Me family of intermetallics, where Me = Fe, Mn, Ni, or Ge.

The irradiation behavior of U_6Me plate-type dispersion fuel has been previously investigated as a candidate for high-density fuel. U_6Fe and U_6Mn have been shown to have poor irradiation behavior in a dispersed thin plate configuration due to breakaway swelling of the fuel phase at relatively low burnup.[3-5] Other U_6Me compounds were not investigated as it was assumed that they would behave in a similar manner. The effect of additives added to the Al matrix on the interdiffusion behavior of the different atomic species in the matrix and in the U_6Me fuel have not been assessed. Consequently, metallic uranium alloys are the only materials with the characteristics required to be used in high-density LEU dispersion fuels for high power research reactors.

It has been shown that γ-stable (bcc crystal structure) metallic fuels are more resistant to swelling than α-uranium-based (orthorhombic crystal structure) fuels under low-burnup, high-temperature irradiation conditions.[2] γ-U is not thermodynamically stable under the fabrication and irradiation conditions of interest, however some alloys of uranium can remain in the γ-phase in a metastable state indefinitely at room temperature, and for long periods of time at elevated temperature. Of particular interest, due to their properties under irradation, are U-Mo and U-Nb-Zr alloys. However, due to extensive fuel/matrix reaction and marginal fuel performance,[6] the class of U-Nb-Zr alloys has been eliminated from further fuel testing, although there is renewed interest to study the behavior of this fuel with an Al matrix with high Si content.

Alloys based on U-Mo are potential candidates because a solid solution of Mo in γ-U has acceptable irradiation properties for reactor fuels. These fuels were irradiated to a maximum burnup of 70% ^{235}U at a temperature of approximately 338 K. Fuels with 6 wt% or more molybdenum content performed well during irradiation, exhibiting low to moderate fuel/matrix interaction and stable fission gas bubble growth. Fuel particles containing 4 wt% Mo reacted extensively with the matrix aluminum during fuel fabrication and irradiation.[7-9]

However, recently, unexpected failures of LEU UMo dispersion plates and tubes in high neutron fluxes during irradiation tests were found.[10] These failures have been ascribed to the formation of ternary compounds and large porosity located at the interface between the (UMo)/Al interaction product and the Al matrix under high power operating conditions.[10-12] As a consequence, fuels based on UMo have not yet been qualified. However, a potential remedy by suitable modification of the matrix by alloying Al, or the UMo fuel by adding selected elements in order to change the properties of the interaction layer and improve its irradiation behavior has been proposed.

Several basic issues must be addressed in order to have a better understanding of the fundamental behavior of the different fuels as, for example, the processes involved in the interaction between the aluminum matrix and the UMo solid solution, or the interdiffusion or interfacial reaction with the Al matrix or cladding. Other basic questions relate to the role of new additives in the stability of the bcc phase, as well as their influence on the thermal compatibility with the Al matrix and on potential porosity formation.

The answers to these questions could arise from experiment, theory, or from a combination of both but, as mentioned above, the nature of the problem at hand has made it particularly difficult to any approach. This is partly due to the fact that, until recently, the method for developing or improving specific alloys has been mostly based on extensive experimental trial and error work, which is both expensive and time consuming. Recently, however, the increasing role of atomistic computational modeling in the development of structural materials has shown promise as a valuable tool to aid the experimental work. If theoretical modeling could also be included in the development of nuclear fuels, then the experimental process could be better directed and, as a consequence, the number of experiments could be reduced to specific ones correlated in number and nature to the guidance provided by the theoretical predictions.

While the virtual design of new materials through complex computer simulations is yet a goal to be achieved, its success will depend on the availability of a unified approach that provides the same level of simplicity and accuracy for any possible application, whether it is directed to surface and/or bulk analysis. Although first-principles (FP) approaches provide the most accurate framework for such studies, the complexity of the tasks and their substantial computational requirements impose limitations that still prevent these approaches from becoming economical predictive tools. In fact, the theoretical description of actinide metals and their alloys poses a severe challenge to modern electronic structure theory and has eluded accurate treatment by semiempirical or quantum approximate methods (qam). As a consequence, theoretical modeling efforts to describe the complex behavior observed experimentally in these systems have been limited, and it is only recently that their role has began to be felt.

The tendency to incorporate atomistic simulations as a standard tool in the analysis of complex systems has imposed high expectations on the range of applicability of qam, their computational efficiency, their ease of implementation, and the type of output that they provide. The purpose of qam is to provide an efficient and accurate way to compute the energy of arbitrary atomic systems in terms of their geometrical configuration. Almost independently of their foundation and formulation, these methods rely on simplifications which, as a result, inevitably require the introduction of parameters. The recent trend of combining FP with qam has created new possibilities in the field of atomistic simulations, as they provide accurate and valuable input for the determination of these parameters when experimental data are not available. In most cases, the applications of existing qam are restricted to a few systems for which a specific (and therefore nontransferable) parameterization is developed, thus limiting their use. Additional restrictions generally apply, resulting in limitations on the efficiency or accuracy of the method in terms of type of lattice structure, and number and type of element.

A recent addition to the growing family of qam is the Bozzolo-Ferrante-Smith (BFS) method for alloys,[13-16] which fulfills several requirements for applicability in terms of simplicity, accuracy, and range of application, as it has no limitations in its formulation on the number and type of elements present in a given alloy. Moreover, it has shown promise in describing diverse problems, particularly in the area of surface alloys[13] and high-temperature ordered intermetallics.[14-16] The BFS method relies on approximations by replacing the exact process of alloy formation with virtual processes where the end result is, or is expected to be, a good description of the result of the real process. BFS is then expected to reproduce the essential features of the equation of state of the solid at zero temperature and, in particular, around equilibrium providing structural information that is, ultimately, contained in the binding energy curve describing the solid under study. Starting from the same initial state, BFS tries to provide an alternative, virtual path leading to the final state with a minimum number of parameters to guide its way. While more flexibility can be gained by letting these parameters vary according to the specific problem at hand, no such degree of freedom is added to the method, and the parameters remain fixed, fully transferable, for any case dealing with the same elements, regardless of their number, type, or structural properties. This restriction implies that in order for the method to be equally valid in a number of diverse situations the parameters must contain all the necessary information to warrant the accuracy of the virtual path chosen for describing the process of alloy formation. If this description is correct, then the method should accurately reproduce the most critical properties of the solid in its final state, including the cohesive energy per atom, compressibility, and equilibrium Wigner-Seitz radius. As long as these properties are sufficient for an equally accurate description of defects, their accuracy is essential for addressing issues such as the site preference behavior of alloying additions in multicomponent systems.[13]

It is also worth noting that the parameterization of the BFS method implies a somewhat different approach for the interaction between different atoms. In general, most approaches introduce some sort of interaction potential, with the parameters describing each constituent remaining unchanged. In BFS, it is precisely the set of parameters describing the pure element that it is perturbed in order to account for the distortion introduced by the nearby presence of a different element or defect. The additive nature of the perturbative theory results in that only binary systems need to be known in the BFS framework. Multicomponent systems are thus studied via binary perturbations: the perturbation in the electron density in the vicinity of any given atom is computed as the superposition of individual effects due to each neighboring atom. This allows for an accurate, but also computationally simple way to detect general trends in idealized multicomponent systems as in, for example, the study of additives to the Al/UMo interface.

In this chapter, we apply the BFS method for alloys to the description of the basic features observed in nuclear fuels that are currently under development. Our aim is to understand the basic behavior of each component of the multicomponent system, and to investigate the possibility of applying the BFS method to the development of new nuclear fuels. Section 2 is devoted to the validation of the parameterization of the BFS method, while section 3 shows the modeling results obtained with Monte Carlo simulations and analytical calculations of the Al-U and Al-UMo systems, while sections 4 and 5 describe the role of Si and Ge, respectively, in the Al/UMo interface. Finally, general conclusions are presented in Section 6.

2. THE BFS METHOD FOR ALLOYS

The BFS-based methodology, used to provide detailed insight on different aspects of bulk or surface alloys properties, assumes no *a priori* information on the system at hand and none of the experimental information is used in the formulation and application of the method. The only input necessary consists of the basic parameterization of the participating elements, their binary combinations, and lattice structures.

The BFS parameters are determined from FP calculations by using the full potential linearized augmented plane wave (LAPW) method, as implemented in the WIEN97 package,[17] but their validation depends on their ability to describe basic known features of the binary combinations of these elements. However, unlike other systems for which there is abundant theoretical or experimental information, the background available for U-X compounds (X = Al, Si, Mo, Ge) is rather limited. Accurate parameterization, coupled with the ability to reproduce experimental behavior is clearly essential. Within the framework of BFS, it is of particular importance because one essential feature on the methodology is the precise role that each atom plays in any given compound. For example, UMo forms a disordered bcc solid solution. The sign of the individual contribution to the heat of formation from U or Mo atoms determines their role in the alloying process. This piece of information should be clearly established and tested so that no information is lost or no error is augmented as the number of participating elements increases. Therefore, as a first step, it is necessary to investigate all the possible binary combinations of elements by testing the BFS predictions against known experimental evidence.

The BFS parameters for the single elements used in this study and their binary combinations are listed in Table 1. Details of the method are described in Chapter 7.

2.1 The U-Al system

LAPW calculations for a metastable B2 UAl binary alloy result in perturbative parameters Δ_{UAl} and Δ_{AlU} characterized by negligible Al perturbative attraction ("attraction") of U atoms, which can be easily offset with the strain of Al atoms, and a rather intense U perturbative repulsion ("repulsion") of Al atoms. Evidence for Al interdiffusion in U can be seen in the theoretical results displayed in Table 2, where the differences in energy (with respect to the configuration where the adatom occupies a bulk site in the substrate) are shown for a number of static configurations for Al (or U) on a U (or Al) (100) surface. Al(L_1) and U(L_2) denote the Al (or U) atom in layer L_1 (or L_2), where L denotes the overlayer (O), surface (S), first plane below the surface (1b), second plane below the surface (2b), or a bulk site (bulk). The subindex denotes the distance (in neighbor layers) between the Al and U atoms. The results predict no barriers for Al Interdiffusion in

Table 1. LAPW results for the lattice parameter, cohesive energy, and bulk modulus for the bcc phases of U, Al, Si, Mo, and Ge, the resulting equivalent crystal theory (ECT)[18] parameters p, α, l and λ and the BFS perturbative parameters Δ_{AB} and Δ_{BA}.

	Lattice parameter (Å)	Cohesive energy (eV)	Bulk modulus (GPa)	ECT parameters p	α (Å^{-1})	l (Å)	λ (Å)
U	3.45012	5.55175	141.38	12	4.8689	0.31322	0.88015
Al	3.23811	3.44327	69.13	4	1.76396	0.36414	1.02322
Si	3.08731	4.07901	90.93	4	1.85741	0.35390	0.99446
Mo	3.16156	6.66707	260.54	8	3.47728	0.26414	0.74223
Ge	3.3847	3.23457	65.14	6	2.5282	0.35562	0.99929

BFS parameters Δ_{AB} and Δ_{BA} (in Å^{-1})

A\B	U	Al	Si	Mo	Ge
U		-0.01356	-0.04831	-0.04373	-0.0253
Al	0.0642		0.02607	-0.03351	-0.02673
Si	0.05004	-0.00526		-0.04488	0.00405
Mo	0.15898	0.10065	0.27565		0.04139
Ge	-0.00482	0.03664	0.00394	-0.02673	

Table 2. Difference in the energy of formation of different configurations with Al or U atoms in a U or Al (100) substrate. The reference energy is that of the Al (or U) atom in the bulk of the slab.

	ΔE		ΔE
Al(O)	0.533	U(O)	-1.285
Al(S)U(O)*l*	0.479	U(S)Al(O)$_l$	-1.189
Al(S)U(O)$_f$	0.364	U(S)Al(O)$_f$	-1.234
Al(1b)U(O)	0.149	U(1b)Al(O)	-0.025
Al(2b)U(O)	0.003	U(2b)Al(O)	0.001
Al(bulk)U(O)	0	U(bulk)Al(O)	0

Table 3. Coordination matrices for the simulated BFS results and the ideal $C11_b$ structure.

	BFS		Ideal $C11_b$	
NN	U	Mo	U	Mo
U	50.7	49.3	50.	50.
Mo	100.	0.	100.	0.
NNN	U	Mo	U	Mo
U	82.6	17.4	83.3	16.7
Mo	35.4	64.6	33.3	66.7

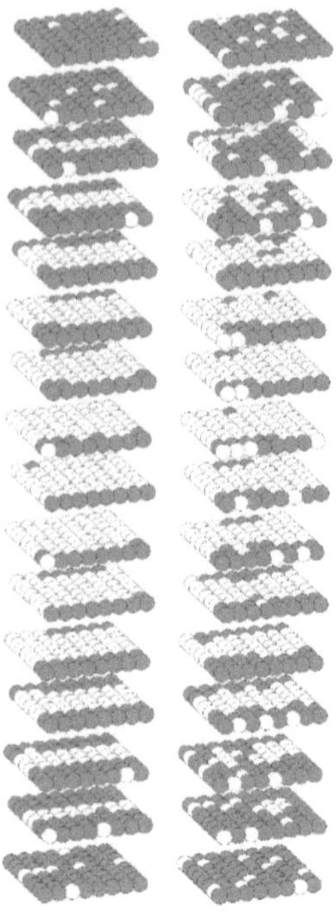

Figure 1. Phase separation of U-Al: $U_{50}Al_{50} \rightarrow Al+U(Al)$ at T = 1 K (left) and the T = 400 K state (right). U and Al atoms are denoted with dark and white circles, respectively. U(Al) denotes the U solid solution.

U(100) and a rather strong barrier for U interdiffusion in Al(100), favoring U atoms in the overlayer or in the surface layer. Additional information on the phase structure of U-Al alloys can be obtained from Monte Carlo - Metropolis simulations (MC) of 3D cells of UAl, where all atoms exchange sites until the lowest energy configuration is found at each temperature. The results for $U_{50}Al_{50}$ alloys predict phase separation, UAl -> Al + U(Al), as shown in Fig. 1, in agreement with the phase diagram.[19,20]

2.2 The U-Mo system

The phase diagram suggests a bcc-based solid solution as well as a possible U_2Mo phase around T = 1000 K for more than 60 at% U. Simulations were done for $U_{30}Mo_{70}$ (where there should be phase separation of Mo and U_2Mo), $U_{67}Mo_{33}$ (which should form an ordered $C11_b$ structure) and $U_{80}Mo_{20}$ (which should form a solid solution in the 800 K < T < 1500 K range). MC simulations indicate a solid solution for the correct temperature and concentration range ($0.6 < x_U < 1$).

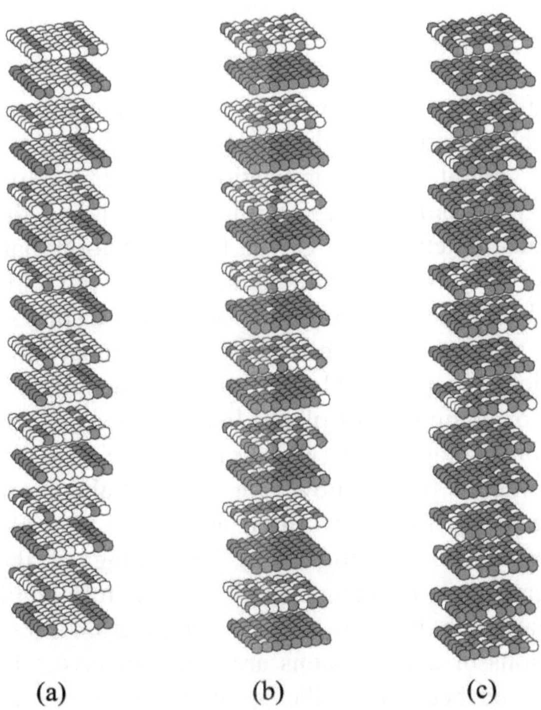

	$U_{30}Mo_{70}$	$U_{67}Mo_{33}$	$U_{80}Mo_{20}$
ΔH	-0.70336	-0.50797	1.18016
a_0	3.3681	3.3249	3.2381

Figure 2. Final state (T = 1 K) of a Monte Carlo simulation of $U_{30}Mo_{70}$ (left), $U_{67}Mo_{33}$, (center) and (c) $U_{80}Mo_{20}$ (right). U and Mo atoms are denoted with solid and white circles, respectively. The table below shows the values of the energy of formation ΔH at T = 0 K and the corresponding equilibrium lattice parameter.

Table 4. Difference in the energy of formation of different configurations with Si/Ge or U atoms in a U or Si/Ge (100) substrate. The reference energy is that of the Si/Ge (or U) atom in the bulk of the slab.

X/U(100)	ΔE (X = Si)	ΔE (X = Ge)	U/X(100)	ΔE (X = Si)	ΔE (X =Ge)
X(O)/U	1.536	-0.196	U(O)/X	-2.427	0.0401
$X(S)U(O)_1$	1.582	-0.017	$U(S)X(O)_1$	-2.348	-0.1092
$X(S)U(O)_f$	1.586	-0.136	$U(S)X(O)_f$	-2.334	0.0585
X(1b)U(O)	0.135	-0.059	U(1b)X(O)	-0.077	-0.0036
X(2b)U(O)	0.001	0.001	U(2b)X(O)	0.015	-0.0003
X(bulk)U(O)	0	0	U(bulk)X(O)	0	0

Low temperature states are shown in Fig. 2, where the negative energy of formation could be due to the slight hints of order present in the cell. These 'strings' of Mo might be an indication of the U_2Mo phase, as the coordination of Mo atoms in these strings is similar to the correlation in $C11_b$. For the stoichiometric case, U_2Mo, the correlation matrix for the final state, shown in Table 3, is comparable with the observed $C11_b$ structure. This is, however, misleading as this coordination matrix displays averages of the number and type of bonds over the whole computational cell. The simulation results do not yield a perfect $C11_b$ structure, characterized by two U (100) planes followed by a single Mo plane. Instead, there are different ordered regions which, on average, satisfy the same correlation seen in the $C11_b$ structure. This ordered phase allows for substantial interplanar relaxation which lowers the energy from -0.0714eV/ atom to -0.1314 eV/atom.

The MC simulations are somewhat misleading: if they allowed for relaxation at every single exchange, the $C11_b$ structure would have a better chance to appear. The rigidity of the computational cell (where only isotropic expansions or compressions are allowed) favors B2 ordering, but $C11_b$ is also favored because of the optimized balance between pure and mixed bonds. As a result, a competition between B2 and $C11_b$ ordering can be expected, resulting in $C11_b$-like correlation that can be obtained by alternative arrangements (all of which can be seen in the simulation for $U_{67}Mo_{33}$). It is unlikely, therefore, that the computational cell will show perfect rigid $C11_b$ distribution if the number of iterations is finite. A strong test of the validity of the parameters is the comparison of the lattice parameter dependence with composition for the U-Mo solid solution.[21] The experimental results yield $a(x)/a_U = 1 - 0.0009021x$ (x in at% Mo) for the range 12.7 to 35.5 at.% Mo at 1173 K. The BFS prediction for the same temperature is $a(x)/ a_U = 1 - 0.0009044x$. In both cases, the data points for the range of concentration studied is used to determine the value of a_U (Dwight[21] extrapolates this value to 3.4808 Å). LAPW calculations, however, predict a slightly smaller value for a_U, 3.4501 Å. Finally, MC simulations show a slight curvature for increasing Mo concentration as it approaches the critical value of 33 at% for which an ordered U_2Mo structure is possible.

2.3 The U-Si and U-Ge systems

An atom-by-atom analysis of the U_3Si ($L1_2$), USi (B2) and USi_3 ($L1_2$) ordered phases shows that the parameters predict negative energies of formation for U_3Si (which exists as a $D0_c$ structure, a distorted bcc). Fig. 3 shows the results of MC simulations for (a) $U_{85}Si_{15}$, where the cell separates into U and U_3Si, and (b) $U_{60}Si_{40}$, where it splits between U_3Si and B2 USi. For $U_{15}Si_{85}$ the set predicts perfect separation of Si + B2 USi.

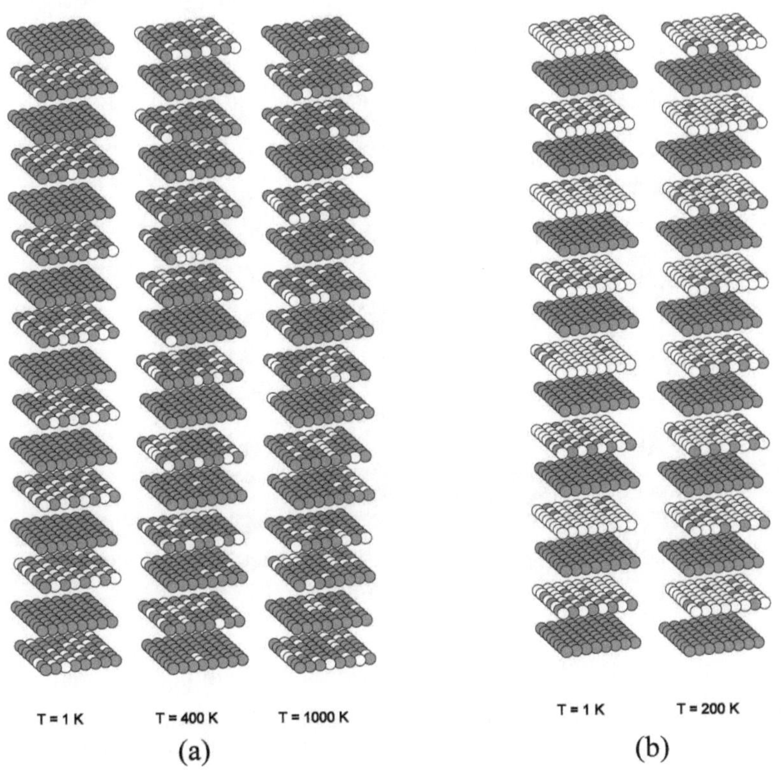

Figure 3. Intermediate states of a Monte Carlo simulation for (a) $U_{85}Si_{15}$, for T = 1, 400 and 1000 K, and (b) $U_{60}Si_{40}$, for T = 1 and 200 K. U and Si atoms are denoted with solid and open circles, respectively.

The results for U_3Si can be explained in terms of the individual atomic contributions of U and Si atoms and their respective strain and chemical components: the negative (attractive) chemical energy contribution of

U atoms (-0.02 eV/at) turns into a net positive (repulsive) total energy contribution (+0.015 eV/at) once the strain component is added. Conversely, Si atoms have a much stronger chemical energy term (-1.76 eV/at) which is not offset by strain, leading to a negative total energy contribution (-0.92 eV/at) and very strong U-Si bonds.

Using this set of parameters for the interdiffusion of Si in U (or viceversa) shows a very small energy barrier, as shown in Table 4. There is no further barriers for Si interdiffusion in U. For U/Si, there is a noticeable barrier for U(O) → U(S) exchanges and a much larger barrier that precludes U(S) → U(1b) exchanges.

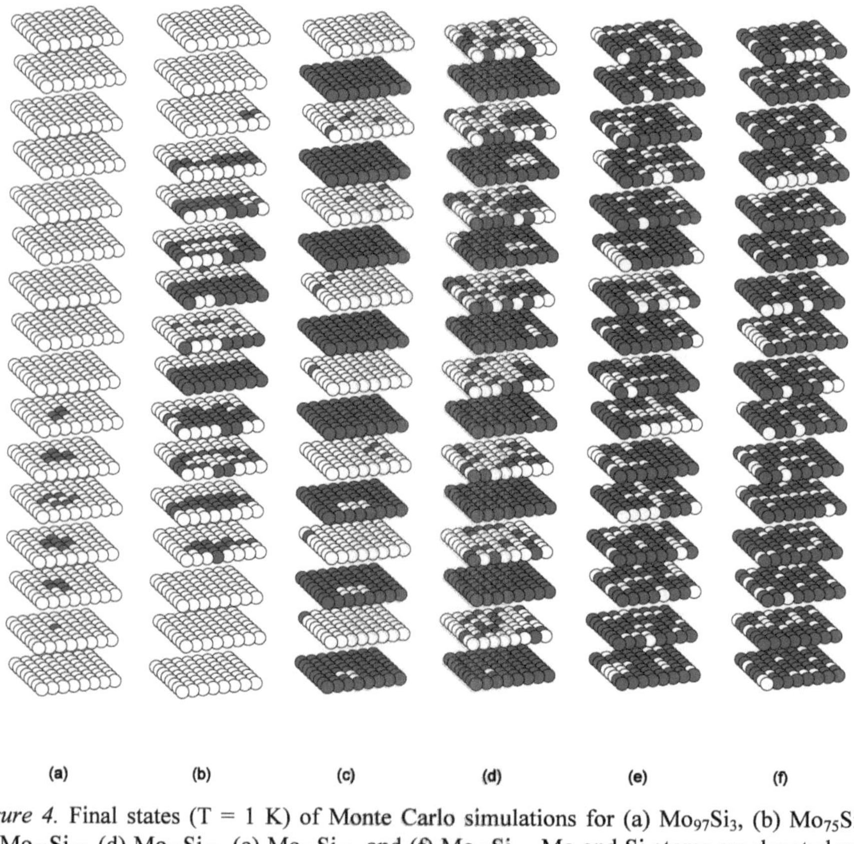

Figure 4. Final states (T = 1 K) of Monte Carlo simulations for (a) $Mo_{97}Si_3$, (b) $Mo_{75}Si_{25}$, (c) $Mo_{50}Si_{50}$, (d) $Mo_{33}Si_{67}$, (e) $Mo_{25}Si_{75}$, and (f) $Mo_{20}Si_{80}$. Mo and Si atoms are denoted with open and dark circles, respectively.

In spite of the similarities between the U-Si and U-Ge phase diagrams, and the (identical) role assigned to Si and Ge by the chosen BFS parameters, there is a qualitative difference in the U/Si and U/Ge interdiffusion analysis, also shown in Table 4. Ge diffusion in U is generally inhibited with mildly increasing energies as Ge penetrates the U(100) slab.

2.4 The Al-Mo system

Simulations of $Al_{90}Mo_{10}$ using the selected set predict isolated Mo atoms surrounded (up to next nearest neighbors) by Al atoms, consistent with the existence of an $Al_{12}Mo$ compound. For higher Mo concentration, Al-Mo alloys evolve into Mo + P, where P denotes an Al precipitate with Mo in solution, or an ordered Al_xMo compound. For $Al_{75}Mo_{25}$, Mo appears encapsulated with Al up to nearest neighbors, and the coordination matrix for the computational cell has the exact distribution of an fcc Al_3Mo $L1_2$ ordered compound ('hidden order', where up to first neighbors the system has an average distribution that exactly matches an ordered distribution in a different lattice). An additional feature for testing the selected set is the somewhat ambiguous behavior seen in Al deposition on Mo(110). Experiment[22] shows layer-by-layer growth, with little evidence for surface alloying, in agreement with the BFS results.

2.5 The Al-Si and Al-Ge systems

The selected set reproduces the main feature of the eutectic Al-Si phase diagram, mostly for what it does not predict (solid solution, intermetallics), rather than for what it predicts: a patchy phase separation. In spite of the similarities between the Al-Si and Al-Ge phase diagrams, the Al-Ge system is better described by parameters where Ge favors alloying and Al does not. Any other alternative régime (in terms of the role of Al, Si or Ge atoms) results in strong ordering tendencies at room temperature.

2.6 The Mo-Si and Mo-Ge systems

The phase diagram shows the existence of a $MoSi_2$ $C11_b$ phase, verified with the selected set of parameters with negative energy of formation. Figure 4 summarizes results for different concentrations, ranging from Mo + Si phase separation for $Mo_{97}Si_3$, to a $C11_b$ for $Mo_{33}Si_{67}$. Similar arguments apply to the Mo-Ge case. Once again, Mo is the element favoring alloying, resulting in simulated results nearly identical to those shown in Fig. 4 for Mo-Si.

3. MODELING RESULTS FOR THE Al/U-Mo INTERFACE

Starting with the analysis of the simple Al/U interface, in this section we study the influence of Si and Ge alloying additions to Al on the evolution of the Al/U-Mo interface. Several effects will be singled out for different Mo concentration in the fuel (5 and 10 wt% Mo), and the alloying addition concentration in Al (5 and 10 wt% Si or Ge). Although the calculations are restricted to a rigid bcc cell with the lattice parameter of the U portion, it is also true that for the purpose of this study, focused on determining qualitative trends and changes in behavior for different additives, the proposed framework is largely appropriate. It should also be noted that the low temperature results are a consequence of the simulation scheme used, based on a Monte Carlo - Metropolis algorithm where atoms 'diffuse' based only on the energetics of the initial and final state, and not due to their ability to overcome diffusion barriers. In other words, the simulation results just say that there are lower energy accessible states to which the system can evolve, but accurate diffusion rates can only be obtained if diffusion barriers are properly taken into account. In spite of these limitations, if the parameterization and general modeling scheme is correct, it is possible to obtain information on general trends and the driving forces for the processes found experimentally.

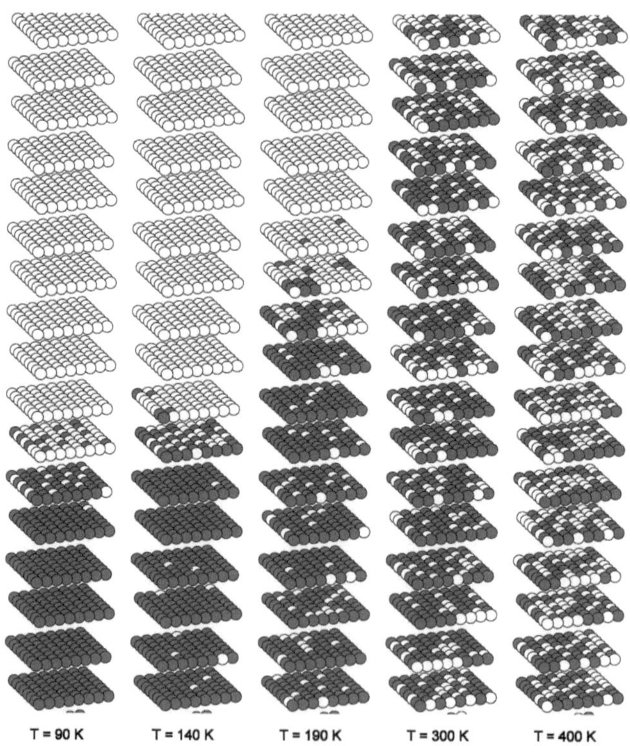

Figure 5. Intermediate states of a Monte Carlo simulation. The computational cell displays the Al (white circles)/U (grey circles) interface for T = 90, 140, 190, 300 and 400 K (from left to right).

3.1 The Al/U system

The experiments show that an $L1_2$ UAl_3 compound forms in this interface.[23-25] The modeling is done in a rigid cell of uniform symmetry (bcc), so it is not possible to observe fcc-based structures but, within the theoretical and modeling framework proposed in this work, the origin of the compound formation tendencies observed in this system can still be understood.

Dealing with a single symmetry in the simulations, as the examples do, is not a limitation of the method. To extend the calculations to multiple symmetries, however, requires a general parameterization scheme dealing with variable reference states which is currently under development. In what follows, a large computational cell (~10000 atoms) is used to simulate the Al/U interface. In the examples, the T = 0 K computational cell is heated steadily up to 400 K (700 K in some cases), in small temperature steps. The MC simulation is performed at each temperature stage until the computational cell is stabilized.

Consistent with the results in Table 2, the simulation results in Fig. 5 show that U has limited "diffusion" in Al, while Al is seen to "diffuse" deeply into the substrate even at low temperatures. As mentioned previously, the simulation results just say that there are lower energy accessible states to which the system can evolve, as opposed to real accessible states which are dominated by diffusion rates at low temperatures. For higher temperatures, Al is observed to mix with U resulting in a composition profile across the interface that is comparable to experiment, as shown in Ref. 9. Performing atom-by-atom analysis of the interface helps us to understand the role of each atom and the driving force for the observed final state. As mentioned in Sec. 2, this system is characterized by negligible Al "attraction" of U atoms and a rather intense U "repulsion" of Al atoms. The balance between chemical and strain of each atom in the simulations results in "diffusion" and compound formation tendencies that are in agreement with the trends observed experimentally.

3.2 The Al/U-Mo system

The experimental results for Al/(U,Mo) show that the U-2 wt% Mo/Al dispersions increase in volume by 26% at 673 K after 2000 h. This large volume change is mainly due to the formation of voids and cracks resulting from nearly complete interdiffusion of U-Mo and Al. No significant dimensional change occurs in the U-10 wt% Mo/Al dispersions. Interdiffusion between U-10 wt% Mo and Al is found to be minimal. The differences in diffusion behavior are primarily due to the fact that U-10 wt% Mo particles are supersaturated with substitutional Mo, more so than the U-2 wt% Mo particles. Al diffuses into the U-2 wt% Mo particles relatively rapidly along grain boundaries with nearly pure U forming UAl_3 almost fully throughout the 2000 h anneal, whereas the supersaturated Mo in the U-10 wt% Mo particles inhibits the diffusion of Al atoms.[8]

In addition, experimental results[11] show that the U–7 wt% Mo alloy previously homogenized retains the γ-(U,Mo) phase, and the formation of $(U,Mo)Al_3$ and $(U,Mo)Al_4$ at 853 K is observed. Also, a very thin band close to the Al side can be assigned to the structure of the ternary compound $Al_{20}UMo_2$. When the decomposition of the (U,Mo) took place, a drastic change in the diffusion behavior was observed. In this case, XRD indicated the presence of phases with the structures of $(U,Mo)Al_3$, $Al_{43}U_6Mo_4$, γ-(U, Mo) and α-U in the reaction layer.

In spite of the rigid cell limitation in these simulations, based on a bcc cell with a lattice parameter characteristic of the U-Mo portion of the cell, it is possible to gather the necessary information to understand the behavior of Al in terms of Mo concentration. One first example can be seen in Fig. 6, where the concentration profiles for Al/U and Al/U-5 wt% Mo indicate reduced diffusion of Al in U-Mo relative to the binary Al/U case described before: up to 190 K, Al "diffusion" is limited to just the first plane. The effect of Mo on Al "diffusion" can also be observed at 400 K, where substantial differences between cells with 5 and 10 wt% Mo can be seen. In the first case, Al "diffuses" to layers far from the interface, with barely 20% in its vicinity. For 10 wt% Mo, however, Al preferentially goes to regions near the interface with low Mo concentration. Far from the interface, Al concentration at 400 K varies from 50% for pure U, 15% for U-5 wt% Mo, to nearly zero for U-10 wt% Mo. The concentration profiles at T = 400 K also reveal that Al and Mo show a tendency toward compound formation. However, this effect could also be understood as a result of Al "diffusion" to Mo-defficient regions which, statistically, populates sites in the vicinity of Mo atoms. Either way, the Al and Mo count would show peaks and valleys of the same nature in the concentration profiles.

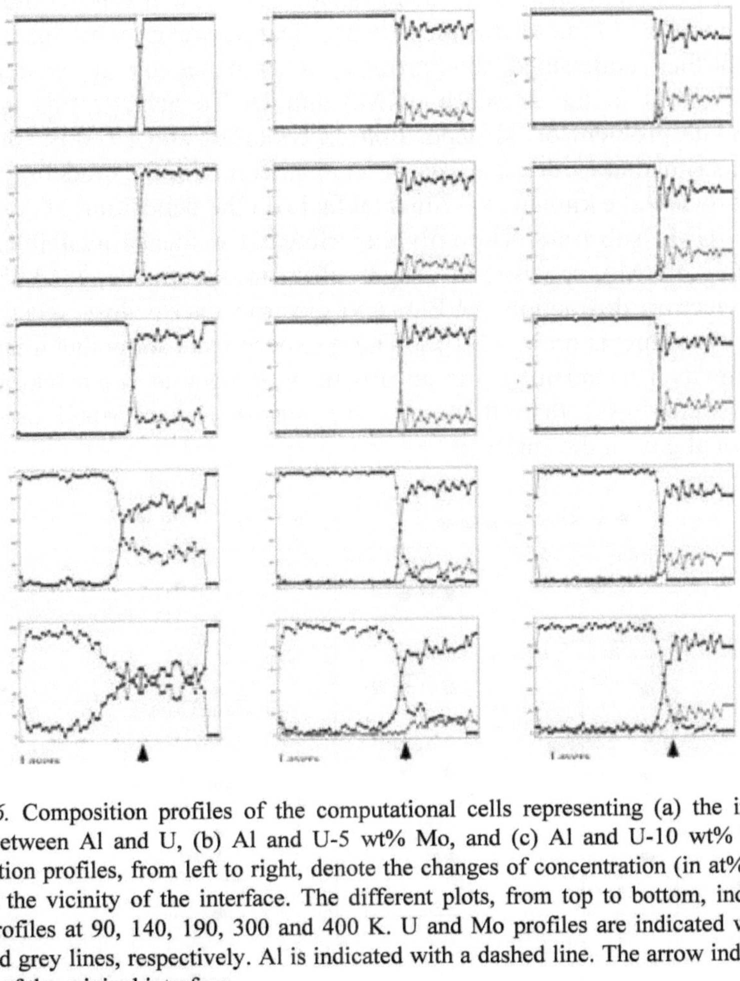

Figure 6. Composition profiles of the computational cells representing (a) the interfacial region between Al and U, (b) Al and U-5 wt% Mo, and (c) Al and U-10 wt% Mo. The composition profiles, from left to right, denote the changes of concentration (in at%) in each plane in the vicinity of the interface. The different plots, from top to bottom, indicate the stable profiles at 90, 140, 190, 300 and 400 K. U and Mo profiles are indicated with solid black and grey lines, respectively. Al is indicated with a dashed line. The arrow indicates the location of the original interface.

3.3 Atom-by-atom analysis of Al deposition

At this point in the discussion, it is important to note that alternative simple modeling schemes can be used to gain further understanding of the problem at hand. While the large scale MC simulations clearly show the qualitative features of interest, it is also useful to exploit the numerical simplicity of the methodology and investigate, in more detail, the atomistic processes leading to the observed behavior. The behavior described above suggests that a necessary next step would be to address the basic nature of the interaction between Al, Mo and U, which can be accomplished by means of an atom-by-atom analysis of this system. One clear feature is the role of Mo in inhibiting Al interdiffusion. It is not clear, however, why this is so.

To further understand this process, it is necessary to examine the behavior of Al in the presence of Mo and U. To achieve this goal, we consider the problem of Al deposition on Mo(100) and U(100), for which some experimental evidence exists for comparison with the modeling results. We first review the known experimental facts on the deposition of Al on Mo, U and a UMo substrate. The only experimental evidence available for Al deposition on Mo consists of Auger electron spectroscopy (AES), low energy electron diffraction (LEED) and electron energy loss spectroscopy (EELS) experiments on Mo(110).[22] These experiments show that a layer-by-layer growth mechanism of the adsorbate is primary at room temperature. For coverage lower than 0.34 ML, the adsorbate is formed as a two-dimensional gas on the surface.

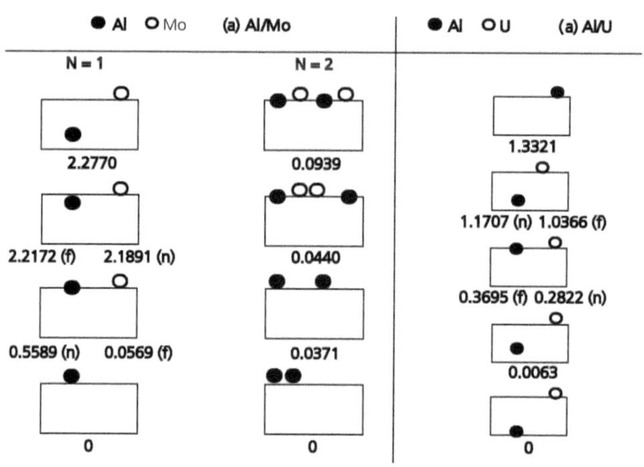

Figure 7. (a) Energy level spectrum for Al deposition on Mo. The first column shows results for the $N_{Al} = 1$ case labeled with the difference in energy (in eV/atom) with respect to the lowest energy state (bottom). The adatom can either be located in the overlayer (on top of the cube), in a surface site (crossing the top edge of the cube), in the first layer (1b) below the surface (immediately below the top edge of the cube) and two layers (2b) below the surface (in the center of the cube). The center column shows similar results for $N_{Al} = 2$. (b) Energy level spectrum for Al deposition on U(100). Al and Mo (or U) atoms are indicated with solid black disks and open circles, respectively.

Al atoms occupy random positions on the surface, a behavior which is probably responsible for the appearance of the (1x1) pattern of the substrate. All the LEED patterns observed at room temperature originate from more or less deformed, hexagonal layers of Al which have different crystallographic orientations with respect to the substrate. Most probably no alloys are formed, though the possibility is not fully excluded.[22] For Al/U, in contrast with the behavior observed in Al/Mo, experimental observations indicate that there is strong interdiffusion of Al and intermetallic formation with different stoichiometries.[8,23-25]

From a modeling standpoint, the deposition of Al on a Mo substrate can be studied by considering a few basic configurations where the Al atom occupies sites in the overlayer, surface, or planes immediately below the surface plane. Figure 7 shows the results for such configurations, displayed as a function of increasing difference in energy with respect to the lowest energy state. From the onset, it is clear that there is no penetration of Al in the inner layers of Mo as there is a noticeable energy barrier preventing Al substitutions in surface sites. The case for two Al atoms, also shown in Fig. 7, indicates that increasing coverage leads to the formation of Al(O) chains along the close-packed direction. Quite the opposite behavior occurs for deposition of Al on U. Figure 7.b summarizes the results for deposition of one single Al atom on a U(100) slab. In this case, the lowest energy state corresponds to Al in the bulk.

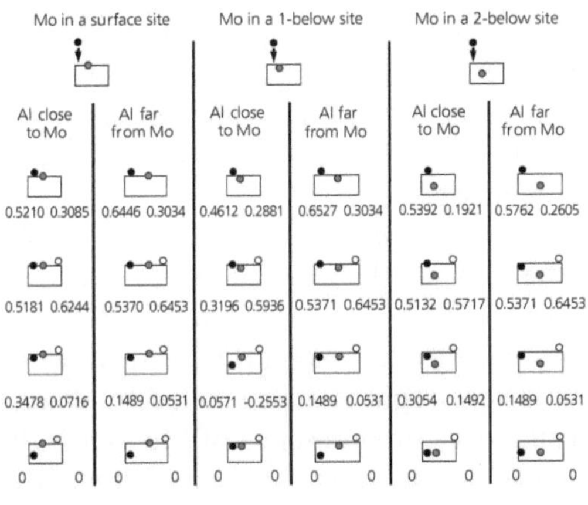

Figure 8. Energy level spectrum (in eV/atom) for Al deposition on a U slab in the presence of a Mo atom a) in a surface site, b) in a 1b site, and c) in a 2b site. In each case, two different situations are examined: when the deposited Al atom is close (left column) or far (right column) from the Mo atom. The configurations are ordered, top to bottom, in terms of decreasing difference in energy with the lowest energy state. Al, Mo and U atoms are indicated with solid black disks, grey disks, and open circles, respectively.

The energy of the computational cell increases steadily as the Al atom approaches the surface, consistent with the high solubility of Al in U. Similar results are observed for two Al atoms, where the lowest energy state consists of two Al atoms in solution in the U bulk.

There is, however, a close low energy state (0.03 eV/atom above the lowest energy configuration) where the two Al atoms are located at third-neighbor distances. This can be seen as an emerging trend for ordering with increasing Al concentration. The results shown in Fig. 7 are, in all cases, in agreement with available experiments.[22-25]

Regarding the Al/U-Mo system, the energy gaps between the configurations in Fig. 7 are of comparable magnitude and, as a consequence, it is difficult to derive the three-element behavior from the binary cases. The competing behaviors observed for Al/Mo and Al/U result in a specific behavior of the ternary system that can only be described by a proper accounting of the interaction between the different pairs and how such interactions are modified by the presence of a third element. To simplify the analysis, we first examine the deposition of a single Al atom on a U-Mo(100) substrate as a function of the distance between Al and Mo and as a function of the location of the Mo atom. The results are summarized in Fig. 8. When the Mo atom is located in a surface site, Mo(S), the lowest energy state is Al(2b)+Mo(S)+U(O). This is true both for the case in which Mo(S) is close (left column) or far (right column) from the deposited Al atom. Clearly, the presence of a single Mo atom, whether it is close or far to Al, does not change the trend observed for Al/U(100), where Al "diffusion" dominates. When Mo occupies subsurface sites, proximity between Al and Mo introduces small energy barriers that would slow, but not inhibit, interdiffusion of Al to deeper layers.

The results in Fig. 8 generalize the analysis of Al on U-Mo indicating that Al diffuses into the U-Mo substrate. However, a close examination of the magnitude of the energy gaps between the configurations whose ordering lead to this conclusion indicates that such effect is facilitated in regions with low Mo concentration. These regions are modeled by configurations where Mo is far from Al. In summary, the presence of Mo favors the location of Al in surface or overlayer sites, thus affecting, but not inhibiting Al interdiffusion in the bulk of U-Mo solid solution for this level of coverage, i.e., in the very dilute limit.

More insight on this behavior can be obtained by analyzing a set of configurations that model the penetration of Al into subsurface layers in regions of high Mo surface concentration. The fifteen configurations in Fig. 9 describe the process in different ways: a) from top to bottom, showing the evolution of an Al atom in the overlayer (left column), surface layer (center column) and 1b layer (right column), in the presence of a cluster of Mo(S) atoms, and b) from left to right, showing the evolution of the Al atom as it moves from the overlayer site, to the surface site, and to the 1b layer. This is shown for different locations of Al relative to the Mo(S) patch: far from the

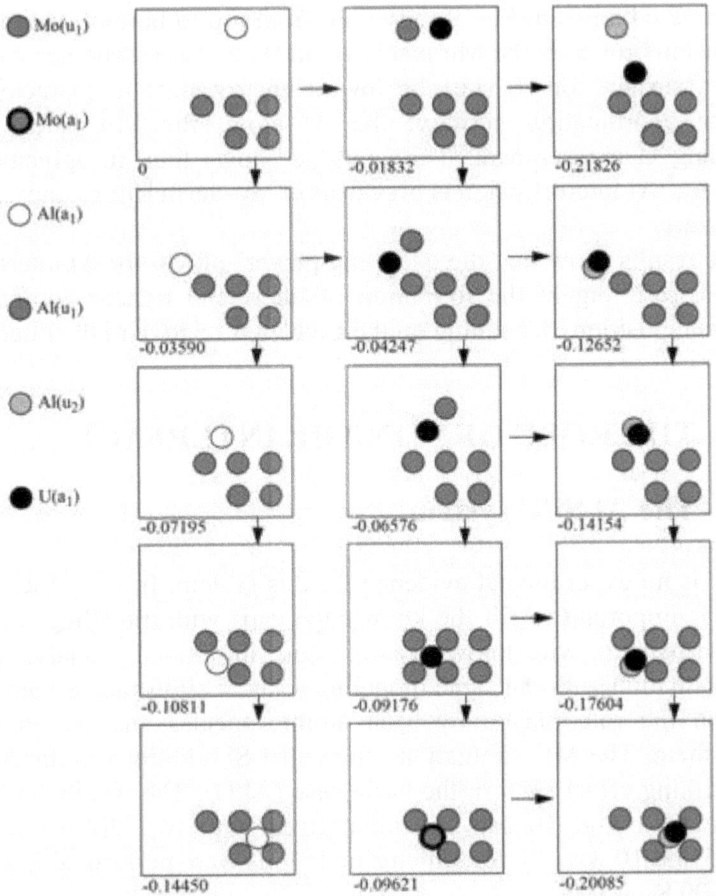

Figure 9. Top view of different configurations showing an Al atom (black disk) in different overlayer (left column), surface (center column) and subsurface sites (right column) in the vicinity of a cluster of Mo atoms (grey disks) in an U substrate. Ejected U atoms are denoted with an open circle. Each configuration is labeled by the difference in energy (in eV) with the reference state in which an Al adatom is located away from the Mo patch (top left corner). The horizontal or vertical arrows indicate those transitions that are energetically favored.

patch (top row), as a nearest-neighbor of a 'corner' Mo(S) atom (second row), with two, three and four (third, fourth and fifth row, respectively) nearest-neighbors in the Mo patch. The top row indicates that the penetration of Al (when deposited far from the Mo patch) in subsurface layers is energetically favored, an effect that is less pronounced as Al(O) approaches the Mo patch. The second row indicates that the process is still possible if Al(O) connects to the Mo(S) patch via one Mo(S) atom, but unlikely when the number of Al(O)-Mo(S) bonds increases (third, fourth, and fifth row). However, regardless of their proximity to the patch, Al atoms tend to migrate to Mo-rich regions.

All three columns in Fig. 9 represent, from top to bottom, the process of Al "diffusion" towards the Mo patch. In all three cases, whether Al is in an overlayer, surface, or 1b site, the lowest energy state is characterized by maximum coordination between the Al atom, the Mo patch and the surrounding U environment. This analysis shows that, in agreement with experiment,[8] Al interdiffusion is prevalent in Mo-defficient regions of the U-Mo substrate.

These results show that the 'stopping power' of Mo for Al interdiffusion is limited, resulting in the formation of interfacial ternary compounds of varying composition, depending on their location relative to the interface.

4. THE ROLE OF SI IN THE INTERFACE

4.1 The Al-Si/U system

There is no experimental evidence for this system. It is in cases like this where it is important to fill the knowledge gaps with modeling results that assure consistency with known cases. More importantly, analyzing every possible situation with the same modeling scheme allows us to gain detailed insight on the role played by each atomic species and the interactions between them. The MC simulations show that Si additions to the Al matrix have a striking effect even in the basic case (Al/U). The results for this system, shown in Fig. 10 can be summarized with two distinct examples, namely, 5 and 10 wt% Si, for which simulations were performed in the range $1 < T < 400$ K.

In contrast with the Al/U system, Al interdiffusion in U is drastically affected by Si. The simulations show higher "diffusion" rates for Si than for Al. As shown in Fig. 10 for T = 190 K, where Al-5 wt% Si/U and Al-10 wt% Si/U results are shown, Si inhibits Al interdiffusion at low temperature.

This effect is more noticeable in regions wih higher Si concentration. As T increases, Si diffuses to deeper layers allowing for Al "diffusion" to the interfacial region. There is a noticeable difference between the 5 and 10 wt% Si in Al cases, as seen in Fig. 10: Al depletion in near-interfacial regions where Si is in the majority leads to the formation of ordered compounds.

For 10 wt% Si in Al, there is a noticeable effect in the diffusion of Al to U, mainly in the regions where Si is in the majority, forming compounds with U. The simulations show that the diffusion path of Al is through Si-poor regions (particularly for T = 400 K), or where U-Si precipitates disappear.

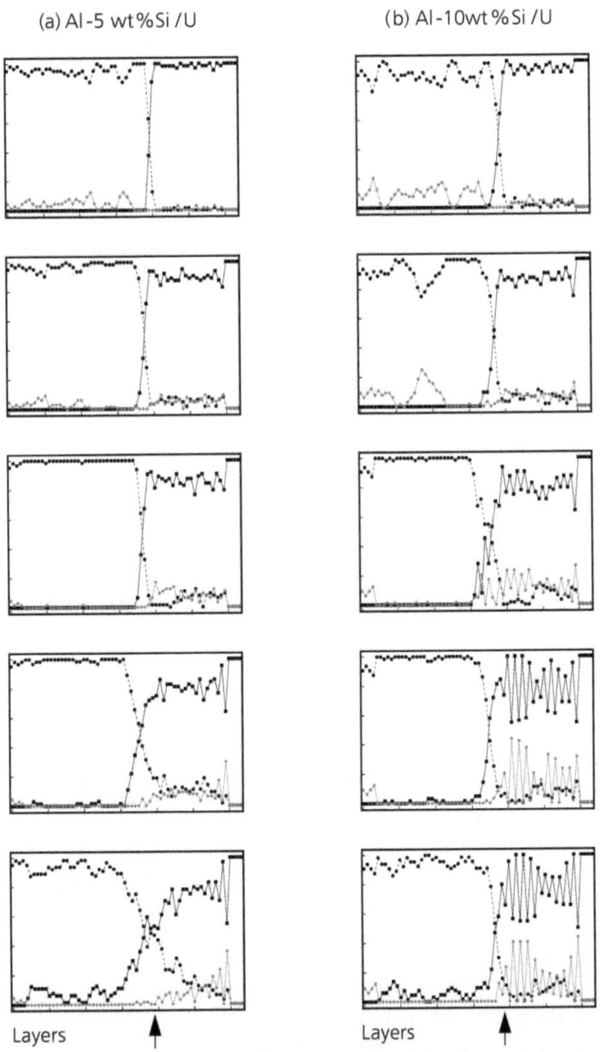

Figure 10. Composition profiles of the computational cells representing (a) the interfacial region between an Al-5 wt% Si and U, and (b) Al-10 wt% Si and U. The composition profiles, from left to right, denote the changes of concentration (in at%) in each plane in the vicinity of the interface. The different plots, from top to bottom, indicate the stable profiles at 90, 140, 190, 300 and 400 K. U profiles are indicated with black solid lines. Al and Si profiles are indicated with black (dashed) and grey (dot-dashed) lines. The arrows indicate the location of the original interface.

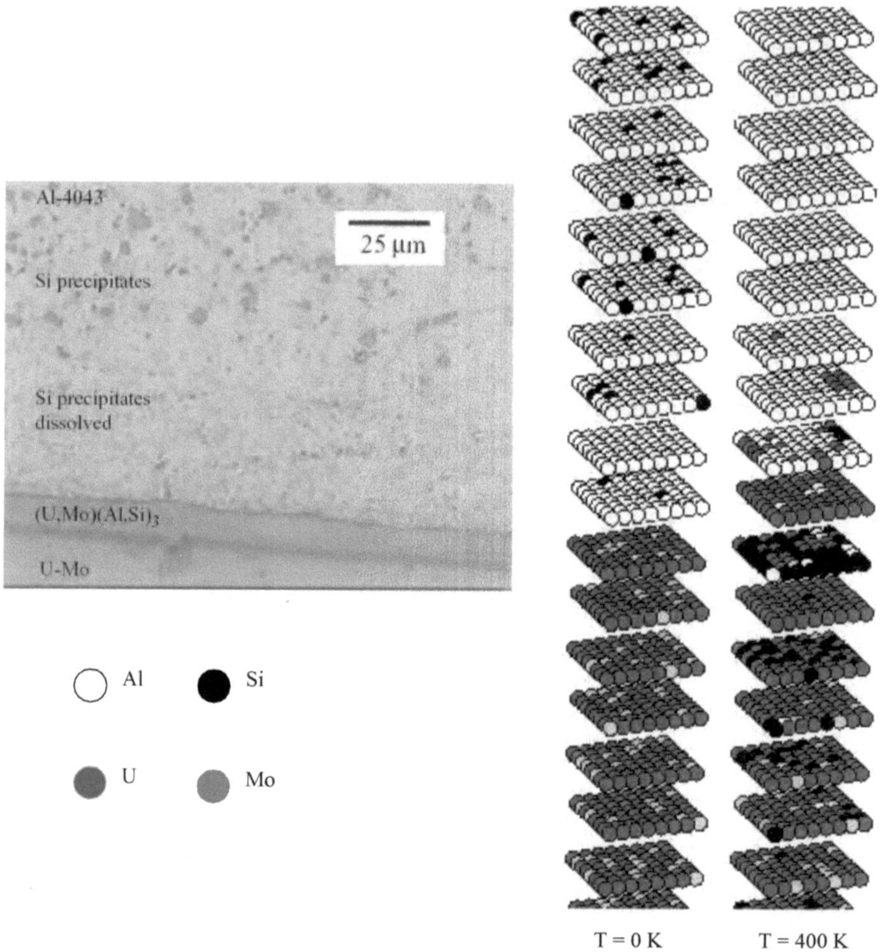

Figure 11. Al-5.2 wt% Si/U- 7wt% Mo sample's interdiffusion layers showing the zone of depletion of Si precipitates near the interdiffusion layer.[26] (Reproduced with the permission of the authors). The right panel shows results of BFS calculations of the Al-Si/U-Mo interface: the initial state (at T = 0 K) on the left, and the T = 400 K, displaying features similar to those observed experimentally.

4.2 The Al-Si/U-Mo system

Experimental results, shown in Fig. 11, suggest that Si additions to Al introduce important changes in relation with the phases found inside the interaction layer: $(U,Mo)(Al,Si)_3$ near the Al alloy, a two-phase zone consisting of $(U,Mo)(Al,Si)_3$ and, probably, $(U,Mo)(Al,Si)_{2-x}$ near U-Mo alloy. No $(U,Mo)Al_4$ was found. $(U,Mo)(Al,Si)_3$, which is the main component of the interaction layer, is the phase reported as the interaction product in dispersion Si fuel elements.[26]

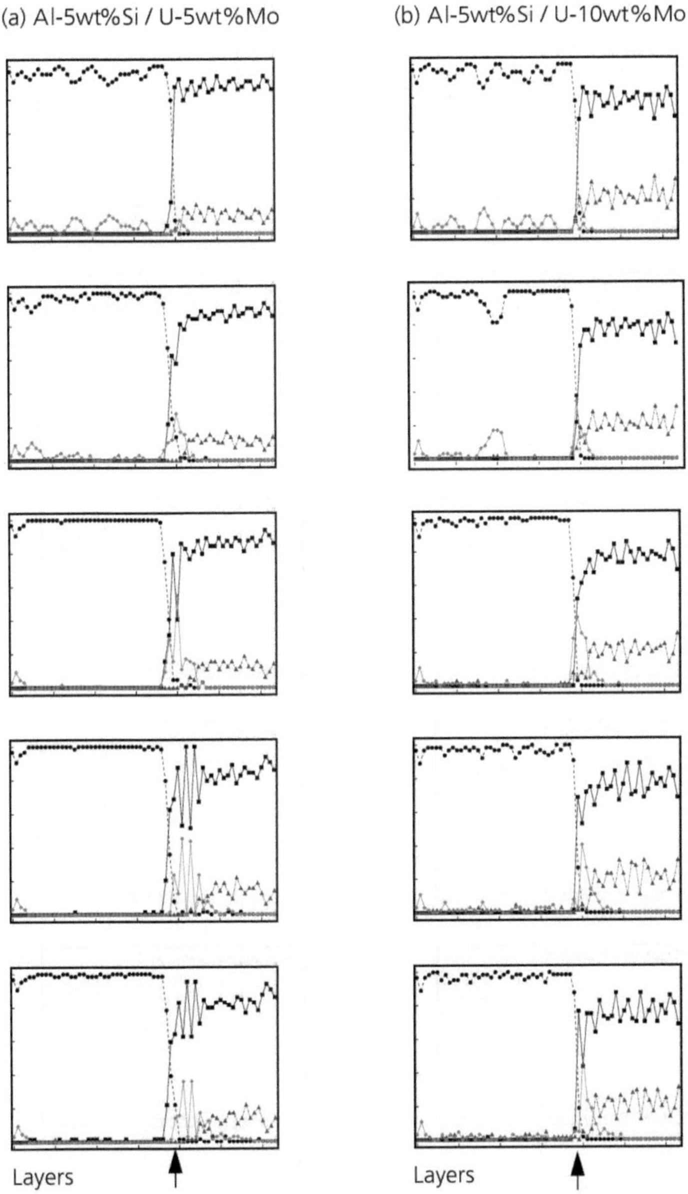

Figure 12. Composition profiles of the computational cells representing (a) the interfacial region between an Al-5 wt% Si and U-5 wt% Mo, and (b) Al-5 wt% Si and U-10 wt% Mo. The composition profiles, from left to right, denote the changes of concentration (in at%) in each plane in the vicinity of the interface. The different plots, from top to bottom, indicate the stable profiles at 90, 140, 190, 300 and 400 K. U and Mo profiles are indicated with black and grey solid lines, respectively. Al and Si profiles are indicated with black (dashed) and grey (dot-dashed) lines. The arrows indicate the location of the original interface.

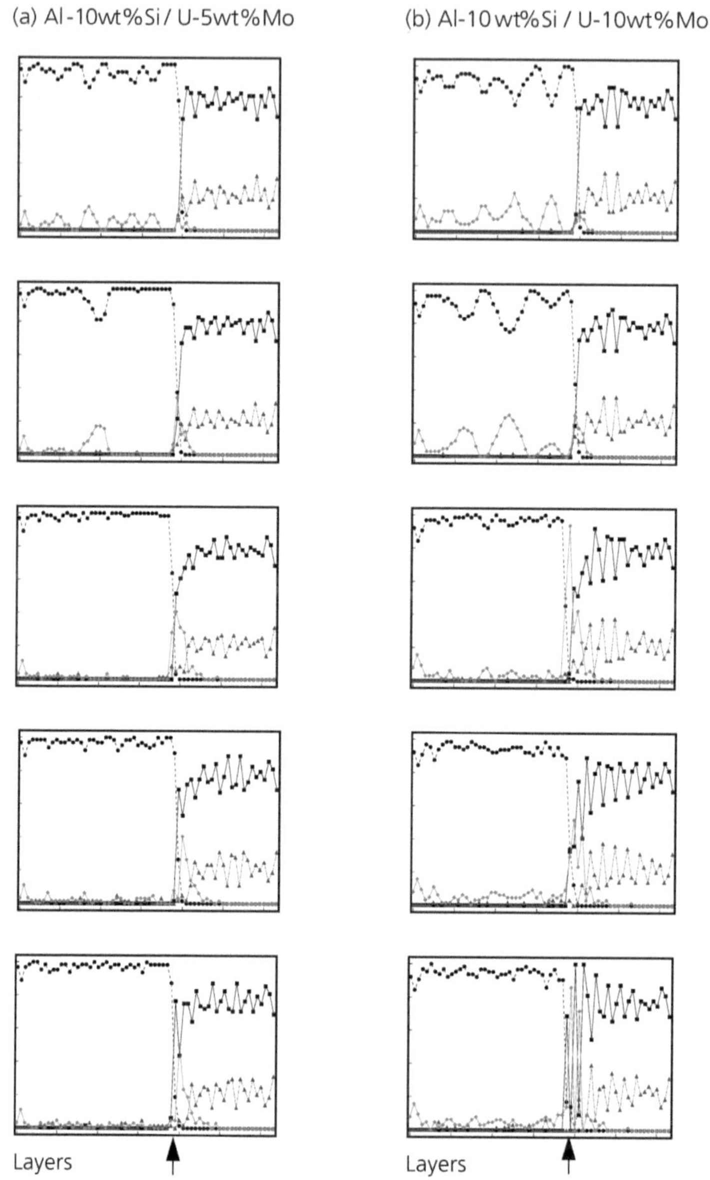

Figure 13. Composition profiles of the computational cells representing (a) the interfacial region between an Al-5 wt% Si and U-5 wt% Mo, and (b) Al-10 wt% Si and U-10 wt% Mo. The composition profiles, from left to right, denote the changes of concentration (in at%) in each plane in the vicinity of the interface. The different plots, from top to bottom, indicate the stable profiles at 90, 140, 190, 300 and 400 K. U and Mo profiles are indicated with black and grey solid lines, respectively. Al and Si profiles are indicated with black (dashed) and grey (dot-dashed) lines. The arrows indicate the location of the original interface.

The following set of simulations, depicted in Figs. 12 and 13, addresses the simultaneous effect of adding Si to Al and Mo to U, their influence in the structure of the interface, and in the interactions between the two alloying additions. The results show that the trends observed in the ternary case (Al/U,Mo) are, to a certain extent, conserved. The most striking result, a consequence of the interaction between all four elements, is the complete depletion of Si in Al regions close to the interface, an effect that is in excellent agreement with experiment.[26]

The simulation results for Al-5 wt% Si (Fig. 12) and for Al-10 wt% Si (Fig. 13), reproduce the main features of the experimental results, highlighting a strong trend towards the formation of interfacial compounds. As mentioned earlier in this discussion, the limitation of working on a rigid bcc cell prevents us from properly determining the structure and composition of these compounds, but does not hinder the fact that their formation strongly depends on the Si and Mo contents. Additional effects due to the interactions between the participating elements are also observed. The most noticeable is the interaction between Mo and Si, resulting in a region free of Mo and Al where Si (in the majority) forms compounds. This effect, where Mo inhibits Si diffusion, is proportional to Mo concentration, allowing for Si-rich planes resulting in changes in the composition of the interfacial compounds from U_3Si to B2 USi. The combined effect of Si and Mo and their interactions is a noticeable decrease in Al diffusion, more noticeable with increasing content of either element.

5. THE ROLE OF Ge IN THE INTERFACE

One expectation when using additives in the Al matrix is that the element or elements added will form weak bonds with Al and a tendency towards compound formation with U and Mo. The opposite would be expected with elements added to the fuel. Ultimately, the goal is to diminish the interfacial growth kinetics and perhaps avoid porosity development and the formation of unwanted compounds. There is one other alternative, namely, coating the fuel particles with one or more elements or compounds with the aim of generating diffusional barriers in the interface. This alternative, however, has been questioned due, first, to the thickness of the deposition necessary before the coating becomes an active and effective barrier under irradiation, and, second, to the negative effects of the rolling process of fuel fabrication on the barrier behavior (if the coating consists of a brittle compound). In both cases, Si is an element that has been proposed as an additive to the Al matrix or as diffusional barriers through the deposition of pure Si or Si-based compounds.

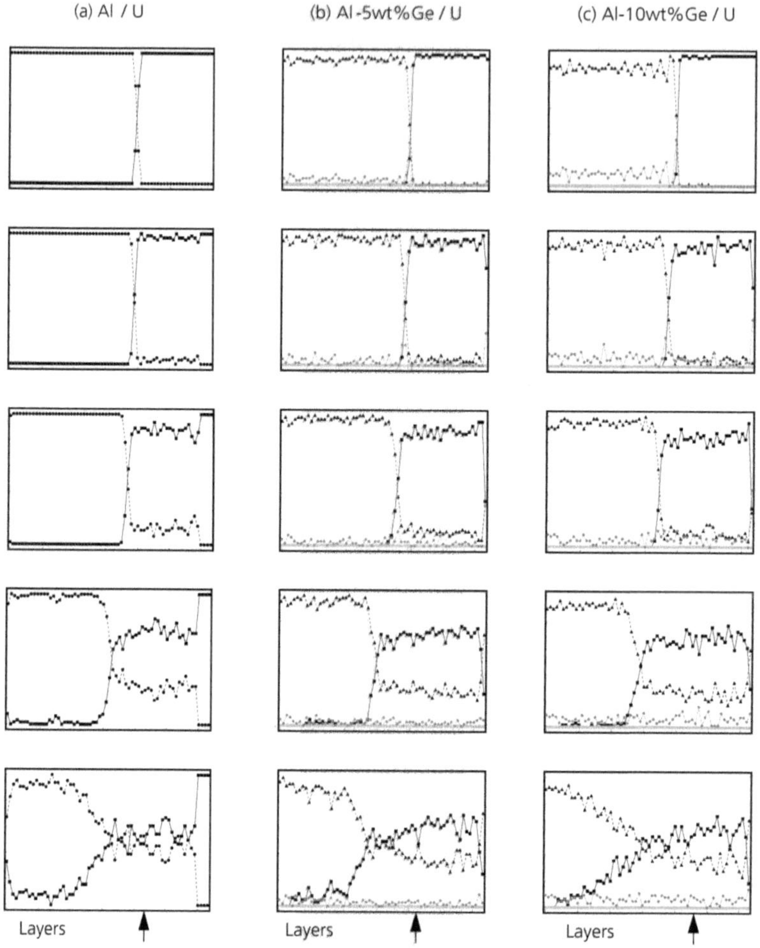

Figure 14. Composition profiles of the computational cells representing (a) the interfacial region between an Al and U, (b) Al-5 wt% Ge and U, and (c) Al-10 wt% Ge and U. The composition profiles, from left to right, denote the changes of concentration (in at%) in each plane in the vicinity of the interface. The different plots, from top to bottom, indicate the stable profiles at 90, 140, 190, 300 and 400 K. U profiles are indicated with black solid lines. Al and Ge profiles are indicated with black (dashed) and grey (dot-dashed) lines. The arrows indicate the location of the original interface.

The observed and computed behavior of Si on Al diffusion raises questions if another element of the same group, as Ge, present the same behavior if it is used as an additive in Al. This is yet one more case in which the lack of experimental results or any guiding evidence motivates the use of modeling in order to preview qualitative changes before proceeding to experimental verification. Therefore, in what follows, we apply the same modeling tools to establish a comparison between the role of Si and Ge, with particular emphasis on the formation of metastable compounds with Al.

5.1 The Al-Ge/U system

To define a proper basis for the comparison between Si and Ge additions to Al, we consider computational cells with the same number of Si and Ge atoms. This amounts to 4 and 10 wt% Ge. The most noticeable result in the simulations for the Al-4 wt% Ge/U case is the trend toward the formation of interfacial compounds and high Ge concentration in the interface at low temperatures, and their disappearance at higher temperatures. In spite of this effect, Ge has little influence on Al interdiffusion. Similarly, little if any Ge diffusion in U is seen. These conclusions are valid for both Ge concentrations, in contrast with Si, where for the Al-10 wt% Si/U case there is a much more noticeable change in Al interdiffusion and the formation of interfacial compounds. Also, little Ge depletion is observed near the interface, as shown in Fig. 14.

5.2 The Al-Ge/U-Mo system

The results for the simulations, shown in Fig. 15, make it clear that there are noticeable differences between the behavior of Si and Ge, for 10 wt% Ge added to Al and 10 wt% Mo added to U. In contrast with the behavior of Si described above, the Ge remains mainly associated with Al, a consequence of the tendency for metastable compound formation observed experimentally in the Al-Ge system.[27-29] In fact, a close examination of the concentration profile for Al shows very little difference with the pure Al case. This is rather surprising, for in spite of the many similarities between the parameterization schemes for each element (see Sec. 2), there are very noticeable differences between the role of each addition in a multicomponent environment, as can be seen in Fig. 15, where the composition profiles for Al/U-Mo, Al-Si/U-Mo, and Al-Ge/U-Mo are shown. The strong tendencies for compound formation between Si and U, magnified by the interaction between Mo and Si, is not observed for Ge which remains almost exclusively in the Al side with little Ge diffusion into U. In spite of the observed weak trends for the formation of interfacial compounds (especially at low temperatures), no major qualitative differences are observed with respect to the pure Al case, casting doubt on the ability of Ge additions to achieve the desired effects. In spite of the apparent failure of Ge to reach the desired goals, the modeling analysis is a good example that shows the synergy between modeling and experiment, guiding choices, and understanding the observed results.

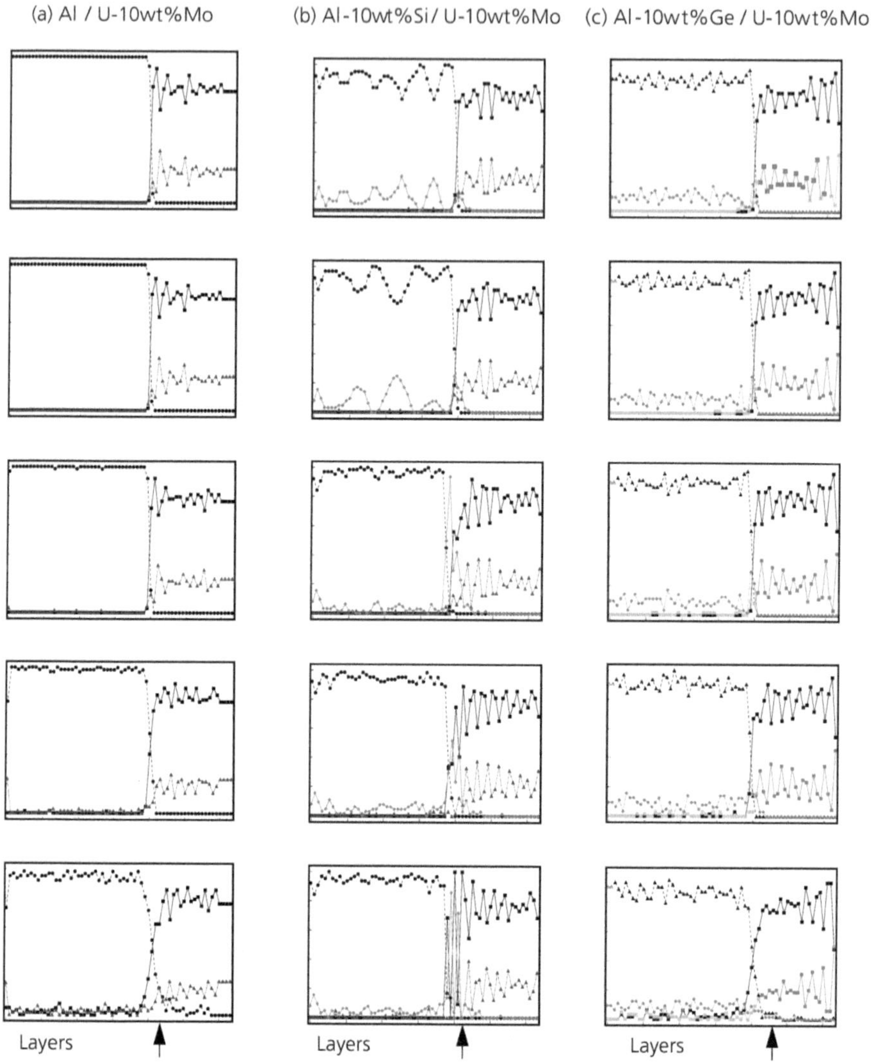

Figure 15. Composition profiles of the computational cells representing (a) the interfacial region between Al and U-10 wt% Mo, (b) Al-10 wt% Si and U-10 wt% Mo, and (c) Al-10 wt% Ge and U-10 wt% Mo.The composition profiles, from left to right, denote the changes of concentration (in at%) in each plane in the vicinity of the interface. The different plots, from top to bottom, indicate the stable profiles at 90, 140, 190, 300 and 400 K. U and Mo profiles are indicated with black and grey solid lines, respectively. Al and Si (or Ge) profiles are indicated with black (dashed) and grey (dot-dashed) lines. The arrows indicate the location of the original interface.

6. CONCLUSIONS

Modeling real systems and their environments with equal treatment of every possible degree of freedom, while desirable, is not always either achievable or as helpful as it could be expected. Freezing some degrees of freedom, or modeling equivalent, simplified systems can provide guidelines when dealing with the growing number of technological problems related with the development of new materials. There are several techniques currently available to narrow the gap between the atomistic theories and experimentation but this is not yet a trivial task, and considerable effort has to be made to improve these methods in order to enhance their predictive power for further development and research. In this chapter we apply the BFS method for alloys to model interfacial phenomena and show that atomistic computational modeling could become a valuable tool to aid the experimental work in the development of nuclear fuels through the understanding of basic features observed in the Al/UMo interface and the effect of additives to Al matrix. However, several questions remain unanswered, as are the crystallographic nature of the interfacial compounds that form or how small quantities of alloying additions both in the matrix or fuel particles can influence the formation of these compounds and their effect in the overall behavior of the system. This chapter provides the first attempt to tackle these issues in an oversimplified modeling framework. However, in spite of the fact that all the calculations refer to a rigid environment where both the Al matrix and UMo particles are represented by a bcc lattice characteristic of the UMo fuel, the basic features characterizing the real system can still be identified: 1) the increased Al/UMo interdiffusion with temperature, 2) the Al "stopping power" of increasing Mo concentration, and 3) the formation of interfacial compounds.

Moreover, significant results were also obtained in the case of Si additions to Al, once again reproducing the main features observed experimentally: 1) the trend indicating formation of interfacial compounds, 2) much reduced diffusion of Al into UMo due to the high Si concentration, 3) Si depletion in the Al matrix, and 4) an unexpected interaction between Mo and Si which avoid the Si diffusion to deeper layers in the UMo solid solution thus improving the stopping power for Al diffusion. Also, the role of another element, Ge, of the same group as Si, was investigated. In contrast with Si behavior, we observed a weak tendencies to compound formation and, basically, no differences were observed in Al difussion with respect to pure Al. As no experimental results or guiding evidence are available for the system Al-Ge/U-Mo, this is a good example of how computational modeling could help to understand the behavior of the system before proceeding to experimental verification.

REFERENCES

1. 10CFR50, Limiting the use of highly enriched uranium in domestically licensed research reactors, Federal Register, vol. 51, no. 377, 1986.
2. J.L. Snelgrove, G.L. Hofman, M.K. Meyer, C.L. Trybus and T.C. Wiencek, Development of very-high-density low-enriched-uranium fuels, *Nucl. Eng. Des.* **178**, 119 (1997).
3. M. Ugajin, A. Itoh, M. Akabori, N. Ooka and Y. Nakakura, Irradiation behavior of high uranium-density alloys in the plate fuels, *J. Nucl. Mater.* **254**, 78 (1998).
4. M.K. Meyer, T.C. Wiencek, S.L. Hayes and G.L. Hofman, Irradiation behavior of U_6MnAl dispersion fuel elements, *J. Nucl. Mater.* **278**, 358 (2000).
5. G.L. Hofman, R.F. Domagala and G.L. Copel, Irradiation behavior of low-enriched U_6Fe-Al dispersion fuel elements, *J. Nucl. Mater.* **150**, 238 (1987).
6. M.K. Meyer, G.L. Hofman, T.C. Wiencek, S.L. Hayes and J.L. Snelgrove, Irradiation behavior of U-Nb-Zr alloy dispersed in aluminum, *J. Nucl. Mater.* **299**, 175 (2001).
7. M.K. Meyer, G.L. Hofman, S.L. Hayes, C.R. Clark, T.C. Wiencek, J.L. Snelgrove, R.V. Strain and K.-H. Kim, Low-temperature irradiation behavior of uranium molybdenum alloy dispersion fuel, *J. Nucl. Mater.* **304**, 221 (2002).
8. D.B. Lee, K.H. Kim, C.K. Kim, Thermal compatibility studies of unirradiated U-Mo alloys dispersed in aluminum, *J. Nucl. Mater.* **250**, 79 (1997).
9. K.H. Kim, J. Park, C.K. Kim, G.L. Hofman y M.K. Meyer, Irradiation behavior of atomized U-10 wt% Mo alloy aluminum matrix dispersion fuel meat at low temperature, *Nucl. Eng. and Design* **211**, 229 (2002).
10. G.L. Hofman, M.R. Finlay and Y S. Kim, Post-irradiation Analysis of Low Enriched U-Mo/Al Dispersion Fuel Miniplate Tests, RERTR-4 and 5, International Meeting on RERTR, November 7-12, 2004, Vienna, Austria.
11. M.I. Mirandou, S.N. Balart, M. Ortiz and M.S. Granovsky, Characterization of the Reaction Layer in U-7 wt% Mo/Al Diffusion Couples, *J. Nucl. Mater.* **323**, 29 (2003).
12. P. Lemoine, F. Huet, B. Guigon, C. Jarousse and S. Guillot, French Development and Qualification Programs for the JHR Project Fuel Element, International Meeting on RERTR, November 7-12, 2004, Vienna, Austria.
13. G. Bozzolo and J.E. Garcés, *Atomistic modeling of surface alloys, Surface alloys and alloy surfaces*, The Chemical Physics of Solid Surfaces, Vol. 10, Elsevier, 2002.
14. G. Bozzolo, J. Khalil, M. Bartow and R.D. Noebe, Atomistic modeling of ternary and quaternary ordered intermetallic alloys, *Mat. Res. Soc. Symp. Proc.* **646**, N6.2 (2001).
15. A. Wilson, G. Bozzolo, R.D. Noebe and J. Howe, Experimental Verification of the Theoretical Prediction of the Phase Structure of a Ni-Al-Ti-Cr-Cu Alloy, *Acta Mater.* **50**, 2787 (2002).
16. G. Bozzolo, R.D. Noebe and H. Mosca, Atomistic Modeling of Pd Site Preference in NiTi, *J. Alloys Compds.* **386**, 125 (2005).
17. P. Blaha, K. Schwartz, and J. Luitz, WIEN97, Vienna University of Technology. Improved and updated Unix version of the copyrighted WIEN code, P. Blaha, K. Schwartz, P. Sorantin and S. Trickey.
18. J.R. Smith, T. Perry, A. Banerjea, J. Ferrante and G. Bozzolo, Equivalent crystal theory of metals and semiconductor surfaces and defects, *Phys. Rev. B* **44**, 6444 (1991).
19. R. Hultgren, R.L. Orr, P.D. Anderson and K.K. Kelley, Selected values of the Thermodynamics Properties of Binary Alloys, Wiley, New York, 1963.
20. T.B. Massalski, H. Okamoto, P.R. Subramanian, L. Kacprzak (Eds.), Binary Alloy Phase Diagrams, ASM International, 1990.
21. A.E. Dwight, The Uranium Molybdenum Equilibrium Diagram Below 900 C, *J. Nucl. Mater.* **2**, 81 (1960).

22. J. Kolaczkiewicz, M. Hochol and S. Zuber, LEED, AES, $\Delta\phi$ and EELS study of Al on Mo(110), *Surf. Sci.* **247**, 284 (1991).
23. D. Subramanyam, M. Notis and J. Goldstein, Microstructural investigation of intermediate phases formation in Uranium-Aluminum diffusion couples, *Met. Trans.* **16A**, 589 (1985).
24. R. Pearce, R. Giles and L. Tavender, Preparation and properties of UAl_x coatings formed on uranium via the electrophoretic deposition of Aluminum powder, *J. Nucl. Mat.* **24**, 129 (1967).
25. J. Buddery, M. Clark, R. Pearce and J. Stobbs, The development and properties of an oxidation resistant coating for uranium, *J. Nucl. Mat.* **13**, 169 (1964).
26. M. Mirandou, M. Granovsky, M. Ortiz, S. Balart, S. Aricó and L. Gribaudo, Reaction layer between U-7wt%Mo and Al alloys in chemical diffusion couples, International Meeting on RERTR, November 7-12, 2004, Vienna, Austria.
27. M.A. Shaikh, M. Iqbal, J.I. Akhter, M. Ahmad, Q. Zaman, M. Akhtar, M.J. Moughal, Z. Ahmed and M. Farooque, Alloying of immiscible Ge with Al by ball milling, *Mater. Lett.* **57**, 3682 (2003).
28. K. Chattopadhyay, X.-M. Wang, K. Aoki and T. Masumoto, Metastable phase formation during mechanical alloying of Al-Ge and Al-Si alloys, *J. Alloys Compds.* **232**, 224 (1996).
29. S. Srikanth, D. Sanyal and P. Ramachandrarao, A re-evaluation of the Al-Ge system, *CALPHAD* **20**, 321 (1996).

INDEX

Note: Citations derived from figures are indicated by an *f*; citations from tables are indicated by a *t*.

MIX
Papier aus verantwortungsvollen Quellen
Paper from responsible sources
FSC® C105338

If you have any concerns about our products,
you can contact us on
ProductSafety@springernature.com

In case Publisher is established outside the EU,
the EU authorized representative is:
Springer Nature Customer Service Center GmbH
Europaplatz 3, 69115 Heidelberg, Germany

Printed by Libri Plureos GmbH
in Hamburg, Germany